14th International Conference on Shot Peening 2022 (ICSP14)

Milan, Italy
4-7 September 2022

ISBN: 979-8-3313-0519-2

Printed from e-media with permission by:

Curran Associates, Inc.
57 Morehouse Lane
Red Hook, NY 12571

Some format issues inherent in the e-media version may also appear in this print version.

International Scientific Committee for Shot Peening
Department of Mechanichal Engineering
Politecnico di Milano Campus Bovisa Sud
via La Masa 1, 20156 Milano, Italy

Phone: +39.02.23998500
Fax: +39.02.23998202

www.shotpeening.org

Additional copies of this publication are available from:

Curran Associates, Inc.
57 Morehouse Lane
Red Hook, NY 12571 USA
Phone: 845-758-0400
Fax: 845-758-2633
Email: curran@proceedings.com
Web: www.proceedings.com

TABLE OF CONTENTS

Further enhancing fatigue properties with HFMI by customized edge treatment

J. Schubnell[1], S. Fliegener[1] and M. Farajian[2]

1 Fraunhofer Institute for Mechanics for Materials (IWM), Freiburg, Germany
2 GSI Gesellschaft für Schweißtechnik International mbH, Duisburg, Germany

Abstract
Free corners and edges are potential weak points for fatigue crack initiation at High Frequency Mechanical Impact (HFMI) treated welded joints even if the treated welded seam itself shows a high fatigue strength. In this study the residual stress state around cutting edges of HFMI-treated notched and welded specimens were investigated. For this, X-ray diffraction (XRD) technique and Finite Element (FE) Analysis were used. It is verified that no compressive residual stresses close to the edges were present. Fatigue tests were performed at notched specimens made of S960QL steel. Some specimens were HFMI-treated at the cutting edges with a concave pin. This additional treatment leads to a shift of the crack initiation location from the edge to the center of the specimen and to a significant increase of fatigue life. HFMI edge-treatment of welded joints showed in some cases that the edge treatment may also cause defects that could lead to crack initiation even from treated edges.

Keywords: High Frequency Mechanical Impact treatment, Cutting edges, Residual stresses, Fatigue

Introduction

Besides welds seams free cutting edges are potential weak points for fatigue crack initiation of steel welded structures [1]. Due this reason, the current IIW-recommendation [2] and Eurocode 3 [3] for the fatigue design of welded structures recommend grinding of corners or edges to remove potential defects and decrease the stress concentration. Experimental investigations show that the fatigue strength of the cutting surface is often related to the condition of the edges or corners, where pre-dominantly fatigue cracks initiate [1], [4]. Furthermore edges of High Frequency Mechanical Impact (HFMI) treated weld seams show as well in many cases fatigue failure from the edges [5]. Therefore, belt or disc grinding was usually performed to avoid fatigue crack initiation from the edges. However, it is obvious that the quality of grinding is strongly related to the user. For this reason, this investigation is focused on the edge-treatment by HFMI with special concave pins to provide an alternative method to improve the edge condition that expect a high reliability compared to grinding methods.

Specimen and mechanical treatment

High strength steel S960QL (1.8933) according to DIN EN 10025-6:2018-07 with a bainitic-martensitic microstructure was used in this investigation. The mechanical properties of the material are summarized in Table 1. Simple Notched and welded specimens (double-sided transverse stiffener with single-layer fillet welds) were manufactured of S960QL. The geometrical notch of the notched specimen was manufactured by electrical discharge machining (EDM) and has a stress concentration factor (SCF) of 2.26. The dimensions are illustrated in Figure 1 (a). Further details are given by Schubnell et al. [6]–[8].
Edge treatment was performed with a pneumatic impact treatment (PIT) device Pitec Weld Line 10, illustrated in Figure 1 (b). The recommended working pressure for steel was decreased from 6 bar to 4 bar to avoid an over-treatment of the edges. The impact frequency was 90 Hz. The impact angle was around 45° and the radius of the concave-pin was 2 mm. After edge-treatment the edges of some notched specimens remain in untreated condition (unt. edge). After edge-treatment defects were removed (chamfering) and the very

rough edges were polished (edge-treatment condition, HFMI-edge). The complete process is illustrated in Figure 1 (c).

Table 1. Mechanical properties of the investigated materials

Material	Yield strength [MPa]	Ultimate tensile strength [MPa]	Elongation [%]	Hardness [HV10]
S960QL	1011	1060	14	316

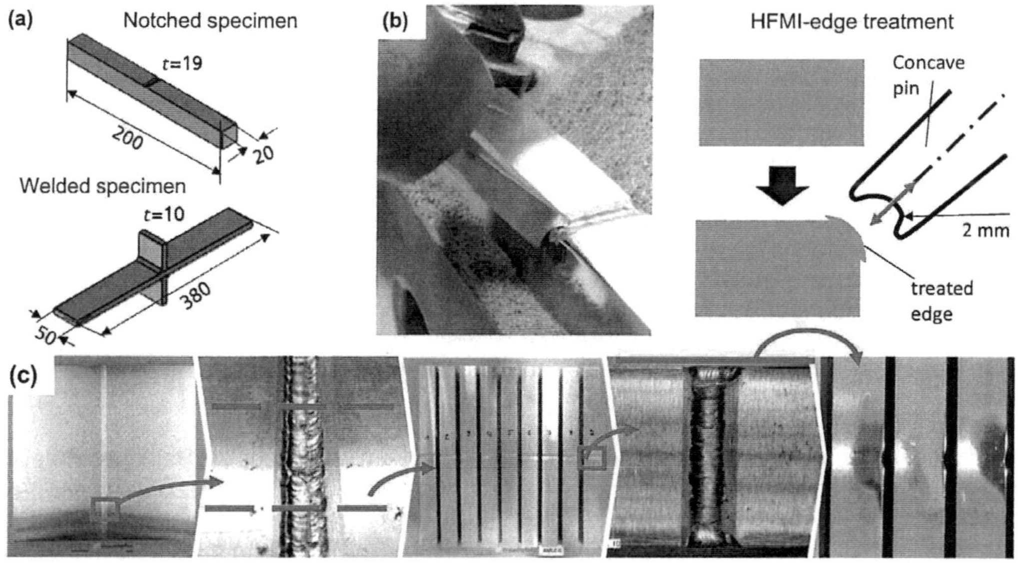

Figure 1. (a) Dimensions of specimens, (b) HFMI edge-treatment, (c) Manufacturing process of notched specimen

HFMI-process simulation

The residual stress state after HFMI-treatment, especially at the edges of the specimen, were analyzed by Finite Element (FE) process simulation. For this, the force-controlled FE-model implemented in ABAQUS according to Ernould et al. [9] and Schubnell [10] was used. With this simulation technique the vertical pin movement of the pin was controlled by a defined force. A combined isotropic-kinematic constitutive model with strain rate dependent flow stress according to Chaboche [11] modified by Maciolek [12] were used. The local material properties of the heat affected zone (HAZ) for the welded specimen from S960QL [13] were taken into account. Details of the process models are given by Schubnell et al. [6] (welded specimen) and by Schubnell et al. [7] (notched specimen).

The process simulation contains two steps: First the simulation of the HFMI-process, see Figure 2. Second, the deletion of the edges to simulate the re-distribution of the residual stress caused by the cut of the single specimen out of the base plate. The deletion of the edge was performed via a VUMAT-subroutine (vectorized user material) that set the stress on defined elements to zero. As shown in Figure 2, compressive residual stress was induced by HFMI-treatment up to a depth of around 1.5 mm for both types of specimens. However, compressive residual stresses are not determined close to the edge of the specimen. Here high residual stresses were induced by HFMI-treatment close to the specimen's edge. High tensile residual stresses were also present after deletion of the edge (simulation of cutting process).

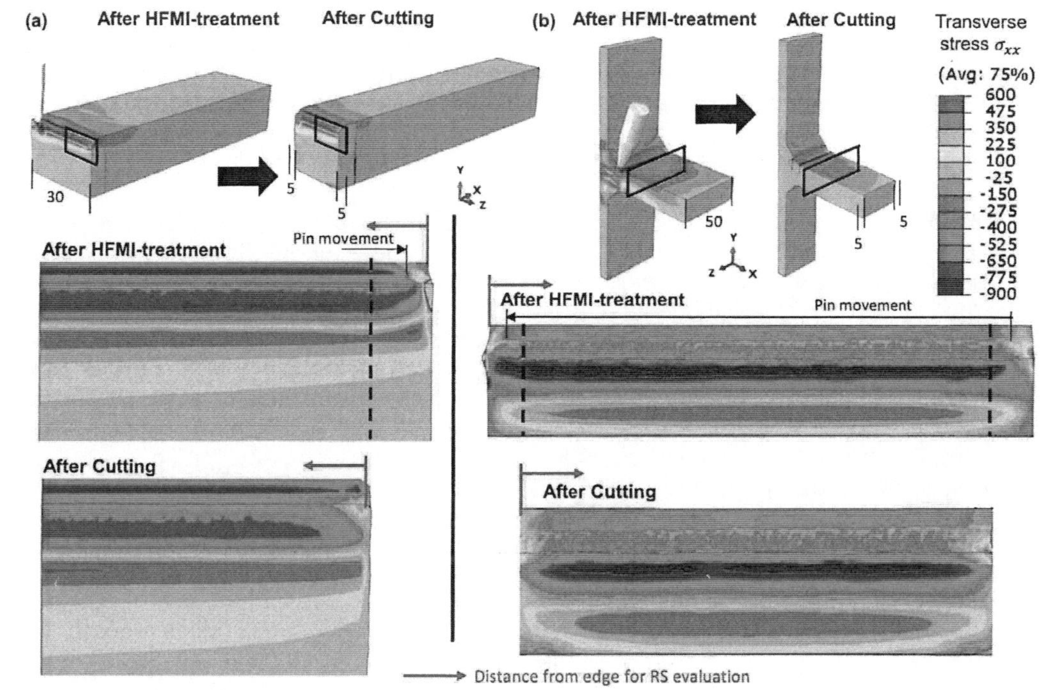

Figure 2. Finite Element Simulation of the HFMI-treatment and cutting of the notched specimen (a) and welded specimen (b)

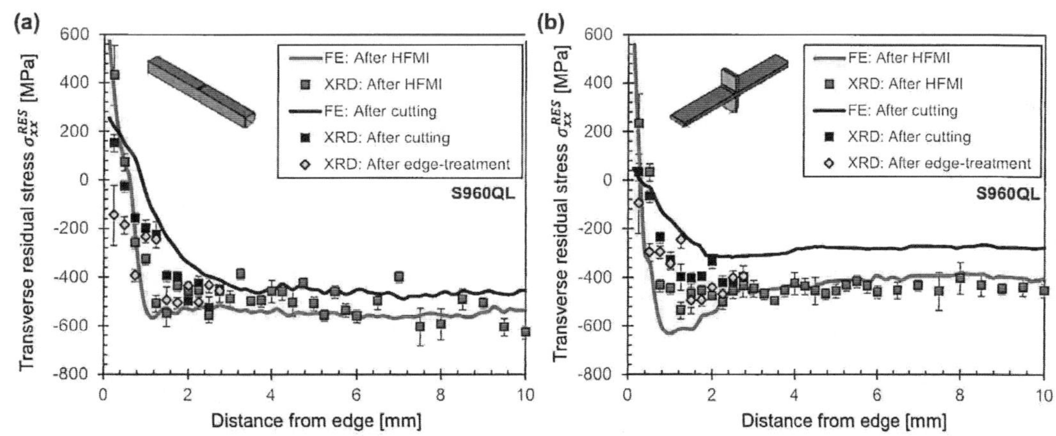

Figure 3. Residual stresses at the surface of the HFMI-treated groove from the distance of the edge (see Figure 2) of the notched specimen (a) and welded specimen (b)

The results of the numerical determined residual stresses were compared to experimental determined values. For this, surface near residual stress measurements of the S960QL specimens were performed using X-ray diffraction with CrKα-radiation at the {211}-ferritic lattice plane. The measurements were performed with a diffractometer type Pulstec µ-360, based on the cosα-method [14], [15].The reliability of this method in combination with the 2D-detector compared to the commonly used $\sin^2\psi$-method was evaluated by Matsuda et al. [16]. The collimator diameter (measurement diameter) was 0.2 mm and the X-ray incident angle was ψ = 35° for all measurements. As elastic constants for the evaluation E = 220 GPa and v = 0.29 were used. The results are presented in Figure 3. As shown, high residual stresses appear in a distance of 0.5 mm from the edge of the HFMI-treated groove by XRD-analysis. After the cutting process (in order to obtain single specimens from the base plate), the tensile residual stresses close to the edge are decreasing. HFMI-edge treatment, the removal of defects and polishing leads to compressive residual stresses. However, it cannot be stated if this is related to the HFMI-edge treatment or the polishing of the edges. Also, the compressive residual stresses after edge treatment and polishing are less significant than the compressive residual stresses in distance of 1 mm away from the edge.

Fatigue tests

Fatigue tests were performed to quantify the influence of the edge-treatment technique on the fatigue strength of notched and welded specimens. All fatigue tests were performed with a stress ratio of R=0.1. The fatigue tests of the notched specimens were performed under 4-point bending load and for the fatigue tests of the welded specimen pure tensile loading was used. The fatigue tests were performed until the fracture of the specimens. Further details are given by Schubnell et al. [17] (notched specimen) and Gkatzogiannis et al. [18] (welded specimen).

The fatigue tests results are summarized in Figure 4. As shown, only a slight improvement was determined from the as-welded condition to the HFMI-treated condition without edge-treatment (unt. edge). Especially for the notched specimen several specimens do not exceed the range of fatigue life compared to untreated condition. However, with an additional HFMI-treatment and polishing of the edges a significant increase of fatigue life was determined for both types of specimens.

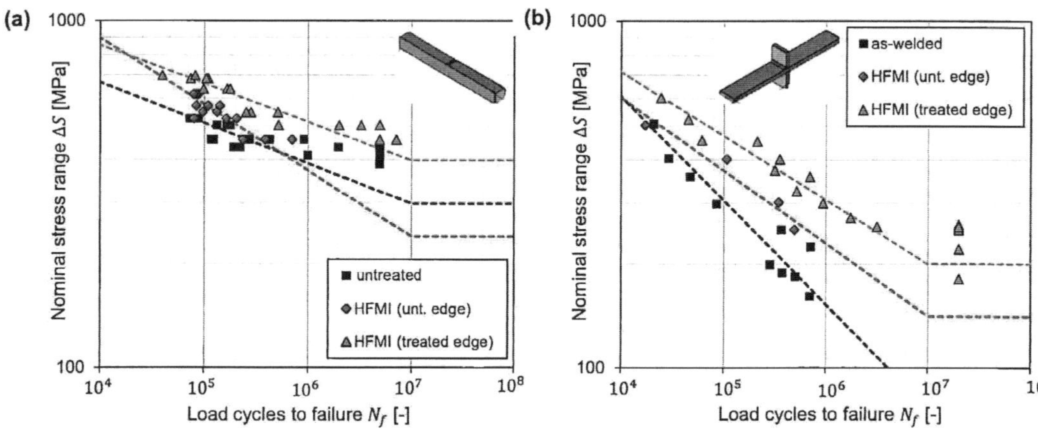

Figure 4. Fatigue test results of notched specimen (a) and welded specimen (b)

So-called beach-mark tests were performed for visual detection of the crack initiation location and crack propagation after the fracture of the specimen. For this, the load ratio was temporarily changed from R=0.1 to R=0.7. The crack propagation and the number of load cycles during R=0.7 was neglected for the evaluation. The fracture surfaces of one specimen of each condition were displayed in Figure 5 (a) (notched specimen). The displayed fracture represents the typical location of crack initiation (crack starter) of each condition. As shown, the location of crack initiation shifts from the middle of the notch (untreated condition) to the edges (HFMI (unt. edge)). After HFMI-edge treatment and polishing of the edge the crack starter shifts again to the middle of the HFMI-treated notch. Similar behavior was investigated for the welded specimen. However, in this case no additional removal of defects or polishing of the edges were performed after HFMI-edge treatment. The investigated fracture surfaces show still a crack propagation from the edge to the center of the specimen in some cases. For further investigation, the location of crack initiation was detected with by scanning electron microscopy. For all investigated specimen a high number of crack starters were determined along the HFMI-treated weld toe, see Figure 5 (b). However, in some cases crack like defects were determined as a reason of

HFMI-edge treatment, see Figure 5 (c). In this case it is strongly assumed that such a crack-like defect is the dominant failure mechanism.

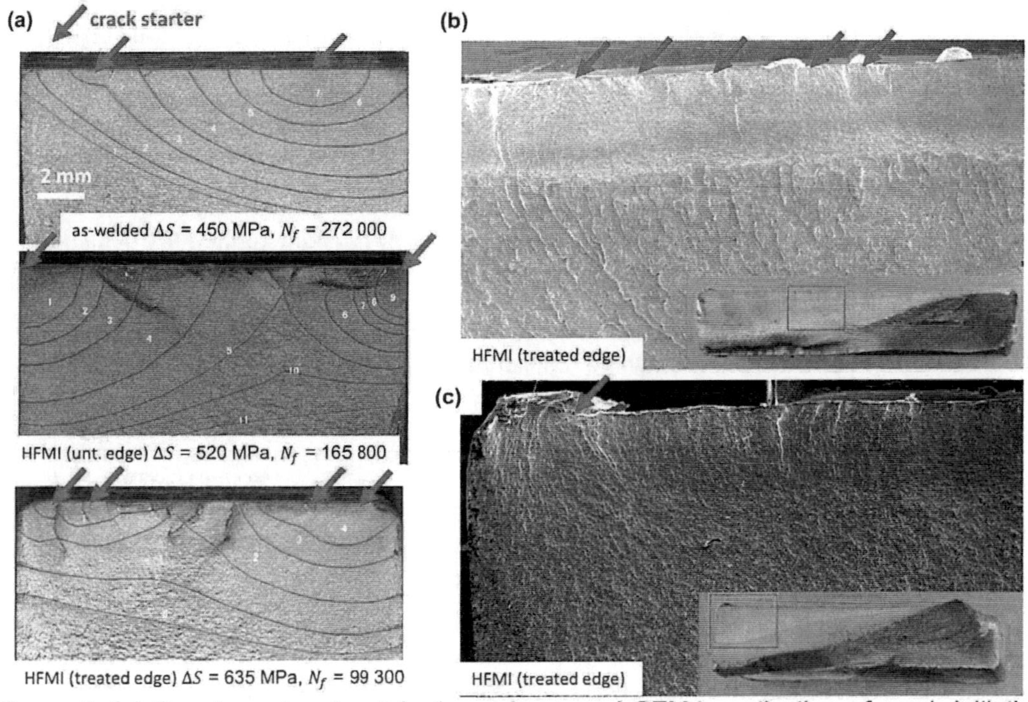

Figure 5. (a) Beach-marks at notched specimen and SEM-investigation of crack initiation locations from specimen center (b) and specimens' edge (c)

Discussion and Conclusions

Geometrical edges and corners are potential weak points in welded and non-welded constructions. This study verifies that no compressive residual stresses are present after HFMI-treatment in the vicinity of the edges based on FE simulation and XRD measurements. Two different specimen types (notched and welded) from S960QL steel were investigated. It is assumed that the lack of compressive residual stresses, the presence of defects and roughness at the edge leads to a comparable low fatigue strength and to crack initiation at these positions. An additional HFMI-edge treatment with special concave-pins was performed to eliminate these weak points. Tensile residual stresses at the edge of the HFMI-treated specimen were removed by this edge-treatment in combination with polishing of the edge but the compressive residual stress level was still low compared to the center of the HFMI-treated specimen. However, a significant improvement of fatigue life was determined for notched and welded specimens. The location of crack initiation shifts from the edge of the center of the specimen. SEM investigations of the fracture surface of the welded specimen shows that cracks also initiate at defects induced by HFMI-edge treatment. Thus, it is assumed that this kind of edge-treatment should be performed with an additional polishing of the edge and a removal of such crack-like defects.

References

[1] P. Diekhoff, J. Hensel, T. Nitschke-Pagel, and K. Dilger, "Investigation on fatigue strength of cut edges produced by various cutting methods for high-strength steels," *Weld. World*, vol. 64, no. 3, pp. 545–561, Jan. 2020, doi: 10.1007/S40194-020-00853-Y/FIGURES/14.

[2] A. F. Hobbacher, *Recommendations for Fatigue Design of Welded Joints and*

Components, 2th illust. Springer, 2016.

[3] *Eurocode 3: Design of steel structures -Part 1-9: Fatigue, 1993-1-9:2005.* Brussels: Europeon Commitee of Standardization, 2009.

[4] T. Stenberg, E. Lindgren, Z. Barsoum, and I. Barmicho, "Fatigue assessment of cut edges in high strength steel – Influence of surface quality," *Materwiss. Werksttech.*, vol. 48, no. 6, pp. 556–569, Jun. 2017, doi: 10.1002/MAWE.201600707.

[5] I. Weich, "Edge Layer Condition and Fatigue Strength of welds improved by mechanical post-weld treatment," *Weld. World*, vol. 55, no. 1, pp. 3–12, 2011.

[6] J. Schubnell *et al.*, "Residual stress relaxation in HFMI-treated fillet welds after single overload peaks," *Weld. World*, vol. 64, no. 6, pp. 1107–1117, Jun. 2020, doi: 10.1007/S40194-020-00902-6/FIGURES/9.

[7] J. Schubnell, J. Preußner, and M. Farajian, "Decreasing and Increasing the value of the compressive residual stresses induced by High Frequency Mechanical Impact during Cyclic Loading," 2022.

[8] J. Schubnell and M. Farajian, "Fatigue improvement of aluminium welds by means of deep rolling and diamond burnishing," *Weld. World*, vol. 1, pp. 1–10, Dec. 2021, doi: 10.1007/S40194-021-01212-1/TABLES/5.

[9] C. Ernould *et al.*, "Application of different simulation approaches to numerically optimize high-frequency mechanical impact (HFMI) post-treatment process," *Weld. World*, vol. 63, no. 3, pp. 725–738, May 2019, doi: 10.1007/s40194-019-00701-8.

[10] J. Schubnell, "Experimental and numercial investigation of the fatigue performance of notches and welded joints after High Frequency Mechanical Impact treatment," Karlsruhe Institute of Technology, 2021.

[11] J.-L. Chaboche, "Time-independent constitutive theories for cyclic plasticity," *Int. J. Plast.*, vol. 2.2, pp. 149–188, 1986, doi: https://doi.org/10.1016/0749-6419(86)90010-0.

[12] A. Maciolek, "Implementierung eines elasto-viskoplastischen Materialmodells zur Simulation des Kugelstrahlens an Komponenten aus 42CrMoS4 Stahl, (Implementation of a elasto-viscoplastic material model of the simulation of shot peening at components of 41CrMoS4 steel," KIT Karlsruhe, 2017.

[13] J. Schubnell, D. Discher, and M. Farajian, "Determination of the static, dynamic and cyclic properties of the heat affected zone for different steel grades," *Mater. Test.*, vol. 61, no. 7, pp. 635–642, Jul. 2019, doi: 10.3139/120.111367.

[14] K. Tanaka, "The cos*α* method for X-ray residual stress measurement using two-dimensional detector," *Mech. Eng. Rev.*, vol. 6, no. 1, pp. 18-00378-18–00378, 2019, doi: 10.1299/mer.18-00378.

[15] S. Taira, K. Tanaka, and T. Yamasaki, "A Method of X-Ray Microbeam Measurement of Local Stress an Its Application to Fatigue Crack Growth Problems," *J. Soc. Mater. Sci.*, vol. 27.294, pp. 251–256, 1978.

[16] M. MATSUDA, K. OKITA, T. NAKAGAWA, and T. SASAKI, "Application of X-ray stress measurement for residual stress analysis by inherent strain method - Comparison of cos<i>α</i> and sin²<i>Ψ</i> method-," *Mech. Eng. J.*, vol. 4, no. 5, pp. 17-00022-17–00022, 2017, doi: 10.1299/mej.17-00022.

[17] J. Schubnell, P. Pontner, R. C. Wimpory, M. Farajian, and V. Schulze, "The Influence of Work Hardening and Residual Stresses on the Fatigue Behavior of High Frequency Mechanical Impact Treated Surface Layers," *Int. J. Fatigue*, vol. 134, pp. 125–138, Jan. 2020, doi: 10.1016/j.ijfatigue.2019.105450.

[18] S. Gkatzogiannis, J. Schubnell, P. Knoedel, M. Farajian, T. Ummenhofer, and M. Luke, "Investigating the fatigue behaviour of small scale and real size HFMI-treated components of high strength steels," *Eng. Fail. Anal.*, vol. 123, p. 105300, May 2021, doi: 10.1016/J.ENGFAILANAL.2021.105300.

Development of Residual Stresses Induced by Deep Rolling at an Additive Structure

E. Mueller

Department of Mechatronics and Mechanical Engineering, Bochum University of Applied Sciences, Bochum, Germany
Steinbeis-Transfercenter for Spring Technologies, Component Behavior and Process, Iserlohn, Germany

Abstract

This paper shows the residual stress distributions over the depth of a dice out of 1.2344 produced as an additive structure. Because of the layer structure (obtained by SLM) it gives two interesting directions (along with the layer and perpendicular to the layer) to be investigated concerning residual stresses. Next a deep rolling process was done. Deep rolling gives different residual stress distribution. Perpendicular to the rolling track higher compressive residual stress is induced in comparison along the track. Combined with the layer structure of the dice, four different combinations are obtained that are investigated concerning residual stresses.

Keywords: residual stress, deep rolling, additive structure.

Introduction

Additive manufactured structures are very interesting for one-piece manufacturing. The wish is to use them in dynamic applications. Compared with conventional components, the durability is not very high. An increase could be done by mechanical surface treatment like shot peening and deep rolling. The characteristic of the compressive residual stresses plays an important role in the fatigue strength. Deep rolling is one common method to induce high compressive residual stresses with a deep penetration [3]. The layer structure of an additive manufactured component gives different compressive residual stress profiles after shot peening [4,5]. Resulting of this fact, an important question is, how are the compressive residual stress profile by deep rolling in dependence of the layer orientation.

Experimental Methods

Specimen

For this experiment, a specimen with the shape of a dice was printed with the SLM (Selective Laser Melting) method. The edge length is 30 mm, and the layer thickness is 30 µm. The material is H13 (1.2344) with a hardness of 52 HRC.

Figure 1: The specimen

Residual Stress Determination

The residual stress was determined via X-ray diffraction. The used equipment was a Strainflex MSF-2M from Rigaku. The tube had a Cr-anode with CrK_{α} radiation and a current of 10 mA. The detector was a photomultiplier with a slit aperture on the tube and detection side. The exposed area was 6 mm * 8 mm. The measurement errors are +/- 30 MPa or 7% (the higher one). To get the full residual stress profile into the depth, the material was removed electrolytically by an area of 8 mm diameter.

Deep Rolling

The specimen was deep rolled with the tool HG 6 from Ecoroll (see: https://www.ecoroll.de/produkte/hydrostatische-werkzeuge.html). This hydrostatic tool has a ball of 6 mm in diameter. The specimen was rolled in a meander shape with a spacing ΔX of the tracks ΔX =150 µm, and the pressure p was p = 400 bar.

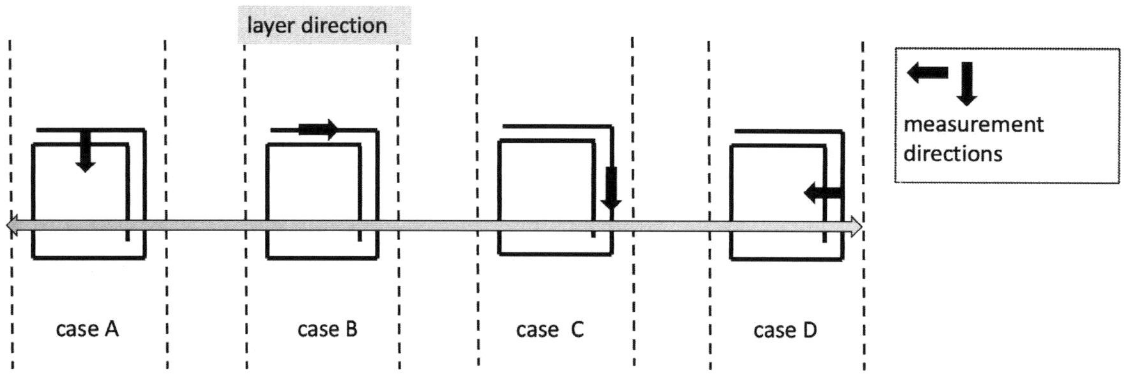

Figure 2: The four different cases between deep rolling and layer structure

Deep rolling induces fundamental different residual stresses along and perpendicular to the rolling track, which means there are four cases:

Table 1: Explanation of the different cases (**≡** = parallel, **⊥** = perpendicular)

case	**rolling track ↔ layer directions**	**measuring direction ↔ trolling track**
A	≡	⊥
B	≡	≡
C	⊥	≡
D	⊥	⊥

These four cases were examined concerning residual stress distribution in dependence of the depth.

Experimental Results

At first, the residual stress profiles for the non-deep-rolled side of the dice are shown what the initial state is (fig. 3 a). Additionally, the full width of half maximum (FWHM) is given as a relative the number of dislocations (fig. 3b). If there are dislocations, the spacing of the atomic layers differs a lot which causes a widening of the reflected beam.

There are different residual stress distributions. All residual stresses are tensile stresses, but perpendicular to the layers, they are very high in opposite to the moderate tensile residual stresses along with the layers. The FWHM is very high, which means lots of dislocation.

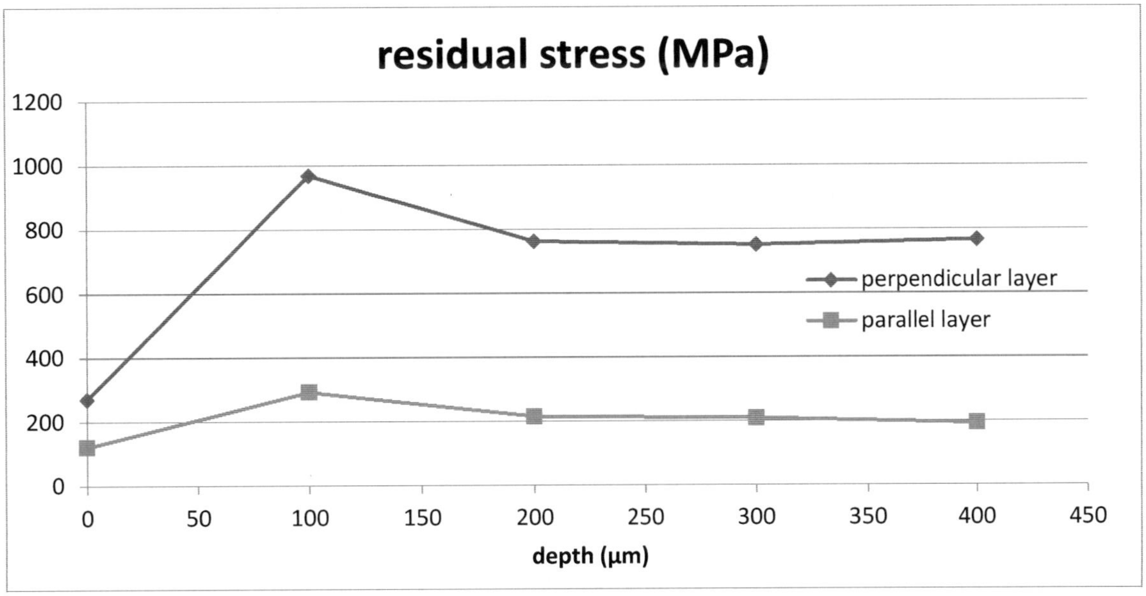

Figure 3a: Residual stress profile in the untreated state

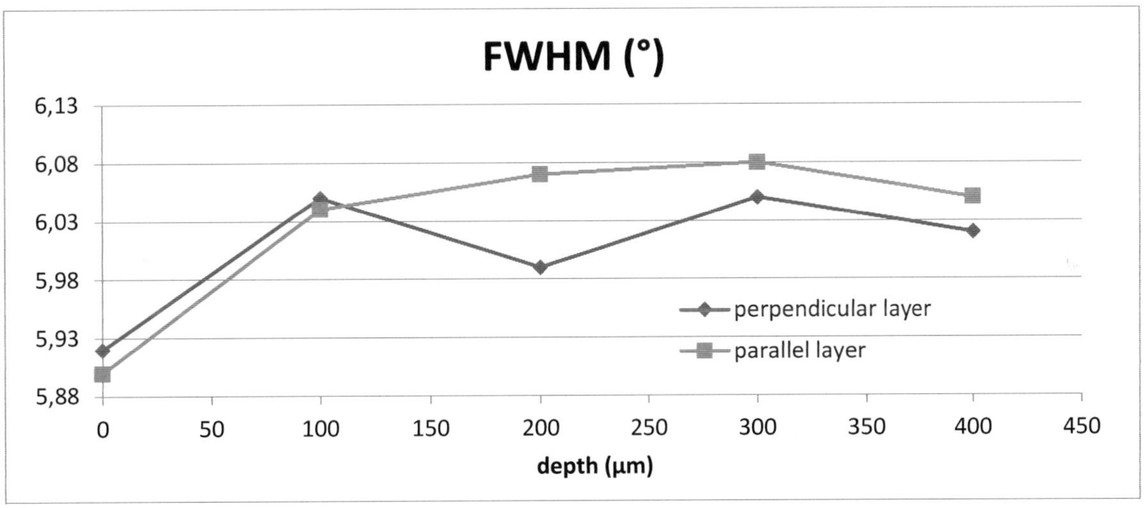

Figure 3b: Development of the FWHM over the depth.

Interesting is, to compare cases A and D respectively cases B and C. At cases A and D, the residual stresses are determined perpendicular to the rolling track. For deep rolling in this direction, higher compressive residual stresses are expected. Cases B and D are both along the track, and the expected compressive residual stresses are significantly lower. In this case, the profile looks like induced by Hertzian pressure. Perpendicular, the compressive residual stresses are decreasing continuously with the depth [6, 7].
Figure 4 shows the comparison of case A and case D in the residual stresses and the FWHM.

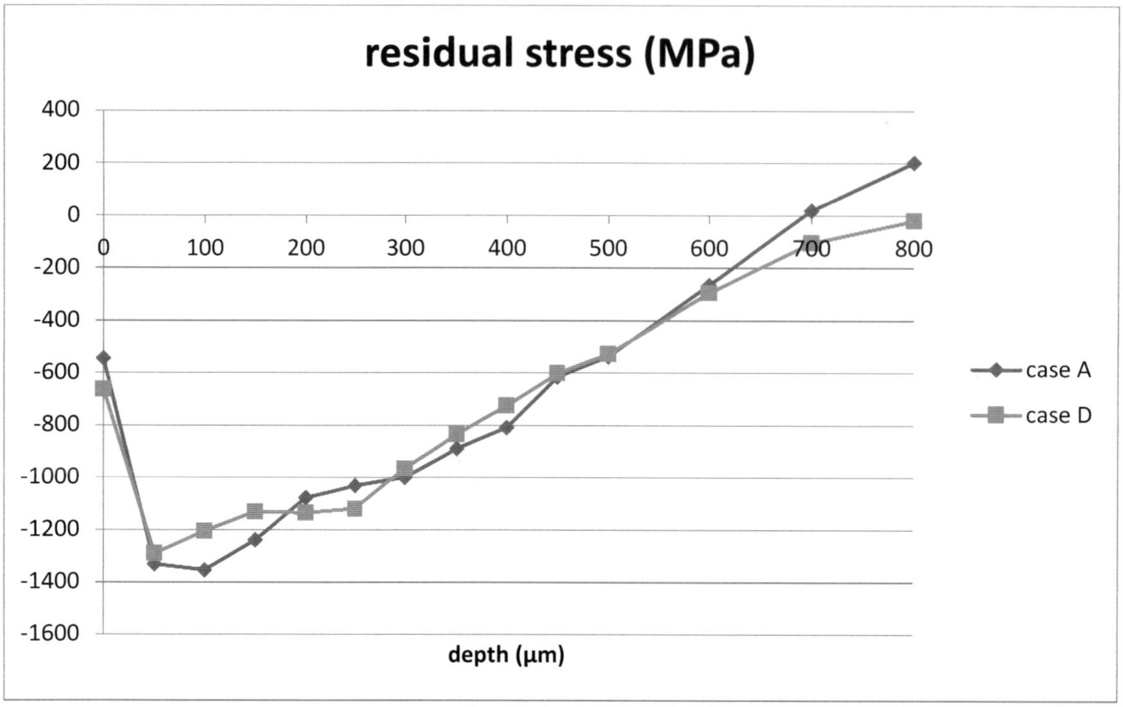

Figure 4a: Residual stress distribution of the cases A and D

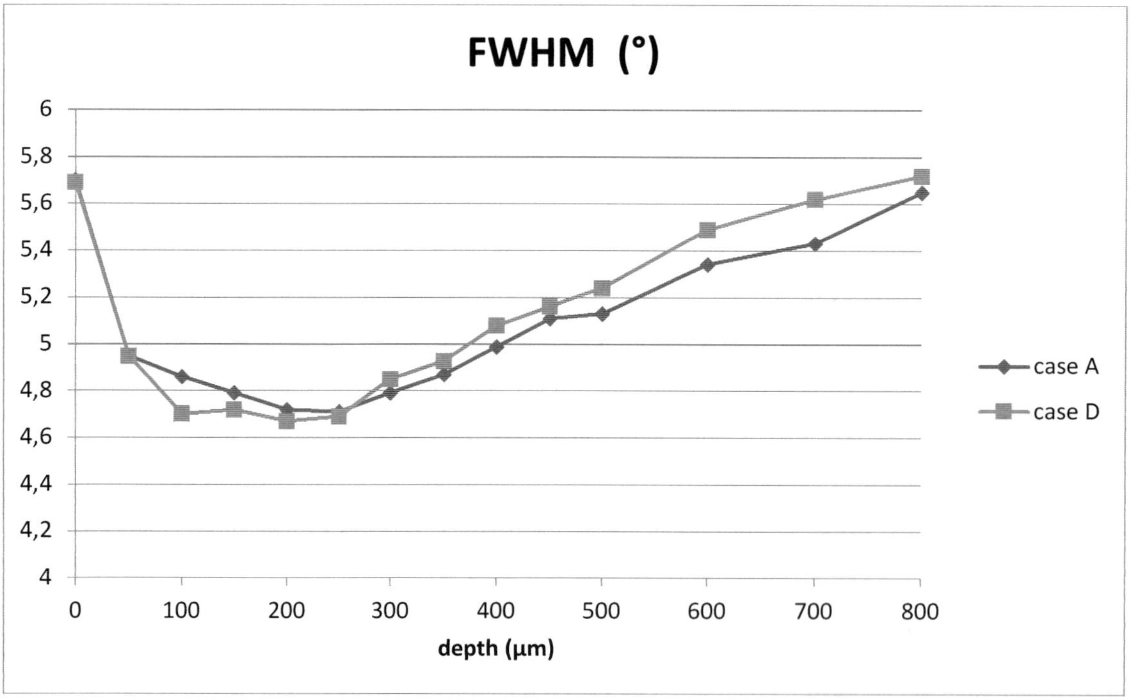

Figure 4b: Development of the FWHM over the depth for cases A and D

The residual stress profile and FWHM for cases B and C are shown in figure 5. The typical residual stress distribution induced by Hertzian pressure is achieved, and the compressive residual stresses are lower compared with the compressive residual stresses perpendicular to the rolling track.

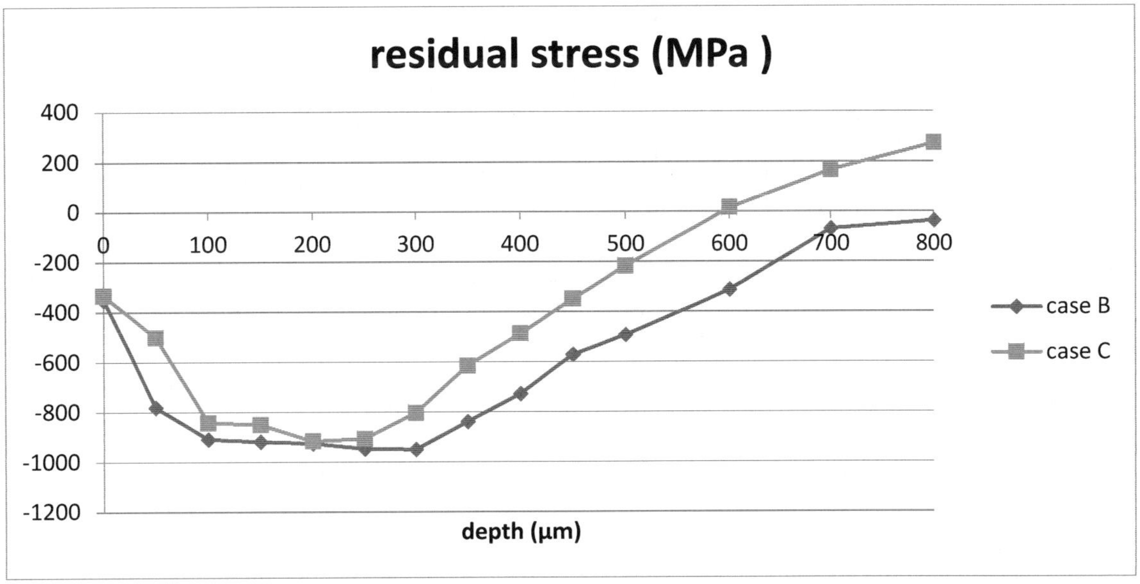

Figure 5a: Residual stress distribution of the cases B and C

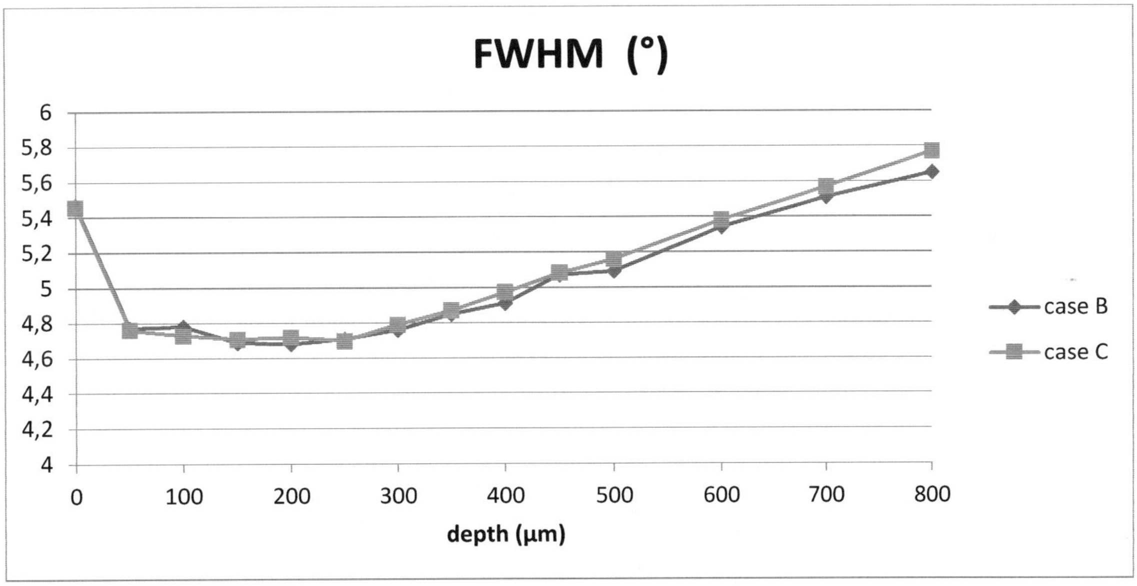

Figure 5 b: Development of the FWHM over the depth for cases B and C

The interpretation of the result will follow now.

Discussion and Conclusion

The different tensile stresses are strongly dependent on the layer orientation at the untreated surface. After deep rolling, this splitting of the profiles is totally gone for the induced compressive residual stresses perpendicular to the rolling direction (cases A and D). The behavior of the residual stresses is as expected surface values [6,7]. Here, surface effects cause the lower values of the compressive residual stress. No dependence on the layer orientation can be seen. The distribution of the FWHM shows that the number of dislocations has been decreased like also at shot peening. [8] At the maximum of the compressive residual

stress, the dislocations are moving to the grain borders, and the scattering of the X-ray beam is less.
Looking at the residual distributions that are along the track, show the distributions obtained by Hertzian pressure and the maximum of the compressive residual stress is lower compared to the stresses perpendicular to the track. The splitting of the profiles started around 300 µm in depth. Is the track along with the layer, a higher penetration depth of the compressive residual stresses can be observed. The FWHM distribution is nearly the same as perpendicular to the track.
The conclusion is that the induced residual stresses by deep rolling are dependent on the layer orientation. For a deep rolled component, it is important to take this fact into account to improve the fatigue strength in an optimal way.

References

[1] D. H. Wohlfahrt, *Einfluß von Mittelspannung und Eigenspannung auf die Dauerfestigkeit*, VDI-Berichte 661: Dauerfestigkeit und Zeitfestigkeit, VDI-Verlag, Düsseldorf, 1988, p 99

[2] M. Habschied, B. de Graff, A. Klumpp, V. Schulz, F*ertigung und Eigenspannungen,* Journal of Heat Treatment Materials, Vol. 70 (2005) 3, p 119

[3] I. Altenberger, *Deep rolling – the past, the present and the future*, in V. Schulze, A. Nikulari: Shot Peening and other Mechanical Surface Treatments, IITT-International, Noisy-le-Grand, 2005, pp 144-155

[4] E. Müller, *Die Ausbildung von Eigenspannungen durch Kugelstrahlen bei einer additiven Struktur*, in H.-J. Christ (ed.), Werkstoffe und Bauteile auf dem Prüfstand, Inventum GmbH, Sankt Augustin, 2019, pp 57-62

[5] E. Müller, *Eigenspannungsverteilungen durch Kugelstrahlen in einer additiv gefertigten austenitischen Struktur*, in S. Brockmann, U. Krupp: Werkstoffe und Bauteile auf dem Prüfstand, Stahlinstitut VDEh, Düsseldorf, 2021, pp 343-346

[6] E. Müller, *Comparison of the Residual Stresses Induced by Shot (Stress) Peening and Rolling in Spring Steel*, in L. Wagner (ed.): Shot Peening, Wiley-VCH-Verlag, Weinheim 2003, pp 441 – 446,

[7] E. Müller, *Evolution of the residual stresses by stress rolling*, in V. Schulze u. A. Nikul-Lari (ed.): Shot Peening and other Mechanical Surface Treatments, IITT-International, Noisy-le-Grand, 2005, pp 436 – 441

[8] E. Müller: *Die Ausbildung von Eigenspannungen nach der Additiven Fertigung und nach dem Kugelstrahlen eines ferritischen Stahls*, in M. Eiber, R. A. Richard (ed.): Additiv gefertigte Bauteile und Strukturen, Bericht 406, DVM, Berlin, 2021, pp. 31-45, https://doi.org/10.48447/ADD-2021-BB

Coverage and Peening Angle Effects in Shot Peening on HCF Performance of Ti-6Al-4V

A. Jochen P. Fuhr[1]
B. Mahmoud Basha[2], C. Manfred Wollmann[2], D. Lothar Wagner[2]
1 Curtiss-Wright Surface Technologies/Metal Improvement Company, Germany
2 Institute of Materials Science and Engineering, TU Clausthal, Germany

Abstract
Shot peening of titanium alloys is known to enhance the HCF performance by inducing in near-surface regions residual compressive stresses and increasing the dislocation density.
The present work aimed at studying the effect of coverage and peening angle effects in shot peening on cyclic fatigue performance on Ti-6Al-4V.
For shot peening a soft and hard type of cast steel shot and a comparable size conditioned cut wire shot have been used.
The coverage was varied from 20 % to 1200 % and the peening angle from 90 to 30°. Fatigue performance of shot peened rotating- and alternating-bending fatigue specimens was studied.
The results indicate that low (20%) coverage peening leads to a loss whereas full (100 %) up to a high (1200 %) coverage was found to result in a marked increase in HCF strength.
The variation of peening angles resulted in a significant beneficial effect on the fatigue performance in direction to increasingly flat angles and high coverages (1200 %) at constant Almen intensities of 0,20 mmA.
These results are even more interesting as those peened surfaces resulted in the highest surface roughness and a topography with a starting waviness.

Key Words – Shot Peening, Cyclic Fatigue, Impact Angle, Ti-6Al-4V, Waviness

Introduction

Ti-6Al-4V is a widely used Ti-alloy in aerospace industry and many of the components made from this are shot peened to increase the cyclic fatigue performance.
Nearly all existing fatigue data of shot peened components are based on the "idealized" conditions of ≥ 100 % coverage and an impingement angle of 90°. But due to complex geometries or the repeating repair of several of these components or areas it cannot be avoided that some of these surfaces will or must be shot peened with a much wider range of coverage and/or impingement angles.
Target of this work was to investigate the effect of shot peening with realistic ranges of coverage and impingement angles on the HCF performance of the Ti-6Al-4V material.

Material & Experimental Methods

Ti-6Al-4V was received as cylindrical rods Ø 12.7 mm. The used material was Grade 5 (3.7165) fulfilling the material specifications ASME SB348, ASTM B348, UNS R56400.

<u>Hour-glass samples for coverage investigations</u>
For fatigue testing the 20, 100 and 1200 % coverage, hour-glass shaped specimens (6 mm minimum cross diameter) were machined. Electrolytically polished (EP) and mechanically polished (MP) conditions were used as references to which the shot peening at various coverage degrees were compared. EP was done to remove about 100 μm from the as-machined surface to exclude any machining effects that could have masked the results. During MP, a surface layer of about 50 μm was removed from the as-machined surface resulting in the same smooth surface condition as observed in EP condition.
Fatigue tests were performed in load controlled rotating beam loading (R = -1) in air at a frequency of 50 s-1.

Flat-shaped samples for impingement angle investigations

For the testing of the 90, 60, 45 and 30 degrees impingement angle at 100 and 1200 % coverage flat-shaped fatigue samples have been machined and shot peened. The fatigue testing was carried out in a load-controlled fatigue machine where the force on the flat specimen is applied by a cam and linkage (R = -1) in air at a frequency of 8 s^{-1}.
All Specimens were considered as run outs after 10^7 cycles.

Shot Peening

Shot Peening was carried out using a direct pressure blast system using a single nozzle.

Three peening medias were used for this study: SCCW14, ASR110 and ASH110 with an average diameter of 0.35 and 0.3 mm (Fig. 1).

Fig. 1: Magnification (200x) of the shot material (a) SCCW14, (b) ASR110 and (c) ASH110

The target Almen intensity for all samples was 0.20 mmA.
Only exception in terms of used Almen intensity was the 30 degrees flat-shaped fatigue sample with a maximum shot peening intensity of 0.16 mmA limited by the maximum available pressure in the specific machine configuration.

All three coverage degrees (20, 100, 1200 %) and impingement angles (90, 60, 45, 30 degrees) represent occurring values in the industry e.g. in areas with very low coverage rates in opposition to areas with e.g. ricochet peening or complicate geometries where coverage rates > 1000 % and low impingement angles can be reached.

To produce 20 % coverage a rotating disc with a slit was used in between the nozzle and the sample ensuring that only for a short time a low amount of shot particles with the appropriate speed was impacting the surface of the rotating hour-glass shaped fatigue sample.

Experimental Results

Hour-glass shaped fatigue samples

Surface roughness

The roughness of all shot peened samples was increased in comparison to the electropolished (EP) and mechanically polished (MP) condition. The highest value was reached by using the SCCW14 at 1200 % coverage.

Fatigue:

The results of the fatigue tests (Fig. 2) confirm the experience [1] that the lowest performance in HCF fatigue can be seen in all samples peened with low coverage of 20 % followed by the stress-free EP condition. All versions covered with 100 % resulted in the highest fatigue performance. The Lowest fatigue strength at 20 % coverage showed the ASR110 with -15 % to the EP fatigue level followed by ASH110 and SCCW14 shot. On the upper side the SCCW14 showed the highest increase with +13 % followed by the ASR110 and the ASH110 in comparison to the EP level of 650 MPa. The obvious decrease of fatigue performance of the 20 % covered samples can be explained by the insufficient residual compressive stress fields which cannot compensate the early crack initiation caused by the high roughness especially around single shot indents.

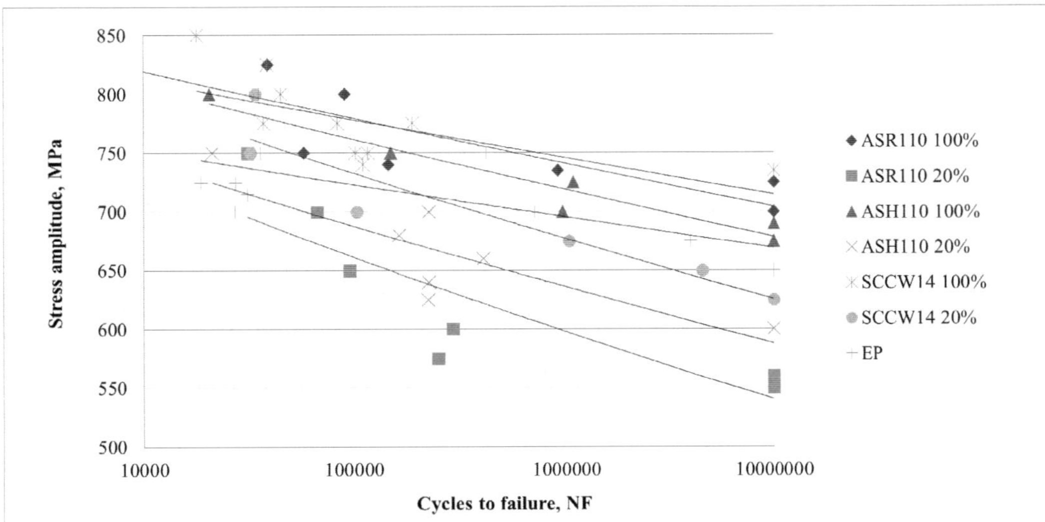

Fig. 2: S-N curves of the fatigue probes in different treatment conditions

Flat-shaped fatigue samples
A complex geometry of a workpiece can lead to a wide range of impact angles others than 90 degrees. Most aircraft and engine manufacturers [12-13] are limiting the angles in a range of 90 to 45 degrees whereas the SAE J443 [9] reduces the lowest possible angel of impact to a minimum of 30 degrees.
For this reason 4 different impact angles (90, 60, 45, 30 degrees) have been investigated in terms of resulting surface topography (roughness, waviness) and fatigue performance.
Important to notice is that for the trials on the flat-shaped fatigue samples peened with different peening angles the pressure was increased with lower angle of attack to keep the Almen intensity to or as close as possible to the target of 0.20 mmA.

Surface roughness/waviness:
Fig. 3 shows the results of roughness over the 4 impact angles at 100% and 1200 % coverage.

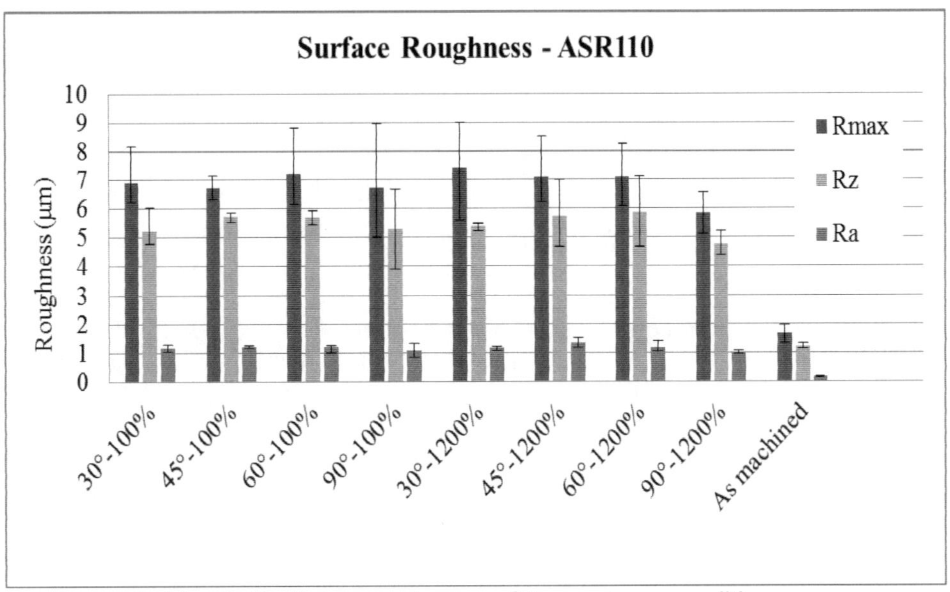

Fig. 3: Surface roughness of the various conditions

Beside the increased roughness with lower impact angles and increased coverage rates the samples showed a parallel increase of waviness which is obviously caused by the pushing movement of the material along the impact direction of the shot. Further amplifier for this deformation and material moving effect is the increasing shot speed at lower impact angles to keep the targeted Almen intensity at the same level.
This waviness has its visual and measured maximum at 45 degrees and coverage of 1200 % (see Fig. 4 and Table 1).

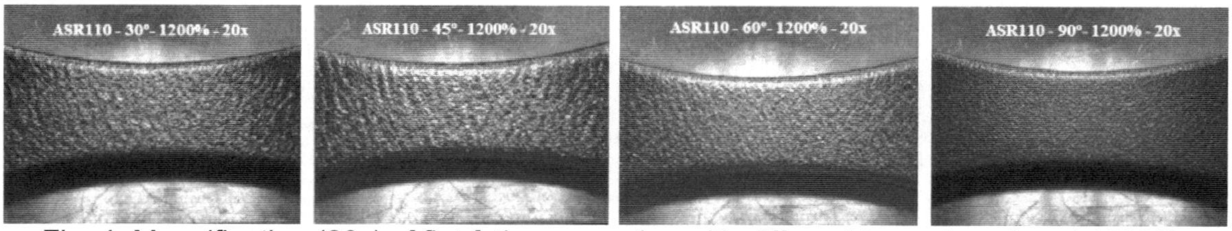

Fig. 4: Magnification (20x) of flat fatigue samples with different impact angles at 1200% coverage

Table 1: Waviness [Wt] of flat-shape fatigue samples with different impact angles

Impact angle [degrees]	Waviness [Wt in μm]	
	100 % coverage	1200 % coverage
90	2.50	3.25
60	2.53	5.93
45	5.00	13.67
30	3.71	6.15

<u>Fatigue:</u>
The results of the fatigue tests on the flat-shape samples can be seen in Fig. 5 for 100 % and 1200 % coverage over the 4 different investigated impact angles.

(a) (b)

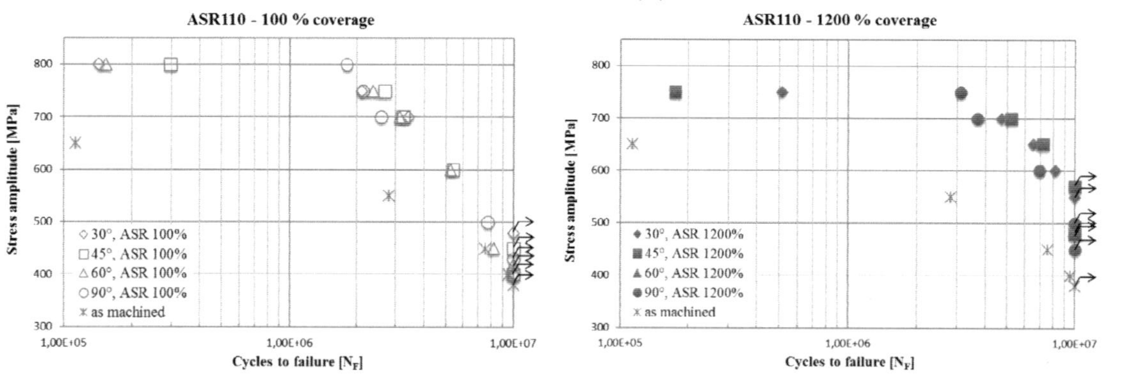

Fig. 5: S/N curves of the flat fatigue probes in the SP condition using ASR110 at 90, 60, 45, 30 degrees impact angle at (a) 100 % and (b) 1200 % coverage

All shot peened samples showed an increase in fatigue performance in comparison to the as machined condition. The samples with 1200 % coverage result in the highest increase.
Looking on the different impact angles the largest improvement can be found at an angle for 45 degrees at 1200 % coverage – but also the 30 degrees impact angle shows a significant increase even so the shot peening intensity here was due to machine reasons lower than for all other investigated impact angles.

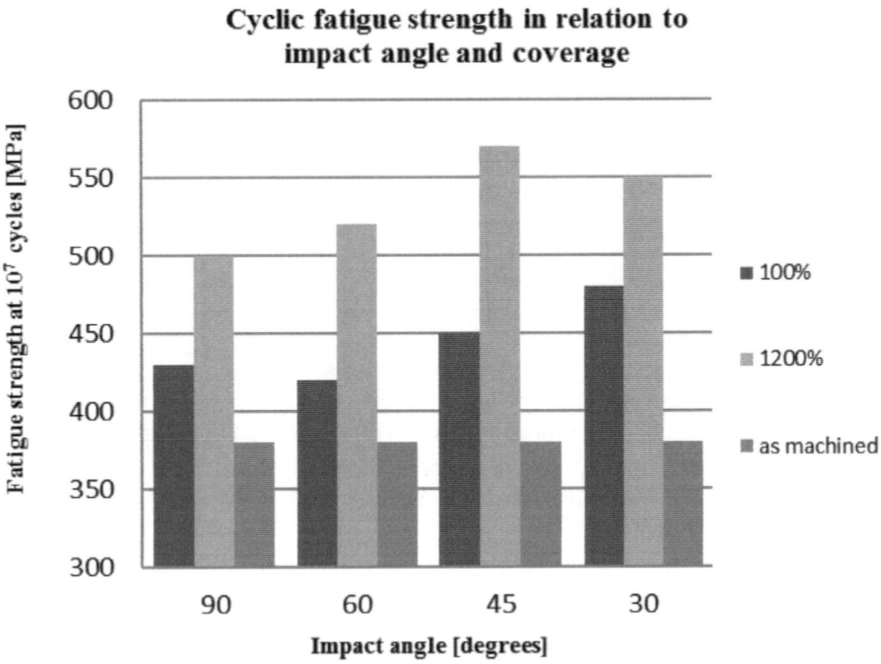

Fig. 6: Cyclic fatigue strength of the flat bending fatigue sample in relation to shot peening impact angle and coverage treated with ASR110

Fig. 6 shows that especially for the high coverages all impact angles smaller than 90 degrees show a larger increase in fatigue strength with a maximum at 45 degrees resulting in a 14% higher value than the shot peened version with same coverage at 90 degrees impact angle and 50 % higher strength than the as machined condition.

This result stands in contradiction to the increased roughness but especially the significant increased waviness (Table 1) caused by the shear plastic flow of the material at lower impact angles.

A possible explanation for the consistent fatigue performance at different impact angles is presumably the targeted constant Almen intensity kept by adjustment of air pressure and the resulting shot particle speed. The resulting compressive stress, dislocation density and material flow in the close surface area under shear plastic loading seems to induce comparable or even higher residual compressive stresses and dislocation densities which are retarding the growth rates of surface cracks increase the resistance to fatigue crack initiation (Fig. 7).

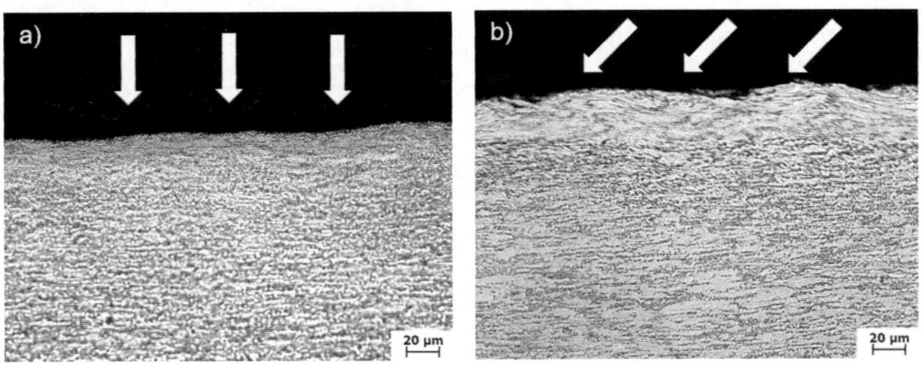

Fig. 7: Microsection of the flat bending fatigue samples shot peened with (a) 90 degrees and (b) 45 degrees

Conclusions

Shot peening increased the surface roughness compared to the reference as-machined conditions. However, the roughness values were still lower than the allowed roughness values as specified by many aerospace shot peening specifications.

Low coverage (20 %) resulted in a significant loss of fatigue performance. High coverages (1200 %) showed in the most cases a slight increase of roughness values and induced compressive residual stresses.

On the flat bending fatigue probes the highest fatigue strength was found with an increasing amount for the high coverages (1200 %) in direction to smaller impact angles than 90 degrees. Significantly increased fatigue strength values have been found for 45 degrees impact angle and 1200 % coverage although theses samples showed the highest roughness values and surface waviness.

The waviness causing deformation of the surface material texture orientated 90 degrees to the preferred cracking direction seems to be a further fatigue improving factor beside the well-known effects of residual compressive stress and increased dislocation density due to the shot peen treatment. Within the investigated range of shot peening intensities and coverages the resulting roughness and waviness seems to be a negligible detrimental factor for the resulting fatigue performance.

In all cases SP led to a marked improvement in the fatigue life compared to the EP and MP conditions. Using different SP media had no significant effects on the resulting fatigue life even a slight better fatigue performance can be seen using cut wire in comparison to cast steel shot.

References

[1] L. Wagner et al, Fatigue Performance of Al 7075-T73 and Ti-6Al-V: Comparing Results after Shot Peening, Laser Shock Peening and Ball-Burnishing, International Journal of Structural Integrity, 2, (2011), 185-199

[2] T. Ludian, L. Wagner, Coverage Effects in Shot Peening of Al2024-T4, ICSP9; Paris, Marne la Vallee, France, (2005), pp 296.]

[3] P.S. Prevéy, J.T. Cammett, The Effect of Shot Peening Coverage on Residual Stress, Cold Work and Fatigue in a Ni-Cr-Mo Low Alloy Stee,l ICSP8, Garmisch-Partenkirchen, Germany, (2002)

[4] S. A. Meguid, Effect of Partial-Coverage upon the Fatigue Fracture Behaviour of Peened Components, Fatigue Fract. Engng. Mater. Struct. Vol. 14, No. 5, (1991) pp 515

[5] L. Wagner, T. Ludian, Importance of Coverage in Shot Peening on Fatigue of Al 2024-T4, Metal Finishing News, 6, (2005)

[6] J. O. Almen, SAE Journal, 51, (1943), 248

[7] SAE Recommended Practice; Procedures for Using Standard Shot Peening Test Strip, SAE J442 Rev, (2008)

[8] SAE International, J2277 OCT2009, Shot Peening Coverage Determination, (2009)

[9] SAE International, J443 JUN2010, Procedures for Using Standard Shot Peening Almen Strips, (2010)

[10] Rolls-Royce, RPS428, Controlled Shot Peening, Iss. 17, (2010)

[11] SNECMA Moteurs, DMP28, Surface Compressive Stressing, Iss. L, (2007)

[12] Boeing Aircraft Corp., BAC5730 (Revision R), p. 28, (2011)

[13] Airbus Industries, AIPI 02-02-004, p. 8, (2011)

Shot-peened induced residual stresses relaxation on steel DIN 34CrNiMo6 under different axial loading

N. Leguinagoicoa[1], J. Albizuri[1]

1 Mechanical Dept., Faculty of Engineering of Bilbao, University of the Basque Country

Abstract
Shot-peening treatment (I_{SP}: 8A & C:200%) was applied to improve the R: -1 axial fatigue strength of DIN 34CrNiMo6 (Q+T) steel specimens, due to the introduction of a compressive residual stress field in the near-surface area. However, this residual stress field tends to reduce or relax when an external stress, quasi-static or cyclic, is applied. Quasi-static tests were conducted to evaluate the role of tensile or compressive applied stress. Results showed that tensile applied stress induced lower relaxation and compressive applied stress required lower magnitude to initiate the stress relaxation. Cyclic tests, with R: -1, were developed to evaluate the influence of the cyclic applied stress magnitude on the surface stress relaxation. In addition, cyclic tests with load ratio $R \neq -1$ ($-0.5 < R < -0.25$), including tensile mean stress, were performed to assess the influence of the tensile mean stresses on the residual stress relaxation. Results showed that, in the range of R studied, higher tensile mean stresses reduce the amplitude of residual stress relaxation.

Keywords shot-peening, residual stress relaxation, axial applied stress, mean stress.

Introduction
Shot-peening is a mechanical treatment widely employed to increase fatigue resistance of the treated parts. This improvement is related with the introduced residual stress field, which reduces the crack growth rate, and the work-hardening, which enhances the mechanical properties of the treated material, due to the increase of dislocation densities, twinning, phase transformations or nano-crystallization [1]. This microstructure distortion is quantified by the parameter FWHM.
However, there are two main factors which reduce the fatigue improvement caused by shot-peening. Surface plastic deformations increase the surface roughness and surface micro-defects or micro-cracks, which facilitate the crack nucleation. In addition, compressive residual stress field tends to reduce when an external stress, quasi-static or cyclic, is applied. If the external applied stress is high enough to overcome the yield point, the onset of plastics strains will produce residual stress relaxation [2]. So, higher applied stress produces higher relaxation rates. When the applied stress is cyclic, cyclic mechanical properties of the treated material should be considered, especially when the material tends to cyclic-softening.

Experimental Methods
Hourglass specimens were manufactured according to ASTM E466-15 with DIN 34CrNiMo6 (Q+T) steel, whose monotonic mechanical properties are presented in Table 1. This material presents cyclic-softening behaviour and the cyclic yield point was set at 825 MPa.

Table 1. Monotonic mechanical properties of DIN 34CrNiMo6.

σ_{yp}	σ_{ut}	ε_{ut}	Z
1,084 MPa	1,209 MPa	12.18 %	60.17 %

Surface roughness of the shot-peened samples were measured according to ISO 4287, obtaining R_a 1.93 µm and R_z 11.3 µm. Surface hardness and the in-depth microhardness according to ISO 4545-1:2018 were measured for shot-peened samples. Shot-peening produced a slight hardness increment, from 38 HRC to 39.8 HRC for a depth of 0.15-0.20 mm. In addition, FWHM and residual stress field in-depth were measured with X-Ray diffraction technique [3] and the correction for layer removal is performed following [4]. Results are shown in Figure 1.

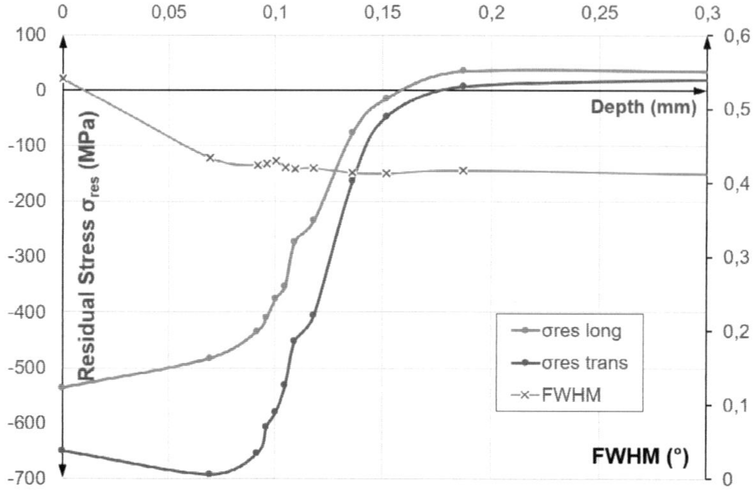

Figure 1. Shot-peened surface properties.

Fatigue tests were carried out with a servo hydraulic test rig to assess the fatigue improvement provided by the shot-peening treatment. With the same set-up, quasi-static and cyclic axial tests were performed to assess the surface residual stress relaxation. Quasi-static tests were applied in tensile or compressive direction and six different stress applied magnitudes were selected. Test data can be seen in Table 2. Surface residual stresses were measured before and after the tests.

Table 2. Quasi-static tests definition.

	Stress applied magnitude (MPa)					
Stress applied direction	Nº 1	Nº 2	Nº 3	Nº 4	Nº 5	Nº 6
Tensile (T)	630	850	978	1,084	1,119	1,154
Compressive (C)	-630	-850	-978	-1,084	-1,119	-1,154

Cyclic applied stress tests with no mean stress (R: -1) were conducted with four different magnitudes of σ_a; A_1: ± 849 MPa, A_2: ±776 MPa, A_3: ± 703 MPa and A_4: ± 630 MPa. After applying a determinate set of cycles, the samples were removed from the test rig and the surface residual stresses were measured. In addition, cyclic tests including mean tensile stress ($-0.5 < R < -0.25$) were performed with three different stress ratio R. Selected stress cases can be seen in Table 3.

Table 3. R≠ -1 cyclic tests definition.

Case	MT_1	MT_2	MT_3
σ_a	733.7	607.5	611.4
σ_m	244.6	244.6	366.8
R	-0.50	-0.426	-0.25

Experimental Results

Axial Fatigue Tests

S-N curves for machined and shot-peened variants were calculated according to ASTM E739-10-2015. Their S-N curves are presented in Figure 2. Shot-peening treatment provides an improvement of 31% for the fatigue limit compared to non-peened condition (machined variant surface condition: R_a 0.8 μm and R_z 4.5 μm)

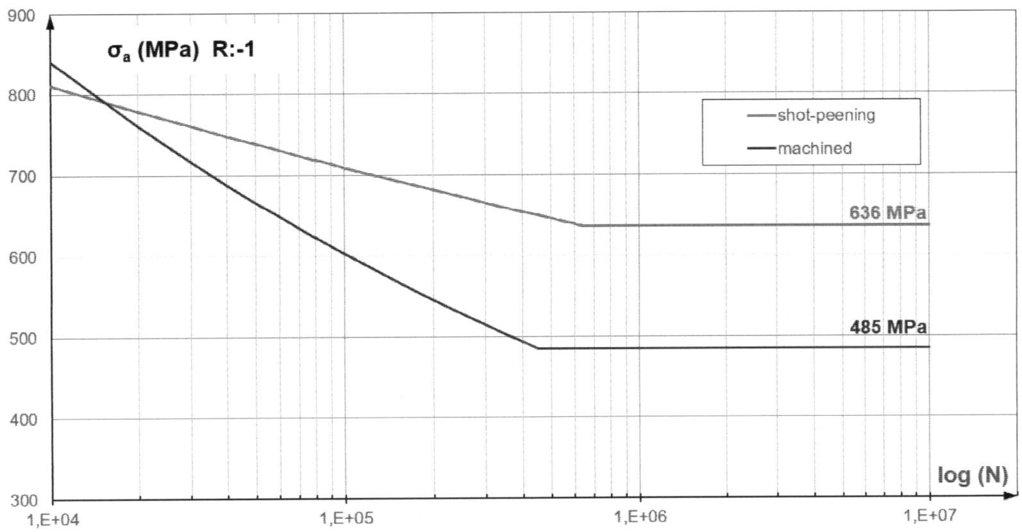

Figure 2. S-N curves for machined and shot-peened variants.

Quasi-static tests

Relaxed surface residual stresses after the quasi-static tests are presented in Figure 3. Compressive applied stresses produce higher residual stress relaxation than tensile applied stresses. Applied stresses with higher magnitude produces higher relaxation.

The onset of residual stress relaxation with tensile applied stress is set at 997 MPa, which is the *effective tensile yield point* for surface residual stress relaxation. In this case, residual stress relaxation is produced by the plastic strains generated in the core of the sample.

The onset of residual stress relaxation with a compressive applied stress can be set at -444,7 MPa, which is an amplitude wide lower than tensile loading. Considering the average longitudinal residual stress after shot-peening treatment is -535.5 MPa, the *effective compressive yield point* can be set as the sum of both stresses in compressive direction, 980.2 MPa, which is quite similar to tensile case. In this case, plastic strains are developed in the external ring, the area where the residual stress field is generated by shot-peening process.

R: -1 cyclic tests

Cyclic relaxation presents higher relaxation rate if the applied stress has higher magnitude, as can be observed in the Figure 4. Higher applied stress magnitudes generate higher plastic strains according to Ramberg-Osgood equation for cyclic loading [5]. In this research, the two highest applied stress levels (A_1: 849 MPa & A_2: 776 MPa) produce residual stress relaxation until the last failure cycle. This is related with the important plastic strain level and the cyclic-softening tendency of the material. In addition, the value of the FWHM gradually tends to reduce until achieve its reference value 0,41°. This reduction is related with the movement and reduction of dislocation densities.

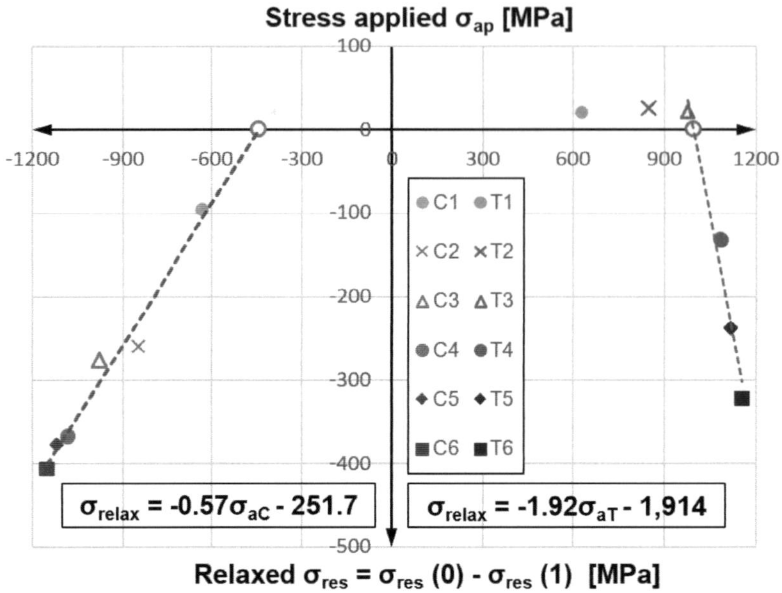

Figure 3. Quasi-static residual stress relaxation.

Logarithmic equation (1) predicts the surface residual stress evolution for the number of cycles of different applied stress magnitudes, where $\alpha = \sigma_a/\sigma'_{yp}$, $\sigma'_{res\,(0)}$ can be considered the average measured value of longitudinal stresses (-535.5 MPa) and N can be calculated from S-N curve.

$$\sigma_{res}(n) = \sigma^0_{res} + ([\alpha \cdot \sigma_a] - \sigma^0_{res} - \sigma'_{yp})\frac{\ln(n+1)}{\ln(N+1)} \quad (1)$$

If the applied stress magnitude decreases, plastic strain level is also reduced. Then, the relaxation rate is lowered, especially after 0.4N cycles, when the relaxation rate is almost null. In this stage, the *stabilized residual stress* on the surface appears. In this study, the stabilized residual stress was set if the stress applied magnitude is lower than 720 MPa. The lack of important plastic strains can be appreciated with the value of FWHM for the two lower applied stress levels, which presents a steady value along the complete fatigue life, as can be seen in Figure 4. The equation (2) correspond to the stabilized value of the surface residual stresses. Due to the cyclic-softening of the material, cyclic yield point must be considered as a threshold for the onset of plastic strains.

$$\sigma_{res\ st} \approx \alpha \cdot \sigma_a - \sigma'_{yp} \quad (2)$$

Stabilized values obtained after cyclic testing are presented in Table 4, where slight deviations between obtained and calculated stabilized residual stresses are obtained. Two more cases of *run-out* specimens are also included in this study.

Table 4. Stabilized residual stress tests results, values in MPa.

Applied stress	Obtained $\sigma_{res\ ST}$	Calculated $\sigma_{res\ ST}$	Difference
703	-233	-226	-7
645	-337	-321	-16
630	-343	-344	-1
630	-348	-344	-4

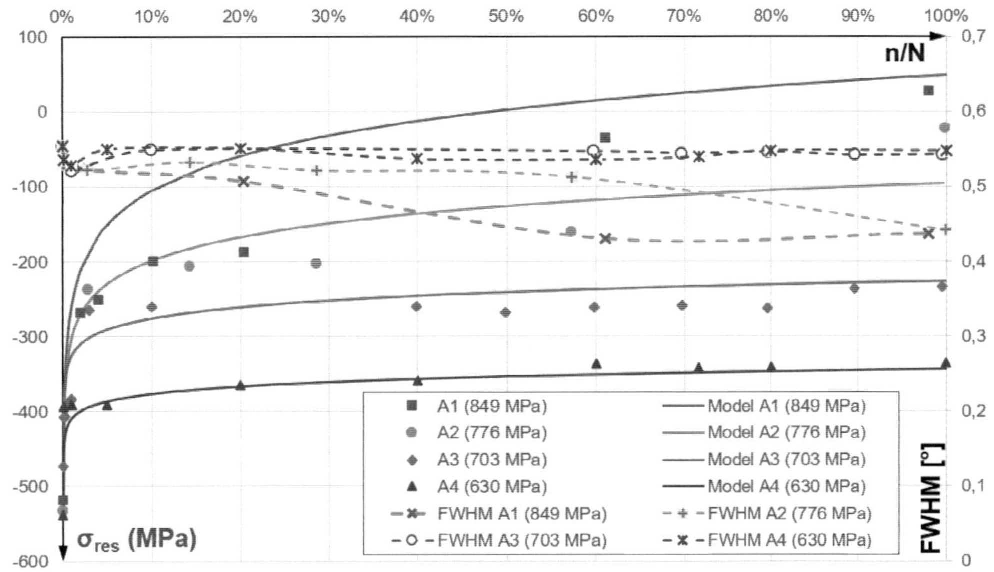

Figure 4. R: -1 cyclic tests results.

<u>R≠ -1 cyclic tests</u>

Test results show the influence of the tensile mean stress on the surface residual stress relaxation. When mean stress magnitude is high enough, it reduces the relaxation rate. The coefficient β is introduced to quantify the influence of the tensile mean stress. Thus, equation (1) can be generalized to equation (3):

$$\boldsymbol{\sigma_{res}(n) = \sigma^0_{res} + ([\alpha \cdot \sigma_a + \beta \cdot \sigma_m] - \sigma^0_{res} - \sigma'_{yp})\frac{\ln(n+1)}{\ln(N+1)}} \qquad (3)$$

The logarithmic prediction equation fits properly with the obtained data, as can be seen in Figure 5. In the Figure 5, the surface relaxation for σ_a 609 MPa, according to equation (1), is represented to better compare the influence of the tensile mean stress on the relaxation rate.

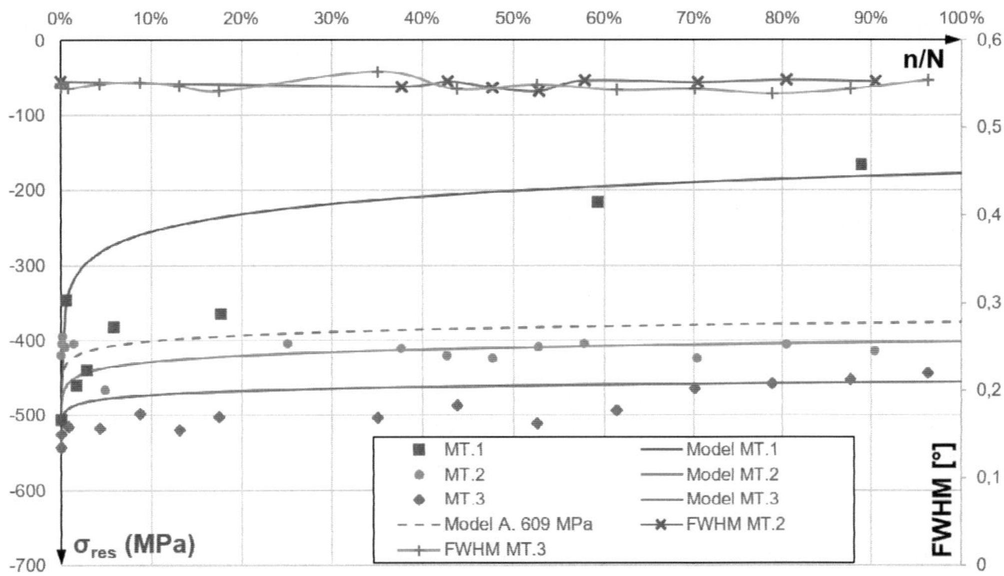

Figure 5. R≠ -1 cyclic tests results.

The relationship between the coefficient β and stress ratio R is presented in the Figure 6. In the studied range of R (-0.5<R<-0.25), higher stress ratio produces higher influence of the mean stress, which reduces the relaxation rate.

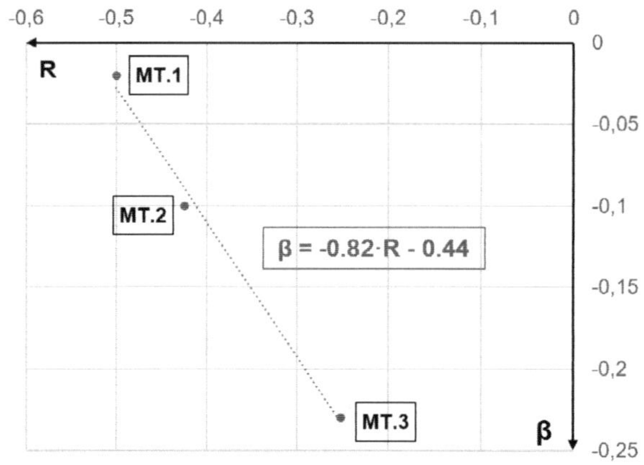

Figure 6. Relationship between β and R.

Discussion and Conclusions

Applied shot-peening increases the fatigue performance of DIN 34CrNiMo6 steel compared to the as machined variant. The residual stress field introduced by shot-peening relaxes when an external stress is applied. The relaxation is related to the onset of plastic strains on the treated material. So, higher stress magnitudes produce higher relaxation rates.

Quasistatic tests showed that the applied stress direction affects the relaxation rate. Tensile applied stresses produce lower relaxation. However, compressive applied stress, due to the compressive residual stress field, require lower magnitude to produce residual stress relaxation. The sum of both compressive stresses achieves a higher plastic strain level.

Cyclic tests with a high magnitude of applied stress showed the influence of plastic strains on the relaxation process, which can be easily identified by the variation of the value of FWHM. Lower applied stress magnitudes maintain FWHM steady and achieve the stabilized residual stress. The introduction of a tensile mean stress, within the stress ratio R range of (-0.5 < R < -0.25), produce lower relaxation rate compared with an R: -1 applied stress.

References

[1] I. Altenberger, B. Scholtes, U. Martin, & H. Oettel, *Cyclic deformation and near surface microstructures of shot peened or deep rolled austenitic stainless steel AISI 304*, Materials Science and Engineering, Vol. 264. (1999), pp 1-16.

[2] H. Holzapfel, V. Schulze, O. Vöhringer, & E. Macherauch, *Residual stress relaxation in an AISI 4140 steel due to quasistatic and cyclic loading at higher temperatures.* Materials Science and Engineering, 248, (1998), pp 9–18.

[3] I. C. Noyan, & J. B. Cohen, *Residual stress: measurement by diffraction and interpretation.* Springer, (2013).

[4] M. G. Moore & W. P. Evans, *Mathematical correction for stress in removed layers in X-ray diffraction residual stress analysis. SAE Transactions*, *66*, (1958) pp 340–345.

[5] C. Boller, & T. Seeger, *Materials data for cyclic loading: Low-alloy steels,* (Vol. 42). Elsevier, (2013).

A New 3D Surface Characterization Method Is More Quantitative and Reproducible Than Traditional Peening Metrology

J. Wheeler, K. Brown, and E. Novak

4D Technology Corporation, 3280 E. Hemisphere Loop, Tucson, AZ 85706, USA

ABSTRACT

Current metrology methods of shot peened surfaces have many shortcomings. 2D stylus methods are directionally sensitive, can't measure corners or edges, often difficult to setup, and offer small overall coverage. 3D microscopes have high resolution and good areal coverage but lack on portability and can take significant time to capture a single data point. This paper presents a high-resolution, extremely portable 3D optical measurement system for comprehensive, accurate, fast shop floor assessment of shot peen coverage. A variety of component measurements will be presented along with studies on accuracy, repeatability, and reproducibility in real-world environments.

Keywords: structured light, polarized structured light, 3D measurements, dynamic interferometry, surface inspection, shot peen, surface roughness, S-Parameters, R-Parameters

1. INTRODUCTION

Surface structure, or surface roughness quantification is a well-known technique for aiding in the qualification of a peened component. This paper will not attempt to correlate surface parameters (S-Parameters) to peening parameters but instead, assume a correlation exists already, based off many published works, and attempt to show that 3D methods, especially portable solutions, are much more comprehensive, repeatable, and reproducible for a given surface structure on the shop floor compared to other methodology traditionally used.

Optical metrology based on interferometry, fringe projection or speckle methods[1] has become indispensable to quantify a variety of surfaces in the production of precision machined parts. The choice of parameters for testing surfaces is large: the field-of-view, lateral and vertical resolution, and the depth of field all can be suited to a particular application. Shop-floor, 3D-measurements of key parameters saves time and improves accuracy. Parts do not have to be moved to a measurement laboratory for characterization, and 3D characterization eliminates the errors usually associated with conventional 2D trace methods. Such systems are also necessary because some objects, such as turbine blades, shafts, or engine cowlings, are simply too large to be tested using a laboratory-based stylus or 3D microscope system. Additionally, 3D optical measurement systems provide non-contact measurements, and the most advanced systems can be immune to vibrations, allowing hand-held operation. Although other fast methods exist[2,3], they often are limited to flat samples, require multiple cameras, or have limited resolution, making the systems unsuitable for hand-held, shop-floor use.

2. CURRENT MEASUREMENT METHODS OF SHOT PEENING ON MACHINED SURFACES

Assessing the peening or surface roughness on machined parts on the production floor commonly relies on visual or stylus-based methods. Alternatively, these assessments can rely on a lab-based microscope system which requires a small sample, often not the part itself, and can take hours to provide any meaningful data. While these methods are commonly practiced, they are often slow and/or difficult to duplicate. Much faster, quantitative measurement can be achieved via a microscope using rubber, replicast material, representative material, or a small section cut off the real part. This process is super time consuming, potentially destructive of samples, and not very reproducible.

Stylus based systems can provide some shop-floor capability for shot peening or general roughness measurements, but again only over a single, 2D line. This can lead to large variations in results from operator to operator or trial to trial even on the same part, as there are both local variation sin shot peening depth and larger variances due to the overall part geometry. They also typically have lateral and vertical resolutions of tens of micrometers and cannot measure fine geometries. This system additionally suffers from alignment errors, is easily affected by vibration, and often cannot measure over sharp edges without breaking the stylus tip. Setup time on a complex part can be quite long as the stylus tip cannot pass over an edge without risk of breaking and finding a vibrationally-stable measurement location on a complex part is often quite difficult. Due to setup time and the need to take multiple scans to have statistically meaningful data, it can take more than 30 minutes to achieve consistent results.

Fringe projection systems can provide the needed precision and portability required for shop-floor characterization of shot peened surfaces. Precision of a few microns is readily achievable when they are designed to have an effective wavelength of the fringes of a few hundred microns. However, typical fringe systems also have limitations. This next section will quickly review the limitations of most fringe projection/structured light setups.

Fringe projection systems consist of three parts: fringe generation and projection on the object, fringe detection, and fringe processing and analysis. Fringe generation and projection used to rely on the projection of a Ronchi grating onto the object while introducing some defocus to make fringes more sinusoidal. Digital micromirror projectors (DMD) or liquid crystal displays (LCD) are now the most common due to their flexibility. The programmability of these systems allowed for the development of techniques based on projection of other than sinusoidal or binary fringes, mainly grey-code light projection or structured light of various spatial patterns. But DMD, and LCD-based systems are not well suited for projection of high frequency fringes that have an effective wavelength of a few hundred micrometers due to limitations in the number of pixels of the devices. Thus, such systems do not have sufficient vertical resolution to assess most shot-peened surfaces with the required accuracy and repeatability. Interferometrically-generated fringes have no such limitations.

The detection system is commonly a CCD or CMOS array for most all fringe-projection systems. However, fringe processing and analysis typically require projection and detection of a few phase-shifted fringe patterns over multiple camera frames. These systems are therefore sensitive to vibration since lateral motion will mean one pattern is not aligned with the next. Single-frame fringe projection systems typically rely on fringes with a high carrier frequency, which in turn are analyzed using Fourier or similar transforms based or spatial carrier phase shifting method. Those methods restrict the range of measurable slopes and discontinuities on the object. However, with a specialized camera using micro-polarizers, a single-frame system without severe slope and discontinuity limits can be achieved.

3. POLARIZED STRUCTERED LIGHT (PSL) METHOD

Systems based on fringes generated by interference with polarization interferometry combined with a detector with a micro-polarizer array for single camera frame detection are well-suited to hand-held shop floor measurements. Four phase-shifted sets of data can be acquired simultaneously.[6] This single camera method is much more compact and easier to align than traditional systems based on three cameras with polarizers, which also can be used for simultaneous shifted fringe detection in polarization systems. By selecting information from pixels of the same polarizer rotation, the phase-shifted images can be displayed and analyzed using, for example, four-frame, phase-shifting interferometry PSI algorithm[7]

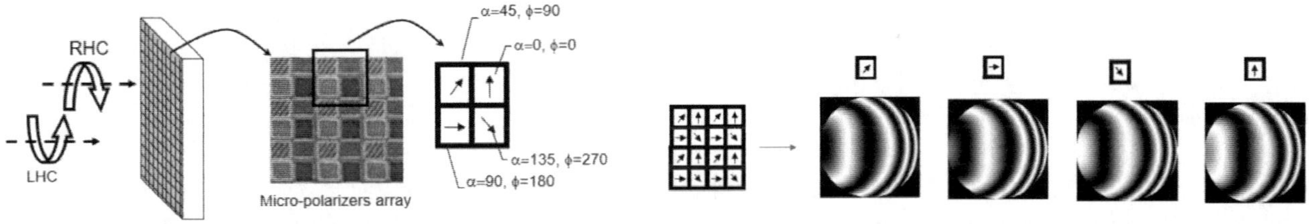

Figure 1. Left, Instantaneous phase shifting method for two orthogonally polarized wavefronts (Right-Hand Circular and Left-Hand Circular) by a camera with micro-polarizer mask with polarizers oriented at angles 0, 45, 90 and 135 degrees and introducing 0-, 90-, 180- and 270-degrees phase shift between beams. Right, fringes extracted for polarizers of the same angular orientation.

To apply this method to fringe projection, the fringes must be created by the interference of two orthogonally polarized beams at the surface of the object. This can be achieved in different ways, such as a Twyman-Green interferometer, Wollaston prism beam splitting, or a special polarization grating. Because the wavefronts are orthogonally polarized, the fringes are not observed on the object with the naked eye but only if the polarizer is placed in front of the detector.

Some caution and system calibration are needed in systems with polarization. For example, if polarization deviates from being circular, some phase ripple may be present in the measurement. Thus, it is important that the polarization is close to being circular. Another potential source of error may be a difference in transmission of the micropolarizers at different polarizations; such differences, however, may be corrected via calibration. Other error sources will vary with the method used for splitting the light into two orthogonally polarized light beams. Another source of error is the variation in brightness across the sample and if this variation is calibrated the error will be significantly reduced. Sample brightness variation also comes into play when determining the optimal projected fringe frequency, since fringe frequency should not be higher than the brightness variation frequency but should not be too low either as it would decrease vertical resolution.

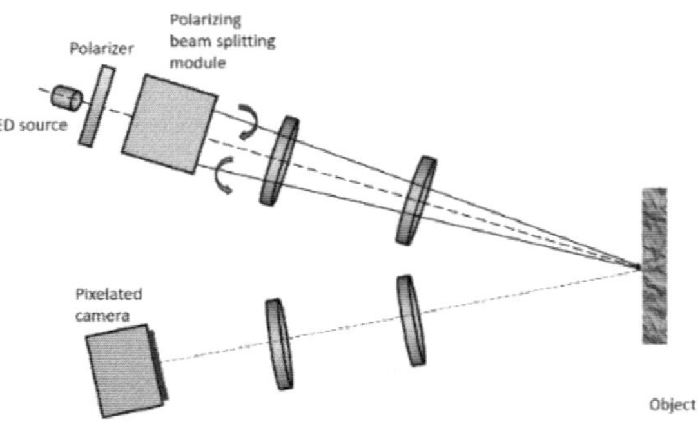

Figure 2. Schematic diagram of single frame polarized structured light system.

The newly developed single shot method based polarized structured light seems to be fitting very well needs of the niche market in aerospace, auto and ship building industry and other large precision mechanical parts builders where testing equipment needs to be brought to the sample.

4. 3D SURFACE CHARACTERIZATION AND MEASUREMENT OF SHOT PEENING

S-parameters, the 3D version of the more commonly known 2D R-Parameters, are ideally suited for determining characteristics of a surface which can be correlated to a given function for use in many applications. Ra of a single trace is problematic for several reasons. First, vastly different surface structures can have similar Ra, as it doesn't consider the spatial scale of any height variations. Secondly, as mentioned previously, traces are directionally dependent and also provide very small overall sampling of the part. Sa, the 3D equivalent to Ra, provides roughness information across the entire 3D measurement, so provides information on a larger area without having to change any drawing parameters. However, in the case of shot peening various S parameters such as those relating to bearing ratio (Smr, Smc, Sdc…) or the ratio of peaks to valleys (Ssk, Sku, Spk…) will likely eventually be adopted by industry since they can correlate with the function of surfaces and not just provide a control number that doesn't relate to how a surface performs such as Ra/Sa. For now, Sa can be logged alongside all the other comprehensive 3D parameters to 'fingerprint' surfaces and ultimately when problems arise that extra information can be used to determine which parameters can help control those problems.

Single trace techniques used to calculate roughness are rife with limitations. The first one is the limited information gleaned from just one trace. Depending on the location, roughness can vary significantly as readily illustrated below using various traces across a 3D shot peened surface. This example was taken on a comparator standard and ideally should be as even as possible across the whole area. Below the error varies between 2 and 30 percent, depending on location of trace relative to nominal result from comparator manufacturer.

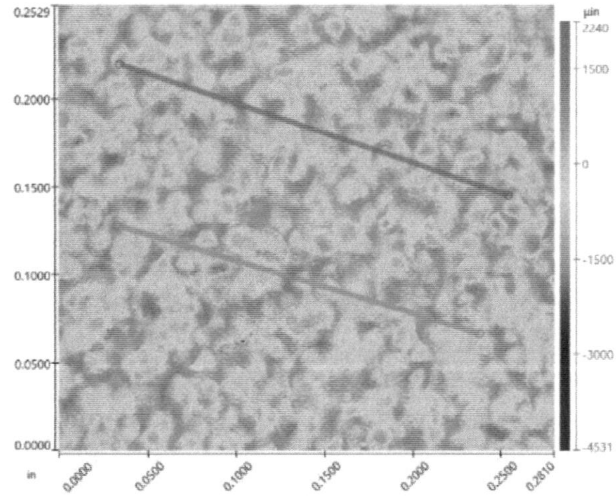

Result	Angle (deg)	ROC (in)	Ra (µin)	RMS (µin)	PV (µin)	Rz (µin)
Line Cursor 1	0.4368	0.1042	492.3	592.7	2955	2909
Line Cursor 2	-0.1349	0.2541	343.9	467.4	2968	2931

Figure 3. R_a varies significantly between 344 and 492 micro-inches, as shown in the table below the surface map. Note: the specified roughness for this sample location, a GAR MICROFINISH COMPARATOR (SH-6), is 500 micro-inches R_a. This demonstrates how the location of a 2D trace can have a large impact on the results.

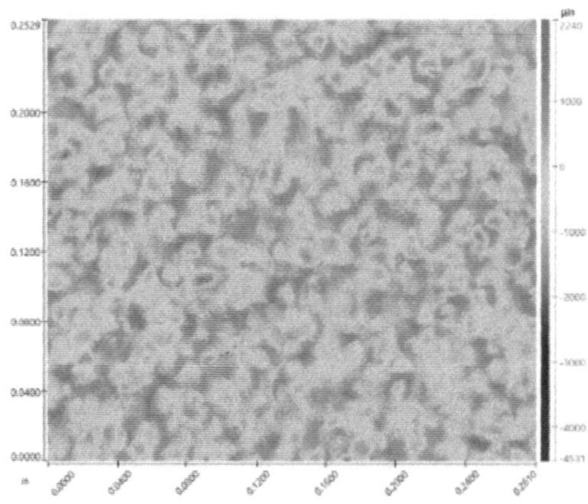

Surface Stats	
Sa	433.0 µin
Sq	542.5 µin
Ssk	-0.04162
Sku	2.977
Sz	6771 µin
Sv	4534 µin
Sp	2237 µin
Sal	0.006176 in
Str	0.8415

Figure 4. Using same data set as in figure 4, 3D results are displayed along with the S-Parameter values. S-parameters are calculated automatically based on the full field of view irrespective of orientation, clocking, or any operator input. 3D analysis of roughness, by its nature, is much more reproducible than 2D trace analysis.

3D Data also can show clear differences visually. See below an example of various peened surfaces and their 3D surface maps.

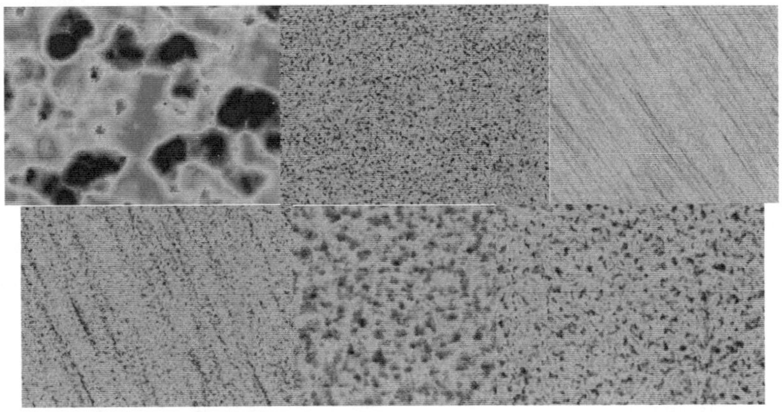

Figure 5. With a variety of peened samples, 3D visual results can vary dramatically.

Finally, the repeatability and reproducibility of 3D metrology is best seen via a gauge study. A gauge study was performed using, for an example of shot peening, a GAR SH-6 MICROFINISH COMPARATOR® Surface Finish Scale. Each of 6 locations was measured 10 times by 3 different operators. Nominal surface roughness varied from 32 to 1000 micro-inches. See results below:

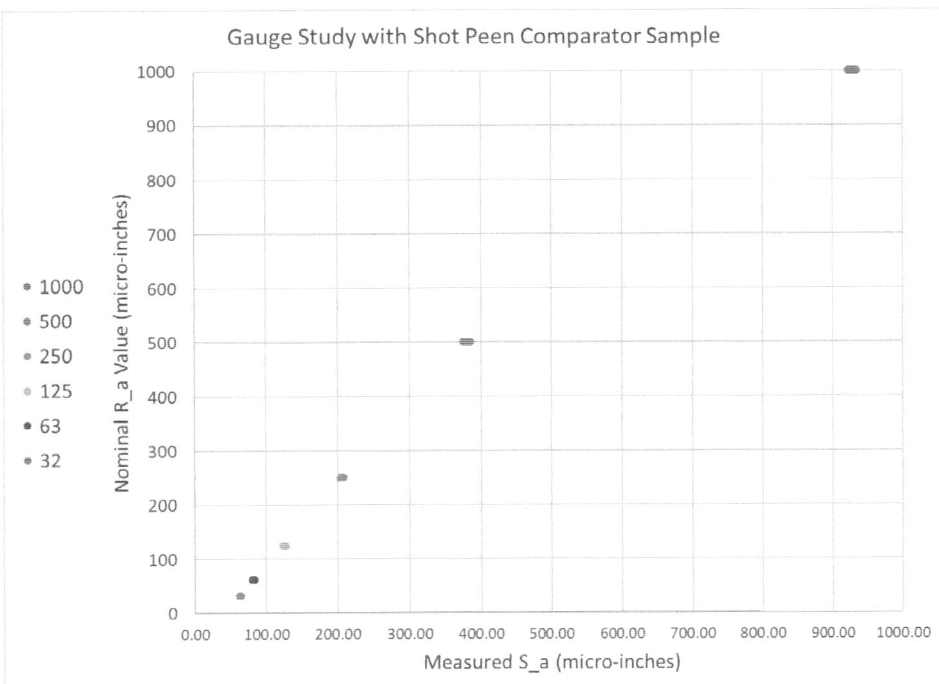

Figure 6: Gauge study results show outstanding repeatability and reproducibility across 3 operators measuring each location 10 times. In plot above, all 30 measurements for each location are overlayed. Highest standard deviation for any location was 1%. Nominal values for standard were 32, 63, 125, 250, 500, & 1000 micro-inches. All of these measurements, with three operators, took a total of ~15 minutes.

5. SUMMARY

Hand-held, 3D shop floor gauges for determination of shot peened surfaces provide numerous benefits for the production of precision machined components. Alignment of the instrument to the part is no longer critical to getting the correct results. Increased amounts of data and measurements over a large field of view allow variations within a part to be readily examined. The large amount of data available, typically millions of points, allows for excellent repeatability and accuracy, allowing inspectors to readily determine if parts need rework, should be failed, or pass specifications.

REFERENCES

[1] Harding, K., Handbook of Optical Dimensional Metrology, CRC Press, Boca Raton, London, New York (2013)

[2] Upputuri, P. K., "Single-shot interferometry techniques for optical testing," Asian Journal of Physics, 24(10), 1317-1338 (2015)

[3] Zhang, Z. H., "Review of single-shot 3D shape measurement by phase calculation-based fringe projection techniques," Optics and Lasers in Engineering, 50(8), 1097-1106 (2012)

[4] Al-Jumaily, G.A., S.R. Wilson, K.C. Jungling et al., "Frequency Response Characteristics of a Mechanical Stylus Profilometer," Optical Engineering, 26, 953-958 (1987)

[5] Kothiyal P. and Delisle R., "Shearing interferometer for phase shifting interferometry with polarization phase shifter," Applied Optics 24(24) 4439–4442, 1985

[6] Millerd, J. E., Brock, N. J., Hayes, J. B., North-Morris, M. B., Novak, M. and Wyant, J. C., "Pixelated phase-mask dynamic interferometer," Proc. SPIE 5531, 304-314 (2004)

[7] J. H. Bruning, D. R. Herriott, J. E. Gallagher, D. P. Rosenfeld, A. D. White, and D. J. Brangaccio, "Digital Wavefront Measuring Interferometer for Testing Optical Surfaces and Lenses," 13(11) 2693-2703; (1974);

[8] Higounenc, O., Correlation of Shot Peening Parameters to Surface Characteristic, ICSP9: Shot Peening, Metal Improvement Company, Montargis, France

[9] Honzghi, Y, Pengfei, Z., Zhi, C., Hui, Z., Yin, Z., Yu Z., Effect of Shot Peening on the Surface Properties and Wear Behavior of Heavy-Duty-Axle Gear Steels, Journal of Materials Research and Technology, March 2022

[10] Gao, Z.; Liao, K.; Chen, J. Surface Characteristic Function of Al Alloy after Shot Peening. Coatings 2021, 11, 160. https://doi.org/ 10.3390/coatings11020160

[11] Cohen, D., www.michmet.com, Michigan Metrology LLC

Preventing Stress Corrosion Cracking and Enhancing Corrosion-Fatigue Performance of Steel and Aluminum by Laser Peening; deployment on nuclear spent fuel canisters.

A Lloyd Hackel[1], B. Jon Rankin[1], C Matt Walter[1], D Brent Dane[1]

1 Curtiss-Wright Surface Technologies, 222 Mountain Vista Parkway, Livermore, California, 94551, USA

Abstract

We report test results showing that laser peening prevents chlorine induced stress corrosion cracking (CISCC) of welded 316L stainless steel and 5000 series aluminums. Our previous work shows similar success preventing cracking in Hastelloy 22, Inconel 600 and Inconel 690. Based on published work identifying the importance of deep (>1 mm) compressive stress to prevent pitting from transitioning into cracking, we provide measurements of the multi-mm depth of compressive stress generated by laser peening in these steel and aluminum materials. The stainless steels tests are important for multi-purpose canisters (MPCs) used for spent nuclear fuel storage and the aluminum panels have important applications in Navy ship and marine applications that have be plagued by sensitization cracking of 5000 series aluminums. Using ASTM G36 (2013) accelerated corrosive testing we decisively show that CISCC does not initiate in weld areas of 316L stainless steel that were laser peened. We further show that for 5083 and 5456 aluminums, laser peening inhibits sensitization and most importantly prevents crack initiation or growth in thermally exposed panels with treatment either before or after sensitization level exposures. The testing of both the stainless and aluminums clearly shows that cracking will initiate and grow in non-peened regions, will not initiate in laser peened areas and that all cracking arrests upon propagating from non-peened areas into laser peened areas. In our work, test panels were exposed to a solution of magnesium chloride at 155°C and rapidly showed extensive cracking in non-peened areas. Based on the results of this work, the laser peening technology was approved by the NRC and used to protect the spent nuclear fuel canisters for the San Onofre Nuclear Power Plant.

Keywords Laser peening, spent fuel storage, stress corrosion cracking, sensitized aluminum

Introduction

Chloride induced stress corrosion cracking (CISCC) is a severe problem impacting many industries and applications including but certainly not limited to nuclear, oil and gas and maritime. For nuclear spent fuel storage, the Multi-Purpose Canister (MPC) design was conceptualized by the U.S. Department of Energy to be a single versatile package for the nuclear industry, equally suitable for on-site storage, transport, and permanent disposal in a future repository. These dry canisters are a temporary approach for storing high-level radioactive waste such as spent nuclear fuel that has been cooled for a required time in a liquid pool. The past and current industry design life of MPCs ranges between 60 and 80 years. The Nuclear Regulatory Commission (NRC) provided initial 20-year license for the facilities using qualified MPCs. Many facilities are applying for license extensions which range from 20 to 40 years.[1,2]

MPCs are typically made of S30400 (304) and 316L stainless steel: although resistant to corrosion, they are known to be susceptible to CISCC. They are roll-formed and welded to cylindrical shape and after loading with spent fuel, sealed with a welded lid. The cylinders are typically filled with an inert gas such as helium and are placed vertically or horizontally in concrete vaults to provide radiation shielding and air flow for passive cooling. The dry canister system is being used in over 20 countries.

CISCC is also of critical importance when evaluating metal alloy dependability. Aluminum derived alloys of 5000 series were introduced due to their high strength and corrosion resistance but have been found highly susceptible to thermally induced sensitization due to precipitation of β phase (Mg_2Al_3) at grain boundaries.[3] Between 2001 and 2002, over 400 commercial marine vessels that were constructed from Al-5083 were found to have experienced severe pitting and stress corrosion cracking.[4] Corrosion cracking on Navy ships presents an enormous economic burden to repair or replace an entire subunit to improve reliability of the vessel.

Chloride Induced Stress Corrosion Cracking

To occur this cracking is generally considered to require the presence of three concurrent factors: 1) material susceptible to CISCC, 2) a corrosive environment and 3) tensile stress above a threshold, for example, tensile stress developed in component welds. With respect to canister's environment, spent fuel storage is predominantly being done at reactor sites which are typically located near a body of water for reactor cooling. The CISCC develops and propagates perpendicular to the vector direction of tensile residual stress. Welding for example is a source of such tensile stress developing along the weld line direction as well as transverse to the weld seam. For welded canisters, the tensile stress developed during the weld cool-down exposes grain boundaries of the sensitive material to the corrosive environment. Parrot and Pits report that fracture mechanics tests have shown that CISCC propagation can begin at low stress intensities in the range of 2 MPa√m to 10 MPa√m and that crack propagation is strongly dependent on temperature but is relatively unaffected by stress intensity.[5] Chen and Kelly have developed predictions of the maximum size of a hemispherical pit in type 304 and 316 stainless steels after exposure to atmospheric conditions.[6] The results of the calculations agree well with several sets of data for near-seacoast exposures on three continents for exposure times out to 26 years. Further evaluations have placed maximum pit size at around 0.2 mm. For structures containing tensile residual stresses, the critical depth of localized corrosion to initiate would be in the range of <1mm.[7]

Deep compressive stress key to prevent pitting and CISCC

It has been observed that a close relationship exists between pitting and stress corrosion cracking in steels.[8] The case for aluminum is the same. Pitting often initiates and advances beyond the surface of a component and then notoriously tends to trigger failures by fatigue or stress corrosion cracking. The electro-chemistry of reactions, such as chloride ions in austenitic stainless steels, drives the pitting corrosion.[9,10] The chlorine ion has been identified as a species which attacks or breaks down a protective film leading to localized dissolution. In both cases of peening as illustrated in Figure 1 the depth of compressive stress reaches the transition to tensile stress at about 300 μm depth. This shallow depth is in the range for pits to grow, thus potentially making shallow peening more susceptible to IGSCC than actually no peening at all. A safety factor thereby dictates that the compressive stress layer extend several times deeper than the potential pit growth depth.

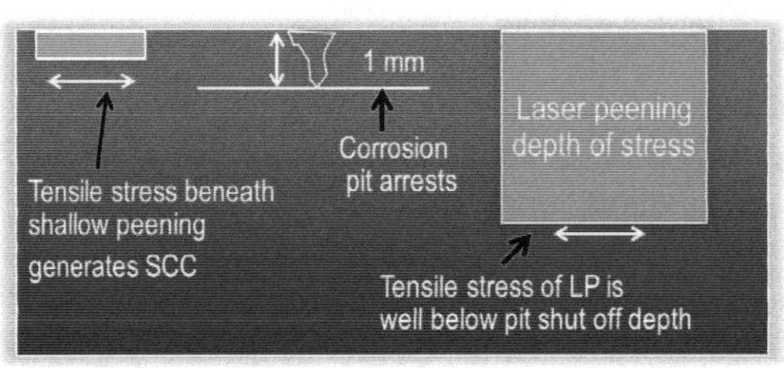

Figure 1. Analyses and tests show corrosion pit depths will only penetrate to somewhat less than 1 mm before arresting. If the pitting penetrates beneath a shallow layer of peening and reaches tensile stress beneath, then cracking can initiate rapidly.

Dry canister test panel preparation, laser peening and residual stress measurement
Using the Holtec International preparation process of roll forming and welding, test panels of 316L stainless steel of 250 mm size by 16.5 mm thickness were fabricated. Figure 2 (5a) shows a plan view of a panel that was roll formed and welded per spent fuel canister fabrication procedures. The weld runs left to right. (5b) shows strongbacks that were lightly stitch-welded to the back of the plate to help provide similar constraints as found in the canisters, thus to hold the cylindrical shape from straining during the laser peening. Stitch welding was used to minimize stress generated during attachment of the stiffeners.

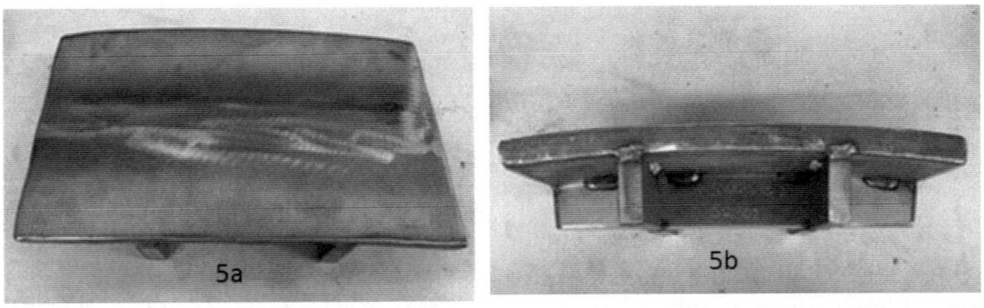

Figure 2. 5a. Canister test panel welded across width. 5b. Side-on view shows stiffening ribs used to help prevent straining as would be the case in the constraint of a canister

One panel was left in the as-welded condition and used for measurements of residual stress for the non-peened state. It was anticipated that the non-peened panel would have tensile stress in the welded area and this was confirmed by the measurement. A second panel was laser peened over the weld and heat affected zone (HAZ) area. Peening was done at 4 GW/cm^2 irradiance, 18 ns pulse duration and with 3 layers of coverage and use of aluminum tape ablative layer. Spot size of each laser pulse was 4.7 mm square (6.6 mm diagonal). Each layer of peening provides fully 100% coverage with the square laser spots robotically positioned within 0.1 mm precision of one next to another and using approximately 3% spot length overlaps. The peening area was 10 cm wide covering the weld and heat affected zones across the width of the panel. Residual stress measurements were performed across the full width of the panel by Hill Engineering using the two-dimensional Contour method [11,12,13] Guided by the FEA predictions, panels were laser peened with variations of process parameters and optimum process for canister peening was selected.

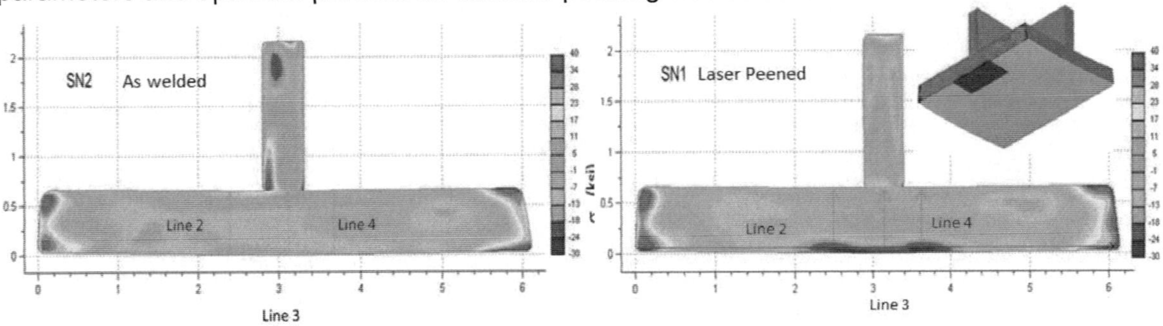

Figure 3. Contour measurement of residual stress in welded panels of 316L Stainless Steel. (SN1) was welded and then laser peened and (SN2) welded and left un-peened.

The left image of Figure 3 shows the stress profile in false-color for an as-welded non-peened panel. The depiction indicates tensile stresses in the 70 MPa (10 ksi) to 140 MPa (20 ksi range as expected in the weld and heat affected zones. These would be of concern for stress corrosion cracking. In contrast, the right image shows measurement after laser peening of an identically welded panel. Three specific areas of importance are indicated with arrows where lineouts of the measured stress are graphically displayed in Figure 4.

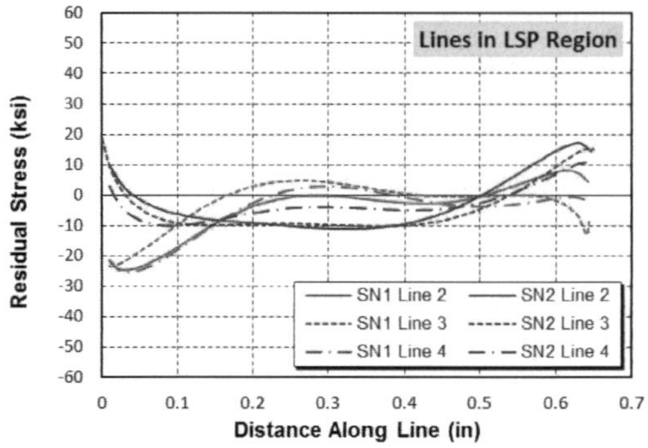

Figure 4. Lineouts of stress profiles of non-peened and laser peened panels measured by Contour. Laser peening converted tensile stress of up to 20 ksi (140 MPa) to 20 ksi (140 MPa) compressive and generated deep compression to 0.16 inch (4 mm) to .25 inch (6.2 mm) depth.

Comparing the lineouts it can be seen that the laser peening transformed these tensile areas to compressive (blue) residual stress in the weld area from approximately 175 MPa (25 ksi) tensile to 140 to 175 MPa (20 to 25 ksi) compressive. In the heat affected areas the transformation was from 140 MPa (20 ksi) tensile to 140 MPa to 175 MPa (20 to 25 ksi) compressive. Depth of compression in the weld zone extended to 4 mm (0.16 inches) to a deeper 5.6 mm (0.225 inches) in the heat affected zones.

But in measuring residual stress it is important to consider geometry. Peening of a constrained component will give more intense and deeper residual stress compared to identical peening of a sample with edges free to strain and thus relax some of the residual stress. Such is the case in measuring and relating the residual stress imparted by laser peening of a 316L unconstrained sample and that of the stress in a treated canister. Using our FE analysis we can calculate corrections. We relate measurements made in isolated panels to the actual stress computed by our FEA in the canister walls. In the methodology blocks 50 x 50 x 16 mm of 316L stainless steel, were fabricated and individually laser peened with 1, 2 or 3 layers of peening. Stress per layer of peening was then derived from slitting measurements. An FEA model of the canister was built and the laser peening results from the blocks applied in the model to specific canister areas. Stress in the canister and then stress in an unconstrained panel were calculated showing the increased stress retained in the canister. Figure 5 shows that the stress retained in the canister is approximately 2 mm deeper than that measured in the panel.

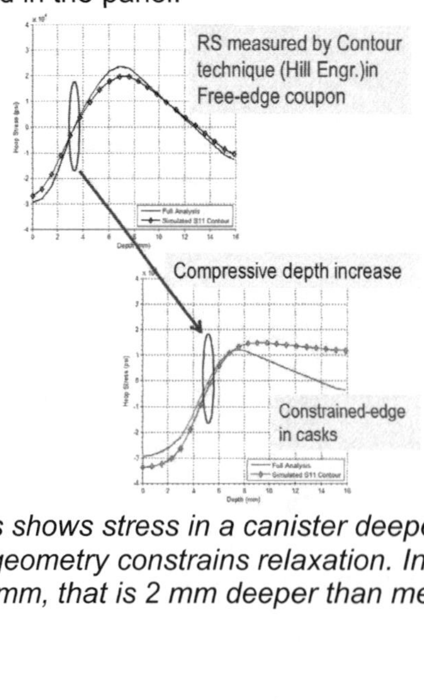

Figure 5. FEA analysis shows stress in a canister deeper than measured in free panel because the canister geometry constrains relaxation. In current case stress depth in a canister is 5.5 mm, that is 2 mm deeper than measured in a free panel.

5000 series aluminum test plates preparation

Aluminum 5083 and 5456 blocks of size 50 mm x 50 mm by 12 mm were fabricated for sensitization testing and 152 mm x 152 mm x 12 mm plates for SCC testing. For this application the peening was done with 4.7 mm square laser beam spots using 4 GW/cm^2 laser irradiance, with18 nanosecond pulse duration and 2 layers of peening, calling this process 4-18-2 in our notation. Aluminum tape was used as the ablative layer. The sensitization test blocks were peened over all six surfaces for the G67 mass loss tests. The plates for SCC were peened on the top surface only and as in the previous tests, laser peening was performed on approximately one half of the plate exposure area leaving the other half non-peened. Because the plates did not contain welds for the SCC exposure, a tensile stress was applied by mechanically loading of individual plates. A steel rod was placed in between a reaction plate and the test plate and outboard bolts were used to load the surfaces to a stress of 80% of the Ultimate Tensile Strength (UTS) of the respective material. This resulted in pre-stressing the Al-5456 to 2597$\mu\varepsilon$ and the Al-5083 to 2200$\mu\varepsilon$ respectively.

Stress Corrosion Cracking test setup

Tests were performed using a setup in the manner of ASTM G36-94[14] (Reapproved 2013) The corrosive exposure of the canisters in actual storage applications will be to their exterior surfaces as will sensitization exposure of 5000 series aluminum on Navy ships. For this reason an acceptably modified version of an ASTM G36 apparatus was used in which a glass cell with flanged open bottom was sealed with rubber o-rings and bolts to the "exterior" face of the test plate. Magnesium chloride crystals ($MgCl_2$-$6H_2O$) were inserted into the glass cell and heat applied to melt the crystals. To accelerate the corrosion exposure rate, a hot plate was placed underneath and heating coils were wrapped around the glass container to provide additional heating of the chloride liquid. A condenser cooled by separated flow from a chiller was inserted into the opening on top of the glass cell. A vapor trap containing a 25 weight percent solution of magnesium chloride was placed on top of the condenser to trap and reflux and thus minimize liquid loss. Power levels were adjusted on the hot plate and heating coils to melt the salt and bring the solution to a boil. The water-salt concentration was adjusted by adding water or allowing it to evaporate by temporarily removing the condenser until the boiling temperature reached a steady 155°C liquid level stabilized. Once the test apparatus was settled in, continuous operation at 155°C was straightforward to maintain and operate for long durations.

Chloride Induces Stress Corrosion Cracking test of 316L

In the initial run, the cell was bolted to a section of plate that had a weld joint running across the diameter. The first test was of a non-peened area, run for 7 days and then cooled and the cell removed: extensive cracks were observed in the non-peened area. For the next test another welded panel was configured with one third of the exposure area partially laser peened and the remaining area left non-peened. This ensured that both peened and non-peened areas were identically exposed to the same conditions.

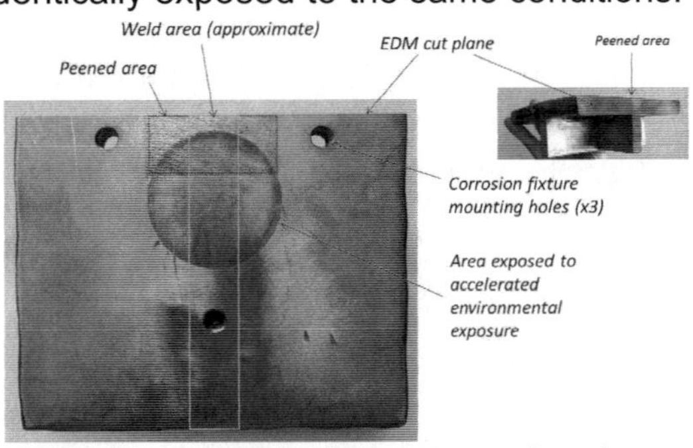

Figure 6. Welded test panel with laser peened and un-peened exposed areas

Figure 6 shows the test panel used. The weld and heat-affected zone ran vertically through the middle of the exposed area. After 57.5 hours exposure to the $MgCl_2$ at 155°C heaters were turned off, the solution solidified and the cell removed allowing inspection of the panel. An array of photos was taken and stitched together with the resultant shown on the left side in Figure 7. The photo with the cracks highlighted is shown on the right. Cracks had extensively developed in the heat affected zone with some cracking in the weld area. No cracks developed in the laser peened region and cracks that propagated into the laser peened region arrested.

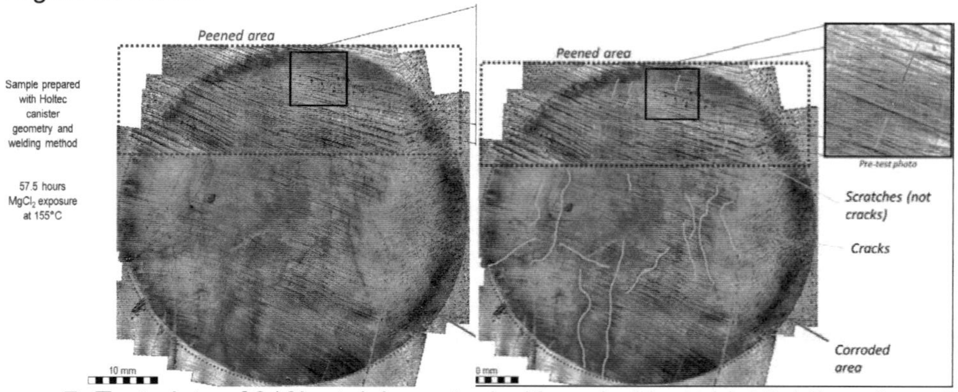

Figure 7. Test plate of 316L stainless steel with and without highlighting of cracks. Observation for cracking was made after 57.5 hours exposure to $MgCl_2$ at 155°C. Cracks extensively developed in non-peened areas, did not develop in peened areas and arrested as they propagated from non-peened into peened areas. Pre-existing scratches in the peened region, identified by the green markings on the right, did not crack even though they would be considered prime candidates for initiation.

To quantify the potential of laser peening to prevent CISCC over extended periods, a panel was laser peened over a 1/3 exposure area, leaving the remaining area non-peened and then similar testing was performed. After 18 hours exposure the panel was inspected to find that extensive panel cracking had already occurred. The apparatus then was reactivated and run to 342 hours total with no crack initiation or propagation into the laser peened area. The results indicate that even in extreme temperature and chlorine exposure, the laser peening will provide in excess of 19 times lifetime increase against CISCC for the welds of 316L spent fuel canisters.

ASTM G-67 sensitization test on 5000 series aluminum

The criterion specified to determine the sensitization level of the aluminum is to expose a sample to an acid solution and measure the mass loss per unit area (mg/cm^2) and compare it to the literature values for degree of sensitization; the greater the mass loss per unit area, the higher the degree of sensitization and the lower the weldability. Results indicate that the laser peened samples did not sensitize. Al-5083 sample that was laser peened on all six sides at 4-18-2 and then exposed for 14 days at 100°C had a mass loss of only 11.61 mg/cm^2 compared to unpeened sample which lost 47.51 mg/cm^2 after a shorter 8 day exposure. Another 5083 sample that was not laser peened but exposed for a total of 15 days lost 21.55 mg/cm^2 whereas a second laser peened sample first exposed for 19 days and the an even longer 21 days lost respectively only 12.39 and 20.82 mg/cm2..

Stress Corrosion Cracking of sensitized 5000 series aluminum

Using the G36 test apparatus exposures were made of both the Al 5456 and Al 5083, testing for cases of sensitized (exposed to sensitizing time and temperature). Samples were then laser peened and the complementary condition of laser peened first and then exposed to sensitizing time and temperature was evaluated. In this case of laser peening first, as for new ship construction, we do not here use the term sensitized because for the severe exposures of our tests, the G67 data indicates the material did not sensitize after laser peening.

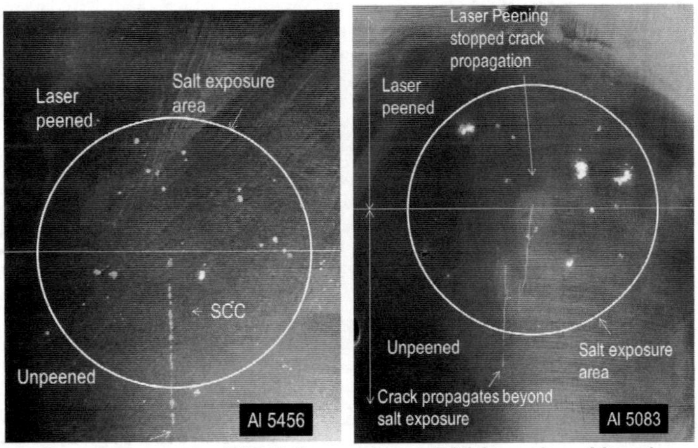

Figure 8. Laser peeingn prevents CISCC cracking 5000 series Aluminums. Al5456 was sensitized and the laser peened. Al5083 was laser peened and then exposed to sensitization environment. Cracks developed and propagated in the non-peened region. No cracks developed in the laser peened region after 300 hours. Cracks in the peened region arrested as they encountered the compressive stress of the laser peened area.

Sensitization cracking tests of Al-5456 were conclusive. Figure 8 left shows the Al 5456 test plate that was peened at 4-18-2 and was then sensitized it by thermal exposure at 100°C for 14 days. The plate was pre-stressed to 597με and run in the G-36 test for 303 hours developing cracks in the non-peened region. Again cracking did not develop in the laser peened region and cracks from the non-peened region arrested as they propagated into the compressive stress of the peened area. A 152 x 152 mm plate of Al-5083 was first sensitized by exposing for 19 days at 100°C. Then approximately one half of its exposure area was laser peened at 4-18-2. Next the sample was loaded with strain gauge monitoring to a tensile strain of 2200με. Considering Young's modulus of 70.3 GPa (10.2 Msi), a tensile loading stress of 155 MPa (22.4 ksi) was applied. Figure 14 right shows this panel cracked in the non-peened region after 96 hours of $MgCl_2$ exposure. Again cracks originating in the non-ypeened region arrested as the propagated into the laser peening.

Deployment of Laser Peening for nuclear spent fuel canisters

Taking the technology beyond the qualifying laboratory tests and evaluations to field deployment was an important step for fully demonstrating the capability of the laser peening process to support nuclear applications. A highly automated laser and robotic system was configured to peen the welds of 75 MPCs for the San Onofre Nuclear Power Plant at the Holtec International canister fabrication facility. The right hand photo of Figure 9 shows the beam delivery hardware and robotics that precisely positioned the laser beam spots onto the canister. A rotation system rotated the canisters to peen the hoop welds and with rotation fixed, the delivery robot translated along a track for peening the longitudinal welds. Individual spots were placed on the canister with 0.1 mm precision and recordings made of laser energy and pulse duration for each shot. The canisters were peened with a 100 mm (4 inch) wide coverage at 4 GW/cm2 irradiance, 18 ns pulse duration and 3 layers of coverage. The photo at right shows the laser peening system as deployed at the manufacturing facility. The UL qualified laser system is housed in a trailer and its output beam is propagated into a light safe tent and to the beam controller mounted on the delivery robot. The beam controller precisely conditions the spot and directs it on to the canister. The water delivery nozzle is attached to and move with the controller. Processing is controlled by operators at the external control station. Canisters were peened at the rate fabricated and the laser system returned to California home base when peening was completed.

Figure 15. Laser peening system configured for treating welds of nuclear spent fuel canisters. At left beam director mounted on the robot (orange) for canister processing. At right system configured for field operation.

Conclusion

Our measurements and analysis for both the 316L and the 5000 series aluminums show that laser peening generates compressive stress to multi–mm depths, well beyond the self-terminating pit depth estimated at about 200 μm due to cathodic current limits. Our accelerated ASTM G36 (2013) tests conducted at 155°C with $MgCl_2$ clearly show that CISCC will not initiate in areas treated with high energy laser peening and that CISCC originating outside of a laser peened zone will arrest upon reaching the peened area. The current work includes stress intensity vs. depth generated from 2-dimensional residual stress measurements made by Hill Engineering using a Contour method. The high energy laser peening offers an excellent safety margin for structural integrity of spent fuel dry storage canisters and alleviates concerns related to cracking prior to moving canisters to a permanent repository facility. For marine applications laser peening alleviates CISCC of sensitized 5XXX aluminums and reopens using this higher strength aluminum for advanced systems.

ACKNOWLEDGEMENT

Work on 5000 series aluminum supported in part by the Naval Shipbuilding Research Program

References (

1. M.R. Wisenburg, "Use of multi-purpose casks for at-reactor spent fuel storage", Proceedings of the Institute of Nuclear Materials management (INMM) spent fuel seminar, p277, 13-15 January 1995, Washington DC.
2. B. Gutherman and D. LeDuc, "Summary of Design Criteria for Dry Cask Storage Systems for ISFSI storage of Shutdown Reactor Spent Nuclear Fuel," U.s. Department of Energy, Nuclear Fuels Storage and Transportation Planning Project, FDRC-NFST-2015-000538, Rev. 0, SRNL-TR-2015-00106.
3. E.H. Dix, Jr., W.A. Andersen, M.B. Shumaker, "Influence of Service Temperature on the Resistance of Wrought Aluminum-Magnesium Alloys to Corrosion", CORROSION, Vol. 15, No. 2, pp. 55t-62t, Feb.1959.
4. H. Bushfield and M. Cruder, "Sensitized marine aluminum plate & ASTM standard specification B928-an update," in SNAME Section Meeting, 2006.
5. R. Parrott, R., and H. Pitts, H., "Chloride stress corrosion cracking in austenitic stainless steel", prepared for the Health and Safety Laboratory, Crown Publishing 2011. Copies can be found online at http://www.hse.gov.uk/research/rrpdf/rr902.pdf.
6. J.J. Truhan Jr., R.F. Hehenamm, "Surface Aspects of Pitting and SCC", NASA CR-135212, April 1977
7. M.T. Woldemedhim, M.E. Shedd, and R.G.Kelly., Evaluation of Maximum Pit Size Model on Stainless Steels under Thin film Electrolyte Conditions, J of the Electrochemical Society, 161(8), E3216-E3224, 2014.
8. J.J. Truhan Jr., R.F. Hehenamm, "Surface Aspects of Pitting and SCC", NASA CR-135212, April 1977

11. M.B. Prime, "Residual Stress Measurement by Successive Extension of a Slot: The Crack Compliance Method," *Applied Mechanics Reviews*, Vol 52, No. 2, pp. 75-96, 1999.
12. M. B. Prime, D. J. Hughes, P. J. Webster, "Weld Application of a New Method for Cross-Sectional Residual Stress Mapping", *Proc. 2001 SEM Annual Conference on Experimental and Applied Mechanics*, June 4-6, Portland, OR, pp. 608-611. (LA-UR-01-1244).
13. M.B. Prime, 2018, "The Two-Way Relationship Between Residual Stress and Fatigue/Fracture," Fracture, Fatigue, Failure and Damage Evolution, Volume 7: Proceedings of the 2017 Annual Conference on Experimental and Applied Mechanics, , pp. 19-23.35.
14. ASTM G36-94(2013), Evaluating Stress-Corrosion-Cracking Resistance of Metals and Alloys in a Boiling Magnesium Chloride Solution, ASTM International, West Conshohocken, PA, (2013), www.astm.org.

Effects of Shot Peening on Surface Layer States and Fatigue Behavior in Experimental Batches of Enhanced Leaf Spring Steels

A. Klumpp[1], R. Elvira[2], D. Dapprich[3], S. Dietrich[1], V. Schulze[1]

1 Institute for Applied Materials (IAM-WK), Karlsruhe Institute of Technology (KIT), Karlsruhe, Germany
2 Sidenor I+D, S.A., Basauri, Spain
3 Stresstech GmbH, Rennerod, Germany

Abstract
Shot peening under pre-stress ("stress peening") is widely used for the mechanical surface treatment of steel leaf springs in heavy truck applications. In this paper, we present the effects of stress peening, applied to experimental batches of four new leaf spring steels. After an examination by means of chemical analyses, microscopy and dilatometry, quenched and tempered specimens of the experimental steel grades were subjected to single and double shot peening under pre-stress at ambient and elevated temperature. Surface layer states were characterized by means of X-ray analyses of residual stresses and full width at half maximum (FWHM) as well as roughness measurements. The fatigue behavior was evaluated using laboratory scale (small) specimens in alternating bending. The shot peening variants gave rise to comparable results, whereas the most pronounced effect could be attributed to the selected steel grades and their respective heat treatment.

Keywords Stress peening, leaf spring steels, surface layer states, fatigue behavior

Introduction
Stress peening is a shot peening variant applied intensively in the steel spring industry, e.g. for different types of leaf springs in heavy truck suspensions [1-3]. Nowadays, leaf springs are typically made of quenched and tempered Cr-V-alloyed steel, such as 51CrV4, 52CrMoV4 or the intermediate grade 52CrV4Mo. This class of steels offers a combination of favorable processability (hot formability, hardenability, decarburization resistance), price and mechanical behavior (yield strength, toughness, fatigue strength), among others. Within the EU-project "LEAFSLIM", the applications of both modified stress peening treatments and improved steel grades for leaf springs have been assessed. The successful implementation of such measures allows for weight reduction (downsizing), cost savings and enhanced fuel efficiency. To cope with those needs, the main design driver for leaf springs during recent years has been the enhancement of strength without affecting other characteristics such as ductility and toughness [4,5].
Steel grades with higher strength are achievable by optimizing the use of strengthening mechanisms applicable to metallic materials, i.e. solid solution, grain boundary and orientation strengthening, precipitation and dispersion hardening as well as work hardening. Full martensitic hardening and subsequent tempering, the heat treatment virtually always applied to leaf springs prior to setting and shot peening, is based on a combination of those effects. Chemical composition, process chain, hardening and tempering conditions interact in a complex manner (see for instance [6]). In this context, several investigations have recently focused on enhancing service life behavior of the classic leaf spring steel 51CrV4 by different means, such as modification of chemical composition [4], microstructure refinement and removal of inclusions [5], nano-particle alloying [7], enhancement of quenching conditions [8] and application of deep cryogenic treatment [9]. All modifications have in common that their implementation must be feasible on an industrial scale at justifiable costs and that they should not affect processability. Although some beneficial treatments have been identified in the past, relationships are complex and the effects can be equivocal. Particularly, all aforementioned studies did mainly concentrate on mechanical

behavior (tensile, impact, fatigue strength), but none of them took reference to any possible interaction between the new steel grade or treatment and the finishing step, i.e. stress peening [2,3]. However, it would not only be interesting to evaluate the behavior of new steel grades, but also the implications of conventional or modified stress peening routes to their surface characteristics and service life behavior, for instance by means of warm stress peening. Warm peening has proven to produce stable surface layer states and is thus considered a promising, though expensive stress peening variant [1,10].

This study aims at providing deeper understanding of such interactions. Referring to the spring steel 52CrV4Mo, an enhanced version of 51CrV4, two different strategies to achieve leaf spring steels with enhanced strength properties are presented. The first strategy is to optimize the hardening behavior of the Cr-V-type steel by an improved balance of the elements C, Cr, V and Mo. The second strategy makes use of Silicon (Si), which in small additions can largely improve mechanical behavior and has been used for very high strength automotive spring steels such as 55SiCr6. Besides the development of new steel grades, improved stress peening treatments have a high potential to enhance service life behavior, such as fatigue strength. Therefore, the present study also aims at providing surface layer states and fatigue behavior of the modified steels after applying a set of modified stress-peening treatments.

Experimental Methods

The low-alloyed spring steel 52CrV4Mo with the chemical composition design shown in Table 1 was used as reference material for the investigation. Based on this material, three experimental steel grades were created under the following basic conditions:

- Cost-effective alloying elements like Mn and Cr must be maintained over a certain limit to not endanger steel hardenability and ensure a fully martensitic transformation.
- No breakthrough modifications are aimed for regarding the current leaf spring upstream manufacturing route: hot rolling and oil quenching and tempering.

The design and actual chemical composition of the four experimental steel casts are shown in Table 1 and 2, respectively. Silicon (Si) is known to increase the yield strength by solid solution strengthening. Furthermore, it delays the transformation of ε-carbide to cementite. However, the element may also cause a shift of the ductile to brittle transition temperature (DBTT) to higher temperature. Moreover, it causes strong decarburization during heating before hot rolling, why its use has been challenging. Nickel (Ni), combined with chromium (Cr), increases hardenability and both impact strength and fatigue resistance. The carbon content directly affects the achievable hardness. Yet, its content has to be limited due to increasing embrittlement and quench crack susceptibility [6]. Therefore, experimental casts of the reference steel and three new steels were created:

- 52CrV4Mo (heat 810): Reference cast
- 45SiCrVMo6 (heat 819): Addition of Si; reduction of C. Goal: Enhanced tensile strength and retained toughness
- 45SiCrV9Ni (heats 812, 816, 817): Addition of challenging amount of Si; reduction of C and addition of Ni. Goal: Enhanced tensile strength and retained toughness
- 55CrVMo6 (heat 818): Addition of C and Cr. Goal: Enhanced hardenability and tensile strength

Table 1. Designed chemical composition of experimental casts.

Grade	C	Mn	Si	P	S	Cr	Ni	Mo	V	Cu	Al	Ti
52CrV4Mo	0.52	1.00	0.35	0.015	0.005	0.95	0.15	0.10	0.10	0.20	0.020	0.005
45SiCrVMo6	0.45	0.80	1.50	0.015	0.005	1.00	0.15	0.15	0.30	0.20	0.020	0.005
45SiCrV9Ni	0.45	0.80	2.25	0.015	0.005	1.00	0.55	-	0.30	0.20	0.020	0.005
55CrVMo6	0.55	1.00	0.35	0.015	0.005	1.50	0.15	0.20	0.30	0.20	0.020	0.005

Table 2. Measured chemical composition of experimental casts.

Heat No.	C	Mn	Si	P	S	Cr	Ni	Mo	V	Cu	Al	Ti
810	0.49	0.94	0.38	0.013	0.010	1.00	0.15	0.09	0.10	0.20	0.010	0.005
812	0.45	0.81	2.31	0.013	0.009	1.03	0.54	0.01	0.29	0.20	0.018	0.006
816	0.44	0.77	2.34	0.015	0.011	1.01	0.56	0.00	0.30	0.20	0.020	0.004
817	0.44	0.78	2.35	0.014	0.009	1.02	0.56	0.00	0.30	0.20	0.020	0.004
818	0.54	0.92	0.39	0.016	0.010	1.53	0.15	0.20	0.30	0.21	0.021	0.004
819	0.45	0.78	1.56	0.014	0.009	1.01	0.15	0.14	0.29	0.21	0.022	0.004

Specimens of all heats were first austenitizied for 7 minutes in a vacuum oven and oil-quenched in an oil bath at 50-55 °C. The austenitizing temperatures for 52CrV4Mo, 45SiCrVMo6, 45SiCrV9Ni and 55CrVMo6 were chosen to 850 °C, 900 °C, 950 °C and 850 °C, respectively. Prior dilatometry experiments were used to determine the A_{c1} and A_{c3} temperatures. After quenching, specimens were tempered for 90 minutes at the following temperatures: 350 °C, 375 °C, 400 °C, 425 °C and 450 °C. For each condition, one tensile sample and three Charpy U-type specimens were tested.
For the stress peening experiments, the flat specimen geometry shown in Figure 1 (left) with a gauge width of 18 mm and thickness of 2 mm was used. Specimens were extracted from the experimental cast blocks by means of EDM. The green drawing will be explained later. Shot peening was carried out with a Baiker SKUK air blast shot peening machine, using the apparatus shown in Figure 1 (right) to apply tensile pre-stress. To realize warm peening, the flow heating device first presented in [10] was used. In Table 2, the test plan of the applied stress peening campaign is shown. Besides conventional stress peening (P1S), warm stress peening (P9S) and double peening (P10D) were applied to the experimental casts. The second shot peening step was carried out using S170 (56 HRC) as shot, an Almen intensity of approx. 0,2 mmA, at room temperature (20°C) and without pre-stress.

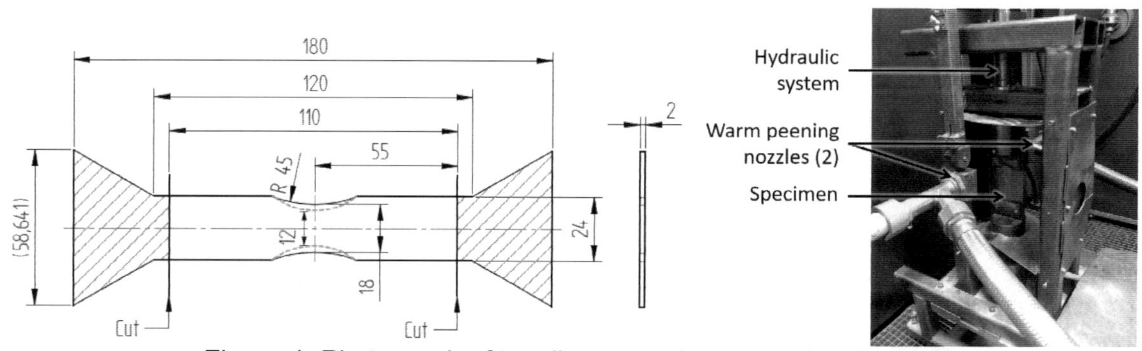

Figure 1. Photograph of tensile warm stress peening device.

Table 3. Parameters for stress peening campaign.

Set	Almen Int.	Shot	Coverage	Tens. pre-stress	Temp.	2nd step
P1S	0.22 mmC	S330 (56 HRC)	100%	1350 MPa	20°C	NO
P9S	0.22 mmC	S330 (56 HRC)	100%	1350 MPa	350°C	NO
P10D	0.22 mmC	S330 (56 HRC)	100%	1350 MPa	350°C	YES

Alternating bending fatigue specimens were extracted from the shot-peened specimens by removing the hatched areas (see Figure 1). Surface states were characterized in terms of residual stress analyses, full width at half maximum values and roughness measurements.
Residual stress measurements were realized with an X-ray diffractometer of the type Stresstech G2R with CrK_{α} radiation, using the $\sin^2\psi$ method. The elastic coefficients C_{11}, C_{12} and C_{44} (for single crystals) were extracted from the literature [11] and are $C_{11} = 2.3*10^5$ MPa, $C_{12} = 1.3*10^5$ MPa and $C_{44} = 1.1*10^5$ MPa. Residual stress profiles with a maximum

depth of 0.5 mm were determined by means of electrolytic layer removal (50 µm steps; circle D 10 mm). Only longitudinal residual stresses (parallel to pre-stress) were measured.
Roughness was measured by means of a Nanofocus µSurf confocal microscope. An area of 1.5*1.5 mm^2 close to the center of the shot-peened surface was evaluated in each case. The Gaussian cut-off wave length was λ_c = 0.8 mm.
Fatigue testing was carried out in alternating bending at a load ratio of -1 and room temperature. The frequency was 25 Hz. Due to the required long heating times, warm peening of fatigue specimens were restricted to 45SiCrV9Ni, using heats 816 and 817.

Experimental Results

Figure 2 shows the dependence of ultimate tensile strength (left) and impact toughness (right) for the experimental casts as a function of tempering temperature. In all experimental steel casts, an improvement of properties with regard to the reference material was achieved. The new steel grades also allow for lower tempering temperatures to achieve satisfactory toughness. The best combination of strength and toughness was found on grade 45SiCrV9Ni. All specimens for the stress peening campaign were tempered as shown in Table 4. By this means, minimum values of ultimate tensile strength (UTS), impact energy (KU) and reduction of area (RofA), as shown in Table 4, could be achieved.

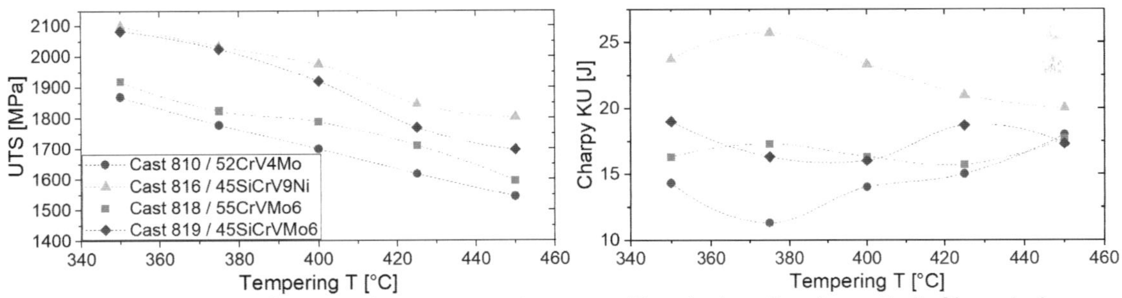

Figure 2. Influence of tempering temperature on ultimate tensile strength (left) and charpy impact toughness (right).

Table 4: Mechanical properties achieved after selected tempering treatment.

Steel grade	52CrV4Mo	45SiCrVMo6	45SiCrV9Ni	55CrVMo6
Tempering T [°C]	425	375	375	375
Min. UTS [MPa]	1600	2000	2000	1800
Min. KU [J]	13.5	15	22.5	13.5
Min. RofA [%]	35	35	40	35

The longitudinal residual stress and full width at half maximum (FWHM) profiles obtained after the different stress-peening treatments are shown in Figure 3. Regardless of the experimental material, the profiles always have a pronounced residual stress maximum with magnitudes up to 2000 MPa beneath the surface. Residual stress penetration depths can be estimated to 0.5~0.6 mm. Warm peening increases the depth of the residual stress maximum. The double peening process, carried out without pre-stress, obviously causes a decrease of the surface residual stresses. After all stress peening treatments, the lowest amounts of residual stress and FWHM can obviously be found on the reference material 52CrV4Mo. The highest amounts of residual stresses and FWHM were found in 45SiCrV9Ni.
Figure 4 shows the roughness values measured on the stress-peened surfaces. All shot peening treatments increase the roughness which was measured to approx. $R_z \approx$ 11 µm in the initial state (EDM cut). In the case of 52CrV4Mo and 55CrVMo6, warm peening causes a higher roughness, which could eventually be reduced by double peening. In the case of the steels 45SiCrV9Ni and 45SiCrVMo6, the effects are less clear. Yet, lower overall roughness was found in the Si-containing steels.

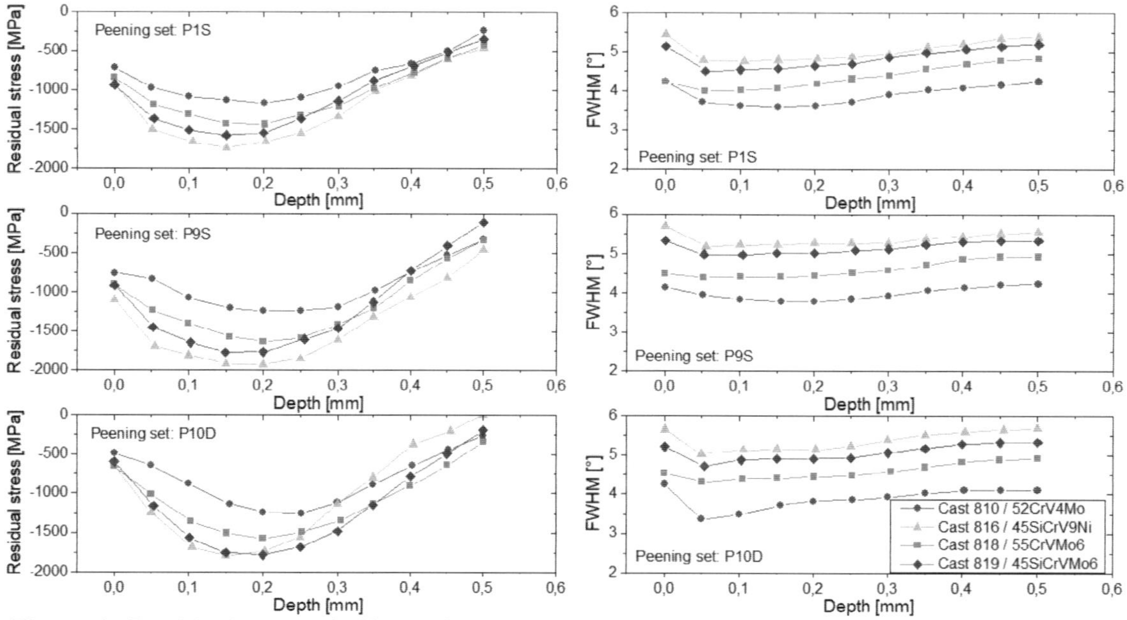

Figure 3. Residual stress (left) and full width at half maximum (FWHM) profiles (right) after different stress peening treatments.

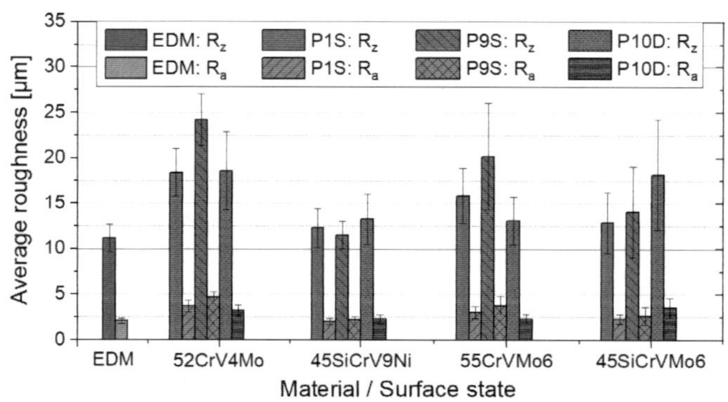

Figure 4: Roughness values after different stress peening treatments.

An overview of the fatigue test results is shown in Figure 5. The tests were commenced with a specimen gauge width of 18 mm. Unfortunately, the bending machines' load maximum was reached with the new steel grades, which is reflected in scatter and a large number of run-outs (Figure 5 (left)). Therefore, the specimen gauge width had to be reduced to 12 mm by means of EDM (drawn in green colour in Fig. 1 (left)). The corresponding fatigue test results are shown in Figure 5 (right) and allow for more interpretation.

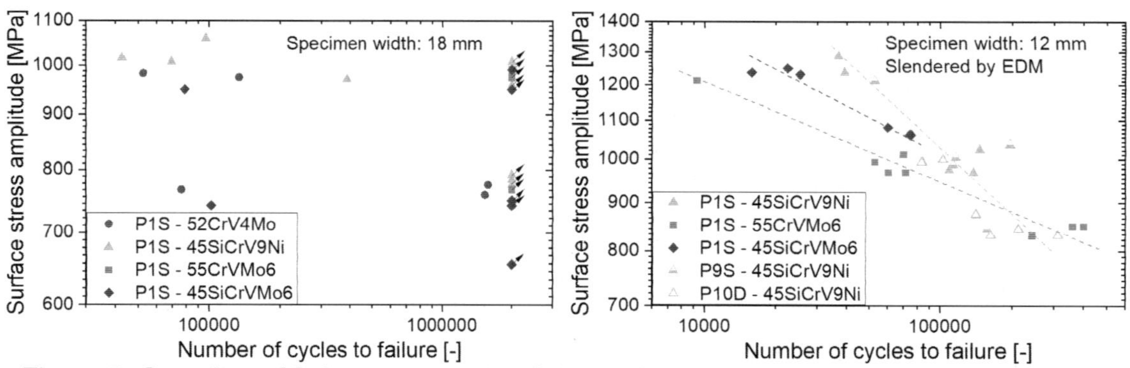

Figure 5. Overview of fatigue test results: Original (left) and slendered specimens (right).

Discussion and Conclusions

Three promising new steel grades have been developed within this study. Particularly the Si-containing grades offer extreme strength at high toughness. The steel grade 45SiCrV9Ni initially has a 25% higher tensile strength than the reference 52CrV4Mo steel. After the standard stress-peening process (P1S) the maximum compressive residual stresses become even ~50% higher as compared to 52CrV4Mo. When comparing both warm peening processes (P9S), the difference is even bigger, improving 45SiCrV9Ni up to a 55% over 52CrV4Mo. The increase in compressive residual stresses from the reference steel and stress-peening process to the warm peening and 45SiCrV9Ni reaches >65%. This growth partially corresponds to its higher strength (~25%), partially to the warm peening process (~10%) and, remarkably, to synergetic effects probably linked to the highest deformability of the 45SiCrV9Ni grade. Obviously, a direct qualitative correlation between fatigue life and residual stresses exists. An advantageous combination of higher residual stresses and lower roughness was found for the steel grades containing Si. However, for the case of 45SiCrV9Ni, no marked difference regarding the different stress peening strategies was identified in terms of fatigue life. Particularly, no beneficial effect of double peening (P10D), as identified by Scuracchio et al. [2], was found. Obviously, the second peening step is not effective and rather detrimental if carried out without pre-stress. Despite the outstanding mechanical behavior, the application of new steel grades requires a thorough analysis of the overall properties. For instance, the decarburization of Si-containing steels within this study could only be limited by means of inductive heating. This is confirmed by the study of Podgornik et al. [4], who discovered a 50% increase in fracture toughness after adding 1,6% Si, while larger contents caused marked decarburization. In summary, promising steel grades and shot peening strategies were identified, though the optimum strategy has still to be identified. In future studies, stress peening parameters such as the ratio of pre-stresses in the first and second shot peening step should be varied more markedly.

Acknowledgement

The authors gratefully acknowledge the financial support of the Research Fund for Coal and Steel under Grant Agreement number: 747346 — LEAFSLIM — RFCS-2016.

References

[1] V. Schulze, *Modern mechanical surface treatment: states, stability, effects,* John Wiley & Sons, 2006.

[2] B. G. Scuracchio, et al., *Role of residual stresses induced by double peening on fatigue durability of automotive leaf springs*, Materials & Design 47 (2013), pp 672-676.

[3] A. Savaidis. *Surface properties and fatigue life of stress peened leaves.* Materials Testing 54.7-8 (2012), pp 529-534.

[4] B. Podgornik, et al., *Effect of modifying the chemical composition on the properties of spring steel,* Metallurgical and Materials Transactions A 49.8 (2018), pp 3283-3292.

[5] B. Podgornik, et al., *Microstructure refinement and its effect on properties of spring steel,* Materials Science and Engineering: A 599 (2014), pp 81-86.

[6] G. E. Totten, ed. *Steel heat treatment: metallurgy and technologies*, CRC press, 2006.

[7] B. Podgornik, et al., *Improving properties of spring steel through nano-particles alloying,* Materials Science and Engineering: A 638 (2015), pp 183-189.

[8] L. Zhang, et al., *Effect of quenching conditions on the microstructure and mechanical properties of 51CrV4 spring steel*, Metals 8.12 (2018), p 1056.

[9] V. R. M. Gonçalves, et al., *Influence of Deep Cryogenic Treatment on the Mechanical Properties of Spring Steels,* Journal of Materials Engineering and Performance 28.2 (2019), pp 769-775.

[10] A. Wick, et al., *Effects of warm peening on fatigue life and relaxation behaviour of residual stresses in AISI 4140 steel,* Materials Science and Engineering: A 293.1-2 (2000), pp 191-197.

[11] H. Wern. *Single crystal elastic constants and calculated bulk properties: a handbook*, Logos-Verlag, 2004.

Experimental Assessment of the Effects of modified Stress Peening Treatments on Surface Layer States and Fatigue Behavior in Leaf Spring Steel 52CrV4Mo

A. Klumpp[1], C. Grau[1], R. Rößler[1], D. Dapprich[2], S. Dietrich[1], V. Schulze[1]

1 Institute for Applied Materials (IAM-WK), Karlsruhe Institute of Technology (KIT), Karlsruhe, Germany
2 Stresstech GmbH, Rennerod, Germany

Abstract
This paper was derived within the EU-project "LEAFSLIM" and deals with surface states and fatigue behavior of quenched and tempered leaf spring steel 52CrV4Mo after conventional and modified stress peening treatments. An experimental setup for warm peening of specimens pre-stressed in bending was developed. A combination of inductive and continuous flow air stream heating was used in conjunction with a hydraulic load device to achieve reproducible surface layer states. Residual stresses and penetration depths in pre-stress direction reached a maximum (by amount) of approximately -1200 MPa and 0.6 mm, respectively. Warm and double peening did not significantly affect the fatigue behavior in the HCF range. Instead, a dominant influence of surface roughness on fatigue lives was found.

Keywords Modified stress peening, leaf springs, surface layer states, fatigue behavior

Introduction

Stress peening is a modified shot peening process in which mechanical pre-stress of the same direction and sign as the future operational load is applied to the work piece during the surface treatment [1,2]. After relieving the pre-load, the shot peening residual stresses are superposed by the reversed pre-stresses. By this means, residual stress amounts and penetration depths can be enhanced markedly in comparison to the conventional shot peening process, which is favorable e.g. in terms of fatigue behavior. Therefore, stress peening has been vastly applied in the steel spring industry, e.g. for parabolic leaf springs in heavy truck applications [3,4]. Leaf springs are typically made of quenched and tempered steel, such as 51CrV4 or the modified grade 52CrV4Mo. In the final stage before anti-corrosion painting, those springs are typically stress-peened under bending pre-load. For the sake of economic efficiency, the interest in optimizing the stress-peening process has increased during recent years. Besides the determination of optimum pre-stress and shot peening parameters, shot peening variants such as warm [5,6] and (warm) double peening [4,6] may be advantageous additions to stress peening.

Warm peening is a modified shot peening process mostly used on steels, in which the work piece exhibits an elevated temperature [1,2]. The latter usually ranges from 170 °C to 350 °C [5,7] and causes static and dynamic strain ageing effects stabilizing the dislocation structure in the work-hardened surface layer. Thus, more stable residual stress states can be achieved [1,8]. Double peening denotes the successive application of shot peening processes to the work piece with different parameters. Peening with coarse and subsequently with fine shot can favorably be used for surface roughness reduction and is therefore promising for fatigue resistance enhancement [2,4,6]. Therefore, a wide range of combinations of stress, warm and double peening (or more peening steps) is imaginable. However, experimental investigations on such combined treatments are rare. Wick et al. [5,8] investigated the residual stress stability and fatigue behavior in quenched and tempered 42CrMo4 after conventional shot peening, warm peening, stress peening and warm stress peening. Varying the temperature between 20 and 290 °C and the tensional pre-stress between 0 and 500 MPa, they found the largest increase of fatigue resistance after warm stress peening. Compared to conventional shot peening, temperature had the most pronounced influence on fatigue strength. However, since warm peening is a very energy- and thus cost-expensive process, it is only applicable if used at optimum conditions. Therefore, Tange and Ando [6] investigated a multiple step warm (torsional) stress peening process on helical compression springs made of quenched and tempered Si-alloyed steel SAE5160. Comparing conventional ambient temperature double peening with warm stress single peening and warm stress double peening, they found the best fatigue behavior after warm stress double peening. In this case, shot velocity, temperature and torsional pre-stress were constant throughout both peening steps and chosen to 76 m/s, 300 °C and 735 MPa, respectively. The shot size was reduced from 0.87 mm to 0.2 mm diameter. Using leaf springs, the double stress peening process at ambient temperature has successfully been applied by Scuracchio et al. [4]. Despite those endeavors, warm stress (double) peening has so far not been carried out on leaf springs. The present study was part of the EU-project "LEAFSLIM" and aims at closing this gap. First, a warm stress peening test bench will be presented. Then, surface layer states will be shown for different parameter combinations in terms of residual stress, full width at half maximum (FWHM) and roughness. Finally, the results of 4-point bending fatigue tests will be presented.

Experimental Methods

The low-alloyed spring steel 52CrV4Mo (or 51CrV4AR, a modified grade of 51CrV4) with the chemical composition shown in Table 1 was used for the investigation. Delivered as hot-rolled bars, it was austenitized in a vacuum furnace at 900 °C for 7 minutes, oil quenched to 50-55 °C and subsequently tempered at 430 °C for 90 minutes. By this means, yield and tensile strength of approx. 1500 MPa and 1600 MPa were achieved, respectively.

Table 1. Chemical composition of spring steel 52CrV4Mo.

C	Mn	Si	P	S	Cr	Ni	Mo	V	Cu	Al	Ti	O	N
0.56	1.00	0.37	0.014	0.003	0.95	0.10	0.087	0.10	0.20	0.025	0.003	0.002	0.011

Bars of the size 740x90x15 mm^3 were used for the stress peening experiments during which they were subjected to 4-point bending as shown in the schematic in Figure 1 (left).

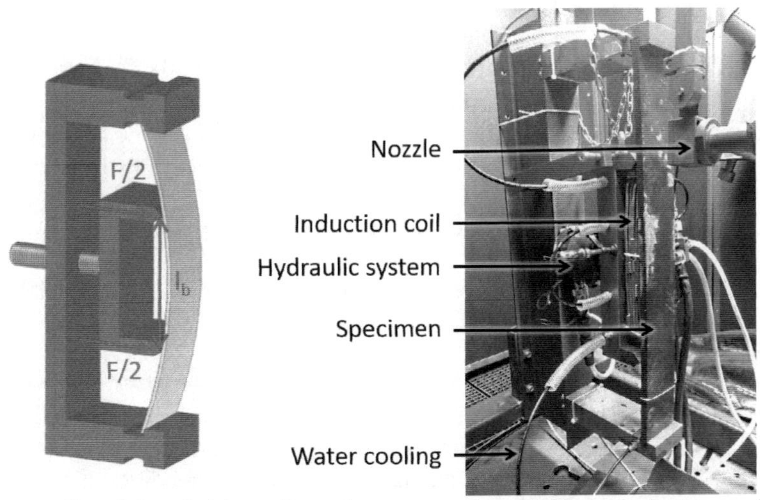

Figure 1. Schematic of 4-point bending stress peening (left) and photograph of realized warm stress peening device (right).

The outer (inner) width between supports was L = 740 mm (l_b = 370 mm). The overall hydraulic force F was limited to 50 kN, corresponding to a maximum bending stress of 1370 MPa. In Figure 1 (right), a photograph of the developed test bench is shown. A Baiker air blast shot peening machine was used. To realize warm peening, the flow heating device first presented in [8] was applied. Due to the large size of the specimens which was chosen to simulate similar stress gradients as compared to leaf springs, an additional inductive heater had to be used to reach the desired temperature of 350°C during the tests. Moreover, a water cooling system was implemented to avoid excessive heating of the stress peening device. In Table 2, the test plan of the applied stress peening campaign is shown. Three bars were shot peened with each parameter set. Almen intensity, pre-stress and temperature were varied. In parameter set P10D, a second peening step was applied (double peening). The latter was carried out using S170 (56 HRC) as shot, an Almen intensity of approx. 0,2 mmA, at room temperature (20°C) and without pre-stress.

Table 2. Parameters for stress peening campaign.

Set	Almen Int.	Shot	Coverage	Surface pre-stress	Temp.	2nd step
P1S	0.22 mmC	S330 (56 HRC)	100%	1370 MPa	20°C	NO
P3S	0.22 mmC	S330 (56 HRC)	100%	685 MPa	20°C	NO
P4S	0.12 mmC	S330 (56 HRC)	100%	1370 MPa	20°C	NO
P6S	0.28 mmC	S330 (56 HRC)	100%	1370 MPa	20°C	NO
P7S	0.22 mmC	S330 (56 HRC)	100%	1370 MPa	290°C	NO
P9S	0.22 mmC	S330 (56 HRC)	100%	1370 MPa	350°C	NO
P10D	0.22 mmC	S330 (56 HRC)	100%	1370 MPa	350°C	YES

Residual stresses were measured using a Stresstech DR45 diffractometer with CrK_{α}-radiation. A primary collimator with a diameter of 3 mm and a secondary 2D detector were applied, allowing for measurement times of approx. 30 seconds per stress value. Cross correlation with a threshold of 85% and the $\sin^2\psi$-method were used for α-{211} peak position determination and residual stress evaluation, respectively. The elastic coefficients C_{11}, C_{12} and C_{44} (for single crystals) were extracted from the literature [9] and are $C_{11} = 2.3\cdot10^5$ MPa, $C_{12} = 1.3\cdot10^5$ MPa and $C_{44} = 1.1\cdot10^5$ MPa. Residual stress depth profiles parallel and perpendicular to the pre-stress direction with a maximum depth of 0.5 mm were determined by means of electrolytic layer removal.

Roughness was measured by means of a Nanofocus µSurf confocal microscope. An area of 1.5*1.5 mm^2 close to the center of the shot-peened surface was evaluated in each case. The Gaussian cut-off wave length was $\lambda_c = 0.8$ mm.

Due to the large size of the bars subjected to shot peening, direct fatigue testing was not possible. Instead, small 4-point bending specimens (sizing 100x20x5 mm^3) were extracted from the bars by milling and EDM. 12 fatigue specimens could be obtained from each bar, yielding 36 specimens per stress peening parameter set. Fatigue testing was carried out in 4-point bending with the maximum stress in the shot peened surface. The outer (inner) width between supports was L = 80 mm (l_b = 40 mm). The load ratio was 0.1. A servohydraulic testing machine was used at a frequency of 30 Hz and room temperature. Residual stress and roughness values were measured on fatigue specimens. Thus it was assumed that a direct link between fatigue life and surface state could possibly be established.

Experimental Results

In Figure 2 (top), the residual stress depth profiles parallel (left) and perpendicular (right) to the pre-stress direction are shown. The full width at half maximum (FWHM) profiles were determined parallel to the pre-stress direction and are depicted in Figure 2 (bottom, left). Anisotropic residual stress states were obtained by stress peening. Shot peening intensity had the largest influence on residual stress profiles in terms of maximum stress amount and penetration depth. The pre-stress during shot peening had only a minor influence on the residual stress profile and only in terms of penetration depth. Double peening caused markedly decreased residual stresses close to the surface. Due to the limitation of in-depth measurements, only values up to 0.5 mm distance to surface were captured. Therefore, the actual penetration depths of residual stresses can in most cases only be estimated. Regardless of shot peening parameters, FWHM values were decreased in near-surface regions. Again, shot peening intensity was the major influencing factor. An overview of the roughness values measured on the stress-peened surfaces is shown in Figure 2 (bottom, right). It is obvious that the shot peening intensity had the most pronounced effect on roughness, while warm and double peening did not affect the roughness markedly. Very favorable small roughness was obtained by stress peening at room temperature and with low intensity.

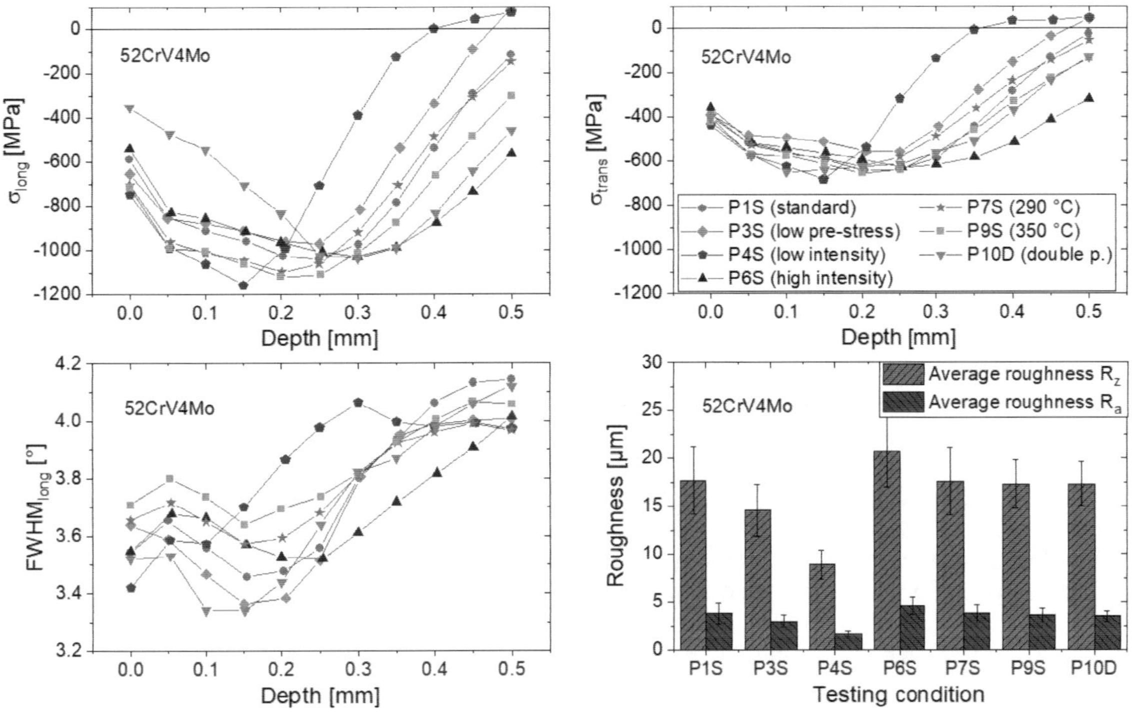

Figure 2. Top: Residual stress profiles; left side: longitudinal measurement (parallel to pre-stress); right side: transverse measurement (perpendicular to pre-stress). Bottom (left side): longitudinal FWHM profiles. Bottom (right side): Surface roughness values.

An overview of the 4P-bending fatigue test results is shown in Figure 3.

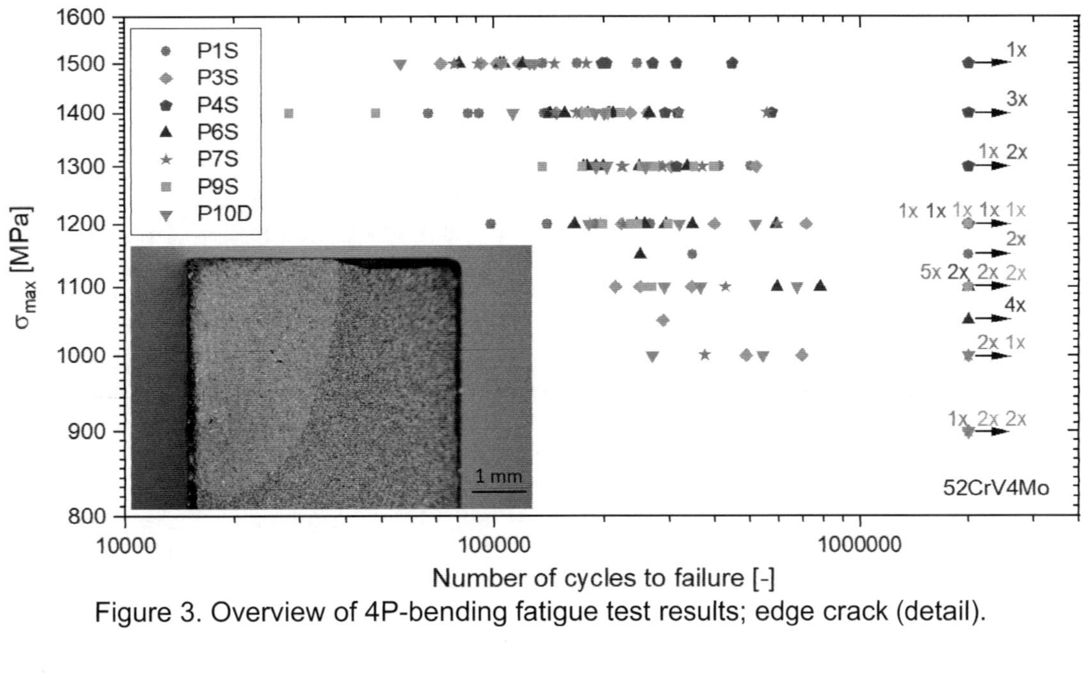

Figure 3. Overview of 4P-bending fatigue test results; edge crack (detail).

Obviously, the fatigue tests were prone to a large degree of scatter and do not allow for meaningful assessment of fatigue lives and limits. Compared to the results derived on the "standard" parameter set, P1S, the only parameter set yielding slightly improved fatigue behavior was P4S, where the shot peening intensity was reduced from 0.22 mmC to 0.12 mmC. While an effect of peening temperature is hardly recognizable, double peening may even have been detrimental to the fatigue behavior.

Discussion and Conclusions

In this study, a test bench allowing for warm stress peening of bars subjected to bending pre-stress was developed. The goal of this study was to assess the effect of modified stress peening treatments on surface states and resulting fatigue behavior on specimens similar to leaf springs. To this end, peening intensity, pre-stress and temperature were varied with reference to "standard" parameters. Moreover, the effects of double peening were investigated, using a second peening step without pre-stress at ambient temperature. Regarding surface states, only the amount of pre-stress, the second peening step and the shot peening intensity markedly affected residual stress and FWHM profiles as well as surface roughness. Different results have been derived in previous experimental studies on warm peening and warm stress peening [5,7,8,10]. Wick et al. [5,8] carried out extensive studies on warm peening, stress peening and warm stress peening of AISI4140. They did not find a significant influence of temperature on residual stresses in studies on stress-free shot peening. Only when pre-stress (tensile; 500 MPa; approx. half of the yield strength) was applied, a marked increase of penetration depth and near-surface residual stress amount was found. In contrast, Schilling-Praetzel et al. [7] and Harada and Mori [10] found increased residual stress penetration depths and slightly reduced amounts of surface residual stress after stress-free warm peening on quenched and tempered spring steels. In each case, surface roughness was increased due to warm peening, which can be explained in terms of decreasing flow stresses at elevated temperatures and thus promoted near-surface plasticity. In contrast, in our study roughness was not markedly affected by the peening temperature. Yet, according to Wick et al. [5], the effect of temperature on roughness might be reduced when shot peening is carried out in the presence of pre-stress [5]. In our study, near-surface FWHM values were decreased regardless of the stress-peening parameters. This is indicative of strain softening and is commonly observed in hardened or quenched and low tempered steels [1]. The affected depth of reduced FWHM values after ambient temperature stress peening can qualitatively be correlated with the resulting roughness values. However, despite comparable residual stress profiles and roughness values, reduction of FWHM values was rather weak after warm peening. Interactions between temperature and pre-stress could prevail and should be investigated in detail in future studies. Regarding double peening, Scuracchio et al. [4] and Tange and Ando [6] found beneficially increased surface residual stresses and roughness values in conventional stress peening and warm stress peening, respectively. Such effects were not identified in this study. Instead, near-surface residual stresses and FWHM values were markedly decreased due to the second shot peening step. The latter was carried out at ambient temperature and without pre-stress. Müller [11] pointed out that a second peening step may only be effective if applied under the same pre-load as the first peening step. In addition, Scuracchio et al. [4] found that the amount of near-surface residual stress could only be increased if the second shot was smaller than the first shot at otherwise same stress peening parameters. In the present study, the second shot was significantly smaller than the first shot and is thus expected to have no marked detrimental effect on residual stresses. Therefore, it can be confirmed that maintaining the pre-stress during the second shot peening step is essential for effective double peening treatments.

Regarding fatigue behavior, the only parameter set clearly leading to enhanced fatigue behavior was P4S (reduced shot peening intensity), which can be attributed to the significantly lower roughness values. Decreasing pre-stress and increasing temperatures, in turn, were detrimental to the fatigue behavior. Any other prevailing effects are obscured by large amount of scatter. With the chosen specimen preparation procedure, the specimen

edges were severely weakened, causing fatigue initiation only at specimen edges. An exemplary edge crack, obtained by macroscopy, is shown within Figure 3. Additional scatter might possibly be caused due to process stability issues. Although no systematic investigations on process stability were involved in this study, repetitive residual stress measurements exemplarily carried out on the parameter set P10D (double peening) showed satisfactory reproducibility of residual stress values. The differences were in most cases within the range of the standard deviation.
In contrast to our findings, warm peening, warm stress peening and warm stress double peening generally enhance the fatigue behavior [4-8]. Particularly, warm peening increases residual stress stability and thus fatigue behavior due to static and dynamic strain ageing effects [1,2]. So far, the warm stress double peening process, proposed by Tange and Ando [6], seems to be the most promising stress peening strategy to improve fatigue behavior. However, process parameters have to be selected with care and must be adjusted to the overall material behavior of specimens or actual parts. In this context, Savaidis et al. [3] carried out fatigue and failure analyses on stress-peened leaf springs made of 51CrV4 and found that roughness may cancel out the residual stress effect with regard to final fatigue lives, particularly when the steel grade is prone to decarburization.
In summary, the development of a bending pre-stress device allowed for modified stress peening with pre-stress conditions similar to industrial leaf spring stress peening treatments. The new test bench allows for the implementation of promising shot peening strategies. Yet, the results derived within this study may have become blurred due to the partly unfavorable choice of peening parameters as well as the specimen extraction strategy. In future studies, further stress peening strategies with larger parameter variations should be tried out on specimens directly subjectable to fatigue tests, as long as the overall procedure is justifiable from both academic and industrial point of view.

Acknowledgement
The authors gratefully acknowledge the financial support of the Research Fund for Coal and Steel under Grant Agreement number: 747346 — LEAFSLIM — RFCS-2016.

References
[1] V. Schulze, *Modern mechanical surface treatment: states, stability, effects,* John Wiley & Sons, 2006.
[2] A. Klumpp, et al., *Mechanical surface treatments,* Proc. ICSP 12 (2014), pp 12-24.
[3] G. Savaidis, et al., *Microstructural, surface and fatigue analysis of stress peened leaf springs*, International Journal of Structural Integrity 6.5 (2015), pp 589-604
[4] B. G. Scuracchio, et al., *Role of residual stresses induced by double peening on fatigue durability of automotive leaf springs*, Materials & Design 47 (2013), pp 672-676.
[5] A. Wick, et al., *Effects of stress- and/or warm peening of AISI 4140 on fatigue life*, Steel research 71.8 (2000), pp 316-321.
[6] A. Tange and K. Ando, *Improvement of spring fatigue strength by new warm stress double shot peening process*, Materials Science and Technology 18.6 (2002), pp 642-648.
[7] M. Schilling-Praetzel, et al., *Influence of temperature of shot peening on fatigue life*, Proc. ICSP 5 (1993), pp 227-238.
[8] A. Wick, et al., *Effects of warm peening on fatigue life and relaxation behaviour of residual stresses in AISI 4140 steel*, Materials Science and Engineering: A 293.1-2 (2000), pp 191-197.
[9] H. Wern. *Single crystal elastic constants and calculated bulk properties: a handbook.* Logos-Verlag, 2004.
[10] Y. Harada and K. Mori, *Effect of processing temperature on warm shot peening of spring steel*, Journal of Materials Processing Technology 162 (2005), pp 498-503.
[11] E. Müller, *Stress Peening—A Sophisticated Way of Normal Shot Peening*, Journal of Materials Science and Engineering A 9.3-4 (2019), pp 56-63.

Prediction of Residual Stress States after modified Shot Peening Treatments using Strain-Rate and Temperature-dependent Material Data

A. Klumpp[1], S. Klaus[1], S. Dietrich[1], V. Schulze[1]

1 Institute for Applied Materials (IAM-WK), Karlsruhe Institute of Technology (KIT), Karlsruhe, Germany

Abstract

Within the EU-project “LEAFSLIM”, the application of modified stress peening treatments to leaf springs for heavy truck applications has been investigated. Part of the project was the development of an adequate Finite Element (FE)-simulation model. Using strain-rate and temperature dependent material data, the predictability of residual stresses after conventional, warm, stress, warm stress and double peening was assessed. Both impact analyses and process-oriented simulations were carried out. The simulation framework shows a good reproducibility of experimentally measured residual stress states and tendencies regarding pre-stress and temperature.

Keywords Modified stress peening, leaf springs, surface layer states, FE-simulation

Introduction

During the last decades, numerical simulations have become a versatile and powerful means to predict, investigate and understand the shot peening process and the hereby generated surface states, determined by measures such as residual stress and roughness, cold work and phase composition. Meanwhile, the efforts undertaken in a large number of original investigations have been summarized in exhaustive literature reviews [1,2], giving an overview about the efforts undertaken in the field. Investigations have focused on feasible model assumptions and simplifications regarding particle dynamics, material behavior of target and shot, boundary conditions, friction, among other aspects, within 2D and 3D micro and macro Finite Element (FE)-simulations [1,2]. Within recent years, a trend towards realistic process models has been recognizable, for instance the model of Gariépy et al. [3] which incorporates the random nature of size and velocity distribution of the shots.

One challenging aspect of shot peening simulation has been the target material modeling. Shot peening causes plastic material deformation at very high strain rates, possibly above 10^5/s [4], causing adiabatic heating and evoking mechanisms relevant to high strain rate deformation such as thermally activated dislocation glide and damping effects [5]. Moreover, the multiaxial Bauschinger effect comes into play, since plastic deformation is confined to very small distances from the target surface and multiple shot impacts cause three-dimensional elastic-plastic strain reversals. Various material-related effects have therefore been incorporated in shot peening simulation, giving rise to different approaches such as complex viscoplastic models with combined kinematic and isotropic hardening rules [4,6,7]. One model presented in the past is the thermal activation continuum flow stress model of Voyiadjis & Abed [6] which was adopted to shot peening simulations by Klemenz et al. [7]. Since this type of model is strain-rate and temperature dependent, its application to shot peening at different temperatures, such as warm peening, a common shot peening variant, would be of high interest. Yet, simulations of shot peening variants are very rare. Besides warm peening, typical shot peening variants are stress peening and double peening, all of which are processes highly relevant to the steel spring industry [8].

Warm peening is a modified shot peening process mostly used on steels, during which the work piece exhibits a temperature usually in the range from 170 °C to 350 °C. Due to static and dynamic strain ageing effects, more stable residual stress states can be achieved [8]. Stress peening is a modified shot peening process in which mechanical pre-stress of the

same direction and sign as the future operational load is applied to the work piece during the surface treatment. After unloading from the pre-stress, the shot peening residual stresses are superposed by the reversed bending stresses and therefore of higher magnitude and penetration depth than after conventional peening [8]. Double peening denotes successively applied shot peening processes with different parameters. Peening with coarse and subsequently with fine shot can favorably be used for surface roughness reduction and individual shot peening variants are theoretically combinable to complex peening operations, such as warm stress double peening [9].
Within the EU-project "LEAFSLIM", the application of such enhanced shot peening variants has been assessed in the context of leaf springs in heavy truck applications (more details in [10,11]). As part of the project, this study aims at presenting an approach toward the simulation of the applied shot peening variants. Besides general tendencies predicted in warm, stress and double peening, some comparisons to experimental data will be shown and subsequently be discussed with regard to other results from the literature.

Experimental and Numerical Methods

The low-alloyed spring steel 51CrV4AR (52CrV4Mo) Q+T was used for the investigation. A detailed description of material and heat treatment can be found in [10].
Shot peening was simulated by means of Finite Element (FE)-Analysis. The FE geometry and mesh were modeled using parameterization. After preliminary mesh convergence and size analyses, a 4*4*4 mm^3 sized cuboid consisting of approx. 200,000 hexahedron elements of type C3D8R was used as simulation model and is shown in Figure 1 (a).

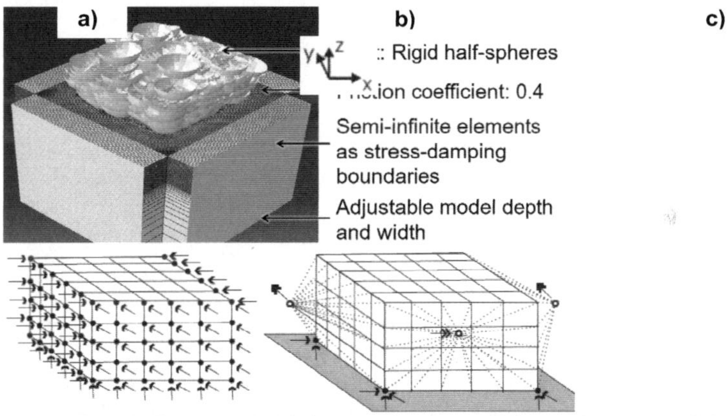

Figure 1. Shot peening simulation model (a); symmetry cell boundary condition (b); kinematic constraint boundary condition (c) [6, 13].

Shots were modeled as analytical rigid half spheres. For the sake of simplicity, mean values of shot velocity and size were assumed using literature data [12]. Shot velocities varying between 20 m/s and 120 m/s were adopted. Since S330 and S170 shots were used in experimental investigations [10,11], the diameters of half spheres in simulation were 0.80 and 0.36 mm, respectively. A friction coefficient of 0.4 between shots and target was used. Semi-infinite elements (CIN3D8) were applied at the laterals to dampen stress waves. Figure 1 (b) and (c) show the applied boundary conditions, namely the symmetry cell and kinematic constraint, as defined by Zimmermann in a previous study [13]. Single end (double end) arrows denote translational (rotational) constraints. Preliminary studies showed that residual stresses generated using the different boundary conditions do not differ markedly.
Stress peening was simulated using the symmetry cell constraint. Pre-stress is applied by releasing the lateral constraint and imposing a pre-defined displacement in x-direction. By this means, tensile pre-stress prior to shot peening is imposed. After the peening step, the displacement is fully reversed. One must bear in mind that full strain reversal is a condition never reached in reality. Yet, it is assumed to be adequate for basic numerical analyses.

Double peening was carried out using two subsequent shot peening steps. Warm peening was simulated by applying an adequate initial temperature condition to the model.
The chosen material model is based on the approach of Voyiadjis & Abed [6] and described in more detail in the work of Klemenz et al. [7]. It makes use of combined kinematic and isotropic hardening. The yield condition reads

$$f \equiv J_2(\boldsymbol{\sigma}' - \boldsymbol{\xi}) - \sigma_G - R - \sigma^*(T, \dot{\varepsilon}),$$

where R and ξ (composed of 2 non-linear variables) denote the scalar isotropic hardening variable and the backstress tensor, respectively. σ_G is the athermal flow stress part of the material. It is temperature dependent only in terms of the shear modulus and independent from strain rate. The thermal flow stress σ* depends on plastic strain-rate $\dot{\varepsilon}^{pl}$ and temperature T. It is given by the equation

$$\sigma^* = \sigma_0^* \left(1 - \left(\frac{k_b T \ln\left(\frac{\dot{\varepsilon}_0}{\dot{\varepsilon}^{pl}}\right)}{\Delta G_0} \right)^n \right)^m,$$

where k is Boltzmann's constant and ΔG_0, σ_0^*, $\dot{\varepsilon}_0$, m and n are parameters which have to be determined experimentally.
Push-pull tests at room temperature and low and high speed tensile tests at varying temperatures were carried out to calibrate the material model. The results of those tests and the model parameters derived hereby are shown in Figure 2 and Table 1, respectively.

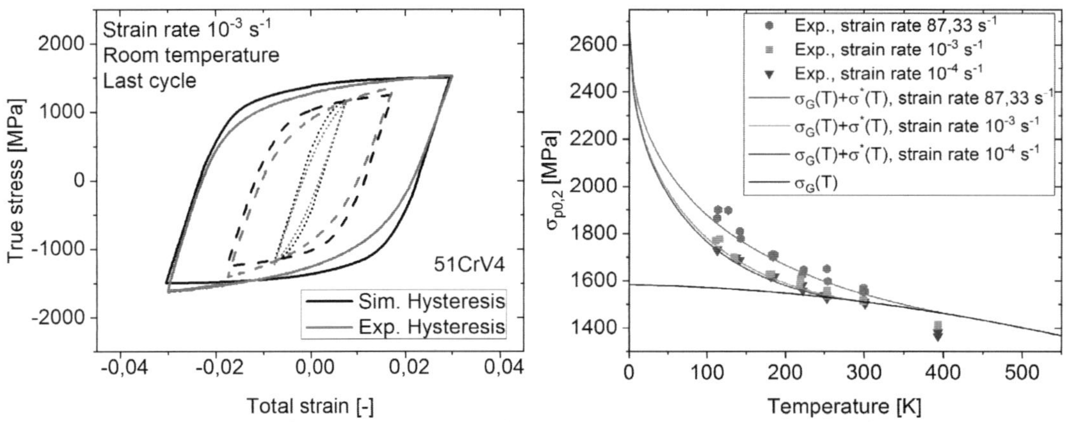

Figure 2. Results of push-pull test (left) and thermal activation model fit (right).

Table 1. Parameters derived for simulation of shot peening variants.

$\sigma_{G,0}$ (σ_G at RT)	ΔG_0	σ_0^*	$\dot{\varepsilon}_0$	m	n
1522 MPa	1.59e-19 J	1088 MPa	5.91e13 s^{-1}	0.5	1.91

Q_1	b_1	C_1	γ_1	C_2	γ_2
-277 MPa	6.93	263050.7 MPa	716.9	24323.3 MPa	83.6

Due to its suitability for nonlinear, dynamic problems, the commercially available equation solver Abaqus/Explicit was used for stress computation. Material data were provided by means of a user material file (VUMAT).
To achieve statistically robust values of residual stresses and field variables after shot peening, an averaging strategy [7] was applied to evaluate the results.

Simulation Results
Figure 3 shows the predicted residual stress states in x-direction after warm peening with different temperatures with (left) and without (right) strain rate dependence of flow stress. In both cases, the simulation predicts lower residual stress magnitudes and higher penetration

depths with increasing process temperature. It becomes obvious that at lower peening temperatures the influence of strain rate is more pronounced. For the simulation, 600 randomly distributed S330-sized shots with a speed of 20 m/s on a surface of 2*2 mm^2 (25% of total model surface) were assumed. This corresponds to a treatment with rather low Almen intensity and high coverage. To achieve absolute comparability of results, shot distribution and shot order were coincident throughout the simulations.

Figure 4 shows the predicted residual stress states after conventional stress peening at 20 °C (top) and warm stress peening at 300 °C (bottom). On the left (right) side, the stress component in x- (y-) direction is shown. For the simulation, 120 randomly distributed S330-sized shots with a speed of 40 m/s on a surface of 2*2 mm^2 (25% of total model surface) were assumed. This corresponds to a treatment with medium Almen intensity and medium coverage (see for instance the investigations of Tange and Ando [9] and Miao et al. [12] for comparison). The tensile pre-stress in x-direction was varied between 0 and 1200 MPa. With increasing pre-stress, both maximum amount and penetration depth of residual stress in x-direction are increased, while the stress in y-direction is not strongly affected. Regarding temperature, the same tendencies as pointed out before prevail.

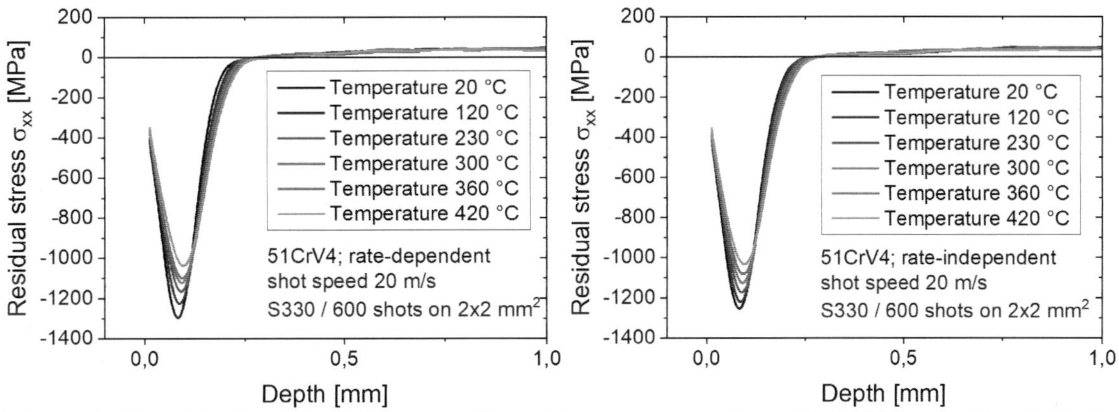

Figure 3. Predicted residual stress profiles after warm peening with (left) and without strain rate dependent material definition (right).

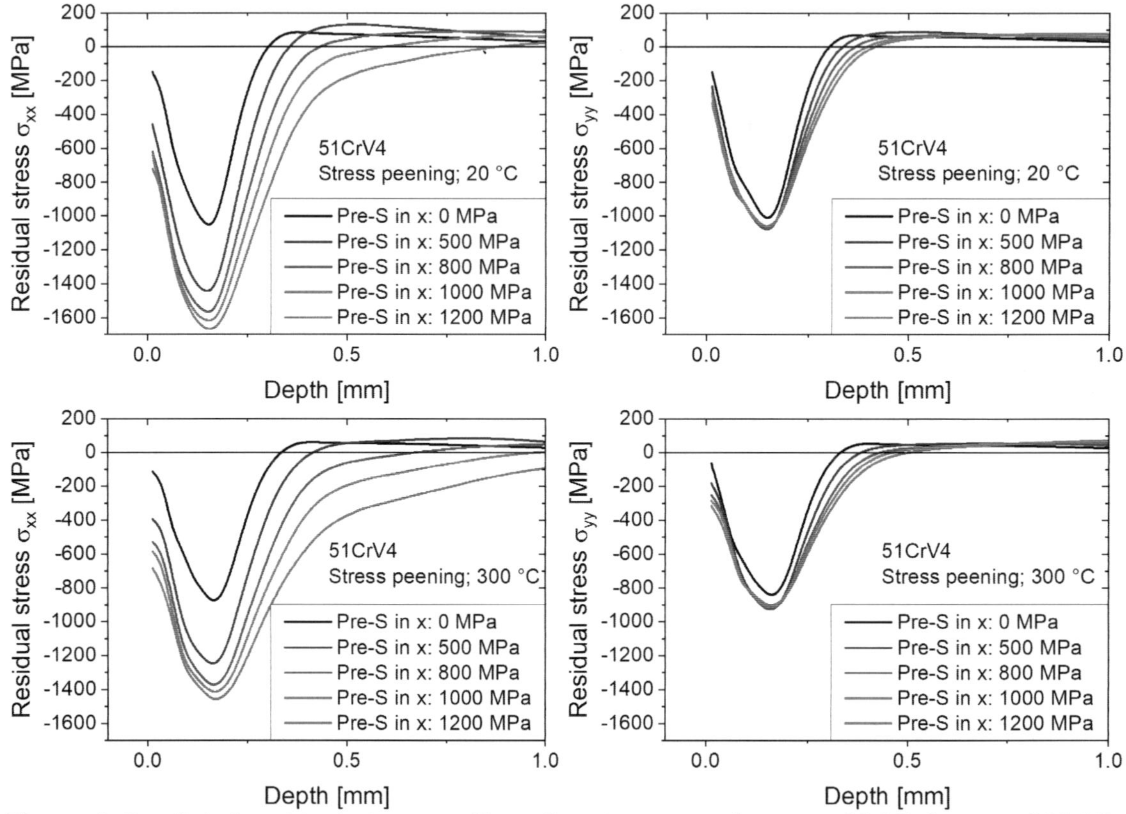

Figure 4. Predicted residual stress profiles after stress peening; top: 20 °C; bottom: 300 °C; left: stress component in x; right: stress component in y.

In Figure 5, the residual stress states after single and double warm peening without pre-stress at 360°C are shown. For the first peening step, 120 randomly distributed S330 shots with a speed of 20 m/s on a surface of 2*2 mm^2 (25% of total model surface) were assumed. For the second peening step, 500 randomly distributed S170 shots were assumed. Shot velocity was varied between 40 m/s and 80 m/s. Obviously, the second shot peening step significantly decreases the near-surface residual stress, while the sub-surface stresses are rather unaffected. This is particularly pronounced for the higher shot velocity.

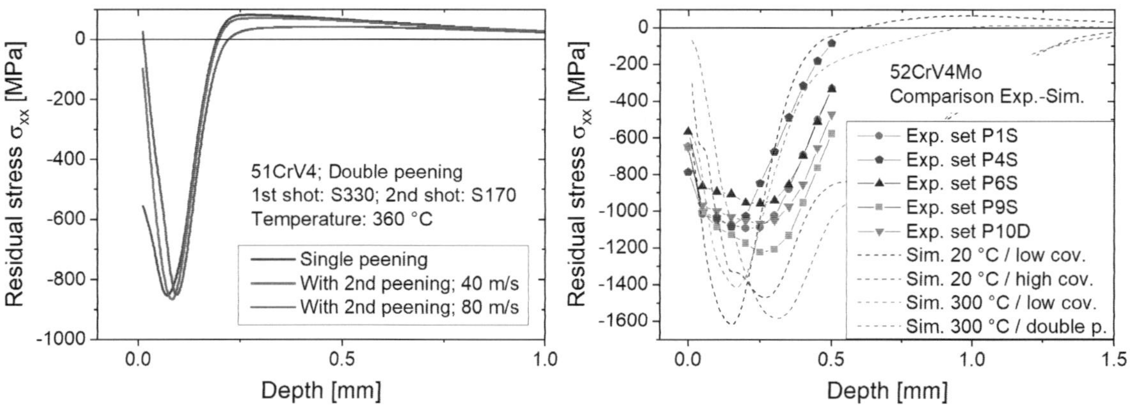

Figure 5. Predicted residual stress states after double peening (left) and comparison with stress peening experimental results (right).

Finally, Figure 5 (right) shows the comparison of numerical with experimental data. Experimental residual stress profiles were derived by stress peening in tension with the procedure described in more detail in [11]. The nomination of parameter sets corresponds to [10]: P1S (standard); P4S (low intensity); P6S (high intensity); P9S (warm); P10D (double).

Experimental stress peening was carried out by shot peening specimens from both sides. It can be seen that despite the fair reproduction of residual stress penetration depth at room temperature, residual stress maxima are largely overestimated.

Discussion and Conclusions

In this study, attention was devoted to the performance of a micro simulation model of conventional shot peening, warm peening and modified stress peening treatments. Besides comparisons of numerical and experimental data, a thorough analysis of the overall predictability of temperature-dependent residual stress formation was carried out. This is necessary since predicted residual stress states markedly depend on model assumptions, particularly regarding material behavior. In this context, Rouhaud et al. [14] compared residual stress prediction with kinematic and isotropic hardening models and showed that a marked influence of the hardening definition exists. Meguid et al. [4] deduced from their investigation on the use of strain-rate dependent flow stress formulation that the definition of the material itself has the strongest influence on the resulting residual stress. Using strain-rate dependent formulations, higher strain rates increase maximum residual stresses and decrease the penetration depth. The same effect prevails using temperature-dependent material data and has experimentally been observed [10,15]. As temperatures are increased, residual stress maxima are reduced and residual stress penetration depths increased. Aside from material definition, the assumption of rigid shots might be questionable when hard targets are treated. Mori et al. [16] deduced from their study that for a yield strength ratio of >2 no plastic deformation of the shot occurs. From target and shot materials used in this study, a yield stress ratio of approx. 2000 MPa / 1600 MPa (56 HRC / 47 HRC) is estimated. Therefore, the assumption of rigid shot in simulations may be a strong overestimation. Rigid shots, in combination with the “hard” strain reversal boundary condition, are expected to yield the strong overestimation of residual stresses after stress-peening, as shown in Figure 5 (right). Regarding double peening, Tange and Ando [9] found beneficially increased surface residual stresses after warm stress double peening. In contrast, decreasing surface residual stresses are predicted by our simulation, as localized deformation – combined with the isotropic softening assumption – softens the material. Yet, strain softening due to shot peening was also observed experimentally and characterized in terms of reduced full width at half maximum values of XRD peaks [10,11].

In summary, strain-rate and temperature dependent material behavior assumptions are an adequate means to predict tendencies in warm peening and warm stress peening. However, other process assumptions regarding the shot and the stress peening conditions might be inadequate and require more realistic modeling. Future studies should thus focus on such process-related assumptions and potential interactions with the material model definition.

Acknowledgement

The authors gratefully acknowledge the financial support of the Research Fund for Coal and Steel under Grant Agreement number: 747346 — LEAFSLIM — RFCS-2016.

References

[1] M. Zimmermann, et al., *Literature review on shot peening simulation*, Int. J. of Computational Materials Science and Surface Engineering 3.4 (2010), pp 289-310.

[2] J. S. Chen, et al., *Literature review of numerical simulation and optimisation of the shot peening process*, Advances in Mechanical Engineering 11.3 (2019), pp 1-19.

[3] A. Gariépy, et al., *Simulation of the shot peening process with variable shot diameters and impacting velocities*, Advances in Engineering Software 114 (2017), pp 121-133.

[4] S. A. Meguid, et al., *3D FE analysis of peening of strain-rate sensitive materials using multiple impingement model*, Int. J. of Impact Engineering 27.2 (2002), pp 119-134.

[5] R. W. Armstrong and S. M. Walley, *High strain rate properties of metals and alloys*, International Materials Reviews 53.3 (2008), pp 105-128.

[6] G. Z. Voyiadjis and F.H. Abed, *A coupled temperature and strain rate dependent yield function for dynamic deformations of bcc metals*, International Journal of Plasticity 22.8 (2006), pp 1398-1431.
[7] M. Klemenz, et al., *Application of the FEM for the prediction of the surface layer characteristics after shot peening*, Journal of Materials Processing Technology 209.8 (2009), pp 4093-4102.
[8] V. Schulze, *Modern mechanical surface treatment: states, stability, effects,* John Wiley & Sons, 2006.
[9] A. Tange and K. Ando, *Improvement of spring fatigue strength by new warm stress double shot peening process*, Materials science and technology 18.6 (2002), pp 642-648.
[10] A. Klumpp, et al., *Experimental Assessment of the Effects of modified Stress Peening Treatments on Surface Layer States and Fatigue Behavior in Leaf Spring Steel 52CrV4Mo*, submitted to Proc. ICSP 14 (2022).

[11] A. Klumpp, et al., *Effects of Shot Peening on Surface Layer States and Fatigue Behavior in Experimental Casts of Enhanced Leaf Spring Steels*, submitted to Proc. ICSP 14 (2022).
[12] H. Y. Miao, et al., *An analytical approach to relate shot peening parameters to Almen intensity*, Surface and Coatings Technology 205.7 (2010), pp 2055-2066.
[13] M. Zimmermann, *Numerische und experimentelle Untersuchungen zur Randschicht-ausbildung beim Druckluft- und Ultraschallkugelstrahlen von IN718*, PhD thesis, Universität Karlsruhe, 2009.
[14] E. Rouhaud, et al., *Finite elements model of shot peening, effects of constitutive laws of the material*, Proc. ICSP 9 (2005), pp 107-112.
[15] Y. Harada and K. Mori, *Effect of processing temperature on warm shot peening of spring steel*, Journal of Materials Processing Technology 162 (2005), pp 498-503.
[16] K. Mori, et al., *Rigid-plastic finite element simulation of peening process with plastically deforming shot*, JSME international journal. Ser. A, Mechanics and material engineering 39.3 (1996), pp 306-312.

Surface nanocrystallisation on Ti-6Al-4V component with a robotic controlled shot peening process

Yijun Liu[1], David Wimpenny[1], Ben Mcgrory[2], Colin Mcgrory[2]

1. Manufacturing Technology Centre, Pilot Way, Coventry, CV7 9JU, UK,

2. Sandwell UK Ltd., 2 Foundry Place, Towcester, Northamptonshire, NN12 6FP, UK,

Abstract

The present study experimentally investigated the influence of shot peening processing on surface nanocrystallisation in terms of the impact kinetic energy as a function of shot peening coverages. The impact kinetic energy of a single shot peening media varied from 1.8J to 77J, while the shot peening coverage changed from 200% to 1600%. In general, nanocrystallization at Ti6Al4V surface has taken place with all shot peening conditions but with different level of surface damages. Shot peening at lower kinetic energy is beneficial to generate nanocrystalline microstructure to a layer thickness of 5-7 μm, without causing harmful damage on Ti6Al4V surface. However, high impact kinetic energies, especially at higher coverage can cause considerable surface damages in the form of delamination parallel to the surface. Delamination was an accumulating result of micro-voids nucleation, growth and their coalescence to form microcracks in the subsurface region where shear localization and severe plastic deformation took place, as evidenced by superimposed high compressive residual stress level and increased microhardness. Shot peening at a low temperature of -21°C also resulted in nanocrystallisation, however, further investigation of the beneficial effects expected from the cooling process is still needed.

Keywords: Shot peening, nanocrystallisation, residual stress, microhardness, shear bands.

Introduction

Surface nanocrystallisation through surface mechanical attrition treatment (SMAT) was reported two decades earlier by Zhu et al in 2004 [1]. The authors developed an appropriate vibration device using an ultrasonic transducer to bombard small steel balls on a commercially pure titanium (CP Ti) surface at high frequency. A nanostructured layer of 50μm thick was produced on CP Ti surface with this ultrasonic peening process. Since then, surface nanocrystallisation (SNC) on metal surfaces has become a hot research topic. Fard et al carried out a comprehensive review on SNC via severe plastic deformation [2]. In recent years, the shot peening process has been extensively investigated by the academic world and researchers found it is possible to generate nanocrstallisation on various metals' surface, such as biocompatible magnesium alloy [3], mild carbon steel [4], and stainless steel [5]. However, so far there is a lack of study on adoption of the shot peening process for nanocrystallisation at a commercial scale. An initial attempt was made in this area by us in 2017, with initial results of surface nanocrystallisation on Ni-based superalloy Inconel 718 presented in ICSP13 [6]. Since then, continuous efforts were made and a robotic controlled shot peening process was developed. This newly developed process has a great potential to produce surface nanocrystallisation at industrial scale for the first time. The new system was also integrated with a cooling device that is capable to keep the peened metal surface at around -21°C because a cooled surface was considered to be beneficial for grain refinement. In the present study, the new system was tested for a parametric study on a commercial Ti-6Al-4V titanium alloy with a wide range of processing parameters.

Experimental Methods

Ti-6Al-4V test coupons were processed with the newly developed Sandwell 5-axis shot peening process at ambient temperature and at temperatures as low as -21°C. Surface temperature was measured with a high speed thermal camera, FLIR X6900sc. Shot velocities (V) were measured using a Progressive Surface/Tecnar G3 Shotmeter. Different masses (M) of cast steel shot were used at a variety of shot velocities in order to produce a wide range of impact kinetic energy (defined as $0.5*M*V^2$ of a single media impact) from 1.8 Joule to 77 Joule at a range of coverage from 200% to 1600%. Coverage is a ratio of the area covered by plastic indentation to the complete surface area impacted by shot peening. In normal production, 98% coverage is taken as full coverage. Coverage beyond 100% is defined as the processing time as a product of the time taken to achieve 100% coverage, as described in SAE J2277_201304. The shot peening process complied with AMS2430U and the shot peening media complied with AMS2431/2F.

The samples for SEM and Optical Microscopy study were polished with Struers's OPS polishing suspension (0.05 microns) and then etched with standard Kroll reagent. Microstructure and deformation bands of the peened samples were examined with Hitachi TM-3000 SEM. Microhardness tests were carried out across the transverse section at a test load of 200g dwelling for 15 seconds. Transverse slices of electron transparent, normal to shot peened surface, were sectioned with a FIB/SEM, the size of a foil specimen is around 5x6 microns. Detailed microstructure of the foil specimen was studied with a transmission electron microscopy (TEM, FEI Talos) at 200kV. Residual stress was measured with Stresstech PRISM ESPI system.

Experimental Results and Discussion

The overall nanocrystallisation results are summarised in Fig.1 for the samples treated at different levels of impact kinetic energy as a function of peening coverage. In general, nanocrystallisation took place on the Ti-6Al-4V surface at all peening conditions investigated in the present study. The main difference observed with SEM is the severity of the surface damages caused by repeated impacts. When the kinetic energy is at a low level, between 1.8J and 3.6J, a uniform nanocrystallisation layer was readily generated on the surface without causing visible damages. It is also interesting to find out that the surface treated at low kinetic energies was maintained at good condition even when the coverage increased from 200% to 1600%. With the increase in the kinetic energy to the range of 14J to 77J, nanocrystallisation continues to take place. However, modest surface damage started to appear at the lowest coverage of 200%. Further increase in coverage beyond 200% resulted in more severe surface damage. Both the modest surface damage and the severe surface damage are undesirable features for commercial engineering components.

Fig.2 shows the effect of the coverage on microstructure in the subsurface region. When the surface was peened at a low kinetic energy of 1.8J at a coverage of 200% (Fig.2a), the grain boundaries of the beta phase (in bright colour) were disintegrated, kneaded together with the alpha phase (in grey colour). Ultrafine microstructure of 5-7 μm thick was formed at the surface. When peening was extended to a coverage of 1600%, the plastic deformation and the microstructural evolution did not seem to extend to a larger depth in the subsurface region (Fig.2b). This may be due to the fact that the surface was worked hardened and mechanically strengthened and any further impacts at low energy level was not sufficient to generate any deeper penetration of plastic deformation in the subsurface.

Fig.3 shows the influence of the kinetic energy on surface microstructural evolution. At the standard coverage of 200%, the increase in kinetic energy to 3.6J did not generate a larger depth of the ultrafine grain structured region than that at 1.8J. However, the subsurface region experienced a more severe plastic deformation at a larger depth than the 1.8J case. A contrasted interface boundary was formed between the top blended region and the subsurface region. This is very similar to the nanocrystallisation reported by other authors in this area [7]. There was no apparent surface damage observed in the low kinetic energy cases of 1.8J to 3.6J.

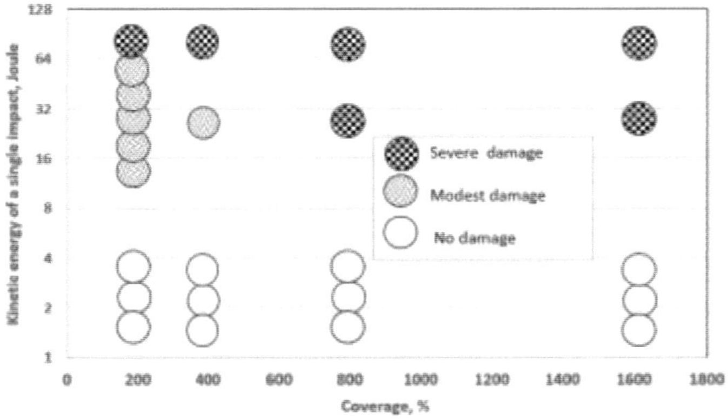

Fig.1: Nanocrystallisation at different impacting kinetic energy as a function of coverage.

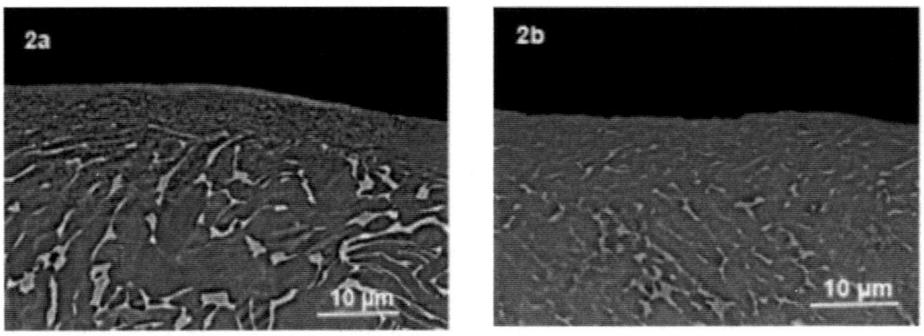

Fig.2: Microstructural evolution in the sub-surface region of the sample peened at kinetic energy of 1.8J at coverages of (a) 200%, (b) 1600% respectively.

When the impact kinetic energy or coverage increased further, a nanocrystalline layer was continuous to form, however, damages at surface and subsurface took place in the form of delamination as shown in Fig.4. When the sample was peened at a higher kinetic energy of 57J for extended period, a single or multiple delamination took place parallel to the metal surface in the subsurface region at a depth of 20-30 µm. Small shear bands or a large single and well developed shear band were formed, as circled with dashed lines in red colour. It looked featureless within the shear band zone. The so-called featureless phenomenon is due to the presence of equiaxed ultrafine submicron or nanocrystalline structure, which provided a direct evidence of dynamic crystallization rather than a boundary migration. Dynamic crystallization took place via a rotational mechanism during a rapid plastic deformation in a very short time at a microsecond scale caused by high velocity impact. This rotational dynamic recovery and recrystallization was similar to what was observed in the Ti6Al4V by Xue et al [8]. Dislocations were accumulated within the shear bands where voids nucleated as a result of high tensile stress in the location. Coalescence of multiple voids then formed microcrack, which was evidenced by the presence of the voids along the edges of cracks in Fig4. Microcrack then grew and propagated to the voids at its tip, leading to a complete delamination parallel to the surface. The mechanism of this kind of damage is also similar to the surface worn mechanism governed by the work hardening law as reported by Zhang et al [9]. Hence, the critical thickness of the nanocrystallisation layer at the surface is a competition between the magnitude of the work hardening within the shear bands in the subsurface region and the dynamic impacting force that applied on it. The increase in microhardness and the high compressive residual stress at the surface and the subsurface zone are the indications of the work hardening and plastic strain.

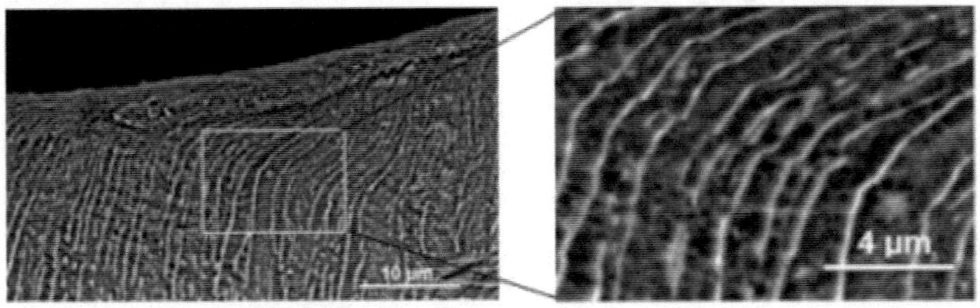

Fig.3: Ultrafine grain structure layer at surface (Left) and severe plastic deformation in the sub-surface region (Right) of the sample peened at 3.6J at a coverage of 200%.

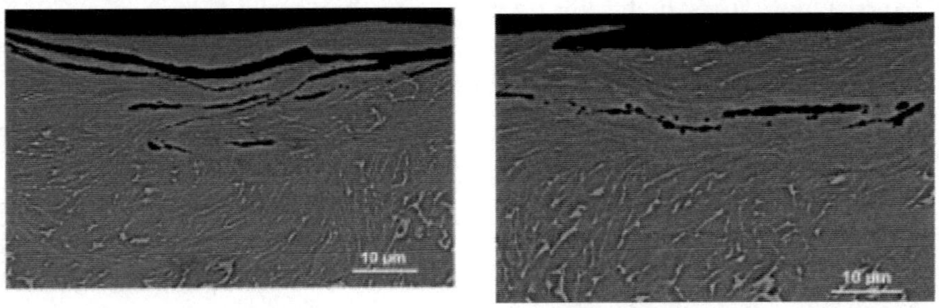

Fig.4: Delamination took place at kinetic energy of 57J at a coverage of 1600%.

Fig.5a shows typical microhardness profiles of the subsurface region shot peened at 200% coverage but with kinetic energies of 1.8J, 14J and 39J, respectively. The surface microhardness increased by around 15-20% in general, but no detectable difference was observed between different kinetic energy levels. It is worth noting that it was not possible to perform the Vickers hardness measurement in the very top surface region of 5-7µm. Hence, the hardness profiles in Fig.5a was not a measurement of the nanocrystalline layer, but they revealed the general work hardening effect in the subsurface zone, which took place at a depth of around 200 µm in the present study. Fig.5b shows the residual stress distributions as a function of coverages when the kinetic energy is 1.8J. A large compressive residual stress was produced by shot peening with all coverages. The depth of the compressive stress area is around 300 µm and was only slightly affected by coverage. The maximum compressive stress falls between 800 MPa and 1000 MPa.

Fig.6 shows the nanocrystalline microstructure observed with TEM when the surface was treated at 1.8J and 2.6J at a coverage of 200%. Microstructure at nanometer scale was generated in the subsurface region at a depth of around 2-3 µm as shown in the bright field images featured by the selected area diffraction ring patterns. The nanocrystallization was not uniform when the surface was peened at low kinetic energy and low coverage. Fine deformation bands mixed with nanocrystals at surface and the subsurface were revealed with EDS mapping in the form of the vanadium rich beta phase boundaries. When the surface was peened for a longer time up to 1600% coverage, the depth of the nanostructured layer was significantly increased up to around 7 µm, as shown in Fig.7a. However, the microstructure still showed the traces of the vanadium rich beta phase boundaries in both situations peened at ambient temperature and when the surface was constantly cooled down to around -21°C. Though the nanocrystallisation at -21°C seemed to be slightly increased for the sample investigated in the present study, the data is insufficient to conclude the benefit that the cooling at -21°C brought on the nanocrystallisation.

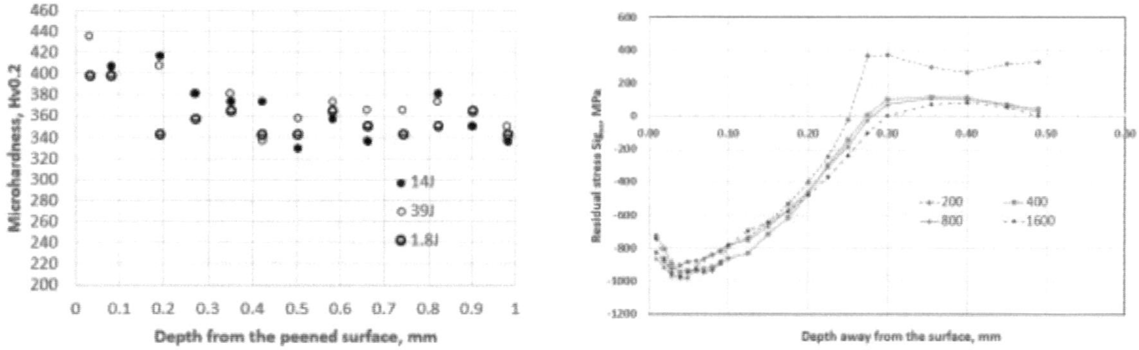

Fig.5: Comparisons of (a) microhardness at different kinetic energy levels at coverage of 200% and (b) residual stress distributions at 1.8J at different coverages of 200%-1600%.

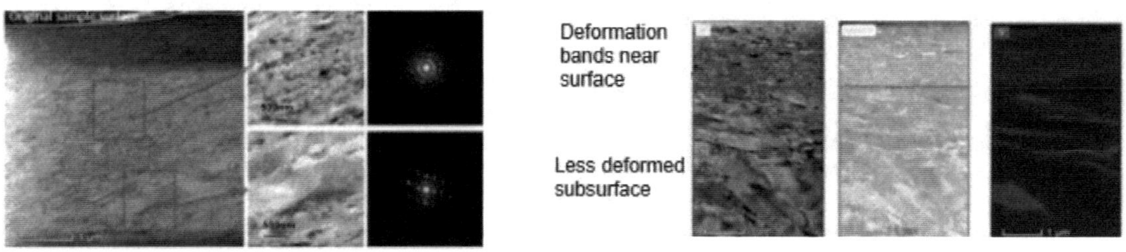

Fig.6: Bright field TEM images and selected area electron diffraction patterns in the subsurface region of the samples peened at 1.8J (Left) and 2.6J (Right) at 200% coverage at ambient temperature.

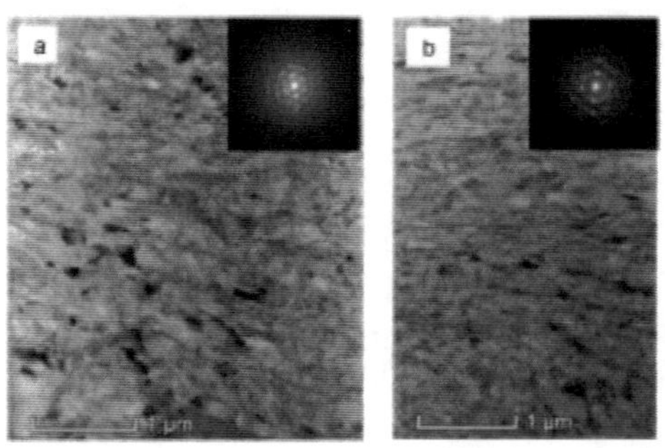

Fig.7: Bright field TEM images and selected area electron diffraction patterns in the subsurface region of the sample peened at 2.6J for 1600% coverage at ambient temperature (a) and constant cooling down to -21°C (b), respectively.

Conclusions

Surface nanocrystallisation on Ti-6Al-4V surface was obtained with the new shot peening process at a depth of at least 5-7 microns in the subsurface region. All shot peening conditions have generated nanostructured surface. However, only the low kinetic energy peening can generate useful nanocrystalline layer without causing harmful damages. Peening at high kinetic energy and high coverage all caused surface delamination. Under severe shot peening conditions, voids were formed initially in the subsurface region, then agglomerated and connected one another causing micro-cracks that led to delamination parallel to the surface. Experimentation on surface cooling influence on nanocrystallisation was carried out, however, further investigation is needed in the future development of this new shot peening approach.

Acknowledgements

The authors acknowledge the support of Innovate-UK fund (70397-487202).

References

1. Zhu, K. et al., 2004. *Nanostructure formation mechanism of α-titanium using SMAT. Acta Materialia,* Volume 52, p. 4101.

2. Fard, S. B. & Guagliano, M., 2009. *Effects of surfaces nanocrystallization induced by shot peening on material properties : a Review. Frattura ed Integrità Strutturale, 7 (2009) 3-16,* Volume 7, pp. 3-16.

3. Bagherifard, S. et al., 2018. *Effects of nanofeatures induced by severe shot peening (SSP) on mechanical, corrosion and cytocompatibility properties of magnesium alloy AZ31. Acta Biomaterialia,* Volume 66, pp. 93-108.

4. Maleki, E., Unalb, O. & Kashyzadeha, K. R., 2018. *Effects of conventional, severe, over, and re-shot peening processes on the fatigue behavior of mild carbon steel. Surface & Coatings Technology,* Volume 344, pp. 62-74.

5. Jayalakshmia, M., Huilgola, P., Ramachandra Bhatb, B. & Bhata, K. U., 2018. *Insights into formation of gradient nanostructured (GNS) layer and deformation induced martensite in AISI 316 stainless steel subjected to severe shot peening. Surface & Coatings Technology,* Volume 344, pp. 295-302.

6. Liu, Y., McGrory, B. & McGrory, C., 2017 pp92-97,. *Nanocrystallisation on the surface of superalloy In718 component with a commercial shot peening process,13th International conference for shot peening,* Montreal Canada, ICSP13 proceedings.

7. Unal, O. & Varol, R., 2017. *Surface Nanostructuring of AISI. Research and Reports on Metals,* 1(3).

8. Xue, Q., Meyers, . M. & Nesterenko, V., 2002. *Self-organization of shear bands in titanium and Ti–6Al–4V alloy. Acta Materialia ,* Volume 50, p. 575–596.

9. Zhang, J. & Alpas, A., 1993. *Delamination wear in ductile materials containing second phase particles. Materials Science and Engineering: A,* 160(1), pp. 25-35.

Analytical-Numerical Model to predict CRS created by Inner Shot Peening Applied on Small ID Tubes

F. Martinez-Guerrero[1], M.G. Ortiz-Lopez[2]

1 CIATEQ, Postgraduate Advanced Manufacturing, San Luis Potosí, Mexico
2 U. Politécnica SLP, San Luis Potosí, México

Abstract
In this work as main aim, the process parameters of inner shot peening by air-blast process combined with special nozzles are analyzed. To guarantee a flow direction, diverting features are used to peen the inner surface of tubular small complex geometry elements with uniform results. The analytical relationships between them are developed and coupled to existing analytical compressive residual stress and Almen Intensity prediction models in order that, by numerical calculation, the resulting process profiles could be used in early stages of product/process development or for the improvement of currently running set-ups.

Keywords Surface treatment, tubular components, inner shot peening, analytical modelling, nozzles for inner shot peening, compressive residual stress profile, Almen intensity.

Introduction

The most recent trends in automotive, aerospace and biomedical product development involve aggressive part weight reduction targets. Therefore, a good and feasible strategy would be to apply the shot-peening process on the inner surface of tubular elements with the purpose to provide them, especially in critical zones or high stress zones, with improved compressive residual stresses, at these zones the wall thickness of tube can be then reduced by subtracting mass to the part, while preserving the same equivalent fatigue life performance or much better than solid elements that are only outer shot peened or than tubulars however not inner treated and with no wall thickness reduction. In [1] it is shown that internal shot blasting increases the life time from 127% in bars with a 48% decrease ratio between the weight and the inner radius of the bar, to 332% in those that have a 36% decrease ratio between the weight and the inner radius of the bar.

The inner shot peening indeed exists for non-complex shape tubes, however applied with blast from outside either into one end of a blind hole or to both ends of a passed hole depending on the L/D ratio or with wheel from outside of the hole, using the projection angle of the shot as shown in Figure 1 or by blasting from inside on the whole surface of straight tubes of large diameters where the peening device runs into the tube as show Figure 2. In [2] is claimed that by blasting, energy consumption is high with not always uniform results, in order to improve this, a method with a shot stream by wheel that with a pre-determined angle to middle point of length axis was there proposed.

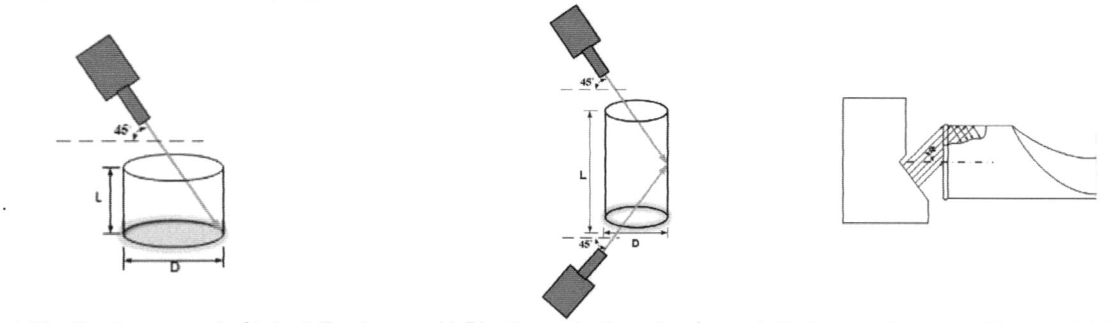

a) Blasting to one end of tube L/D <1 b) Blasting to both ends of tube L/D>1 c) by wheel from outside
Figure 1. Inner shot peening from outside

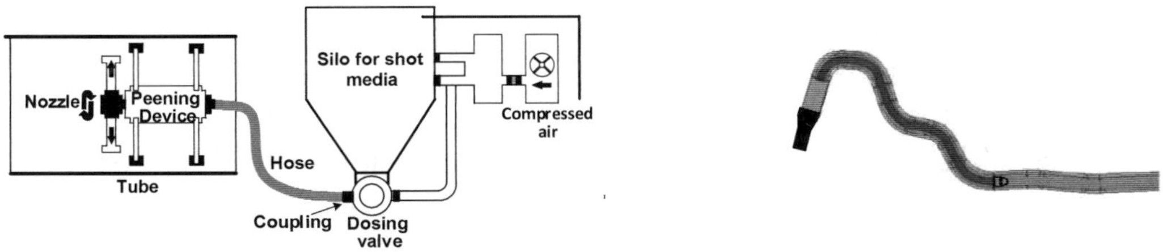

a) By blasting inside a tube of large diameter b) By complex tubular bent bar peened at localized zones

Figure 2. Inner shot peening from inside

Nevertheless, the application of the inner shot peening is under study for very small diameter tubes and complex geometries, such as a tubular stabilizers or other small diameter bent ducts. Main difficulties are the length and thickness of the nozzle. that cannot be freely introduced and travel through the bends of the small diameter tube as well as the hose or lance that stuck or twists then making it hard to guarantee a full coverage-controlled peening of the inner surface of the tube. To improve the profile of compressive residual stresses several adaptations are proposed in the type of nozzle used in the internal blasting process.

The most feasible solution to inner-peen small diameter tubes is to use compressed air blasting. In a conventional process specifically with a straight nozzle, the flow direction of the compressed air and shot mixture inside the hose, and after the outlet of the nozzle well after a stable flow is achieved coincides with the axis of the tube. This means that the angle of impact of the particles against the inner surface of the tube would be very small, so the kinetic energy transferred to it would be as well low. So, if the flow direction of the compressed-air shot mixture is diverted somehow, the inner surface of the tube could be impacted with larger angles, let say with an angle close to 90°degrees.

Development.

There are multiple parameters that influence the results of a shot blasting process and therefore the generation of the compressive residual stress profile σ [3]: the dimension and hardness of the shot (HVs), the hardness of the workpiece (HV), the exposure time (t), the application pressure (P) and the speed of the shot at the moment of impact (v). The latter is very decisive, since its effect is related to the kinetic energy developed in the process and influences both the value of the maximum compressive residual stress that is reached as well as the depth (z) at which this maximum stress is given, as shown in Figure 3.

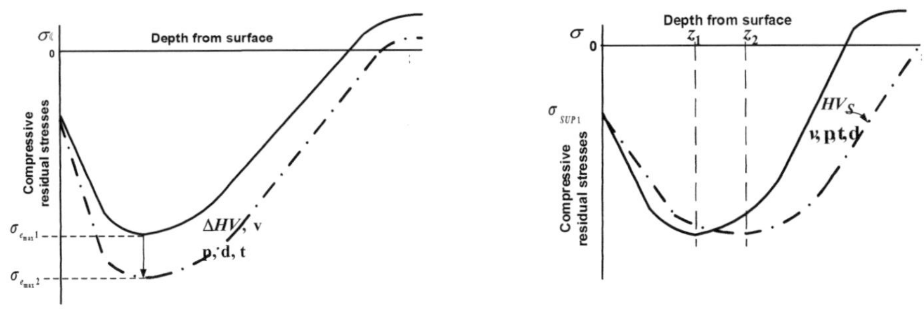

Figure 3 Influence of main process parameters on the residual stress profile.

Among all these process parameters, shot speed is also one of the more difficult parameters to measure, for the particular case of outer shot peening of parts, several methods have been proposed to estimate [4, 5] or measure their value [6], and once known it is possible to analytically perform [7, 8] or by means of numerical methods such as FEM analysis [9], a fairly approximate prediction of the compressive residual stress profile to be achieved. In the case of inner shot peening, once having an estimation of the shot speed, it may be possible to predict compressive residual stress too, so it is important to establish a mathematical relationship between the involved process parameters in order to calculate the shot speed.

A mathematical relationship, that describes the normal pressure acting on two elastic revolution bodies in the absence of friction is based on Hertz theory equation [10].

$$p(r) = p_0\left[1-(r/a_e)^2\right]^{1/2}, \quad a_e = \pi p_0 R / 2E_H \tag{1}$$

where $p(r)$ is pressure distribution along radial direction, r is radius from coordinates origin, p_0 is maximum pressure at r=0, a_e is contact radius, R is relative radius of the interphase R_1 (shot) / R_2 (work piece) and E_H is the equivalent Young's Module.

When it is considered that for the particular case of shot peening, the impact of the shot seen as developed kinetic energy converted into elastic work, the equations for determining maximum elastic pressure and maximum contact radius are transformed into:

$$p^* = \frac{1}{\pi}\left[40\pi k\rho E_0^{\,4}(V\sin\theta)^2\right]^{\frac{1}{5}}, \quad a_e = 5/4[\pi k\rho(V\sin\theta)^2/E_H]^{1/5}(D/2) \tag{2}$$

where Vsinϴ is shot impact speed normal component and K is eficiency coefficient related with elastic and thermal dissipation.

The magnitude of the shot speed to introduce in the former equation for the specific inner shot peening process needs to be dynamically analyzed. The machine of the study is focused on fulfilling process capability, full coverage and consistent compressive residual stress profile, based on a compressed air shot peening machine. This machine consists of several systems that ensure the continuity of the shot flow when applying the process on the workpiece, which are shown in the diagram in figure 4. The inner diameter of the workpiece ranges from a minimum of 15 mm up to a but not limited of maximum 24 mm. Flexible lancers are pushed into the tubular work pieces with servo control of displacement and speed, of course possible to synchronize several lancers in order to maintain an acceptable production rate.

The total flow path of the working materials required for the treatment of the part, impeller (compressed air) and driven (shot) is shown in Figure 4. In it, three interfaces can be identified and their analysis is important to consider. As for the transport stage, there are only changes in the diameters of the conductive hoses in order to increase the speed of the mixture.

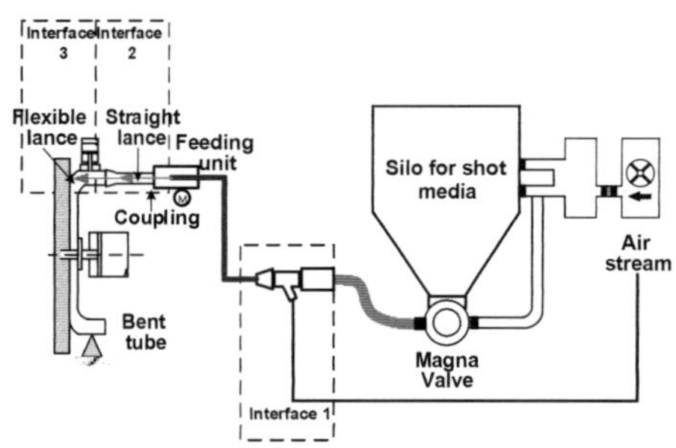

Figure 4 Diagram inner shot peening machine showing material streams and main interfaces.

At interface 1 "incorporation of shot-flow into compressed air stream", the speed of the impact particles (shot) incorporated into the jet is determined by the conservation of the mechanical Momentum, considering that the speed of the abrasive is zero before incorporation with the air flow and that the mixture of air and abrasive have the same speed after incorporation.
At the mixing point of shot and air, assuming momentum conservation, the equation is:

$$(V_{a/s})_2 = (V_a\dot{m}_a)_1 / (\dot{m}_a + \dot{m}_s)_2 \tag{3}$$

where $V_{a/s}$ is air/shot speed, V_a is air speed, $\dot{m}_a$ is mass flow air and $\dot{m}_s$ is mass flow shot.

For steady state and mass conservation at the hoses sections and not having a density change of the air/shot mix between any given point P1(inlet) and P2 (outlet):

$$(V_{a/s})_2 = V_{a/s\ 1}(D_1 / D_2)^2 \quad (4)$$

where D_1 is Diameter inlet and D_2 is Diameter outlet.

Using a cone arrangement like shown in Figure 5 a) one can divert the air-shot flow to the inner surface of the work piece with a suitable angle to get better peening results.

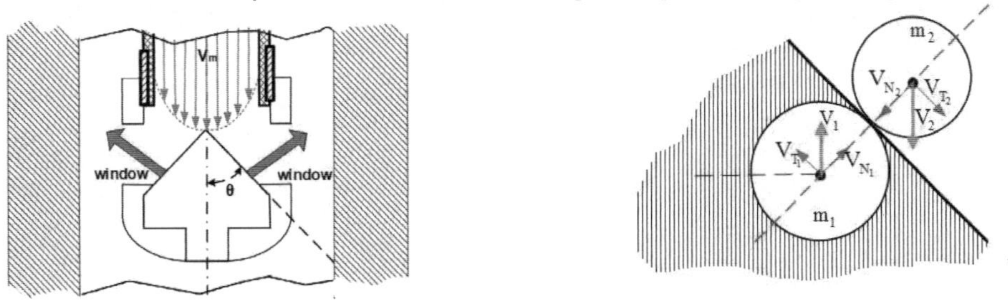

a) Diverting Cone   b) Interaction single shot and cone at impact

Figure 5. Cone arrangement for proper peening direction and impact shot/cone

The impact speeds of a shot particle and the cone are denoted with N sub-index for the normal direction and with T sub-index for the tangential as is shown in Figure 5 b):

$$m_1V_{11} + m_2V_{21} = P \quad (5)$$

$$m_1V_{1N1} + m_2V_{2N1} = m_1V_{1N2} + m_2V_{2N2}, \qquad m_1V_{1T1} + m_2V_{2T1} = m_1V_{1T2} + m_2V_{2T2} \quad (6)$$

where P is the Momentum.

Assuming smooth surface of cone and double rounded shot, then with small surface roughness tangential speeds are negligible for the present study. Solving (5) and (6)

$$V_{1N2} = V_{1N1}(m_1 - m_2) + 2m_2V_{2N1} / (m_1 + m_2), V_{2N2} = V_{2N1}(m_2 - m_1) + 2m_1V_{1N1} / m_1 + m_2 \quad (7)$$

In order to consider a purely elastic impact, the restitution coefficient mainly ruled by material hardness as well as surface state of the impacting materials will be later assigned with value of 1.

$$\varepsilon = V_{rN2} / V_{rN1} = (V_{2N2} - V_{1N2}) / (V_{2N1} - V_{1N1}) \quad (8)$$

where sub-index r represent restitution

Due that speed of the lance and the attached nozzle with inner cone (m_1) is very small with regards to the shot speed (0.005 m/s/100 m/s), the cone is practically considered fixed in the system, then V_{1N1} y V_{1N2} are nearly zero, then $m_1 >> m_2$:

$$V_{2N1}\varepsilon = V_{2N2} \quad (9)$$

As shown in Figure 6 $V_{2N1} = V_{21}\sin\theta$ where θ is half the angle of the cone, then:

$$V_{2N2} = \varepsilon V_{21}\sin\theta \quad (10)$$

The particle's normal speed component of the shot and cone impact determines the impact angle on the inner Surface of the tube. A perpendicular impact on the inner surface of the tubes means the higher amount of kinetic energy used to create the deeper and of higher magnitude compressive residual stresses.

$$V_{2N3} = V_{2N2}\cos\theta \quad (11)$$

Given that $V_{21} = V_2$ Equation (10) is rewritten as:

$$V_{2N3} = \varepsilon \, V_2 \, \sin\theta\cos\theta \quad (12)$$

Figure 7 shows speed factor $\sin\theta\cos\theta$ with angle variation of the nozzle´s cone. The factor is maximum at 90° angle.

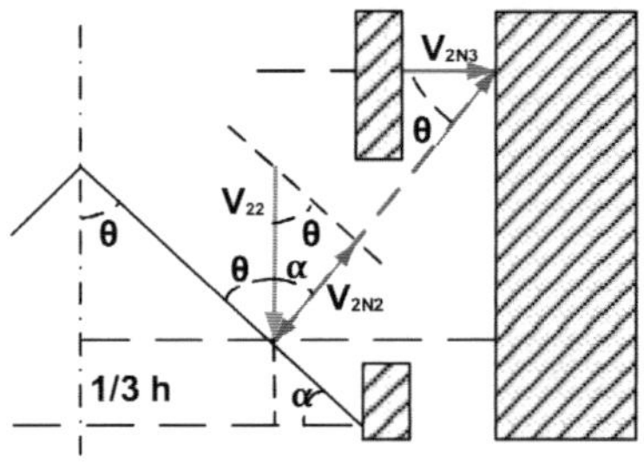

Figure 6. Components of V_2 at impact points 2 and 3

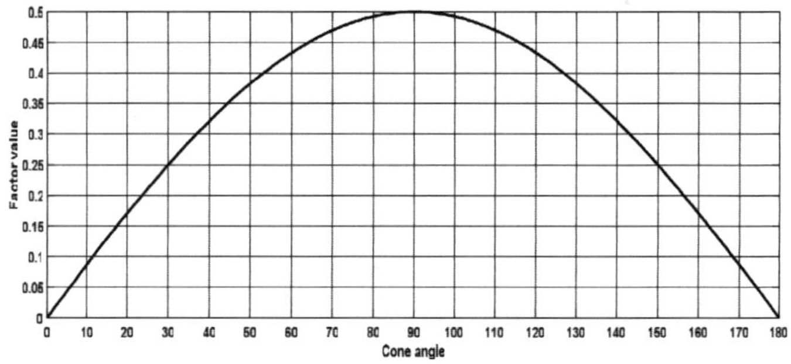

Figure 7. Speed factor $\sin\theta \, \cos\theta$

By Introducing the actual calculated speed V_{2N3} (12) impacting the work piece material at an angle of ca. 90º. Eq. (2) transforms in Eq. (3). Here V_{2N3} replaces V sinɵ because normal component is already considered.

The coupling of the modified Hertz equation (including the V derived for inner shot peening) with the analytical model proposed in [8], [11] in order that the compressive residual stress profile chart for inner shot peening can be predicted, was written in form of a numerical algorithm calculating this value each delta depth of 1 micron trough the thickness of an Almen type "N" strip.

Results

The inner shot peening process were applied, using the machine shown in Figure 4, in critical zones of a tubular element used in the automotive suspension, which has general dimensions of 1200 x 450 x 300 mm and Φ 32 x 5.1 wall thk. mm. Material is 34MnB5 quenched and tempered to Rm = 1650-1800 MPa. To provide fatigue life enhancement it is necessary to achieve an CRS profile with the following specifications in the mentioned zones: -450 MPa at 3 to 5 µm and -600 MPa at 10 to 20 µm.

Using the expression (12) coupled with existing mathematical models of shot peening process and by means of numerical calculation for different sets of input parameters, air Pressure (bar) and shot flow (kg/min), selected from DOE matrix with better output response, the CRS profiles by simulation were obtained, as shown in charts 8 and 9.

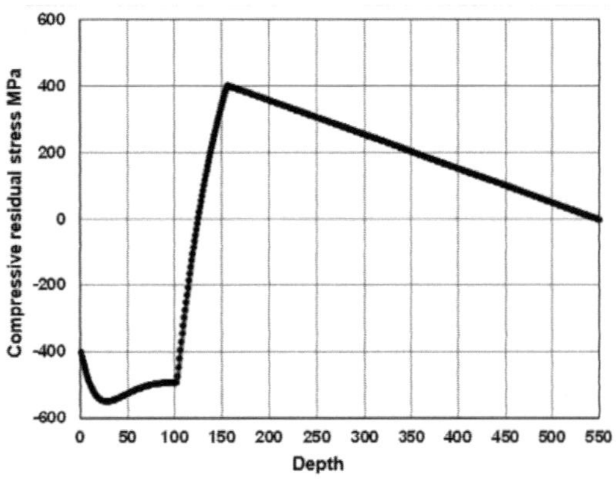

Figure 8 Simulation Param. Set 1
P= 6 bar, Shot flow=1400 kg/min

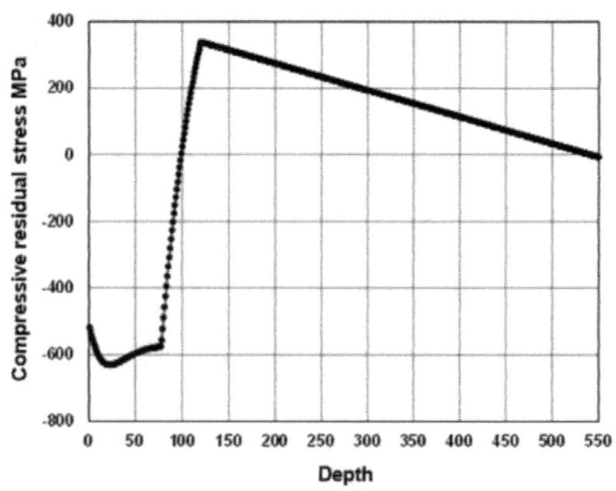

Figure 9 Simulation Param. Set 2
P=5 bar Shot flow=2000 kg/min

In order to verify the correspondence of the predicted results with actual measure values, tubular bars samples were processed with same parameter sets. CRS were measured on these samples by X-R diffractometer according to [12] as reference, at depths of 0, 5 and 20 µm. The results are shown in charts 10 and 11.

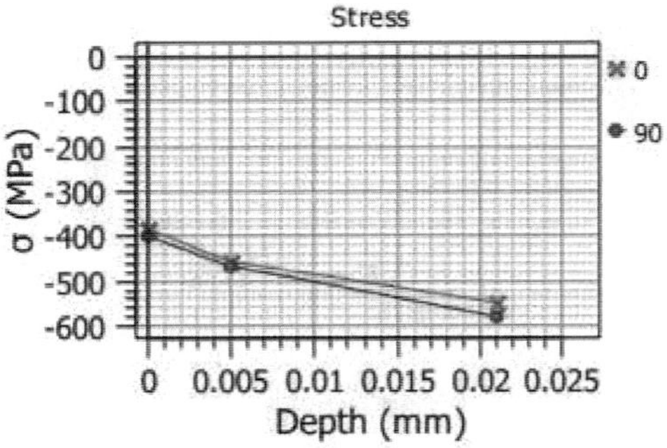

Figure 10 X-R Diffr. Meas. Set 1
P=6 bar, Shot flow=1400 kg/min

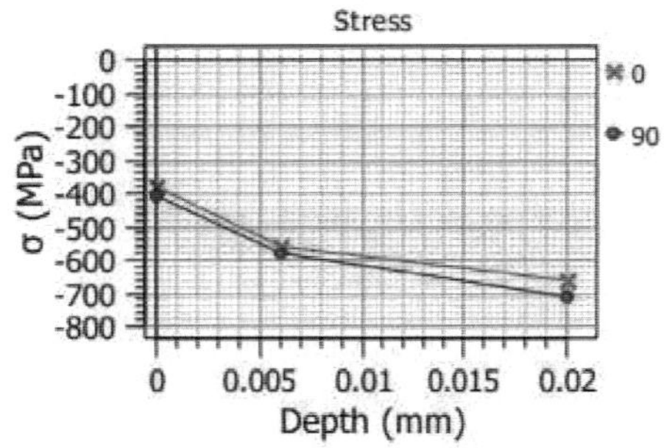

Figure 11 X-R Diffr. Meas. Set 2
P=5 bar, Shot flow=2000 kg/min

Observations and Conclusions.

As initially stated, it was possible to get good peening results on the inner surface of a small tube, when implementing nozzle with a diverting cone in order to profit as much as possible from the kinetic energy, targeting a c.a. 90° angle. This assumption was brought into the an already analytical/numerical proposed model, for peening an "A" Almen Strip peening to full coverage; however here changing to Almen "N" strip due to the range of peening intensity. The given specification of -450 MPa, to comply with the targeted Fatigue Life Cycles for the component of study was well achieved for both proposed parameter sets. Furthermore, is evident that the higher rate of shot mass flow effectively pushed the compressive residual stress profile deeper into the material.

References

[1] Y. Won Seo, J. Won Sur, W. Ki Lee, Y Kim, *Effect of Inner Shot Penning Process for Tubular Stabilizer Bars*, Trans. of the Korean Society of Mechanical Engineers A, Vol 41, num. 11, (2017) pp. 1119-1124.

[2] H.D. Schlick, (2005*), Verfahren zum Innenstrahlen von Rohren*, (*EP 1 591 203 A1*): European Patent Office

[3] R. Herzog, W. Zinn, B. Scholtes, H. Wohlfahrt, *The Significance of Almen Intensity in the Generation of Shot Peening Residual Stresses*, Proceedings of 6th ICSP. San Francisco, Cal., USA, 1996.

[4] H. Guechichi, M. Benhattab, *Predicting of shot velocity in shot peening process*: Proceedings of 13th ICSP. Montreal, Canada, 2017

[5] H. Y. Miao, M. Lévesque, *Effect of shot peening on the fatigue life and crack propagation prediction with experiments and Navarro-Rios Model*, [Internet] The Shot Peener. 2011. [cited date March 2022]. Avail. from: https://www.shotpeener.com/library/pdf/2011.001

[6] R. Clausen, J. Stangenberg, *Device for the Determination of Impact Velocities in Shot Peening*: Proceedings of 8th ICSP. Munich, Germany, 2002.

[7] S. Shen, S. N. Atluri, An Analytical Model for Shot-Peening Induced Residual Stress. Computers, Materials and Continua, Vol.4, num 2, (2006) pp. 75-85.

[8] H. Y. Miao, S. Larose, C. Perron, M. Lévesque, *An analytical approach to relate shot peening parameters to Almen intensity.* Surface and Coatings Technology, Vol. 205, (2010) pp. 2055-2066.

[9] M. Guagliano, Relating Almen intensity to residual stresses induced by shot peening. A numerical approach. Journal of Materials Processing Technology, Vol. 110, (2001) pp. 277-286.

[10] Hertz H. Über die Berührung fester elastischer Körper, Journal für die reine und angewandte Mathematic. 1881. Vol. 92, pp 156-171.

[11] F. Martínez-Guerrero, Estudio sobre el Proceso de Granallado Interno a Barras Estabilizadoras Tubulares para Mejorar el Perfil de Esfuerzos Residuales Compresivos [Doctoral thesis], Posgrado CIATEQ A.C.; 2019. Avalaible from: http://ciateq.repositorioinstitucional.mx/jspui/handle/1020/595

[12] PV1005, 2005-08. Klass.-Nr.51 03 1. Determining the Internal-Stress Depth Characteristics. Ferritic Iron Materials. Revision August 2005.

Application of shot peening to enhance precipitation hardening effect in aluminium alloys

Libor Trško[a], Michal Jambor[b], Ivo Šulák[b], Sara Bagherifard[c], Mario Guagliano[c], Otakar Bokůvka[a]
[a] Research Centre of the University of Zilina, www.researchcentre.sk, Zilina, Slovak Republic – libor.trsko@uniza.sk, otakar.bokuvka@fstroj.uniza.sk
[b] Institute of Physics of Materials, Czech Academy of Sciences, www.ipm.cz, Zizkova 22, Brno, Czech Republic – jambor@ipm.cz, sulak@ipm.cz
[c] Politecnico di Milano, Department of Mechanical Engineering, www.mecc.polimi.it,Via la Masa 1, Milano, Italy - mario.guagliano@polimi.it, sara.bagherifard@polimi.it

Keywords: *aluminium alloy, precipitation, fatigue, shot peening*

Introduction

Positive effect of the severe shot peening (SSP) on the fatigue properties of Aluminum alloys, especially in the very high cycle fatigue region was proved in many studies [1-4]. Improvement of the fatigue properties is achieved by a combination of several factors, however the most significant can be considered to be surface hardening due to the severe plastic deformation. The mechanism which causes the hardening of the material is accumulation of lattice defects by the intensive plastic deformation. Besides the material strength, also other material properties are affected by the number of lattice defects and in case of aluminum alloys, lattice defects play important role in formation of precipitates during age-hardening, thus this arises a question, if the SSP can be used to induce pre-aging plastic deformation to enhance precipitation of strengthening phases.

Al-Cu-Li alloys offer very high mechanical strength while maintaining sufficient ductility. These properties are achieved by the applied age-hardening heat treatment – during which strengthening precipitates are formed in aluminum matrix. The main strengthening phase in advanced Al-Cu-Li alloys is T_1 (Al_2CuLi), whose particles preferentially heterogeneously nucleate on the dislocations. Therefore, to obtain optimal combination of mechanical properties, T8 class heat treatment (plastic deformation applied between solution annealing and aging) is often employed. However, such treatment is hard to apply in case of e. g. welded structures.

In the past, few studies were performed focused on the possibility of precipitation enhancement by the pre-aging severe plastic deformation [5-7]. Increase in the dislocation density resulted in precipitation of denser populated finer particles. Besides that, as a result of dislocation interactions, vacancies are emitted enhancing diffusion rate, thus affecting kinetics of precipitation processes [5,8,9]. On the other hand, excessive severe plastic deformation results in the formation large number of high-angle grain boundaries, which can act as a sink for the solutes and decrease supersaturation of the solid solution [8]. Despite unambiguous precipitation enhancement due to the pre-aging plastic deformation, resulting effects on the mechanical, and especially fatigue properties are questionable. Only a few studies were focused on the fatigue performance of Aluminum alloys with pre-aging severe plastic deformation applied [5]. Application of such surface treatment enhanced precipitation in the affected region, on the other hand, it creates significant surface roughness and can cause creation of various discontinuities on the surface and in subsurface layers. Another aspect is residual stress distribution. In many severe plastic deformation techniques, the residual stresses are the key factor responsible for the improvement of the high cycle fatigue performance. When the pre-aging plastic deformation is applied, the following aging treatment can partially or even fully relieve residual stresses. Increased surface roughness together with relieving of introduced residual stresses can easily outbalance positive effects on enhanced precipitation. This underlines necessity for the careful choice of the pre-aging severe shot peening parameters, to balance plastic deformation intensity and level of surface damage.

In the present study, severe shot peening was applied as a pre-treatment before aging to promote precipitation of the heterogeneously nucleating precipitates. Four series of samples were prepared per Fig. 1. Two different heat treatments were chosen – low and high-temperature regimes. For both heat treatment types, two sample series were prepared. In the first series, the SSP was applied after solution annealing and before aging step. In the second (control) series, the SSP was applied after completed age-hardening. In the low-temperature regime, the strengthening is mostly governed by copper-rich precipitates as GP zones and Θ". These precipitates are formed homogeneously and therefore their precipitation does not depend on the local dislocation density. On the other hand, the high-temperature heat treatment regime promotes precipitation of T_1 precipitates which are predominantly nucleated on the dislocations [10,11]. High dislocation density formed during SSP treatment can promote precipitation of the denser population of finer precipitates in the affected surface zone. The fatigue properties of all four series in the high cycle region were then examined, to evaluate if application of the SSP before aging can improve fatigue properties. The results were then discussed in the light of the microstructural characterization of the subsurface affected layers and the evaluation of the residual stresses state.

Material and Experimental methods

In the present study, commercially available AA 2055 alloy was used as experimental material. The alloy was delivered in the form of extruded bars with a diameter of Ø 66 mm. In total, four series of specimens were machined and subjected to the severe shot peening process. In two series, SSP treatment was applied after solution annealing and before aging (one of the specimen series was subjected to low-temperature heat treatment and the second one to the high-temperature heat treatment). In the other two series, the SSP treatment was applied after finished heat treatment (again, one series subjected to the low-temperature heat treatment and the other one subjected to the high-temperature heat treatment). SSP treatment of Almen intensity 8.9N and coverage of 650% was applied in all cases. The SSP parameters were chosen based on our previous studies [1,2], with the aim to induce intensive plastic deformation of the surface layer, while the surface damage (cracks and other discontinuities) was kept as low as possible.

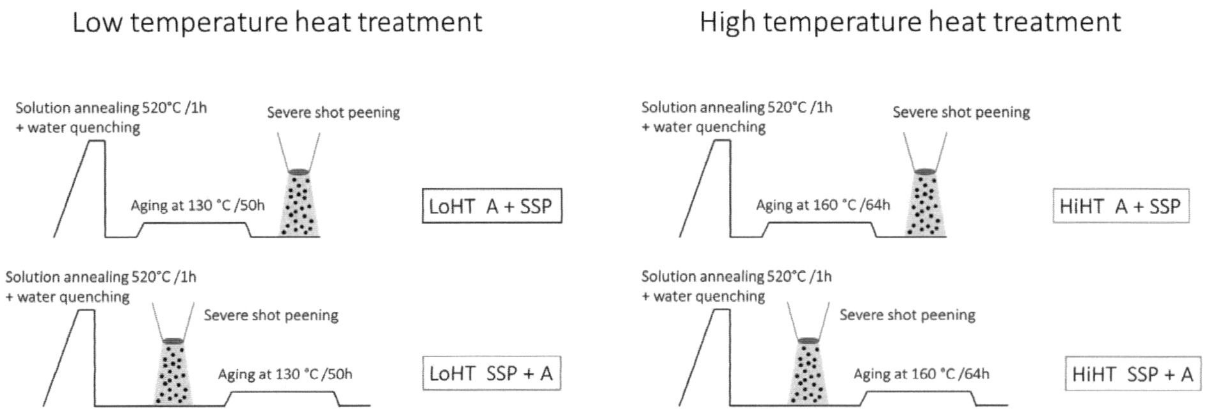

Figure 1. Schematic view of the tested sample series. Four different series were tested, differing in the aging temperature and the application of the shot peening process before/after aging treatment.

Heat treatments were carried out with parameters as follows:

Low temperature (LoHT) - solution annealing at 520°C for 1 hour followed by water quenching, aging at 130°C for 50h.

High-temperature heat treatment (HiHT) - solution annealing at 520°C for 1 hour followed by water quenching, aging at 160°C for 64h.

Residual stress depth profiles were examined using X-ray diffraction. Proto iXRD diffractometer equipped with Cr X-ray source was used and measurements were performed using $\sin^2\psi$ method, with 6 inclinations between ± 40°. Diffraction peaks from {222} planes were collected at 156.9°. In order to reveal residual stress levels at different depths, electrolytic polishing was used for gradual removal of the thin surface layer. Due to the strong crystallographic texture of the base material, the residual stress levels could be measured only in the very thin, heavily deformed layer where the texture was suppressed by the deformation process.

To describe microstructures in subsurface region, lamellas were prepared from all four specimens' series with use of focused ion beam technique (FIB). Approximate dimensions the lamellas were 25×15 μm, and they were oriented in a way, that they were showing microstructural changes from the surface to 25 μm depth below the surface. Microstructure was then observed using Thermofisher Talos F200i transmission electron microscope.

Fatigue tests were carried out on MTS Acumen electrodynamic test system with linear motor. Tests were performed under stress control regime, with fully reversible loading cycle (R = -1). Testing frequency was 30 Hz, and tests were terminated if there was no failure in 10^7 loading cycles.

Results and Discussion

Application of SSP significantly modified microstructure in the near-surface region. All tested series were examined in terms of microstructure evaluation, however only HiHT series are included because of the range limitation. Observation of the LoHT series did not reveal any significant differences neither in the affected zone morphology and the precipitating phase size and distribution. Microstructures of the surface and subsurface regions of both HiHT series are shown in Fig.2. Application of the shot peening on the series after aging (HiHT A+SSP) form heavily refined nanocrystalline zone (zone 1) in depth of 2-3 μm. There were no intragranular precipitates in this zone, what can be explained by the mechanical dissolution due to the strong dislocation activity and the solute segregation on the grain boundaries. Deeper in the material, partially grain refined zone was resolved, with the T_1 particles present. Bended shape of the particles suggests multiple shearing events; however, size of the particles was not affected. The third zone with increased dislocation density but without introduced new grain/subgrain boundaries was present deeper in the structure. There was no sharp interface but rather a continuous transition. In the third zone, average size of the T_1 particles of 122.4 ± 39.4 nm was measured.

Comparison with microstructure of HiHT SSP+A series revealed significant differences. Again, three zones were revealed, however, especially in zone 3, much higher dislocation density appeared. This can be attributed to the absence of strengthening precipitates in initial state (before SSP application), allowing easier dislocation slip within grains. Subsequent aging results in formation of very fine T_1 particles. Particles measurement in the same distance from surface as in the case of HiHT A+SSP series revealed average size of 37.1 ±10.2 nm. Introducing of SSP process before aging thus promotes nucleation of the precipitates resulting in formation of very fine and disperse strengthening particles.

Residual stress distribution in the surface layer of all examined states is shown on Fig. 3. Additional measurement was performed on the sample after solution annealing and SSP, without any following aging treatment. This allowed monitoring of possible relaxation during applied heat treatment. In all examined series, application of SSP process resulted in formation of the compressive residual stresses in the subsurface region. Fig. 3a shows residual stress depth profiles measured on LoHT series. There is no significant difference between LoHT SSP+A, LoHT A+SSP and reference Solution annealed + SSP series. Based on which it can be concluded, that during application of LoHT (series LoHT SSP+A), no significant relaxation occurred. Different behavior was recorded for the HiHT series. In case of the HiHT A+SSP, the maximal residual stress value is similar like in the reference state (solution annealed + SSP) but the affected zone is shallow. It is most likely due to much higher initial hardness of the HiHT A+SSP series. Residual stress profile of the HiHT SSP+A series shows a significant drop compared with the reference ones. It is expected that aging temperature of 160°C is sufficient for partial relaxation of induced residual stresses.

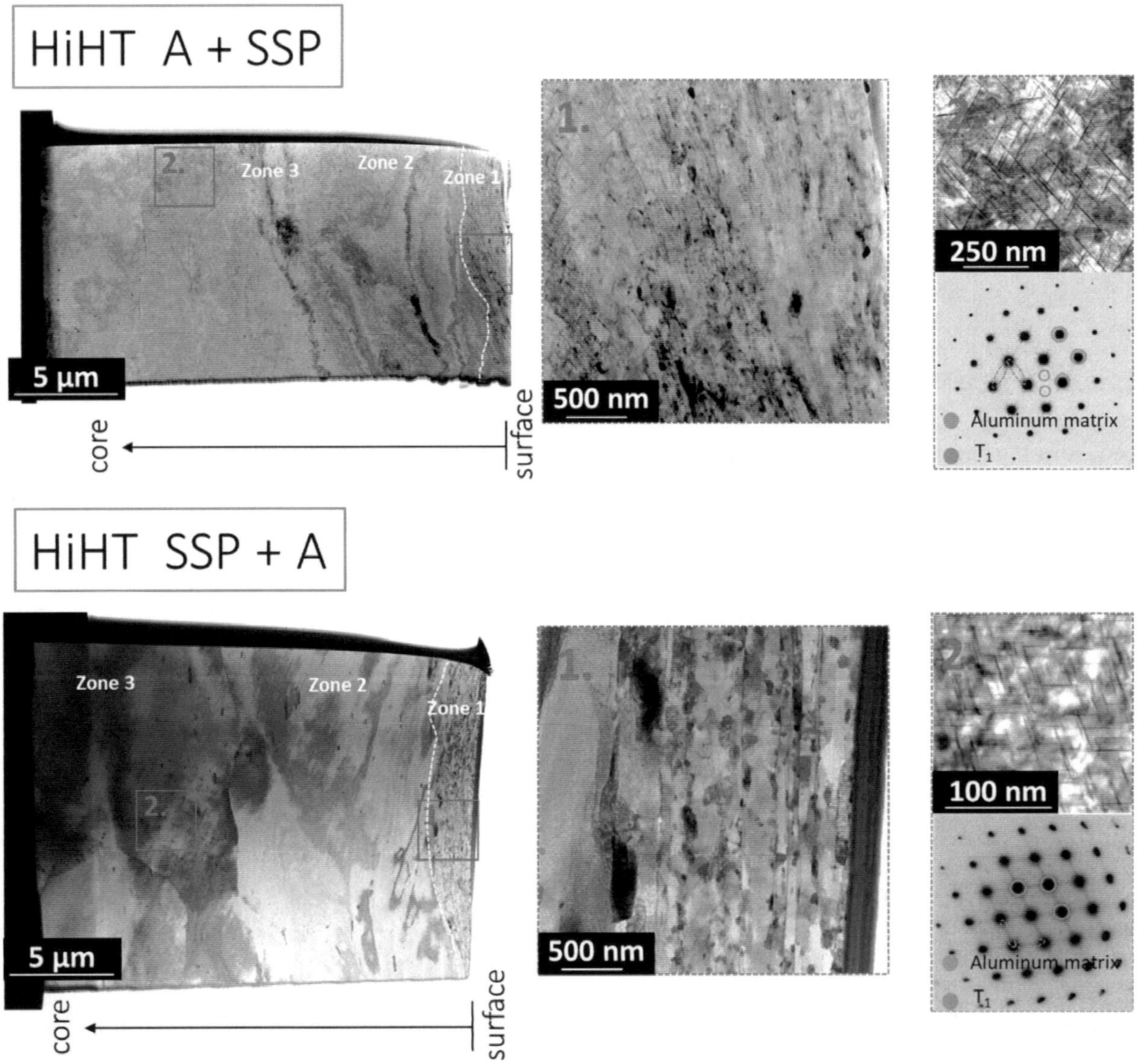

Figure 2. Microstructure of HiHT series (BF-STEM). Images showing overall view on the surface and subsurface region. Images (1) show detailed view on the nanocrystalline layer. Detailed images (2) show T1 plate-like precipitates with corresponding SAED patterns (ZA<110>).

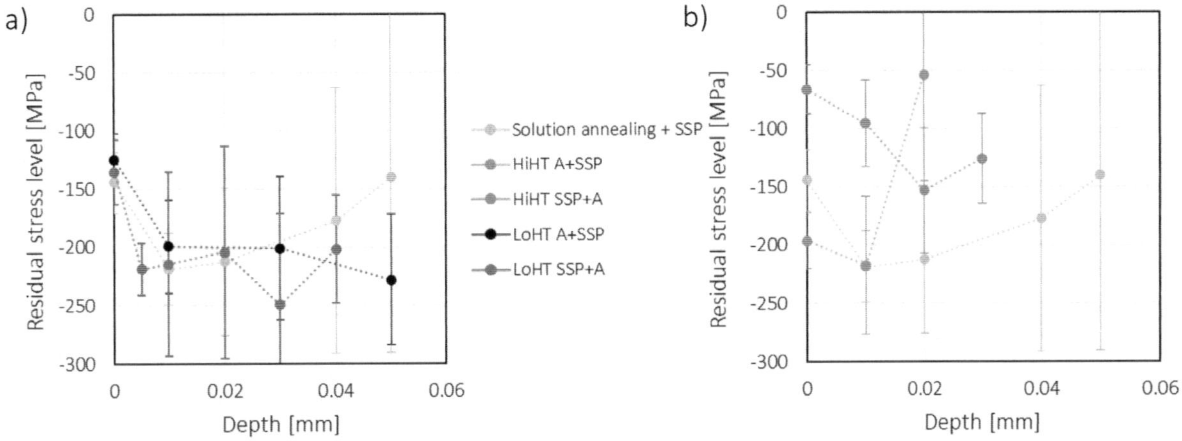

Figure 3. Residual stress depth profiles of all examined series. Compressive residual stresses in the subsurface region induced by SSP process were recorded in all series.

Results of the fatigue tests are shown in Fig. 4. Both series of LoHT exhibited similar behavior at the short and medium fatigue lifetimes. However, at longer fatigue lifetimes, series with SSP process conventionally applied after completed age-hardening show superior fatigue properties than series with the SSP applied before aging. Based on this, it can be stated, that in case of LoHT, there is no beneficial effect of application of severe plastic deformation before aging treatment. Different behavior was recorded for HiHT series. In this case, series with SSP applied before aging shows superior fatigue properties at short and medium fatigue lifetimes. At long fatigue lifetimes, the differences between both HiHT series are diminished. Application of SSP before aging in case of HiHT is therefore beneficial at least in certain regions of the fatigue lifetimes.

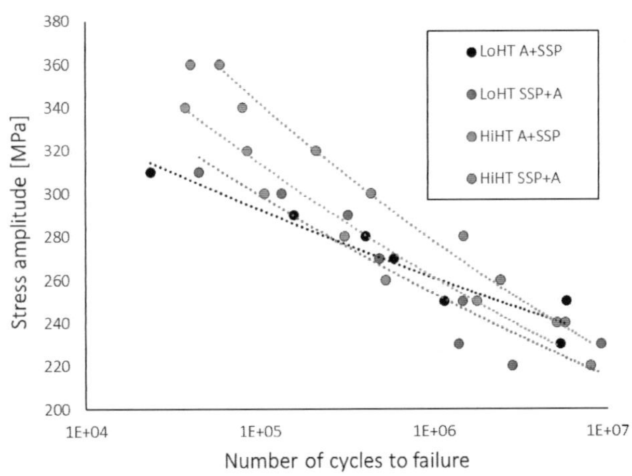

Figure 4. Results of the fatigue test in form of S-N curves. There can be seen, while in case of LoHT, no beneficial effect of introducing SSP process before aging is visible, for the HiHT series

Microstructural observations did not reveal any notable differences between two series after LoHT. Similarly, residual stress depth profiles of both states were comparable. Possible explanation of the worse fatigue properties of LoHT SSP+A series could be the lower initial hardness, which typically result in the higher surface roughness. Such measurements, however, were not included in the study and at this point, the statement cannot be confirmed.

In case of the HiHT series were recorded major microstructural differences. The high dislocation density induced by SSP process result in the precipitation of the very fine T_1 particles in HiHT SSP+A series. Average size of the particles was 37.1 ±10.2 nm at distance of 15 µm from the surface. In the same area, the average size of T_1 particles in HiHT A+SSP series was 122.4 ± 39.4 nm. Significant refinement of the strengthening particles could be beneficial in terms of the strength of particular zones. Another important aspect affecting the fatigue properties is the residual stress distribution. Compressive residual stresses suppress crack initiation and have positive effect especially in the area of long fatigue lifetimes. Partial residual stress relaxation in the HiHT SSP+A series could outbalance the positive effect of the refined T_1 particles, explain equalizing properties of HiHT SSP+A and HiHT A+SSP series at long fatigue lifetimes.

Conclusions

Application of the severe shot peening process as a pre-treatment before aging can be beneficial for the fatigue properties, when optimal combination of the shot peening process parameters and the aging conditions are chosen. Application of SSP before aging results in formation of the finer and denser population of T_1 strengthening particles, in case of high-temperature heat treatment regime. In contrary, no beneficial effect regarding particle distribution was recorded in case of low-temperature heat treatment.

Series of specimens with SSP process applied before aging show superior fatigue resistance in short and medium fatigue lifetimes, compared to the series with shot peening conventionally applied after finished age-hardening. At long fatigue lifetimes, superiority of shot peened + aged samples diminished, and both series exhibited similar properties.

Acknowledgement

The research was supported by projects: ITMS2014+ code 313011ASK8, VEGA 1/0117/21 and APVV-20-0427.

References

[1] L. Trško; et al. *Influence of Severe Shot Peening on the Surface State and Ultra-High-Cycle Fatigue behavior of an AW 7075 Aluminum alloy.* Journal of Materials Engineering and Performance, 26, 2017, 2784-2797.

[2] L. Trško; et al. *Study of Relation between Shot Peening Parameters and Fatigue Fracture Surface Character of an AW 7075 Aluminium Alloy.* Metals, 8(2), 2018, 111.

[3] B. Hadzima; et al. *Shot peening as a pre-treatment to anodic oxidation coating process of AW6082 aluminum for fatigue life improvement.* The International Journal of Advanced Manufacturing Technology, 93, 2017, 3315-3323.

[4] M. Jambor; et al. *Effect of Severe Shot Peening on the Very-High Cycle Notch Fatigue of an AW 7075 Alloy.* Metals, 10(9), 2020,1262.

[5] P. Maurel et al. *How does surface integrity of nanostructured surfaces induced by severe plastic deformation influence fatigue behaviour of Al alloys with enhanced precipitation.* International Journal of Fatigue, 140, 2020, 105792.

[6] J. M. García-Infanta; et al. *Enhanced grain refinement due to deformation-induced precipitation during ambient-temperature severe plastic deformation of an Al-7%Si alloy.* Journal of Alloys and Compounds, 478,2009,139-143.

[7] M. Vaseghi; H. S. Kim. *A combination of severe plastic deformation and aging phenomena in Al-Mg-Si Alloys.* Materials and Design, 36, 2012, 735-740.

[8] Z. Feng; et al. *Surface severe plastic deformation induced solute and precipitate redistribution in an AL-Cu-Mg alloy*. Journal of Alloys and Compounds, 773, 2019. 585-596.

[9] Y. Zhao; J. Liu; T. D. Topping; E. J. Lavernia. *Precipitation and aging phenomena in an ultrafine grained Al-Zn alloy by severe plastic deformation.* Journal of Alloys and Compounds, 851, 2021, 156931.

[10] N. E. Prasad et al. *Aluminium-Lithium Alloys: Processing, Properties and Applications.* Elsevier, 2014.

[11] T. Dorin et al. *Quantification and modelling of the microstructure/strength relationship by tailoring the morphological parameters of the T1 phase in Al-Cu-Li alloy.* Acta Materialia, 75, 2014, 134-146.

Application of on-site positron annihilation lifetime spectroscopy system as non-destructively shot peening evaluation technique

N. Uesugi[1], M. Yamawaki[2], K. Hattori[1], Y. Watanabe[1*]

1. Toyo Seiko Co., Ltd., Japan
2 National Institute of Advanced Industrial Science and Technology, Japan

* Corresponding Author: y_watanabe@toyoseiko.co.jp

Abstract
Positron annihilation lifetime spectroscopy (PALS) is a non-destructively technique to study open-volume lattice defects such as vacancies or dislocations, which is used for fatigue, aging and shot peening inspection. In this work, we developed a portable PALS apparatus that has the ability for on-site measurement. We developed the light-shielding technique which enable us to do on-site measurement. Furthermore, we downsized the apparatus size using Monte Carlo simulation technique. We validated that the developed portable apparatus have the same shot peening potential as the conventional system. We concluded that the developed portable apparatus is sufficiently able to capture the change caused by shot-peening time.

Keywords Non-destructive inspection, Positron annihilation analysis, Lattice defects

Introduction

Positron annihilation lifetime spectroscopy (PALS) is a technique that can detect lattice defects in solids. Lattice defects affect various properties (e.g., mechanical properties), therefore PALS is used for investigating fatigue damage or ageing. PALS technique is also effective in inspecting shot-peened parts. Shot peening produces plastic deformation, and the amount of induced lattice defects is dependent on a shot peening quality. In other words, the amount of lattice defects induced by shot peening increases with the intensity of shot peening increasing. Therefore evaluating the lattice defects leads us to inspect shot peened parts. One of the most significant advantages of PALS is measurement depth. A positron penetrates as ten times deeper than the X-ray generally used in industry, which means that PALS can measure areas non-destructively deeper than the X-ray.

In previous work, we developed the anti-coincidence (AC) method, which made it possible to measure with a single specimen [1]. We also developed a desktop-type PALS apparatus implementing the AC method. However, it is necessary to place the specimen into a dark box in the desktop apparatus for accurate measurement. Therefore a larger specimen needs to be cut, which leads to destructive inspection.

We developed the portable PALS apparatus, which has the ability for on-site measurement [2]. The portable apparatus has a "dark-box-less" structure and small-sized gamma detectors, which leads to convenience for various on-site applications. In this research, we optimized the light-shielding technique, which enables the "dark-box-less" structure to be capable of sufficiently accurate measurement. Furthermore, we optimized the detector layout to improve the counting rate. We applied an MPPC® device (Hamamatsu Photonics K.K.) for the positron detector, which allowed the gamma detectors to be more efficient layout. We conducted Monte Carlo simulation to evaluate the counting rate for various detector layouts with the MPPC.

Positron annihilation lifetime spectroscopy

A positron is an anti-particle of electrons. When a positron and an electron collide, they annihilate each other with two gamma rays emission. The time between the detection of gamma rays due to positron emission and positron annihilation corresponds to the lifetime of positron. The positron lifetime is inversely proportional to electron density at the site where positrons are annihilated. If lattice defects exist in a specimen, positrons are trapped and annihilated there. Since electron density is lower in such defects, positron lifetime becomes longer than a defect-free specimen.
Fig. 1 shows the schematic diagram of our developed PALS system. The system consists of two gamma-ray detectors and a positron detector. The gamma-ray detectors are used to detect positron emission, while the positron detector is used for the AC method. The positron detector comprised a plastic scintillator plate, a photodetector, and a sealed Na-22 positron source with two 7.5 μm thickness Kapton films. Placing a specimen on the positron detector leads to irradiation of positrons into the specimen.

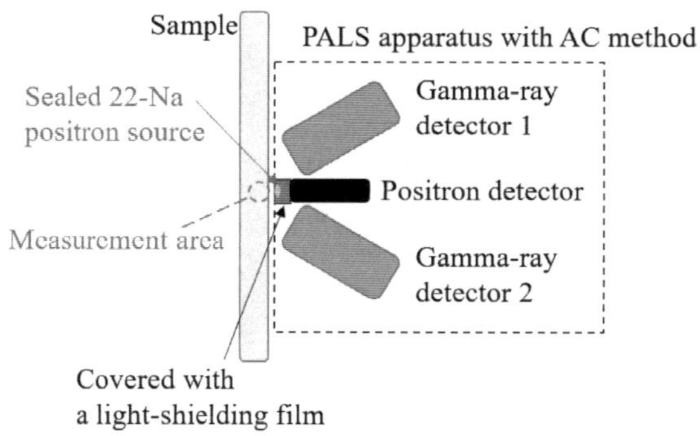

Fig. 1 The diagram of PALS with the AC method

Experimental Methods

The positron detector needs to be light-shielded to prevent the internal photodetector from ambient light. In order to develop a "dark-box-less" structure, we covered the positron detector with light-shielding films. We applied an Al-deposited Kapton film as a shielding film instead of a simple shielding film (e.g., Ti foils). The deposited film makes it possible for the shielding layer to be thin, maintaining a shielding ability. Moreover, Kapton film is of the same material as the sealant of the positron source. Positrons have a definite lifetime in Kapton; thus, the noise signals from Kapton can be separated by analysis. We evaluated various Al-deposited films to thicken the Al deposited layer.
To downsize and lighten the apparatus, we conducted Monte Carlo simulations to evaluate the counting rate for various detector layouts. We used Geant4 (ver. 10.4) [3], a simulation tool kit of the passage for particles through matter. Like our PALS system described above, we placed two gamma-ray detectors and a positron detector. The material information was based on the database of Geant4. In addition, we experimentally evaluated the time resolutions for some layouts and compared them with the simulation results.

We prepared the shot-peened stainless steel SUS304. The specimen size was 15 mm × 15 mm × 2 mm to use the desktop apparatus (with a dark box system) for the measurement. Table 1 shows the shot peening conditions.
We used the developed apparatus to conduct a PALS measurement on the shot-peened specimen and compared the results with those obtained from the desktop apparatus to validate the developed apparatus. We used the analysis software, IPALM [4], to identify the positron annihilation lifetime. Furthermore, we compare the PALS results with the full width

at half maximum (FWHM) of Bragg peak from the (311) plane. The FWHM was measured by μ-X360 (Pulstec Industrial Co., Ltd.).

Table 1 Applied shot peening conditions

Shot media	Conditioned cut wire Φ0.3 mm, HV500 (typ.)
Air pressure [MPa]	0.20
Coverage [%]	50, 90, 95, 100, 200

Experimental Results and Discussions

Fig. 2 shows three Al-deposited Kapton films irradiated with white light from the back. The deposit thickness and the number of deposit layer affected the shielding efficiency, and the 500 nm deposited film had higher shielding efficiency than the 300 nm one. Nevertheless, some irradiated light passed through the film, which indicates that it is not sufficient to use for the positron detector. On the other hand, the film deposited on each side had the highest shielding efficiency despite the thinnest thickness (total 200 nm). These results indicated the presence of the pinholes, which were randomly distributed on a deposited layer. The pinholes could not shield the irradiated light, which caused the film to degrade the shielding efficiency. In the case of the each side deposited film, the light could pass through the film only when the pinholes are present at same position on each layer. Therefore it seemed that the each side deposited film had the highest shielding efficiency. The results indicated that increasing the number of thin deposited layers is more efficient than a single thick deposited layer to make the amount of the thickness of deposited layer thinner. Finally, we adopted the total Al-deposit thickness to 400 nm by performing 200 nm Al deposition on each side of the Kapton film.

300 nm Al on one side   500 nm Al on one side   100 nm Al on each side

Fig. 2 Image of Al deposited Kapton films irradiated with white light from the back.

Table 2 shows the positron lifetime measured by various shielding methods. There was nothing between the positron source and the specimen in the desktop apparatus, which means that contamination did not occur. In the case of shielding with simple Ti foils (thickness of 6 um), the positron lifetime was higher than the desktop one, which means that such thick foils affected the measurement accuracy. On the other hand, the positron lifetime was almost the same as the desktop one in the case of shielding with the Al-deposited film. This result shows that the deposition enabled the degradation from the shielding film to be negligibly small. Based on these results, we developed the "dark-box-less" system enabling accurate positron lifetime measurements using the Al-deposited shielding film.

Table 2 Mean positron lifetime using various shielding methods

	Positron lifetime [ps]
With dark box (the desktop apparatus)	105.4 (±1.2)
Shielding with Al-deposited film (totally 300 nm)	104.9 (±1.2)
Shielding with Ti foils (totally 6 μm)	120.2 (± 1.2)

Fig. 3 shows an image during the counting rate simulation by Geant4. The results identify that it is effective for improving the counting rate to place the positron source near the center of gravity of gamma-ray detector. On the other hand, the positron source is integrated with the positron detector. Therefore, it is limited to bringing the gamma-ray detectors close to each other due to collision. We applied an MPPC device for the positron detector. MPPC is a solid state photomultiplier. An MPPC leads the positron detector to be thinner than the one with a conventional photomultiplifier (Shown in Fig. 3). We varied some layout parameters (e.g., an angle between gamma detectors) and evaluated the counting rate.

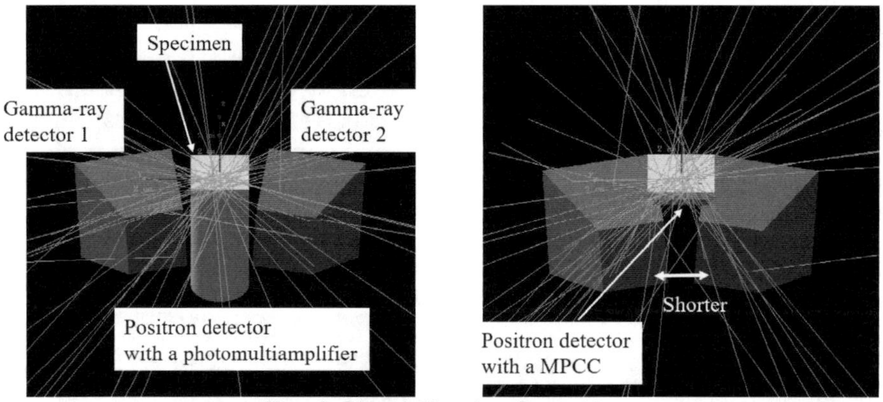

Fig. 3 Simulation by Geant4
(Left: Simulation with a photomultiamplifier, Right: Simulation with a MPPC)

Fig. 4 shows the developed portable apparatus implementing the above works: the light-shielding technique and the detector layout. We validated that the developed portable apparatus has the same potential for shot peening evaluation as the desktop apparatus.

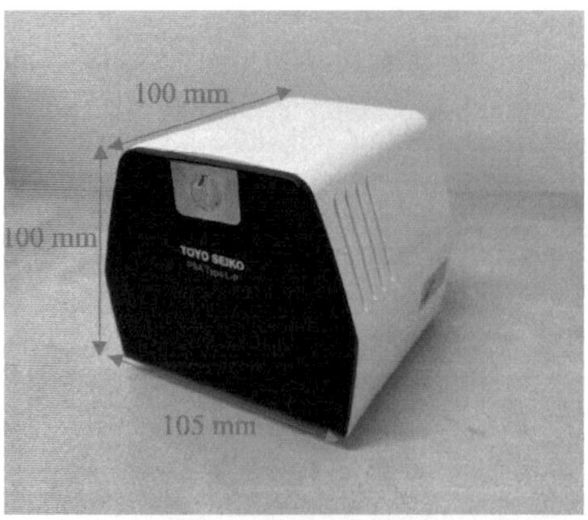

Fig. 4 An overview of the developed portable apparatus

Fig. 5 shows the plot of the change in the 'mean' positron lifetime based on the coverage. The 'mean' positron lifetime measured by the portable apparatus was monotonously increased with the increase in shot-peening time. The 'mean' positron lifetime measured by the desktop apparatus was also monotonously increased with the increase in shot-peening time. The elongation of the 'mean' positron lifetime implies that more positrons are trapped by defects, as shot-peening introduces defects, increasing the defect density. The 'mean' positron lifetimes measured by the portable and desktop apparatus agree well. Therefore, the developed portable apparatus has a sufficient ability to capture the change caused by shot-peening time.

Fig. 6 shows the plot of the FWHM of XRD profile on the coverage as well as Fig. 5. Mean positron lifetime is also showed in Fig. 6 to compare the both. The FWHM increased with the increase in shot peening time, resulting in the dislocations induced by shot peening. In the comparison of the PALS result, there were difference around the full coverage region. The positron lifetime increased more explicitly than the FWHM values in the region. PALS has a high sensitivity for vacancies, not only dislocations. Furthermore, the measurement depth of PALS is ten times deeper than XRD. These difference seemed to result in the difference on the shorter shot peening specimens.

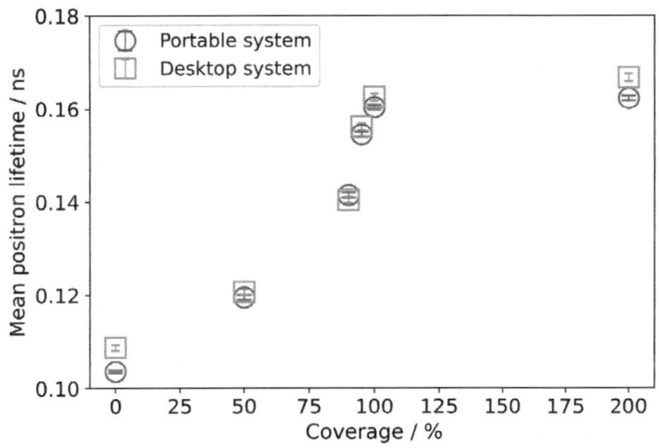

Fig. 5 The mean positron lifetime of shot-peened stainless steel obtained by the portable and desktop apparatus.

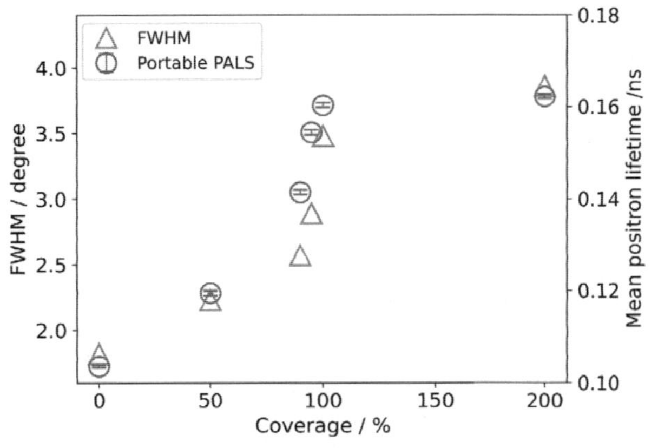

Fig. 6 The FWHM of X-ray line profile of the shot-peened specimens.

Conclusions

We developed the light-shielding technique, which enabled the apparatus to be "dark box less" and optimized the detector layout. We developed the portable PALS apparatus implementing the above works. The apparatus has an ability for on-site measurement, which means that PALS measurement is available for any size of target parts.
We evaluated the performance of the developed apparatus using shot-peened stainless steel; the apparatus could capture the difference in the average positron annihilation lifetime due to increased defect density by shot-peening as well as the conventional desktop apparatus.
We are developing several techniques for short-time measurement [5] and applying them to the developed portable apparatus. It is expected that the ability of short-time measurement with the portable apparatus allows PALS to be widely applied in industry, e.g., fatigue inspections or shot peening inspections.

Acknowledgement

This research was supported by METI Monozukuri R&D Support Grant Program for SMEs Grant Number JPJ005698.

References

[1] M. Yamawaki, Y. Kobayashi, K. Hattori, and Y. Watanabe, *Novel System for Potential Nondestructive Material Inspection Using Positron Annihilation Lifetime Spectroscopy*, Jpn. J. App. Phys., Vol. 50, (2011) 086301

[2] M. Yamawaki; N. Uesugi, Y. Kobayashi, *Development of a small and light portable positron annihilation lifetime measurement equipment for on-site measurements,* Jpn. J. App. Phys., Vol. 61, (2022) 066503

[3] S. Agostinelli et al., *Geant4—a simulation toolkit* , Nucl. Inst. Meth. A, Vol. 506 (2003) 250-303.

[4] M. Yamawaki, K. Ito, K. Hattori, and N. Uesugi, *Development of a commercial positron annihilation lifetime measurement system*, Journal of Physics: Conference Series, Vol. 791, (2017) 012038.

[5] M. Yamawaki; N. Uesugi, Y. Kobayashi, *Feasibility study of non-destructive defect analysis by positron annihilation lifetime measurement*, Jpn. J. App. Phys., Vol. 59, (2020) 066502

Shot peening and deep cold rolling of a single crystal nickel superalloy

I. Bogachev[1], K.M. Knowles[1], G. Gibson[2]

[1] University of Cambridge, Department of Materials Science and Metallurgy, 27 Charles Babbage Road, Cambridge, CB3 0FS, UK – ib358@cam.ac.uk, kmk10@cam.ac.uk
[2] Rolls-Royce plc, PO Box 31, Derby, DE24 8BJ, UK – grant.gibson@rolls-royce.com

Abstract

The present research aims to further the understanding of the effects of mechanical shot peening (MSP) and deep cold rolling (DCR) on the single crystal nickel-based superalloy CMSX-4, used in the manufacture of jet engine turbine blades. Under the thermomechanical conditions of a turbine, the main resistance to fatigue crack growth is thought to come from the surface cold work generated by these procedures. Cold work in samples treated by MSP or DCR was characterised by scanning electron microscopy, both in the as-hardened state and following heat treatment at 900 and 1100 °C. Centre hole drilling was used to find the depths of residual stress also induced by the procedures. DCR was found to produce a deeper layer of relatively low cold work in comparison with MSP. Slip bands were induced by both procedures and were especially dense following MSP. After heat treatment at 900 °C for 500 h, topologically close-packed precipitates formed on slip bands, and discrete recrystallisation occurred in the cold worked layer, with little or no recovery. At 1100 °C for 100 h, complete surface recrystallisation was observed in both samples. DCR resulted in shallower recrystallisation in both cases. The microstructures of both samples underwent rafting at 1100 °C, with the depth levels of the rafting below the surface correlating with the depth levels of compressive residual stress, both being significantly greater than cold work depths.

Keywords: *nickel-based superalloys, single crystal, mechanical shot peening (MSP), deep cold rolling (DCR)*

Introduction

There is rising interest in the aerospace industry in using surface hardening techniques, such as mechanical shot peening (MSP) and deep cold rolling (DCR), to improve the fatigue resistance of jet engine turbine blades. Traditionally optimised to resist creep, turbine blades are typically manufactured from single crystal nickel-based superalloys, which possess excellent creep resistances and high-temperature strengths. These properties are due to an ordered precipitate phase, often the γ' phase, which forms cube-shaped precipitates inside the γ matrix phase and can make up as much as 70 vol% of the material [1].

DCR, in which the surface of a sample is systematically rolled with a hard roller under hydrostatic pressure, is known to produce deeper residual stresses and a better surface finish than MSP. However, DCR is more challenging to apply to complex sample geometries [2].The effects of MSP and DCR on polycrystalline superalloys have been widely studied [3,4], but their effects on the single crystal blade alloys are less well understood.

Surface hardening procedures are known to induce a layer of cold work and a residual compressive stress at the surface of the material. In high stress, high temperature environments, such as in a turbine, the cold worked layer is thought to play the chief role in increasing the resistance to fatigue crack growth [5,6]. The present work, therefore, is concerned primarily with characterising the cold work produced in samples of a single crystal superalloy prior to their use in a high stress, high temperature environment. The microstructures of two samples, treated by DCR or MSP, are analysed and compared, both in the as-hardened state, and following heat treatments at 900 or 1100 °C.

Experimental Methods

Samples of the superalloy CMSX-4 were cast and machined into square cross-section rods, with surface hardening performed on each long face of a sample. One sample was treated by MSP using 110H shot, 0.20–0.25 mmA intensity and 1600 % coverage. Another sample underwent DCR with a 6 mm diameter roller at 30 MPa pressure. Sections of these samples were subjected to two different heat treatments under vacuum, with no externally applied stress: 900 °C for 500 h or 1100 °C for 100 h.

Scanning electron microscopy was performed on sample cross-sections using a Zeiss GeminiSEM 300. Microstructures were examined using backscattered electron imaging. Surface cold work was studied using electron backscattered diffraction (EBSD). Local variations in material crystallographic orientation are dependent on dislocation density, and therefore on the degree of cold work. By calculating these local variations for each pixel in an EBSD scan, maps showing the distribution of cold work in each sample were plotted. The kernel average neighbour misorientation (KANM) parameter [7] (3×3 kernel), in degrees of misorientation, was used to perform this quantification. Depths of cold work were found by plotting profiles of average KANM against depth in each map and finding where this became less or equal to a far-field baseline value, as detailed in ref. 8.

Centre hole drilling (CHD) was used to assess the depths of compressive residuals stresses. It should be noted that current CHD methodology is not adapted to single crystal materials, and therefore stress magnitudes should be not be viewed as quantitative.

Experimental Results

As-hardened material

Significant cold work was produced by both DCR and MSP in the surfaces of the samples (Figure 1), causing a layer of heightened KANM at the surface (Figures 1a and 1b), and bulk variations in material orientation, revealed through channelling contrast (Figures 1d and 1e). The average amount of cold work decreased more or less monotonically with depth in both samples, down to a bulk baseline (Figure 1c).

Average cold work depths were measured to be 330 ± 14 and 107 ± 10 µm for the DCR and MSP samples, respectively. However, cold work in the MSP sample was of a greater magnitude, with a denser set of slip bands. This is evident from comparisons of Figures 1a and 1b, and of Figures 1c and 1d. Numerically, this is shown by the average increase in KANM per depth within the cold worked layer in both samples – 0.26 ± 0.05 and 0.74 ± 0.07 °/µm, respectively. Significant lateral variations in local cold work were also observed in both samples. In DCR, in particular, deformation can be seen to be concentrated in narrow streaks on the surface.

Heat treatment at 900 °C for 500 h

Samples were heat treated at 900 °C for 500 h; this is similar to the total time a superalloy would spend at this temperature in service. The nominal depths of cold work were found to be unaffected by the heat treatment: 340 ± 20 and 108 ± 13 µm in the DCR and MSP samples, respectively. Similarly, no recovery of the bulk cold work was observed in either sample, as shown by the similar KANM values in Figures 1 and 2.

However, recrystallisation (RX) – the formation of new grains with a lower dislocation density [9] – within the cold worked layer was observed. RX grains are seen in KANM maps as dark areas on the surfaces of both samples (Figures 2a and 2b). Large RX grains containing a developed γ/γ' structure are also distinguishable in BSE images (Figure 2e). The maximum depth of RX, which determines to what depth the resistance to fatigue damage provided by the cold work is undermined, was considerably greater in the MSP sample than in the DCR sample – 24 and 8 µm, respectively, despite the latter sample having a much greater cold work depth. This is due to the greater magnitude of cold work in the MSP sample, which provides more driving force and faster diffusion for RX grains to form.

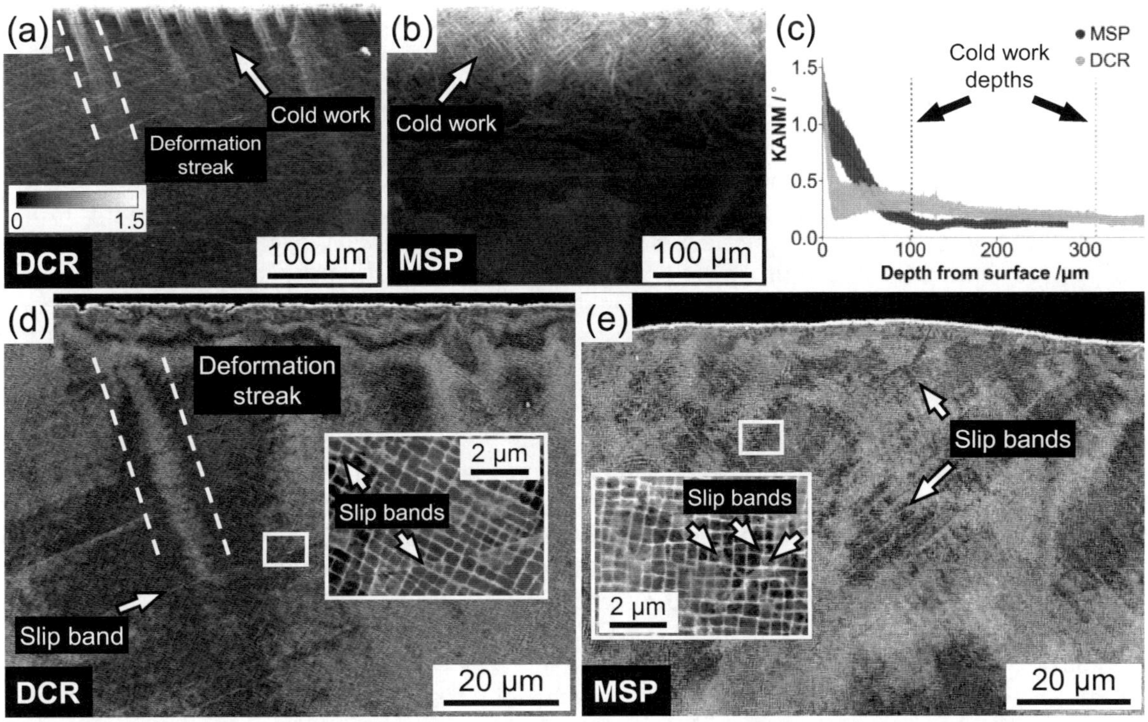

Figure 1: *As-hardened DCR and MSP samples. (a), (b) are KANM maps, with depth profiles of these maps plotted in (c). Sample microstructures are shown in (d), (e). The white boxes give the scales of the enlarged microstructure views, in which γ' is darker and γ is lighter.*

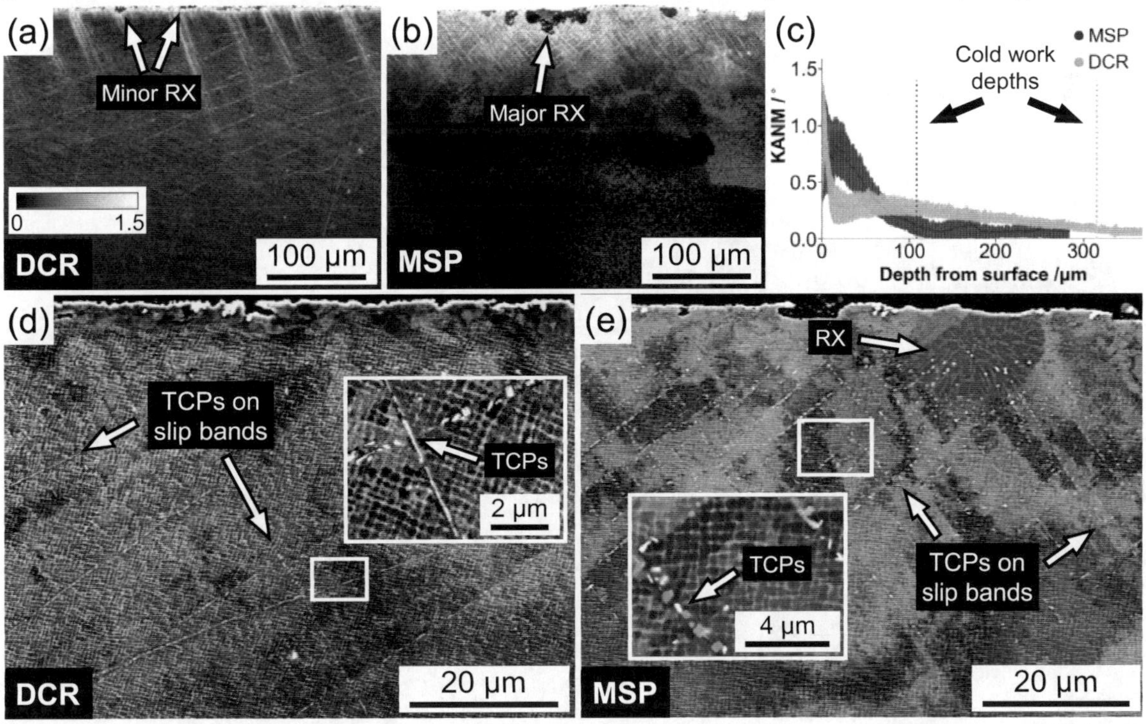

Figure 2: *DCR and MSP samples after 500 h at 900 °C. (a), (b) are KANM maps, with depth profiles of these maps plotted in (c). Sample microstructures are shown in (d), (e). The white boxes give the scales of the enlarged microstructure views.*

Precipitation of topologically close-packed (TCP) phases was observed in both samples (Figures 2d and 2e). These phases are rich in heavy alloying elements and can act as crack nucleation sites [10]. Their formation was seen to begin after just 10 h of heat treatment (images not shown). Superalloys are designed to be kinetically stable with respect to TCP precipitation, and slip bands clearly played a key role in their formation in the present samples. TCPs, seen in Figures 2d and 2e lying predominantly in straight lines, clearly adhere closely to slip bands. TCPs were slightly larger in the DCR sample than in the MSP sample (partly due to their greater elongation, see Figure 2d), with average sizes of 0.30 ± 0.06 and 0.24 ± 0.03 µm, respectively, but were much more sparse in the DCR sample, in which the density of slip bands is lower. TCPs formed up to depths of ~160 and ~60 µm in the DCR and MSP samples, respectively, roughly coinciding with the depths to which slip bands could be visibly distinguished in each sample.

Heat treatment at 1100 °C for 100 h

To test the response of the cold work in a more extreme environment, the two samples were heat treated at 1100 °C for 100 h. In both samples, this promoted the formation of RX grains across the entire sample surfaces, to a greater depth than after 500 h at 900 °C, with the MSP sample again having deeper RX than the DCR sample (Figures 3a and 3b). Indeed, most of the cold worked layer in the MSP sample recrystallised, leaving only a lip of remnant cold work. After this heat treatment, the DCR sample now actually had greater cold work magnitudes at all depths (Figure 3c).

This heat treatment also resulted in rafting – the directional coarsening of γ' precipitates, forming elongated plates or "rafts" (Figure 3d). Rafting is known to occur in single crystal superalloys under an applied external stress, but reports of rafting following surface treatments are rarer [11].

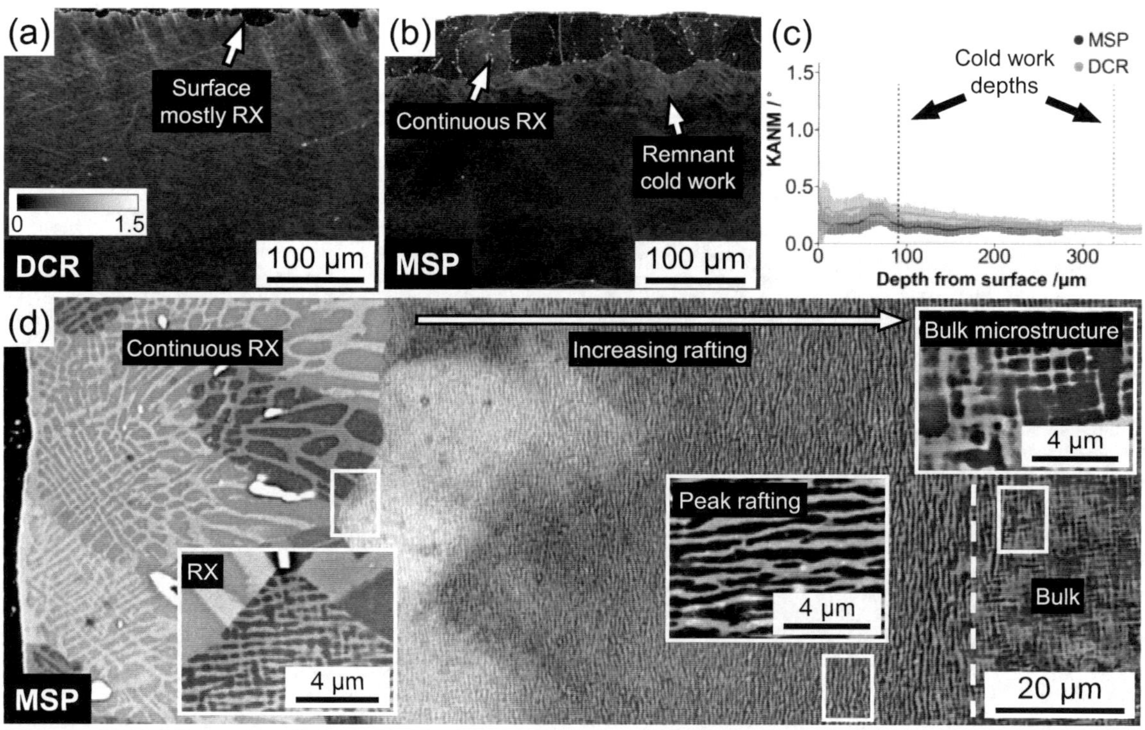

***Figure 3**: DCR and MSP samples after 100 h at 1100 °C. (a), (b) are KANM maps, with depth profiles of these maps plotted in (c). Sample microstructures are shown in (d), (e). The white boxes give the scales of the enlarged microstructure views.*

The extent of rafting, characterised by the "raft ratio" – the ratio of the average lengths of straight raft segments parallel and perpendicular to the raft direction – is plotted against depth for the two samples in Figure 4a. A raft ratio of 1 indicates an unrafted material state where the γ' precipitates have sides of equal length. In both samples, the degree of rafting increased with depth, reaching peak values of ~2.5–3, and then decreased again to ~1, when the material adopted a bulk microstructure (compare this with Figure 3d).

Since rafting behaves in an opposite sense to cold work, which decreases monotonically from a maximum at the surface, it must be occurring under the influence of the residual compressive stresses generated by MSP and DCR. It is known that rafting may continue, once initiated, even if the stress is subsequently removed [12]; therefore any relaxation of the residual stress does not mean rafting would cease. Depths of compressive residual stresses were obtained from residual stress profiles for the two samples, measured by CHD (Figure 4b; magnitudes not quantitative: see Experimental Methods). The depths of cold work, peak rafting, total rafting and compressive residual stress for both samples are given in Table 1.

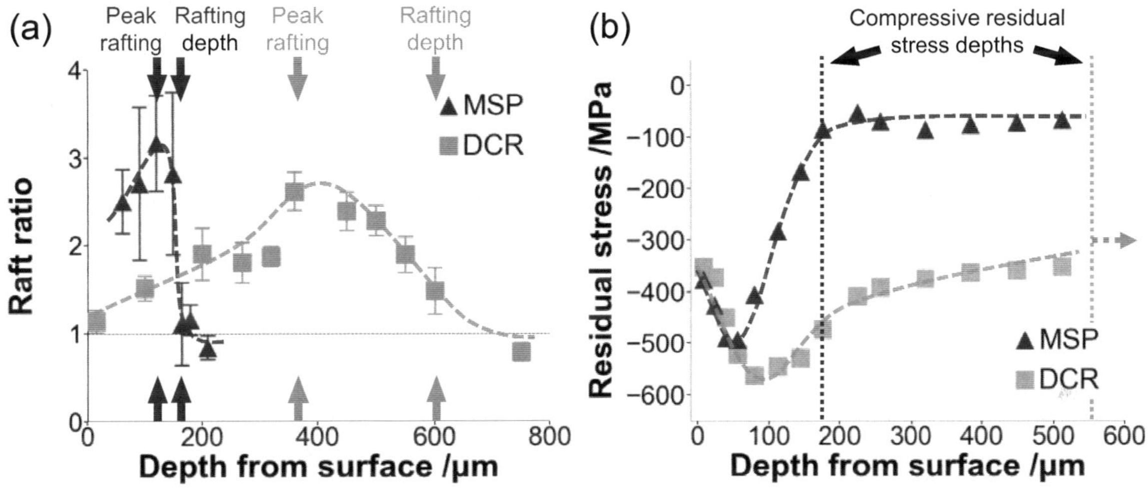

***Figure 4**: (a) Extent of rafting versus depth in MSP and DCR samples after 100 h at 1100 °C, with peak rafting and total rafting depths marked for each sample. (b) Residual stress profiles for each sample in the as-hardened state. Compressive residual stress depth for the DCR sample is outside the measured range*

Sample	Cold work depth	Peak rafting depth	Rafting depth	Compressive residual stress depth
MSP	107	120	165	176
DCR	330	360	600	>500

***Table 1**: As-hardened cold work depths and compressive residual stress depths, and peak and total rafting depths after 100 h at 1100 °C, in μm, in the DCR and MSP samples.*

In both samples, it can be seen that peak rafting occurs beyond the depth of cold work, whereas the total raft depth correlates fairly well with the depth of compressive residual stress. From this, it seems that cold work actually hinders rafting, which is instead driven by the surface compressive residual stresses.

This observation also highlights the fact that the depths of compressive residual stress and cold work do not have to coincide. Residual stress arises from a macroscopic change of the sample shape, whereas cold work is the microscopic change in the material microstructure, resulting in greater dislocation density. The EBSD method measures the depth to which there is a significant increase in dislocation density. However, it can be used to set a lower bound on the depth of compressive residual stress. This is because the depth of the deformed plastic layer below the surface will be equal to or greater than the depth below the surface of the layer where there is a significant increase in dislocation density relative to the base material.

Conclusions

A comparison has been made of the cold work and microstructures of samples of a single crystal superalloy treated by MSP and DCR, both in the as-hardened state and after heat treatments at 900 and 1100 °C.

DCR was found to produce a depth of cold work greater by a factor of three. It was approximately three times less severe per unit of depth than that produced by MSP. Both procedures resulted in multiple sets of slip bands, with those arising from DCR being much more widely spaced.

RX grains formed on the sample surfaces after both heat treatments, with DCR displaying lower RX depths than MSP, and the treatment at 1100 °C producing deeper RX. At 900 °C, TCP precipitates formed on slip bands in both samples, up to ~half the depth of cold work.

After the 1100 °C heat treatment, the microstructures of both samples underwent rafting, with peak rafting occurring beyond the depth of cold work. The depth at which the rafting occurred below the surface correlated well with the depth of compressive residual stress in each sample, showing that residual stress promotes rafting, while cold work retards it.

References

[1] H. Long, S. Mao, Y. Liu, Z. Zhang, X. Han, *Microstructural and compositional design of Ni-based single crystalline superalloys — A review*, Journal of Alloys and Compounds, Vol. 743 (2018) pp. 203–220.

[2] C.C. Wong, A. Hartawan, W.K. Teo, *Deep cold rolling of features on aero-engine components*, Procedia CIRP, Vol. 13 (2014) pp. 350–354.

[3] G.J. Gibson, K.M. Perkins, S. Gray, A.J. Leggett, *Influence of shot peening on high-temperature corrosion and corrosion-fatigue of nickel based superalloy 720Li*, Materials at High Temperatures, Vol. 33 (2016) pp. 225–233.

[4] B. Nagarajan, D. Kumar, Z. Fan, S. Castagne, *Effect of deep cold rolling on mechanical properties and microstructure of nickel-based superalloys*, Materials Science and Engineering A, Vol. 728 (2018) pp. 196–207.

[5] J. Hoffmeister, V. Schulze, A. Wanner, R. Hessert, G. Koenig, *Thermal relaxation of residual stresses induced by shot peening in IN718*, 10th International Conference on Shot Peening (2008) pp. 157–162.

[6] H.L. Cockings, B.J. Cockings, W. Harrison, M. Dowd, K.M. Perkins, M.T. Whittaker, G.J. Gibson, The effect of near-surface plastic deformation on the hot corrosion and high temperature corrosion-fatigue response of a nickel-based superalloy, Journal of Alloys and Compounds, Vol. 832 (2020) art. 154889.

[7] S.I. Wright, M.M. Nowell, D.P. Field, *A review of strain analysis using electron backscatter diffraction*, Microscopy and Microanalysis, Vol. 17 (2011) pp. 316–329.

[8] I. Bogachev, K.M. Knowles, G.J. Gibson, Electron backscattered diffraction analysis of cold work in a shot peened single crystal nickel superalloy, Materialia, Vol. 14 (2020) art. 100860.

[9] D.C. Cox, B. Roebuck, C.M.F. Rae, R.C. Reed, *Recrystallisation of single crystal superalloy CMSX-4*, Materials Science and Technology, Vol. 19 (2003) pp. 440–446.

[10] Z. Zhang, Z. Yue, TCP phases growth and crack initiation and propagation in nickel-based single crystal superalloys containing Re, Journal of Alloys and Compounds, Vol. 746 (2018) pp. 84–92.

[11] H. Biermann, U. Tetzlaff, H. Mughrabi, B. Grossmann, S. Mechsner, T. Ungár, *Investigations on the origin and effect of anomalous rafting*, Ninth International Symposium on Superalloys (2000) pp. 525–534.

[12] N. Matan, D.C. Cox, C.M.F. Rae, R.C. Reed, *On the kinetics of rafting in CMSX-4 superalloy single crystals*, Acta Materialia, Vol. 47 (1999) pp. 2031–2045.

Improvement of fatigue property of A7075 aluminum alloy by laser peening with handheld laser device

K. Masaki[1], Y. Sano[2,3,4], Y. Mizuta[3] and S. Tamaki[4]

1 Okinawa College, National Institute of Technology, JAPAN
2 Institute for Molecular Science, Okazaki, JAPAN
3 SANKEN, Osaka University, JAPAN
4 LAcubed Co. Ltd., Yokohama, JAPAN

Abstract
In recent years, new handheld pulse laser oscillators have been developed that are much smaller than currently-used pulse laser devices. In this study, in order to investigate the effectiveness as a laser source of laser peening (LP), rotating bending high cycle fatigue test was conducted on referenced pristine and laser-peened aluminum alloys of Al-Zn-Mg-Cu system. As a result, the rotational bending fatigue properties were significantly improved by the LP treatment. The fatigue strength is improved by about 1.5 times, and the fatigue life is extended by about 100 times compared with the fatigue property of the pristine material. It was found that the developed new handheld pulse laser oscillator can be used for the LP treatment.

Keywords Laser shock peening, Handheld pulse laser oscillator, Fatigue, Aluminum alloy

Introduction
Laser peening (LP) processing is one of peening technologies in which a metal material placed in water is irradiated with a pulse laser and peened by the generated shock wave [1]. Since conventional LP requires large-scale and high-cost laser equipment, smaller and cheaper pulse laser oscillators have been desired for further industrial applications [2],[3]. In recent years, new handheld pulse laser oscillators have been developed by ImPACT program in Japan that are much smaller and cheaper than currently-used equipment [4],[5]. In order to investigate the effectiveness of the newly-developed handheld pulse laser oscillator for LP treatment, rotating bending fatigue tests were conducted on Al-Zn-Mg-Cu system aluminum alloys.

Experimental Methods
Material
The material used in this study is A7075BE-T6511 aluminum alloy. After extruding into a bar with a diameter of 15 mm, solution heat treatment, residual stress relief treatment and artificial aging treatment were applied. Chemical composition and mechanical properties of the material are shown in Table 1 and 2.

Table 1. Chemical composition of the material. (%)

Si	Fe	Cu	Mn	Cr	Zn	Ti	Ti+Zr	Al
0.08	0.19	1.6	0.04	2.6	5.8	0.01	0.02	Bal.

Table 2. Mechanical properties of the material. (Experimental data)

Tensile strength σ_B MPa	0.2% proof stress $\sigma_{0.2}$ MPa	Elongation δ %	Reduction of area Ψ %	Hardness Hv
685	631	12.6	23.4	189

Observation result of microstructure with an optical microscope is shown Fig.1. In this material, an oriented texture is well developed in the axial direction. The result of crystal orientation analysis (EBSD analysis) is in Fig.2. It is clear also from the IP map that the oriented texture in the axial direction is well developed in this material.

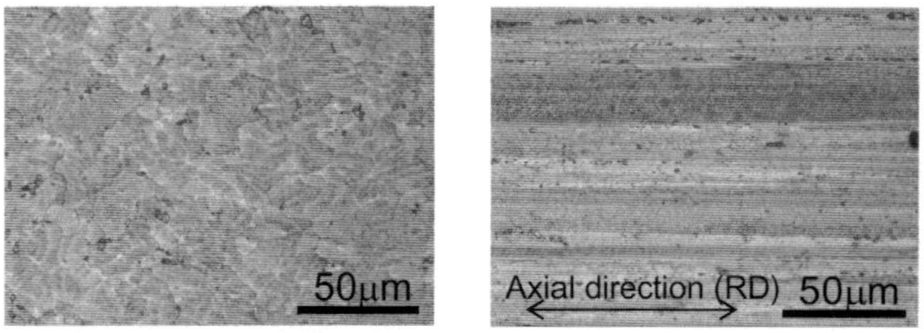

(a) Cross section (b) Longitudinal section

Fig.1 Observation result of microstructure.

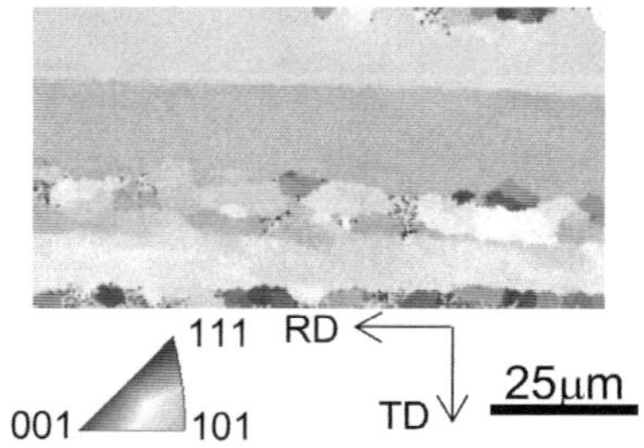

Fig.2 EBSD analysis result (IP Map)

Specimen and fatigue test

The shape of test specimens in this study is an hour glass type as shown in Fig.3. The stress concentration factor of this shape is 1.074 [6]. The LP treatment was performed only to the notch area. The laser device used in this study was a prototype handheld pulse laser oscillator provided by OPTOQUEST Co., Ltd. An over view of the laser peening system is shown in Fig.4. The LP treatment was performed with the following conditions; an irradiation energy of 1.68mJ, a spot diameter of 0.3mm, a repetition rate of 50Hz and laser irradiation densities of 400 and 800 pulse/mm^2. Cantilever type rotating bending fatigue test was carried out with a rotating speed of 4000 rpm in the air at room temperature.

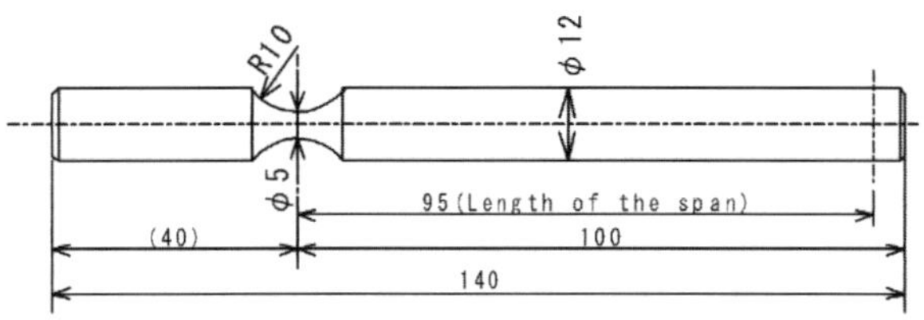

(unit:mm)

Fig.3 Shape of fatigue test specimen

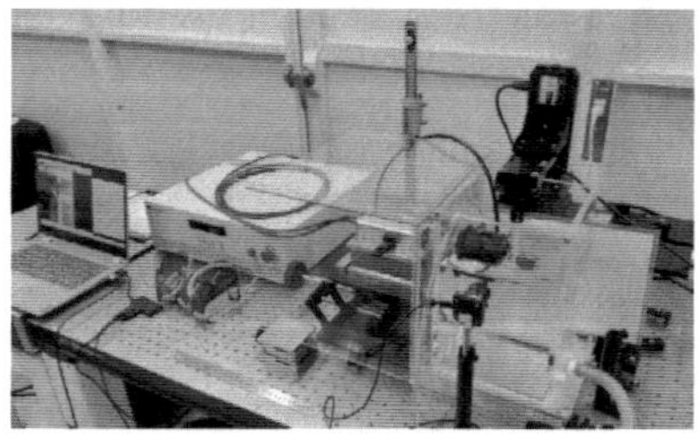

Fig.4 Laser peening system

Experimental Results and Discussion

Peening effect

Surface roughness profiles and surface roughness values are shown in Fig.5. Although the surface roughness increases by handheld LP treatment, the surface roughness values are smaller than conventional SP and LP treatment values [7]. That is, fatigue strength reduction effect due to surface roughness is not significant.

	LP800pls	LP400pls
Ra	0.459μm	0.409μm
Rz	2.75μm	2.47μm

Ra : arithmetic average roughness
Rz : maximum height of irregularities

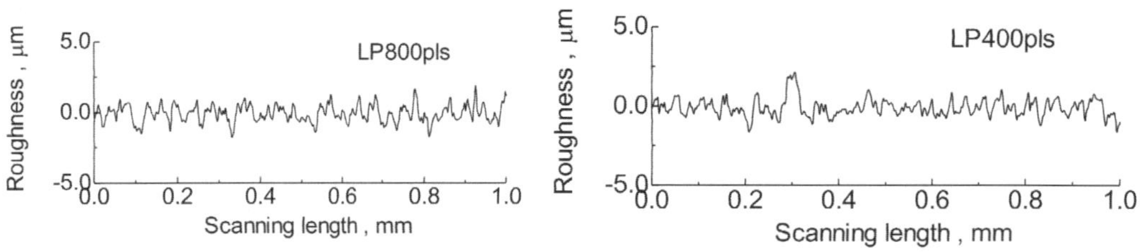

(a) LP800pls (b) LP400pls

Fig.5 Surface roughness profiles and values.

Hardness distributions of LP treated specimens are shown in Fig.6. Strain hardening due to the handheld LP treatment is slight. The hardness distribution of the LP400pls is even from 180 to 190Hv which is the same value as the base metal, 189Hv, as shown in Table 2. On the other hand, the surface hardness value was slightly increased by the LP800pls condition. The thickness of hardened layer is about 50μm and its hardness value is about 210Hv.

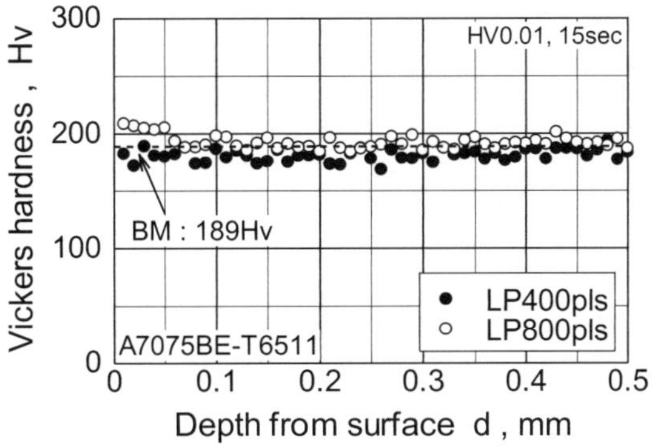

Fig.6 Hardness distributions

At this time, the residual stress distributions have not yet been measured. However, it is expected that the distributions will be similar to that of A7075 material shown in the previous report [5].

Fatigue testing results

S-N curves of fatigue testing result are shown in Fig.7. The rotational bending fatigue properties were significantly improved by the handheld LP treatment. The difference in fatigue properties between different pulse densities was not remarkable. The fatigue strength is improved by about 1.5 times, and the fatigue life is extended by about 100 times than non-peening material. It was found that the newly-developed handheld pulse laser oscillator can be used for LP processing. Typical fatigue fracture surfaces are shown in Fig.8. There was a big difference between the non-peening specimen and LP specimens. The fracture surface of all non-peening specimens were on a flat plane perpendicular to the axial direction as shown in Fig.8(a). On the other hand, most of fracture surface shapes of the LP specimens were a pointed one as shown in Fig.8(b) and (c). As the result of optical microscope observation, a lot of black oxide powder was observed on the inclined surface. That is, the inclined surface was not formed in the final fracture process, but in the fatigue process.

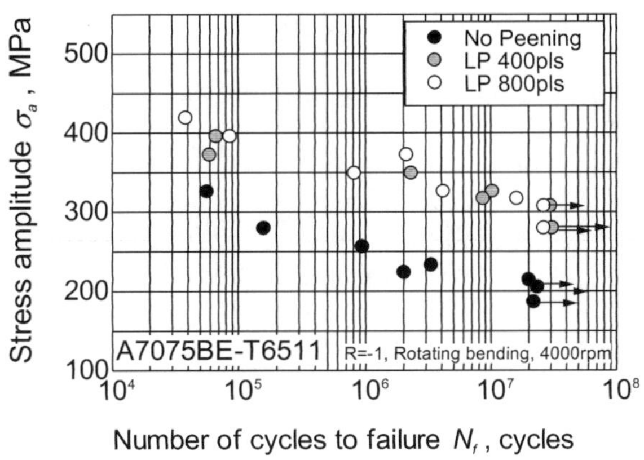

Fig.7 S-N curves of all specimens.

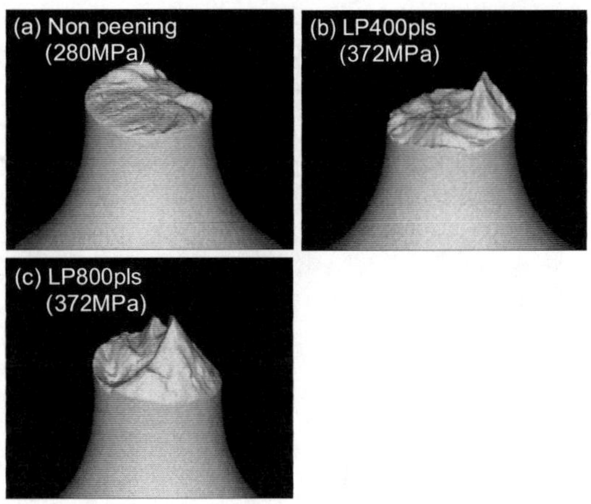

Fig.8 Typical fatigue fracture surfaces

The angle of the inclined surface to the cross section of the specimen is about 55 to 60 degree. The material used in this study has an oriented texture in the axial direction in which the (001) plane of fcc crystal structure coincides with the cross section of the specimen. The angle of slip system {111} <110> for the cross section of this specimen is 54.7 degrees. This value is the same as the angle of the inclined surface. The cause of this phenomenon is considered to be the residual stress due to the LP treatment. It is considered that the generation and propagation of surface cracks was suppressed by the high compressive residual stress. Even if the fatigue crack can be generated at surface, the fatigue crack cannot grow in mode I which is crack open mode. More than that, the large-scale glide plane decohesion occurred inside the test specimen due to the influence of the developed texture in a triaxial stress state. In any case, the fatigue property of A7075BE-T6511 material is significantly improved by LP processing with a handheld pulse laser oscillator. This new device can be widely used in the industry in the future.

Conclusions

In order to investigate the effectiveness of newly-developed handheld pulse laser oscillators as a laser source of LP processing, rotating bending high cycle fatigue test was conducted on A7075BE-T6511 material used for aircraft components. Although the fatigue fracture mechanism was significantly affected by the well-developed oriented texture, the fatigue property was significantly improved by LP processing. The fatigue strength is improved by about 1.5 times, and the fatigue life is extended by about 100 times than non-peening material. It is clarified that the newly-developed handheld pulsed laser oscillator can be used for LP processing and that it can be used in a wide variety of industrial fields.

Acknowledgments

This work was partially supported by the JST-MIRAI Program (grant number JPMJMI17A1), the Supporting Industry Program of the Small and Medium Enterprise Agency (grant number 202041907014), the Amada Foundation (grant number AF-2020239-C2), and the Impulsing Paradigm Change through Disruptive Technologies Program (ImPACT) of the Council for Science, Technology and Innovation.

References

[1] [1] Y. Sano, N. Mukai, K. Okazaki and M. Obata, *Residual stress improvement in metal surface by underwater laser irradiation*, Nucl. Instrum. Methods Phys. Res. B, Vol. 121 (1997), pp 432–436.

[2] A.H. Clauer, *Laser shock peening, the path to production*, Metals, Vol. 9 (2019), p 626.

[3] Y. Sano, *Quarter century development of laser peening without coating*, Metals, Vol. 10 (2020), p 152.

[4] Y. Sano, T. Kato, Y. Mizuta, S. Tamaki, K. Yokofujita, T. Taira, T. Hosokai and Y. Sakino, *Development of a portable laser peening device and its effect on the fatigue properties of HT780 butt-welded joints*, Forces in Mech., Vol 7 (2022), p 100080.

[5] Y. Sano, K. Masaki, Y. Mizuta, S. Tamaki, T. Hosokai and T. Taira, *Effects of laser peening with a pulse energy of 1.7 mJ on the residual stress and fatigue properties of A7075 aluminum alloy*, Metals, Vol. 11 (2021) p. 1716.

[6] D. P. Walter, and F. P. Deborah, *Third Edition - Peterson's Stress Concentration Factors,* John Wiley & Sons, Inc. (2007)

[7] K. Masaki, T. Tsuji and Y. Kobayashi, *Effect of peening treatment for out-of-plane fatigue test on A7075-T651 alloy*, Trans. of the JSME (in Japanese), Vol.81,(2015), DOI: 10.1299/transjsme.15-00003

Fatigue strength improvement of welded joints by means of shot peening and clean blasting

Thomas Nitschke-Pagel, Klaus Dilger, Hamdollah Eslami

Institut für Füge- und Schweißtechnik, Technische Universität Braunschweig,Germany

ABSTRACT

Shot peening is a well established state of the art mechanical surface treatment to improve the fatigue strength of metallic components, clean blasting a familiar process and used frequently in welding practice. The main difference is the peening media used and the minor or missing parameters of clean blasting. The clean blasting process is not yet qualified to optimize fatigue enhancement although it holds similar potential as regular shot peening. Clean blasting is usually applied with the purpose of surface preparation for application of corrosion protection. This article presents results of regular shot peened DV-butt welds made from construction steels S355N and S960QL, The peening parameters are varied widely. The effect of coverage and intensity is investigated to test the robustness of the peening processes. The data is completed with industrially clean blasted welds representing typical workshop conditions. The overall objective of this work is to demonstrate the fatigue strength potential of several variations of peening processes. The results indicate a high robustness of the applied process variations. This indicates high potential of shot peening and clean blasting for fatigue strength enhancement. If clean blasting is accompanied by a simple process control in terms of coverage and intensity the fatigue strength benefit is obvious-

Keywords: Welded joints, fatigue strength, shot peening, clean blasting, residual stresses

1. Introduction

The fatigue strength of welded joints is usually low related to the base material. Since the fatigue strength loss increases with increasing tensile strength due to the stronger notch sensitivity applicable stress amplitudes cannot be improved by changing to a higher-strength base material. Therefore the potential of modern high-strength steels can only be exploited in a meaningful way when high mean stresses are present, since the stress limit is then determined by the level of the upper stress. Extensive investigations over the last 50 years [1,2,6,8,10] have shown that several post weld treatment methods are capable of significantly increasing the fatigue by reducing the notch effect of the critical weld toe fracture. This includes methods such as grinding the seam transitions flat on the sheet or local, or remelting with additional wire-free welding processes like TIG or plasma processes.

Alternatively, mechanical surface treatment methods like hammer-, needle-peening or shot peening are established [4], in which the fatigue strength-increasing effect is achieved through strain hardening of the surface and the generation of favorable residual compressive stresses. While high-frequency hammer-peening processes, which work with high local intensity have recently become very popular and found their way into existing guidelines [5,13], shot peening is still not very common for welded joints, although there is sufficient evidence that it is at least as effective the mentioned methods. Nevertheless a high general scepticism is present about post-weld treatments because their use always involves additional work and costs that one would like to avoid if possible. Another obstacle is the rigid limitation of the usable fatigue strength by the existing set of rules, which is based on the fatigue design rules [3,4], which include a strong limitation of the fatigue strength on the basis of low-strength structural steel. The potential of post-treated welded joints, especially with high-strength materials, cannot currently be exploited in accordance with the regulations, although extensive studies have shown that considerable reserves are available [2,11]. This is particularly important in the current situation in view of the sharp rise in raw material prices and the need for economical and climate-friendly design principles. Better utilization of the strength reserves of construction materials seems urgently required.

In welding practice, clean blasting is a widespread standard procedure integrated into existing production chains. It is used to remove impurities and surface layers from welded or thermally cut components, which would prove to be a quality-reducing factor, e.g. The method is similar to shot peening except a less detailed control of the peening parameters. Since the cleaning effect is the main focus mostly simple shot as quartz sand, gravel or steel shot is used. Anyways clean blasting may also produce improving beneficial changes in surface layers. The aim of the presented work is to demonstrate the potential of different peening treatments and thus to create the prerequisite for an implementation in existing regulations for the fatigue design of welded components.

2. Shot peening and clean blasting of welded joints

Shot peening is a flexible method to modify surfaces of various metallic and non-metallic components. Typical applications are enhancement of surface roughness, cleanliness or enhancement of mechanical parameters and corrosion resistance [9]. The most important peening parameters affecting the fatigue performance are the shot type and size, shot geometry, peening intensity and coverage [10]. The user can influence the kinematic energy of the peening impacts by the choice of the shot type, its size and the impact velocity and additionally the coverage. Steel components are commonly shot peened using steel balls while glass beads are used in case of demands for low peening intensity or with regard to surface finishing. The shot peening process and its intensity is usually adjusted by peening parameters like impact velocity and coverage and the result is controlled by the Almen strip test [11]. Coverage of the surface is defined as the percentage of the peened surface related to the un-peened surface. A coverage of 98 % is the highest coverage value that can be experimentally determined as the indentation spots of the shots are here still distinguishable. A coverage of 98 % is normally denoted as “full” coverage. Higher or lower values of coverage are normally adjusted by a control of the peening time per surface area.

As mentioned the fatigue strength improvement due to shot peening relies on the generation of near surface compressive residual stresses and cold work hardening. According to [5,12] the significance of the generated residual stresses increases with the yield strength of the peened material while in low strength materials the effect of the cold hardening of surface layers is more important due to the higher stability of the residual stress under the particular load conditions which may reduce the residual stresses significantly in weak materials during the first load cycles. Reduction of surface roughness by plastic deformation also becomes more important with increasing ulitimate strength and that is to say in notch sensitive materials like high strength steels and Al-alloys [9]. While the induced compressive residual stress field may retard fatigue cracks or slow down crack growth considerably cold hardening increases the resistance against crack initiation. The residual stress field can be described by the magnitude of the induced compressive residual stress and their depth

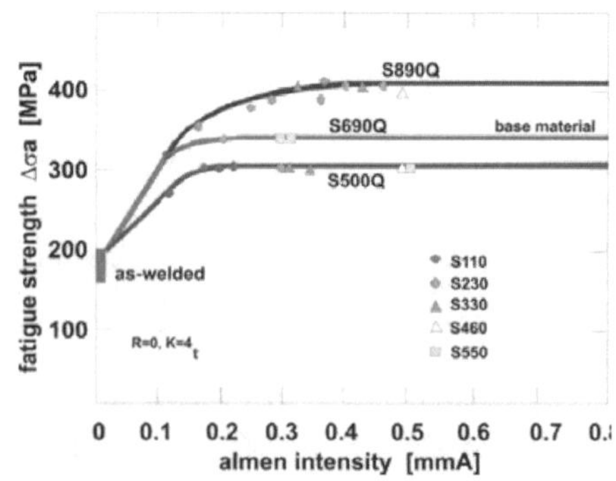

Figure 1: Relationship of peening intensity and expected fatigue strength for welds of different steel grades [1].

profile. Investigations on fatigue strength improvements by shot peening can be found in the literature [2,8,10], where the general benefit is proven clearly. Early investigations [1] have demonstrated clearly the possible benefit of the shot peening procedure, if the intensity is well adjusted to the ultimate strength of the used steel grade (Fig.1).

3. Experimental programme

The experiments where carried out using structural steels S355N (normalized) and S960QL (quenched and tempered), see Tab. 1. The peening media were varied reflecting regular shot blasting (steel shots S280) as well as clean blasting (glass beads and corundum). Further, the coverage (meaning peening time) was varied between 0.25...98% (under coverage) and 2 x 98% (double coverage). Coverage of 25 % reflects the useful lower boundary of clean blasting while 2 x 98% coverage is used in conventional shot peening processes. Another parameter investigated was the influence of the peening intensity which is in pratice described by means of the Almen intensity. The peening intensities used here were 0.3...0.4 mmA, mmN and mmC. Specimens prepared by means of combinations of the aforementioned parameters were used for fatigue testing. The metallographic condition, surface roughness, hardness and the near surface residual stresses were documented.

	R_e [MPa]	R_m [MPa]	A_5 [%]	CEV [%]	T [mm]
S355N	376	518	27	0.41	10
S960QL	995	1033	16	0.56	10

Table 1: Properties o the used steels

The samples were prepared as MAG (metal active gas welding) -welded DV-butt welds. Macrographs and the hardness distributions (HV1) are shown in Fig.2. Typically, steels show heat affected zones with fine and coarse grained zones. Welded steels both show an increase in hardness to approximately 300 HV1 (S355N) and 400 HV1 (S960QL) respectively. In S960Q an area of slight softening from 320 HV1 (base metal) to 280 HV1 is present in the zone, where the temperature was between A_{c1} and the annealing temperature of the base material (Fig.2). All samples were shot peened or clean blasted before testing according to industrial standards at the facilities of an industrial company. Further specimens made from S355N were clean blasted at workshops of different steelwork companies (Details see [14]).

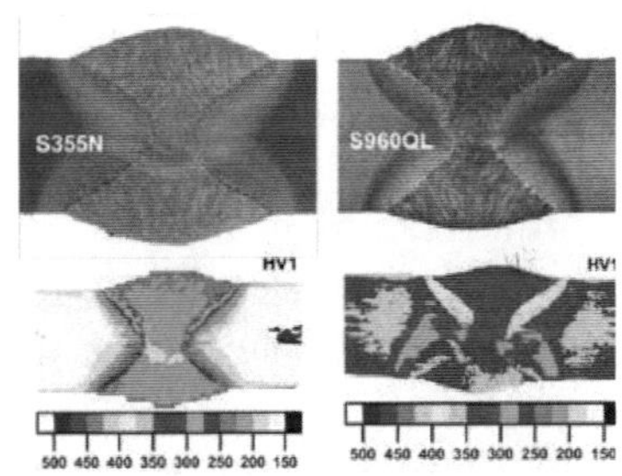

Figure 2: Macrographs of the investigated DV-butt welds.

Residual stress measurements were carried out mainly by means of XRD). The {211}-diffraction lines were measured using CrK$_\alpha$-radiation at 11 ψ -tilts from 0...45°, the collimator size was ø2 mm. Residual stresses where calculated with the $\sin^2\psi$-method using the x-ray elastic constant $1/2s_2$=6.08 mm²/N. The presented results are focused on the component in load direction (transverse residual stresses). Incremental hole drilling method was applied using an air driven milling turbine and 0°/45°90°-strain gage rosettes (HD). Near surface residual stress depth profiles were determined through incremental surface removal by means of electrolytic polishing and additionally applying the hole drilling method.

Fatigue testing was performed under axial loading with a uniform stress ratio of R=0.1. The fatigue strength of the related base material and of the untreated welds was used as reference. Within the test series the peening media, the intensity and the coverage were varied. Further specimens made from S355N were clean blasted at workshops of different companies from the structural steelwork industry The peening conditions of all test series can be found in detail in [14]. The fatigue data were evaluated by linear regression without consideration of run-outs. Stresses are described in the IIW-nomenclature $\Delta\sigma_a = 2\sigma_a$ [2].

4. Results

4.1 Initial residual stress condition of the test samples

Fig.3 represents the initial transverse residual stresses in the as-welded condition and after shot peening. Rather low residual stresses could be found in the as-welded condition, typical for relatively small sized samples. After shot peening compressive almost uniform residual stresses were generated at the surface while no significant differences of the amount was found here due to fifferent peening conditions. The profile of the measured integral width values represent qualitatively the effect of cold hardening of the near surface layer. More detailed information is given by the local residual stress depth profiles at the weld toe (Fig.4). As expected peening with steel shot generates the highest cmpressive residual stresses below the surface in the low strength steel S355N. The total amount of the maximum in the steel S960Ql is more uniform. The higher intensity furthers that the maximum is shifted to a deeper layer in both steels. Similar results were observed after variation of the coverage.

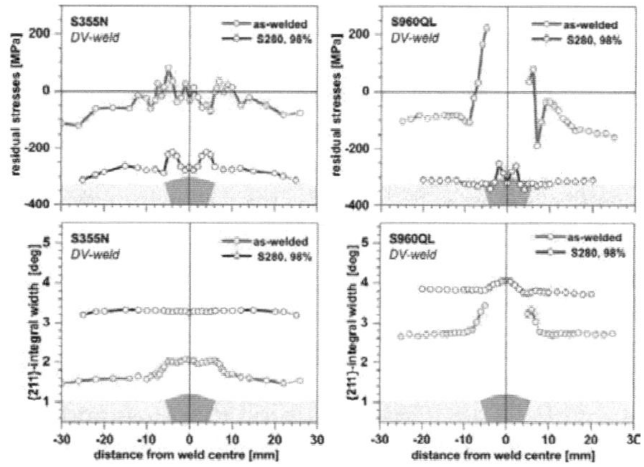

Figure 3: Initial residual stresses (XRD) and {211}-integral widths.

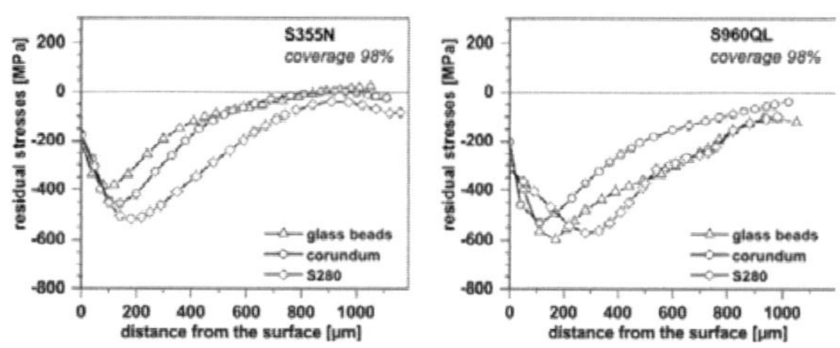

Figure 4: Residual stress depth profiles (Hole Drilling) at the crack sensitive weld toe.

4.2 Stability of the residual stresses during cyclic loading

The most popular argument against the practical use of shot peening for the fatigue strength improvement is the mistrusted stability of the generated beneficial residual stresses under different load conditions. It is assumed that they may be relaxed quickly during the first load cycles (constant amplitude loading) or in practice under service conditions (e.g. variable amplitude loading) due to unexpected single overloads. This narrow perspective is focused on the general major influence of the generated residual stresses. It disregards that their stability depends strongly on the material properties. As described in [5,12] the residual stress stability increases with increasing yield strength. Thus the effect of the residual stresses increases also with the yield strength while in weak materials the positive effect of near surface cold hardening becomes more important. As Fig.5 shows exemplarily for the low strength S355N-steel a static load leads

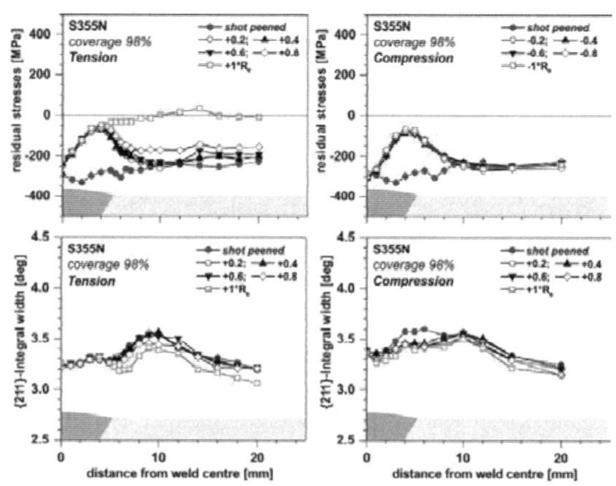

Figure 5: Residual stress surface profiles after different tensional and compressive loads.

to a residual stress relaxation in the weld toe surrounding. The base material is affected only after a nominal tensile at an amount of the yield strength. Similar results were found after different numbers of load cycles while the main relaxation ocurred after the first cycle. On the other hand the integral width distributions of the diffraction lines indicate a stable near surface hardening condition.

4.3 *Fatigue test results*

The results of the fatigue tests are summarized in Fig.6, (S355N). The SN-curves are always representing a lower confidence limit of 97.5% survival probability which is the reference for the design curves in the design rules (i.e. FAT.classes). The results reveal that the fatigue strength of the low strength S355N-steel is shifted almost to the level of the base material. The benefit is almost independent from the individual peening conditions (clean blasting). Controlled shot peening finally furthers the best performance. The highest improvement is given at high numbers of load cycles. The high strength S960QL shows the expected higher dependency of the fatigue strength improvement from the applied intensity, see Fig.7. The

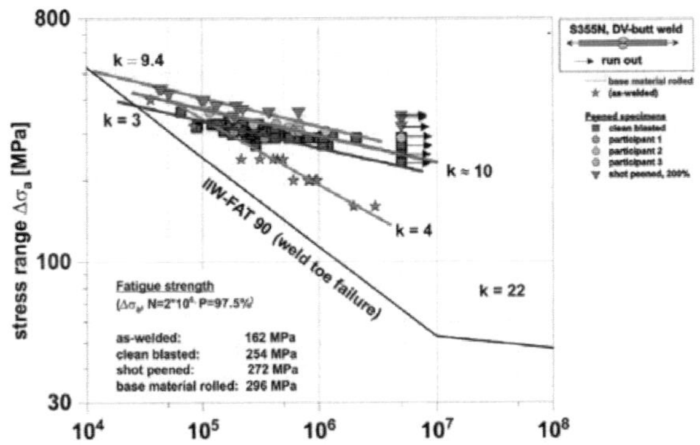

Figure 6: Fatigue test results of the different test series (low strength steel S355N).

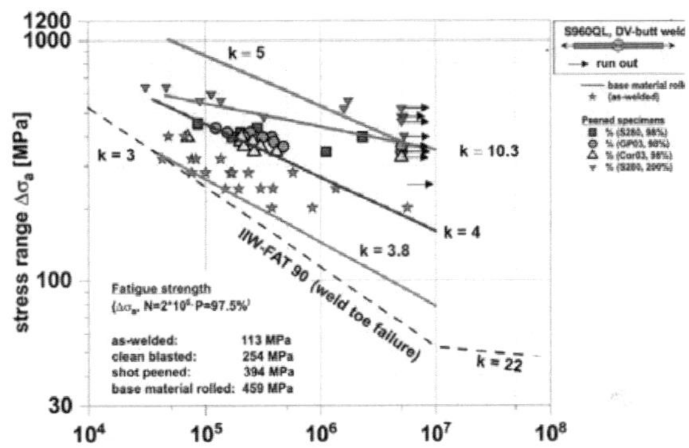

Figure 7: Fatigue test results of the different test series (high strength steel S960QL).

uncontrolled clean blasting treatment also furthers significant shift of the applicable stress amplitudes but the best performance is achieved after shot peening with a sufficient intensity reaching almost the level of the base material at stress levels in the high cycle regime.

5. Conclusions

In accordance to elder research results the presented investigations show the high potential of the application of shot peening as ar recommended post weld treatment procedure for the fatigue strength improvement of welded joints. A comparison with the current design curves evidently shows demonstrates the potential of shot peening procedures. The presented results indicate SN-curves reepresenting a fatigue strength potential significantly above the existing design limits (Fig.7)

1. The shot peening process furthers a strong shift of the applicable stress amplitdes, even if the conditions are not precisely controlled and the procedure is primarily focused on the cleaning effect of he surface.
2. In low strength steels it can be expected that cold working induced surface hardening improves the fatigue crack resistance while the beneficial effect of the generated compressive residual stresses is mainly present at higher lifetime with lower load stress

levels. Clean blasting under common industrial conditions can be almost effective as a precisely controlled shot peening process.

3. In high strength steels clean blasting furthers a higher fatigue strength but the highest performance requires a well adjusted peening intensity. Here the effect of the generated compressive residual stresses is the main improvement factor. Shifting the fatigue strength on the level of the base materiali is generally possible.
4. Shot peening enables the sophisticated consideration of the mechanical properties of the used steel grades which is not yet accepted in a satisfying manor in the design codes.
5. It must be taken into account that the procedures can only be effective in welds without inner defects, i.e. in welds where the crack initiation is located at the weld toe.

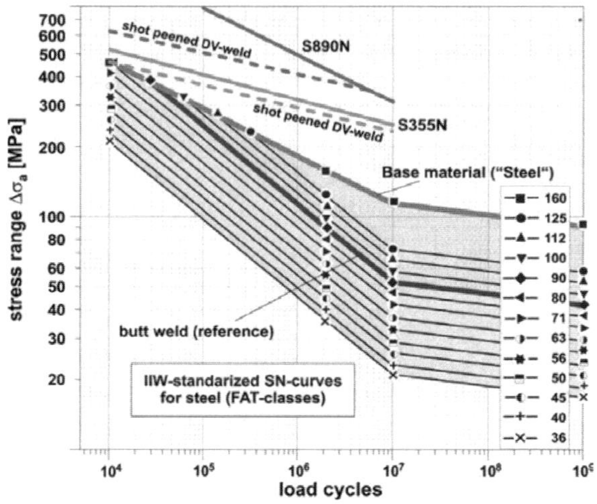

Figure 7: Comparison of the test results on shot peened DV-welds and the design curves fixed in the current design concepts according to [2].

6. References

[1] Müsgen,B.: Komm. der Europ. Gemeinschaften, Technische Forschung Stahl, EUR 7634, 198

[2] Heeschen,J.: Dissertation Universität Kassel (1986)

[3] Hobbacher,A: Recommendations for Fatigue Design of Welded Joints and Components, New York: Welding Research Council, 2009.

[4] Haagensen,P.J.; Maddox,S.: IIW Recommendations on Post Weld Improvement of Steel and Aluminium Structures, The International Institute of Welding, Paris, 2011.

[5] Marquis,G.; Barsoum,Z.: Procedia Engineering, vol. 66, pp. 98-107, 2013.

[6] Macherauch,E.; Wohlfahrt,H.: in *Ermüdungsverhalten metallischer Werkstoffe*, Oberursel, Deutsche Gesellschaft für Metallkunde, 1985, pp. 237-283.

[7] Nitschke-Pagel,Th. et.al.: *Welding in the World* , vol. 51, no. 11/12, pp. 65-75, 2007.

[8] Cheng,X. et.al.: *International Journal of Fatigue,* vol. 25, no. 9-11, pp. 1259-1267, 2003.

[9] Kirkhope,K. et.al.: *Marine Structures,* vol. 12, no. 6, pp. 447-474, 1999.

[10] Maddox,S.: *Metal Construction,* vol. 17, no. 4, pp. 220-224, 1985.

[11] Nitschke-Pagel,Th.; Eslami,H.; Dilger,K.: IIW-Doc. XIII-2483-13," International Institute of Welding, Paris, 2013.

[12] Wohlfahrt, H., Krull, P. (Ed.): Mechanische Oberflächenbehandlungen: Grundlagen – Bauteileigenschaften – Anwendungen. WILEY-VCH-Verlag Weinheim 2000.

[13] Weich,I.: Dissertation Technische Universität Brtaunschweig, 2008

[14] Nitschke-Pagel, Th.; et.al.: IIW Document XIII-2733-18, 2018

Optimization of shot peening parameters on beta titanium alloy Ti10V-2Fe-3Al in order to improve fatigue behaviour

M.R. Chini[a], C. Dides[b], T. Billot[c], N. Guillemot[d]

[a, b] IRT M2P, 4 rue Augustin Fresnel, 57070 Metz, France - maria-rita.chini@irt-m2p.fr,
[c] SAFRAN LANDING SYSTEMS, 9 Rue Guynemer, 64400 Bidos, France - thomas.billot@safrangroup.com
[d] AIRBUS HELICOPTERS, Aéroport International Marseille Provence, 13700 Marignane, France – nicolas.guillemot@airbus.com

Abstract

Shot peening is a mechanical surface treatment widely used in industry to improve the fatigue properties of components. It appears that the parameters of this process must be adjusted to new alloys and/or manufacturing techniques to reach even better fatigue properties. The purpose of this study is to determine the optimal shot peening conditions for the beta titanium alloy Ti10V-2Fe-3Al in order to optimize fatigue life, taking into account different machining conditions. This development was carried out during CONDOR project.

Keywords: Shot peening, Fatigue, beta titanium, Ti 10V-2Fe-3Al, fractography, surface roughness

Introduction

The near-β titanium alloy Ti 10V-2Fe-3Al is highly used in airframe forging applications such as landing gears and rotor systems. It's been widely demonstrated that the introduction of residual compression stresses near the surface through shot peening increases the components fatigue lifetime [1-3]. This process is widely used in industry. However today the parameters are not always adapted to the specific alloy. In particular, titanium alloys, which are generally used in aeronautical applications, are often shot peened like steel alloys. The aim of this study is to determine the optimal shot peening conditions for the titanium alloy Ti10V-2Fe-3Al in order to enhance fatigue properties.

Experimental Methods

Specimens of near-β Ti10V-2Fe-3Al presenting bimodal microstructure were machined from a 135mm diameter billet. A large fatigue test campaign was performed in order to determine the best parameters combination regarding fatigue lifetime. A Design of Experiment (DOE) was set up to take into account different initial surface conditions generated by 3 different sets of milling parameters (see table 1) as well as different shot peening parameters (size and hardness of shots, intensity and coverage - see table 2). The optimization of the DOE was performed by the statistical program minitab. This reduced the number of fatigue tests from 1100 to 245.

Table 1. Surface roughness after machining for the three conditions of study

Milling conditions	Roughness after machining (Ra)
A	0,8 µm
B	1,6 µm
C	3,2 µm

The 4-point bending fatigue tests were carried out at a frequency of 10 Hz at room temperature according to standard EN 6072 [4]. Three levels of fatigue were evaluated (R = 0.1): a level close to the endurance limit (HCF), a stress level above the elastic limit of the material (LCF regime) and an intermediate level. The condition for stopping the tests is $2x10^6$ cycles. The analysis of the results was carried out using the minitab statistical software. Afterwards, fracture surfaces of failed samples were analyzed to determine the failure mechanisms.

Table 2. Surface roughness after machining for the three conditions of study

Shot peening parameters	Values
Shot's size	AS 130 and AS 230
Shot's hardness	Regular (R) and High (H) hardness
Almen intensity	Low and High AI
Coverage	Low and High Cov

In addition, analyses were carried out to characterize roughness and residual stress profiles generated by shot peening. The 3D roughness parameters S_a, S_v, S_p and S_z were measured without cut off in a GT contour microscope from Bruker. Residual stress profiles were determined by ray diffraction using Cu-Kα radiation.

Experimental Results and Discussion

The test campaign results are shown in figure 1. To evaluate the impact of the different parameters on fatigue life, it was chosen to evaluate the residuals of each point compared to the minimum Wöhler curve. This curve is determined from the average Wöhler curve of all the tests of the experimental campaign, from which 3 times its standard deviation is withdrawn.

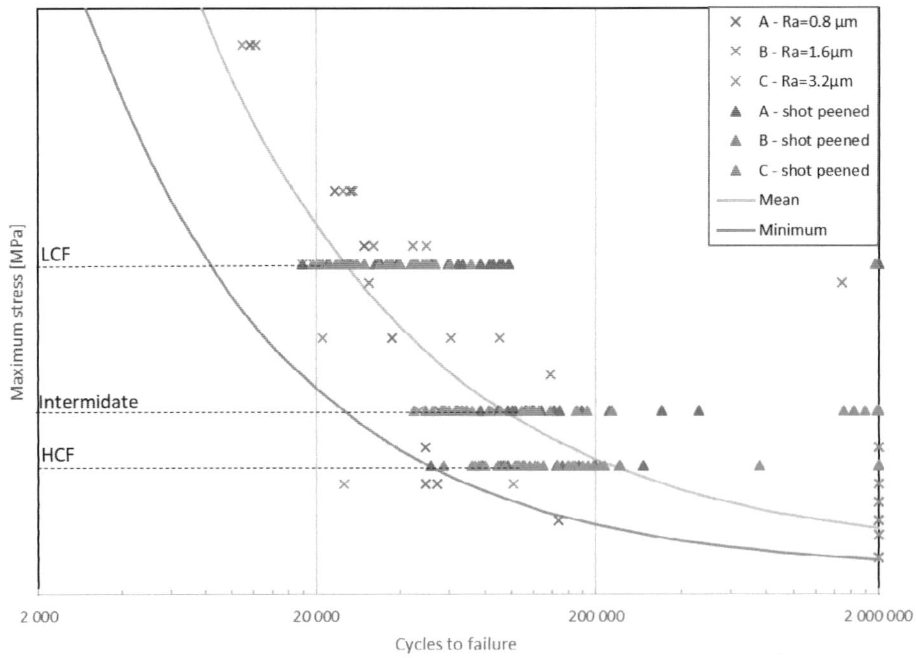

Figure 1. Fatigue results of all tested coupons (reference and shot peened following different conditions). This graph also displays the mean and minimum fit in S-N curves for fatigue data.

The Pareto diagram in figure 2 shows the effect of the different parameters on the fatigue life. All parameters with a normalized effect greater than 1,968 are considered to have a significant impact on the fatigue behaviour. This corresponds to the P value of the experiment plan using an alpha coefficient of 5%. The only impacting parameters without being combined are the Almen intensity (D), the diameter of the shot (C), coverage and roughness before shot peening (A). The Almen intensity alone and coupled with the coverage are the two elements having the most effect on the fatigue life. The shot hardness alone has no significant effect.

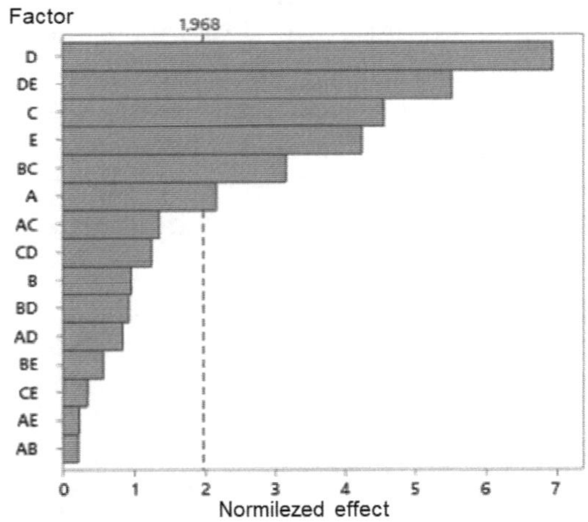

Factor	Parameter
A	Surface roughness before SP
B	Media hardness
C	Media diameter
D	Almen Intensity
E	Coverage

Figure 2. Pareto diagram of DOE on shot peening

Fractography analysis identified two types of cracking initiation sites: surface initiation sites (94% of analyzed sites) and sub-surface initiation sites (6% of analyzed sites). Test coupons that have the highest number of cycles before failure exhibited sub-surface sites that are between 50 µm and 100 µm below the surface. They also exhibited limited surface roughness as a consequence of low Almen intensity during SP. On the other hand, the shortest fatigue lives are consequence of a crack initiation from a surface defect created by shots impact. An example is shown in Figure 3. In these cases, the surface defects were due to the high intensity used during shot peening. This type of defect acts as a stress concentrator, reducing the fatigue life.

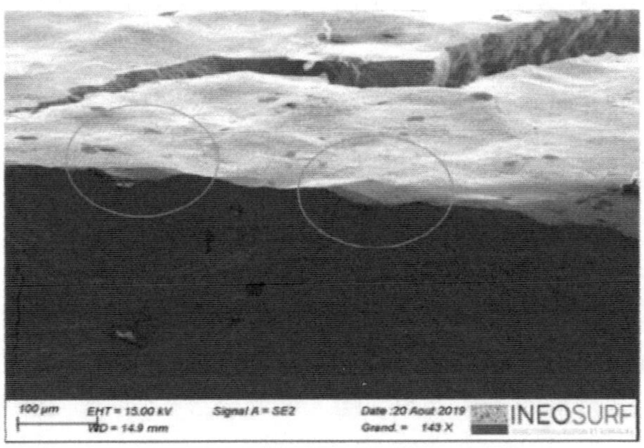

Figure 3. Crack initiation sites as a consequence of a surface defect introduced by shot peening

The impact of the different machining and shot peening conditions on the surface topography is shown in figure 4. In general, roughness measurements confirm that Almen intensity has the predominant role on surface finish. For the same initial topography (A, B or C), using a high intensity leads to higher values of S_a, S_p, S_v and S_z compared to a low almen intensity. As expected, the media hardness also has an influence on the final surface roughness. Regular hardness shots are generally associated with lower roughness values. The influence of the size of the shots is more relevant when using a hard media. In general, higher roughness values are observed when using smaller shots for the same intensity and hardness.

All residual stress profiles measured in shot peened specimens are shown in Figure 5. This graph highlights that the different shot peening and machining conditions induce very different residual stress profiles. As expected, the Almen intensity is the parameter that has the most influence on the affected depth and the maximum residual compressive stress (MRCS). The hardness and size of shots also affect the value of the MRCS and affected depth, but their influence is more moderate and not always obvious. The association of a large size and high hardness of shots induces the highest residual stresses and affected depths at isoalmen. However, it should be noted that there is a great dispersion between profiles corresponding to the same shot peening conditions. This is due to variations of the initial mechanical state caused by the different machining conditions. This implies that a good knowledge of milling conditions is important when choosing shot peening parameters for the same material.

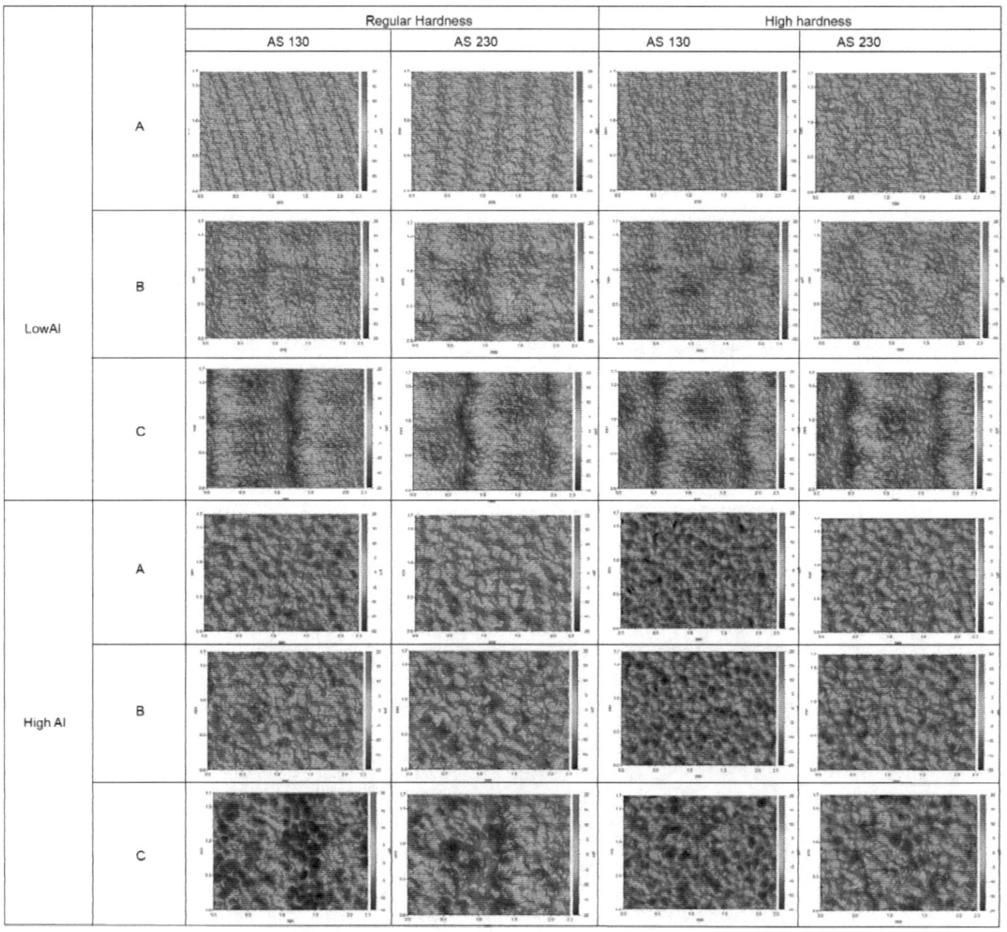

Figure 4. Topography of surfaces issued from different machining and shot peening conditions

The improvement in fatigue life due to the presence of residual compressive stresses introduced by shot peening has been widely demonstrated in the literature [5,8]. However, the graph (a) in Figure 6 shows that there is almost no correlation between the MRCStress and the number of cycles to failure (N). Table 3 shows that is also the case for the affected depth. This is explained by the fact that the peening conditions which introduce the most intense residual stress profiles also produce the most degraded surface finish. On the other hand, the conditions which produce the smoothest surface finish, do not always introduce enough residual stress to delay crack initiation. However, the correlation coefficient between Sa and N is also low (table 3). Coupons with very different surface roughness (S_a or R_a) have similar fatigue lives for a given level of stress. Consequently, the value of S_a nor MRCStress alone does not make it possible to predict the number of cycles to failure.

To evaluate the combined contribution of residual stresses and surface roughness on fatigue life, the correlation coefficient between different combinations of parameters and N was compared. An exponential equation was fitted to the data each time, each point represents the mean life and mean parameter for each set of conditions (machining and shot peening). Table 3 shows the evaluated parameters and the correlation coefficient for all the HCF tests.

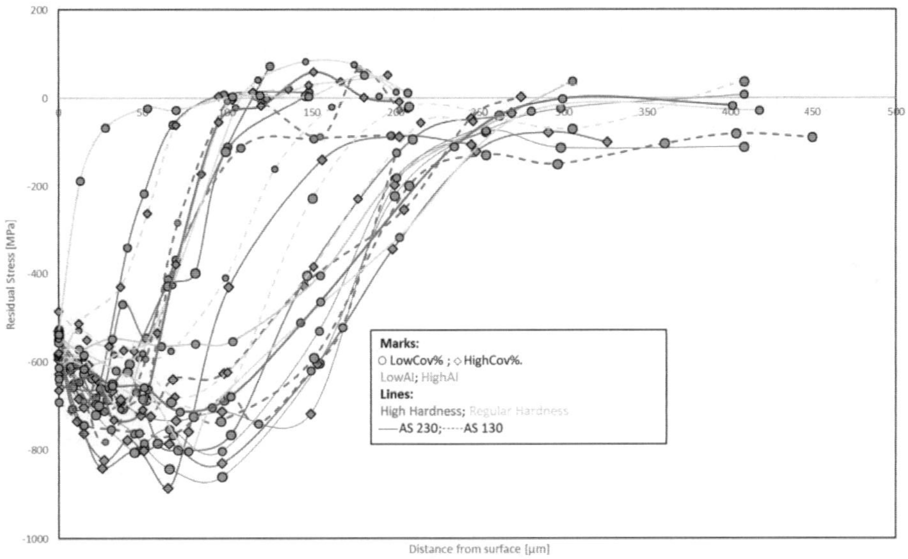

Figure 5. Residual stress profiles induced by different shot peening conditions.

Here, a new parameter (R_{local}) is proposed to characterize the maximum height of defects introduced by shot peening in comparison to the mean roughness, given by: $R_{local} = 0.5{*}S_z{-}S_a$, where $0.5{*}S_z$ is equal to the mean of the absolute values of S_p and S_v. This new parameter presents the highest correlation coefficient of all surface parameters.

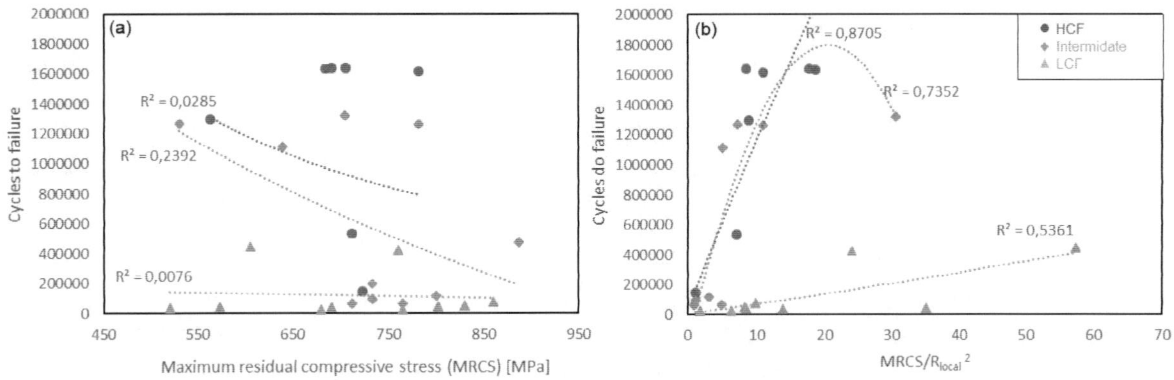

Figure 6. Service life as a function of (a) the maximum residual stress and (b) the ratio between the maximum residual stress and the square of the local roughness (difference between the mean of Sp and Sv (0.5Sz) and Sa - 0 , 5 Sz-Sa). Values correspond to means for each set of parameters

The ratio $\frac{\mathrm{MRCS}}{(\frac{1}{2}S_z - S_a)^2}$ presents the best correlation coefficient (see table 3 and Figure 6b) of all evaluated parameters. This brings out the existing competition between the surface finish and the residual stresses, being the local roughness the most influential parameter. Different studies [9,12] associate a stress concentration factor K_t with the roughness parameter R_z. Therefore, certain shot peening conditions neutralize the beneficial effects of the residual compressive stresses by creating surface defects which increase the stress concentration factor. This demonstrates the importance of reducing dispersion, as generating a surface as uniform as possible is more effective to enhance fatigue life, regardless the average roughness (R_a or S_a).

It should be noted that the correlation coefficient increases with the reduction of fatigue stresses, which indicates that the surface condition and residual stresses are more important when coupons approach the endurance limit (HCF), which agrees with the literature [7,8].

Table 3. Correlation coefficient (R^2) between different parameters and N for HCF solicitations.

Evaluated parameter	R^2
N=f(Sa)	0,318
N=f($S_{a\ normalisé}$= (S_a/S_a before SP))	0,0198
N=f($S_{a\ normalisé}$= ((S_a-S_a before SP)/S_a before SP))	0,0512
N=f(S_z)	0,7979
N=f(S_v)	0,7421
N=f(S_v-S_a)	0,8292
N=f(R_{local}= 0,5*S_z-S_a)	0,8463
N=f(MRCS)	0,0285
N=f(affected depth)	0,2618
N=f(MRCS/S_a)	0,3382
N=f(MRCS/S_z)	0,7683
N=f(MRCS/S_a*S_z)	0,5686
N=f(MRCS/R_{local})	0,8473
N=f(MRCS/R_{local}^2)	0,8705
N=f(affected depth/R_{local})	0,103

Conclusions

The test campaign demonstrated a significant increase in fatigue life for the sets of peening conditions which preserve a regular roughness through samples surface, while introducing residual compressive stresses at depth. This was the case when low Almen intensity was applied with AS H 230 shots. Analyses showed the great influence of local defects, which can be assessed through the roughness parameter S_z. Fractography analysis showed that the shot peening parameters optimization lead in a shift of the crack initiation sites from the surface to the core of the samples, and subsequently increasing the number of cycles to failure. These results show that the competition between the final surface finish and the residual stresses introduced by shot peening is a key element to enhance fatigue life.

References

[1] Fuchs, H. O., *Shot Peening, Mechanical Engineers' Handbook*, Cincinnati: Wiley (1986), pp 941-51.

[2] Hetram, L. S., Om, H., Hetram, L. S., & Om, H., *Shot Peening Effects on Material Properties: A Review.* Int. J. Innov. Res. Sci. Technol, 1, (2015), pp 480-484.

[3] Almen, J. O., *Shot blasting to increase fatigue resistance.* SAE Transactions, (1943), pp 248-268.

[4] EN 6072, *Aerospace series: Metallic materials – Test methods, Constant amplitude fatigue testing*, AECMA standard (2010)

[5] Neuber, H., *Theory of notch stresses: Principles for exact stress calculation* (Vol. 74). JW Edwards(1946).

[6] Roushdy, E. H., & Kandeil, A. Y., *Influence of surface finish on fatigue life of steel specimens subjected to pure bending* (1990).

[7] ASM International Handbook Committee. *ASM Handbook, Volume 19-Fatigue and Fracture.* ASM International(1996).

[8] Withers, P. J., *Residual stress and its role in failure. Reports on progress in physics*, (2007). 70(12), 2211

[9] Koster, W. P., & Field, M., *Effects of machining variables on the surface and structural metals.* PROCEEDINGS OF THE NORTH AMERICAN MANUFACTURING RESEARCH CONFERENCE, SME. (2001, January)

[10] Siebel, E., *Influence of surface roughness on the fatigue strength of steels and non-ferrous alloys.* Engineers Digest, 18, (1957) pp 109-112.

[11] Javidi, A., *Influence of machining on the surface integrity and fatigue strength of 34CrNiMo6 steel.* (2008).

[12] Neuber, H., *Theory of notch stresses: Principles for exact stress calculation* (Vol. 74). JW Edwards. (1946).

[10] D. Arola, C.L. Williams, *Estimating the fatigue stress concentration factor of machined surfaces,* International Journal of fatigue 24 (2002) pp 923-930.

Fatigue performance at 550°C of shot peened and oxidized Inconel 718 additively manufactured by Laser-Powder Bed Fusion (L-PBF)

Charles Bianchetti[a], Pascale Kanouté[a], Louise Toualbi[a]

[a] **DMAS, ONERA, Université Paris Saclay, F92322, Chatillon, France –** charles.bianchetti@onera.fr

Abstract

As-machined, shot peened, oxidized, and shot peened-oxidized samples were fatigue tested at elevated temperature. Shot peening increases fatigue life by 6. Fractographic analyses reveal that cracks mostly initiate on lack of fusion. For the as-machined and oxidized samples, cracks are located at the surface or subsurface within a 120 µm depth range. While for the shot peened samples, cracks are located subsurface at about 850 µm depth. Finally, results show that oxidation does not affect fatigue life even after shot peening.

Keywords

Shot peening, Additive manufacturing, fatigue, high temperature, IN718

Introduction

Metal additive manufacturing is a disruptive technology that is likely to replace some traditional manufacturing processes. However, this process creates undesirable high surface roughness [1], porosity [2], and tensile residual stress [3]. These side effects imply that as-built (AB) test pieces have drastic inferior fatigue life in comparison with that of the conventional counterparts [1,4]. Consequently, additive manufactured parts must undergo surface finish to increase their fatigue life.

Shot peening is a process used to increase fatigue life. Bagherifard et al. [5] showed for the AlSi10Mg alloy that the roughness of the AB parts decreased after shot peening, which consequently increased the fatigue strength by 270% in comparison with the AB condition. However, the effect of shot peening on fatigue life of IN718 additively manufactured is yet to be studied at elevated temperature.

Objectives

The objective of the present study is to evaluate fatigue properties enhancement by shot peening of the IN718 manufactured by Laser-Powder Bed Fusion (L-PBF). In addition, the impact of pre-oxidation on fatigue life before and after shot peening is also studied. This paper presents fatigue results performed on as-machined condition.

Experimental Methods

The studied material was a nickel-based superalloy Inconel 718 elaborated by Laser-Powder Bed Fusion (L-PBF). The L-PBF process conditions cannot be disclosed.

Axial fatigue tests were performed in the built direction on a MTS 318.25 machine under a stress ratio of 0.1 and a maximum stress of 650 MPa, at a temperature of 550°C and a frequency of 10 Hz.

Herein, results for four surface conditions are presented, namely, 1/ as-machined (AM), 2/ as-machined and oxidized (AMO), 3/ as-machined and shot peened (SP), 4/ as-machined, shot peened, and oxidized (SPO). Three samples were fatigue tested per surface condition. All fractured samples were observed under a scanning electron microscope to identify the crack initiation causes.

Samples were shot peened using CW400 media at an intensity Almen of F12 (in millimeter) with a coverage of 150%. Samples were oxidized at 550°C during 1000h in an air environment. Surface roughness profiles of as-machined and shot peened samples were measured to evaluate the effect of shot peening on roughness. Four profiles were measured per surface condition. Finally, microhardness profiles were measured in the plane perpendicular to the built direction at 100 gf.

Experimental Results

Table 1 provides the fatigue results for the studied surface conditions. Shot peening increased fatigue life by 6 when compared to the as-machined condition. Oxidation unaffected the average fatigue life both before and after shot peening. However, oxidation increased fatigue life dispersion both before and after shot peening.

Fractographic failure analyses revealed that most of the cracks initiated at a lack of fusion. Figure 1 shows an example of an observation of a lack of fusion for an as-machined sample. In addition, fractographic analyses of the AM and AMO conditions showed that cracks initiated either at the surface or below the surface at an average depth of 100 µm. Interestingly, surface cracks initiated on lack of fusion located at the surface. These results suggested that surface roughness had a minor impact on crack initiation. Finally, similar analyses performed for the SP and SPO conditions revealed subsurface crack initiations at an average depth of 850 µm.

The roughness parameters of the AM and SP surface conditions are presented in Table 2. Shot peening increased R_a and R_z of the as-machined condition by 10 and 8, respectively.

Finally, microhardness profiles were measured for all the studied conditions. Table 3 provided the microhardness values at the surface and at the end of the zone affected by shot peening. Shot peening hardened the surface within a 200 µm depth. However, oxidation after shot peening did not affect the microhardness peak value. This was surprising, since thermal residual stress relaxation was expected to decrease this value. Finally, the microhardness value of the base material (525 Hv) increased after oxidation to 560 Hv for both the AM and SP conditions. This could result from an additional precipitation of the γ' and γ'' phases.

Table 1. Fatigue results at a temperature of 550°C for a 0.1 stress ratio at a maximum stress of 650 MPa.

Surface condition	Fatigue life: Mean [cycles]	Fatigue life: Coefficient of variation (standard deviation / mean) [%]
As-machined	44 020	3
As-machined + oxidized	49 660	36
As-machined + shot peened	295 161	31
As-machined + shot peened + oxidized	279 862	49

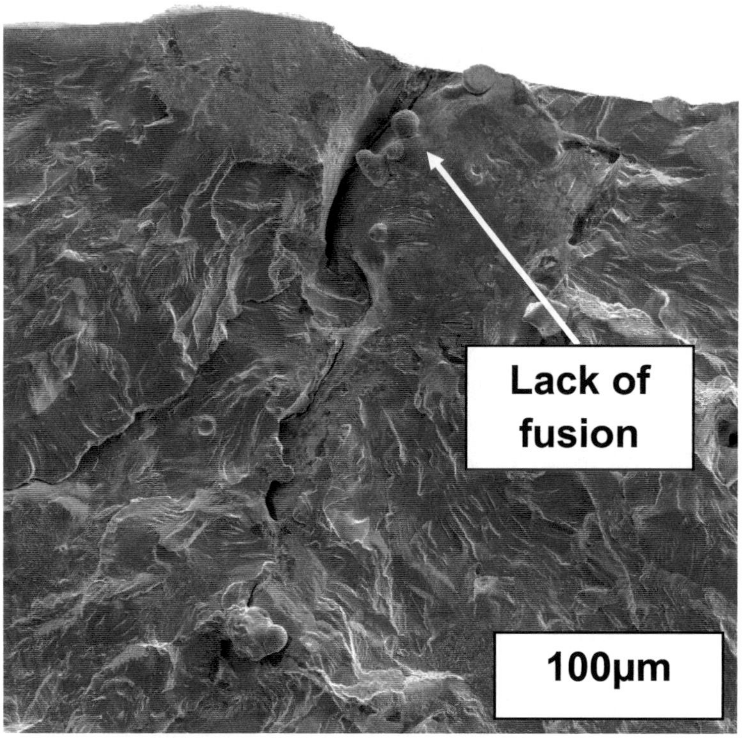

Figure 1. Typical fractographic observation obtained under a scanning electron microscope showing a crack initiation on a lack of fusion.

Table 2. Arithmetic average (R_a) and average maximum peak to valley height (R_z) of the measured roughness profiles.

Surface condition	R_a		R_z	
	Mean [cycles]	Coefficient of variation (standard deviation / mean) [%]	Mean [cycles]	Coefficient of variation (standard deviation / mean) [%]
As-machined	0.1	16	0.8	16
As-machined + shot peened	1.1	5	6.5	12

Table 3. Microhardness measurements

Surface condition	Microhardness, Hv			
	At the surface		At 0.2 mm depth	
	Mean	standard deviation	Mean	standard deviation
As-machined	540	13	525	7
As-machined + oxidized	560	4.5	560	7
As-machined + shot peened	603	7	520	8
As-machined + shot peened + oxidized	601	10	560	10

Discussion and Conclusions

Shot peening increased fatigue life of the as-machined condition by 6. All the cracks that initiated in the AM and AMO samples were contained within the shot peening affected zone, characterized using microhardness measurements. In addition, fractographic analyses of the SP and SPO samples revealed subsurface cracks at a depth of about 850 µm. These results showed that shot peening prevented crack initiation around the surface, which in turn increased fatigue life.

Oxidation increased microhardness values most likely caused by a secondary precipitation of the γ' and γ'' phases.

Oxidation unaffected fatigue life of the AM condition. Oxidizing the material at 550°C during 1000h led to a nanometric oxide layer. Fractographic analyses revealed that cracks initiated either at the surface or below the surface. Moreover, cracks at the surface were initiated by lack of fusion. As a consequence, surface roughness and geometric discontinuity induced by oxidation had no effect on fatigue life.

Oxidation unaffected fatigue life of the shot peened samples. While oxidation barely affected the surface, fatigue life was expected to decrease because of thermal residual stress relaxation that should occur during oxidation. However, the microhardness measurements showed that oxidation unaffected microhardness in the shot peened affected zone. Assuming a correlation between residual stress and microhardness, residual stress did not relax after exposing the material at 550°C during 1000h. This is in opposition with the observations made by Cammett *et al* [6] on a conventional shot peened IN718. In fact, the authors observed that residual stresses decreased by 50% after exposing the material at 525°C during only 10h. Additional work is required to confirm this observation using direct residual stress measurements.

Acknowledgement

The authors acknowledge the Additive Manufacturing Hub (AFH) for partially supporting this study.

References

[1] A.R. Balachandramurthi, J. Moverare, N. Dixit, R. Pederson, Mater. Sci. Eng. A 735 (2018).

[2] E. Hosseini, V.A. Popovich, Addit. Manuf. 30 (2019).

[3] K.S. Sidhu, J. Shi, V.K. Vasudevan, S.R. Mannava, in: ASME 2017 12th Int. Manuf. Sci. Eng. Conf. MSEC 2017 Collocated with JSME/ASME 2017 6th Int. Conf. Mater. Process., 2017.

[4] T.M. Mower, M.J. Long, Mater. Sci. Eng. A 651 (2016).

[5] S. Bagherifard, N. Beretta, S. Monti, M. Riccio, M. Bandini, M. Guagliano, Mater. Des. 145 (2018) 28–41.

[6] J.T. Cammett, P.S. Prevey, N. Jayaraman, 9th Int. Conf. Shot Peen. 9 (2005).

Development of a peening device with a handheld laser on a collaborative robot

Y. Sano[1,2,3], Y. Mizuta[1], S. Tamaki[3], K. Yokofujita[4], K. Masaki[5], T. Hosokai[1] and T. Taira[2]

1 SANKEN, Osaka University, Ibaraki, Japan
2 Institute for Molecular Science, National Institutes of National Sciences, Okazaki, Japan
3 LAcubed Co. Ltd., Yokohama, Japan
4 Unitac Co. Ltd., Onomichi, Japan
5 National Institute of Technology, Okinawa College

Abstract
A compact laser peening (LP) device has been developed for application to large structures such as bridges in service. The device consists of a thumb-sized handheld laser mounted on a collaborative robot, a power supply, a water circulation system, and a PC as a control unit. Various materials such as aluminum alloys and high-strength steels were laser-peened by using this device. Results of residual stress (RS) evaluation and fatigue experiments showed that LP with this device significantly improved the RSs in the near-surface layer and fatigue properties of materials.

Keywords Microchip laser, robotic arm, laser peening, residual stress, fatigue.

Introduction
LP introduces compressive RS on the surface of a water-covered component by exposing it to successive laser pulses [1]. LP is effective in preventing fatigue cracking of components under repetitive loading [2-8]. Existing LP devices [9] utilize high-power lasers assembled on large vibration-proof optical benches. Due to their volume and weight, the application of the existing LP devices has been limited to indoor research or production [10].

There is microchip laser technology [11-14] that can drastically reduce the size and weight of laser oscillators. If a microchip laser could be used as the energy source for LP, the scale of LP devices can be reduced, making it feasible to apply LP not only to production but also to on-site maintenance of existing infrastructure such as bridges in service. Based on this idea, we have developed a compact movable LP device with a handheld-type microchip laser.

Laser Peening Device with a Handheld Laser
The configuration and appearance of the prototype movable LP device are shown in Figs. 1 and 2 [15]. The device consists of two parts: a small 6-axis robotic arm equipped with a thumb-sized handheld-type laser and a power supply including a water circulation system. The power supply incorporates a built-in laser diode, which pumps the handheld laser via an optical fiber cable. The water circulation system recovers and reuses the water for LP. The device operates at 100-220 VAC and have a maximum power consumption of 400 W. The robotic arm weighs 4 kg and the device can be carried as two pieces of airline baggage. The use of the collaborative robotic arm eliminates the need for fences or other safety measures.

Effects on Surface Residual Stress of A7075
Samples of A7075-T73 aluminum alloy plates were laser-peened by the device [16]. Surface RS was analyzed by X-ray diffraction (XRD) and the RS depth distribution was obtained by alternating XRD and electrolytic polishing. The same operation was also performed on the unpeened material to evaluate the effect of LP.

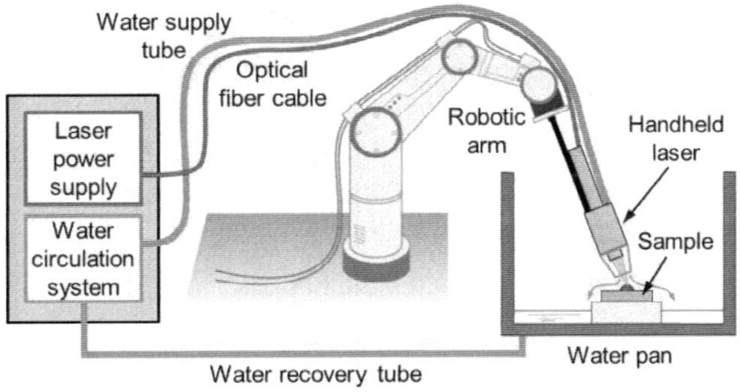

Fig. 1 Configuration of movable laser peening device

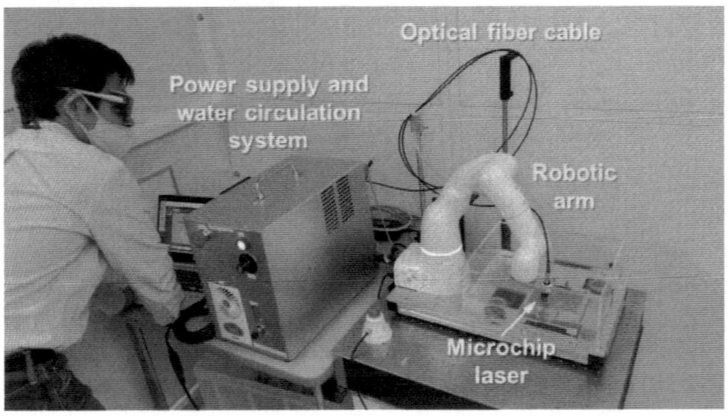

Fig. 2 Overall appearance of movable laser peening device

Figure 3 shows the results of RS evaluation for 3.2 mm (1/8 inch) thick A7075-T73 aluminum alloy plates. Here, σ_x and σ_y are the RS components parallel and perpendicular to the laser sweep direction, respectively. The irradiated laser pulse energy was 7.5 mJ, which is orders of magnitude smaller than that in the usual process with existing LP devices. The standard deviation of the regression in XRD was about ±10 MPa. The depth of compression by LP reaches about 0.3 mm from the surface for an irradiation pulse density of 100 pulses/mm^2 and more than 0.5 mm for 1600 pulses/mm^2.

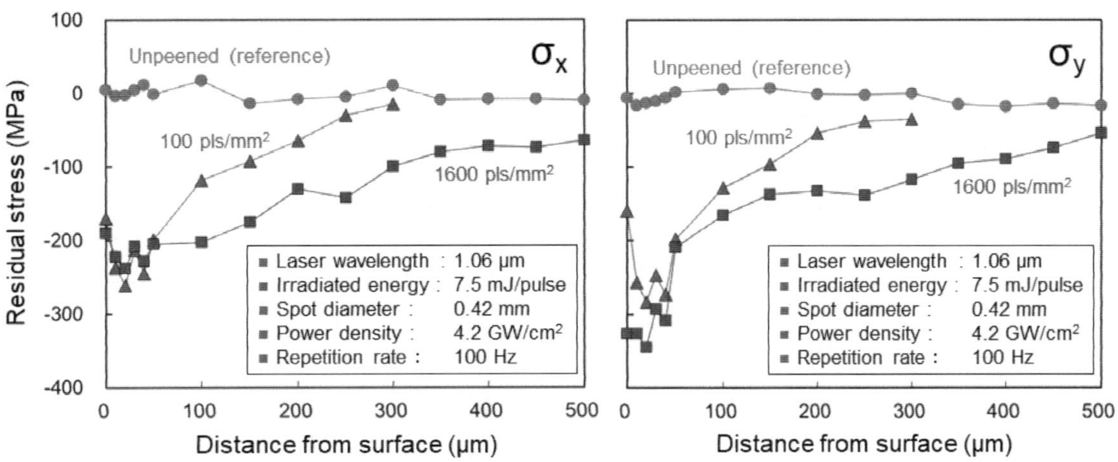

Fig. 3 Residual stress depth profiles of A7075-T73 in x-direction (left) and y-direction (right)

Effects on Surface Residual Stress of HT780

LP was applied to 9 mm thick HT780 high-strength steel plates with mill-scale on the surface. Figure 3 shows the RS profiles of HT780 obtained by the same procedure as A7075-T73. The standard deviation of the regression in the XRD measurement was less than ±10 MPa.

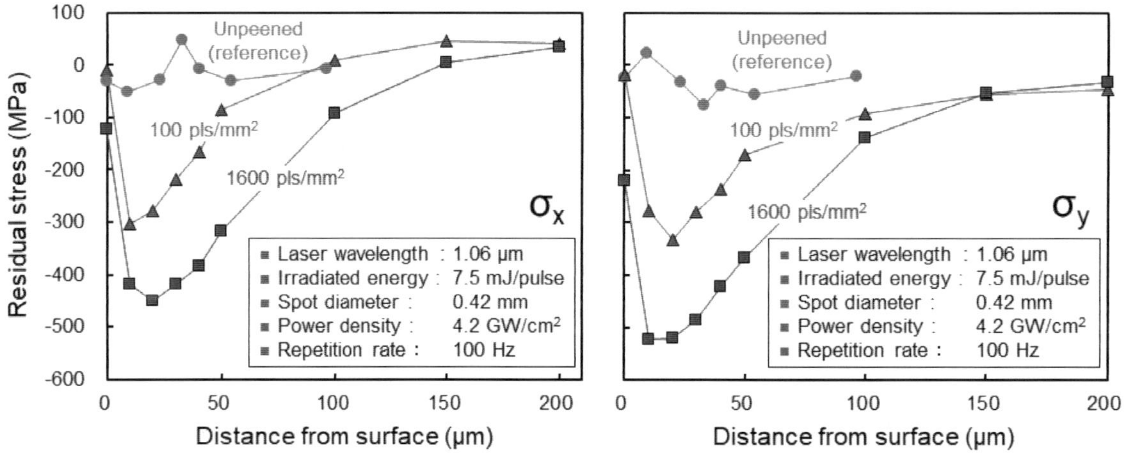

Fig. 4 Residual stress depth profiles of HT780 in x-direction (left) and y-direction (right)

On the top surface of the laser-peened material, the absolute values of the RSs are fairly smaller than those just below the top surface. This may be due to the fact that the mill-scale still remains after LP, affecting the XRD results. When the surface was slightly removed by electrolytic polishing, large compressive RSs appeared. Further experiments are needed to better understand this phenomenon.

Effect on Fatigue Properties

Fatigue experiments were performed for A7075BE-T6511 aluminum alloy samples after LP with a pulse energy of 1.7 mJ [16] and HT780 high-strength steel welded joint samples with 7.7 mJ [15]. As a result, it was confirmed that LP using a microchip laser with pulse energies less than 10 mJ significantly improved fatigue strength and life. The details are provided in the accompanying papers by Masaki and Mizuta at this conference.

Conclusions

A newly-prototyped movable LP device was applied to samples of A7075 aluminum alloy and HT780 high-strength steel with pulse energies of less than 10 mJ. The results showed that RSs in the near-surface layer were significantly improved despite the pulse energies being several orders of magnitude lower than those of the existing LP devices.

The existing LP devices are bulky and cannot be used outdoors, thus their applications have been limited to indoor production [9,10]. With the commercialization of compact and high-power microchip lasers, movable LP devices could be used for a variety of applications. In particular, when combined with a small collaborative robotic arm, LP can be easily applied in a variety of situations, such as post-weld processing at construction sites and infrastructure maintenance.

Acknowledgments

This work was partially supported by the JST-MIRAI Program (grant number JPMJMI17A1), the Supporting Industry Program of the Small and Medium Enterprise Agency (grant number 20334933), the Amada Foundation (grant number AF-2020239-C2), and the Impulsing Paradigm Change through Disruptive Technologies Program (ImPACT) of the Council for Science, Technology and Innovation.

References

[1] Y. Sano, N. Mukai, K. Okazaki and M. Obata, *Residual stress improvement in metal surface by underwater laser irradiation*, Nucl. Instrum. Methods Phys. Res. B, Vol. 121 (1997), pp 432–436.

[2] Y. Sano, M. Obata, T. Kubo, N. Mukai, M. Yoda, K. Masaki and Y. Ochi, *Retardation of crack initiation and growth in austenitic stainless steels by laser peening without protective coating*, Mater. Sci. Eng. A, Vol. 417 (2006), pp 334–340.

[3] K. Masaki, Y. Ochi, T. Matsumura and Y. Sano, *Effects of laser peening treatment on high cycle fatigue properties of degassing-processed cast aluminum alloy*, Mater. Sci. Eng. A, Vol. 468–470 (2007), pp 171–175.

[4] K. Masaki, Y. Ochi, T. Matsumura, T. Ikarashi and Y. Sano, *Effects of laser peening treatment on high cycle fatigue and crack propagation behaviors in austenitic stainless steel*, J. Power Energy Syst., Vol. 4 (2010), pp 94–104.

[5] Y. Sakino, Y. Sano and Y.-C. Kim, *Application of laser peening without coating on steel welded joints*, Int. J. Struct. Integr., Vol. 2 (2011), pp 332–344.

[6] Y. Sano, K. Masaki, T. Gushi and T. Sano, *Improvement in fatigue performance of friction stir welded A6061-T6 aluminum alloy by laser peening without coating*, Mater. Des., Vol. 36 (2012), pp 809–814.

[7] Y. Sakino and Y. Sano, *Investigations for lowering pulse energy of laser-peening for improving fatigue strength*, Q. J. Jpn. Weld. Soc., Vol. 36 (2018), pp 153–159.

[8] T. Kato, Y. Sakino and Y. Sano, *Effect of laser peening with a microchip laser on fatigue life in butt-welded high-strength steel*, Appl. Mech., Vol. 2 (2021), pp 878–890.

[9] A.H. Clauer, *Laser shock peening, the path to production*, Metals, Vol. 9 (2019), p 626.

[10] Y. Sano, *Quarter century development of laser peening without coating*, Metals, Vol. 10 (2020), p 152.

[11] T. Taira, *Domain-controlled laser ceramics toward giant micro-photonics*, Opt. Mater. Express, Vol. 1 (2011), pp 1040–1050.

[12] Y. Sato, J. Akiyama and T. Taira, *Process design of microdomains with quantum mechanics for giant pulse lasers*, Sci. Rep., Vol. 7 (2017), p 10732.

[13] L. Zheng, A. Kausas and T. Taira, *Drastic thermal effects reduction through distributed face cooling in a high power giant-pulse tiny laser*, Opt. Mater. Express, Vol. 7 (2017), pp 3214–3221.

[14] L. Zheng, A. Kausas and T. Taira, *>30 MW peak power from distributed face cooling tiny integrated laser*, Opt. Express, Vol. 27 (2019), pp 30219–30224.

[15] Y. Sano, T. Kato, Y. Mizuta, S. Tamaki, K. Yokofujita, T. Taira, T. Hosokai and Y. Sakino, *Development of a portable laser peening device and its effect on the fatigue properties of HT780 butt-welded joints*, Forces in Mech., Vol 7 (2022), p 100080.

[16] Y. Sano, K. Masaki, Y. Mizuta, S. Tamaki, T. Hosokai and T. Taira, *Effects of laser peening with a pulse energy of 1.7 mJ on the residual stress and fatigue properties of A7075 aluminum alloy*, Metals, Vol. 11 (2021) p. 1716.

Effect of Soft Particle Peening Treatment on Fatigue Strength of Copper Alloy

Koichiro NAMBU [a)] and Ryuji KUNIMATSU [a)]

a) Osaka Sangyo University, Department of Mechanical Engineering, 3-1-1, Nakagaito, Daito, Osaka, Japan
Email : knambu.mech@ge.osaka-sandai.ac.jp

Abstract

This study examined the effect of soft particle peening treatment on the fatigue strength properties of copper alloys. Walnut, apricot, peach, and polyethylene particles were used as the projectiles in the soft particle peening treatment. The hardness of these projectiles was lower than that of copper alloys, and the particles were softer. Plane bending fatigue tests showed that the number of cycles to failure increased at stresses of 140 MPa and 120 MPa regardless of the projectile material. In addition, it is considered that apricot and polyethylene particles are particularly effective in improving fatigue strength when compared with each projectile material. The results of hardness measurement showed that the hardness increased in all the projectile materials. In addition, apricot and polyethylene showed particularly large increases in hardness. Surface roughness measurements showed that surface roughness increased for all projectile materials. The increase in surface roughness was the smallest for polyethylene. The results show that polyethylene particles are effective in improving the fatigue strength of copper alloys among the particles used in this study. The results of this study suggest that the soft particle peening treatment is effective in improving the fatigue strength of copper alloys.

Keywords Shot peening, Soft Particle, Fatigue, Copper alloy

1. Introduction

In recent years, the electrification of automobiles and other transportation equipment, such as EVs and hybrids, has been promoted. From this perspective, the use of copper alloys is increasing, not only in connectors, but also through the use of joining techniques such as dissimilar metal joining. Copper alloys have excellent heat transfer and electrical properties, but they have lower strength than alloys used for structural components such as steel and aluminum alloys. In addition, copper alloys have not received much attention for use as structural members, and little research has been conducted on methods to increase the strength of copper alloys because of the difficulty of increasing strength through heat treatment and other methods. In addition, the methods for strengthening copper alloys have not been studied extensively because it is difficult to increase strength by heat treatment.

In this study, we focused on shot peening, which utilizes work hardening by particle impact, as a method for improving the fatigue strength of copper alloys.[1-3]

Shot peening improves fatigue strength by increasing hardness and imparting residual stress through work hardening, and is thought to be effective in improving fatigue strength even in copper alloys with low melting points.

On the other hand, conventional shot peening treatments generally use ceramic or steel particles as the projectile. When such hard particles are used, there is a possibility that the impacted particles may penetrate the surface of the work piece, as reported by the authors and Kameyama et al.[4,5]

This impinging effect may cause microcracks and other factors that reduce fatigue strength. In addition, the increase in surface roughness caused by the use of hard particles is a factor that reduces fatigue strength. Therefore, in this study, a "soft particle peening treatment" using softer particles than those used in the conventional shot peening treatment was performed. The objective of this study was to clarify the effect of the soft particle peening treatment on the fatigue strength of copper alloys by using naturally occurring particles with lower hardness than copper alloys as the projectile material.

2. Treatment and Experimental Methods

2.1 Specimens

Oxygen-free copper (JIS C1020) was used as the test material. The chemical composition of the material is shown in Table 1. The specimens were plane bending fatigue specimens as shown in Figure 1, and were machined by electrical discharge machining. The specimen surfaces were not polished.

Table 1 Chemical Compositions

	Cu	Pb	Fe	Sn	Zn	Al	Mn	Ni	P
C1020	99.96%over	−	−	−	−	−	−	−	−

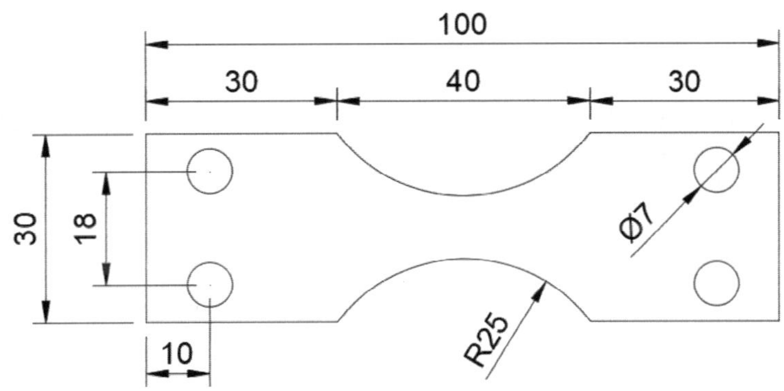

Figure 1 Bending fatigue specimen

2.2 Soft particle peening treatment

A pneumatic sandblasting machine was used for the soft particle peening treatment. Walnut, apricot, peach, and polyethylene were used as the projectiles. The particle size and hardness of each projectile are shown in Table 2. Figure 3 shows an optical micrograph of each particle. The projectile pressure in the soft particle peening treatment was 0.5 MPa, and the treatment time was 120 seconds.

2.3 Test method

The fatigue test was performed by plane bending fatigue test. The number of cycles was 10^6, and the number of specimen fractures was measured at stresses ranging from 160 to 100 MPa. Hardness tests were conducted using a Shimadzu dynamic hardness tester. DHT-115 was used as the unit of hardness measurement. Surface roughness was measured by non-contact measurement using a laser microscope.

Table 2 shot particles

	Walnut	Polyethylene	Peach	Apricot
Diameter (μm)	80	180	150	110
DHT-115	1.6	5.6	2.4	0.6

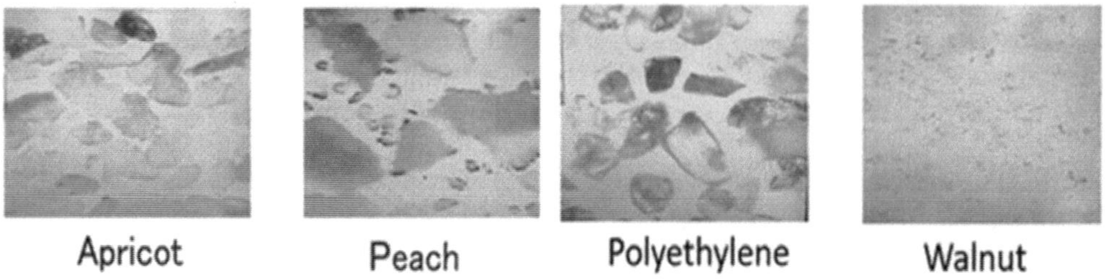

Figure 3 Optical micrographs of each particle

3. Experimental Results and Discussion

3.1 Results of Plane Bending Fatigue Tests

Figure 2 shows the fatigue test results. The fatigue limit was 110 MPa for both the untreated material and the material treated by soft particle peening. Next, let us focus on the 140 MPa limit. Compared with the fracture frequency of the untreated material, the fracture frequency of the soft-particle peened material increased regardless of the projectile material, indicating that the time strength increased. The time strength of the soft-particle peening material at 120 MPa was also higher than that at 140 MPa. These results suggest that the soft-particle peening treatment is effective in improving the time strength of copper alloys.

At these stresses, differences in the soft-particle peening material projectile are discussed. At 140 MPa, the order of the soft particle peening materials was apricot, peach, walnut, and polyethylene. At 120 MPa, the order was polyethylene, peach, apricot, and walnut. In terms of the hardness of the projectile, polyethylene has the highest hardness of 5.6 DHT115, but the hardness of the copper alloy is sufficiently small compared to 140 DHT115. In addition, since there was only one measurement point in the fatigue test, it is considered that there is some variation in the test results.

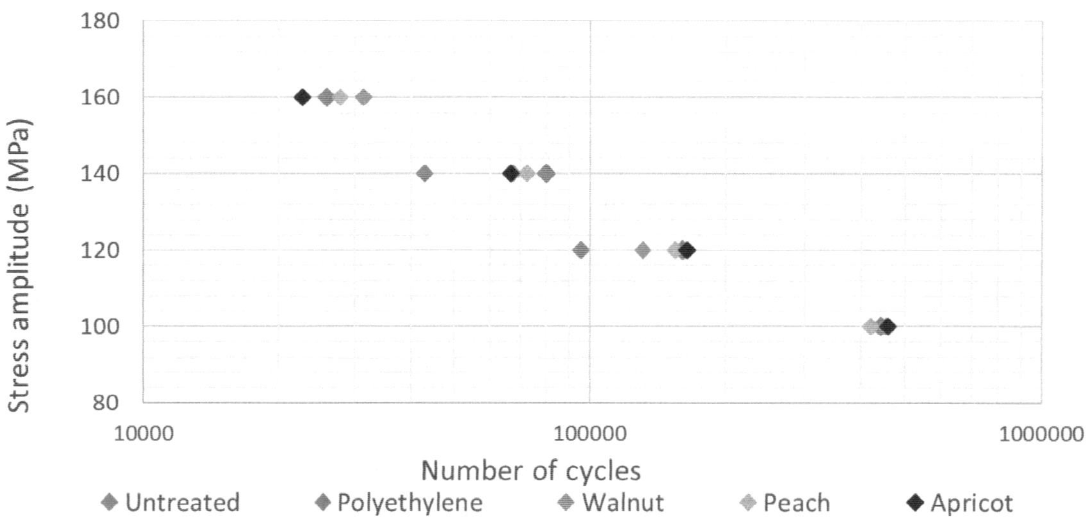

Figure 2 Result of fatigue experiment

3.2 Hardness Measurement Results

Figure 3 shows the results of hardness measurements. The hardness of the untreated material decreased from the surface to a depth of 20 μm, which is thought to be due to work softening that occurred during cutting or rolling.

No decrease in near-surface hardness is observed for any of the soft particle peened materials. This result indicates that near-surface hardness is not reduced by soft particle peening in the vicinity of the surface where the hardness is reduced by work softening.

This result suggests that the soft-particle peening treatment improved the hardness in the vicinity of the surface where the hardness decreased due to work softening.

On the other hand, as mentioned above, the hardness of the projectile materials used in the soft-particle peening treatment is lower than that of the copper alloys. It is believed that the hardness is conventionally increased by work hardening due to plastic deformation in the shot peening treatment.

The hardness of the shot peening treatment is said to be improved by work hardening due to plastic deformation. In the present study, plastic deformation is not considered to have occurred in the soft-particle peening treatment. Therefore, the mechanism of hardness improvement by soft particle peening treatment may be different from that by conventional shot peening treatment. Therefore, the mechanism of hardness improvement by soft particle peening treatment is expected to be different from that by conventional shot peening treatment. The mechanism of hardness improvement should be further investigated.

Next, a comparison of the projectile materials showed that the hardness increased the most when polyethylene particles were used, followed by polyethylene, apricot, walnut, and peach.

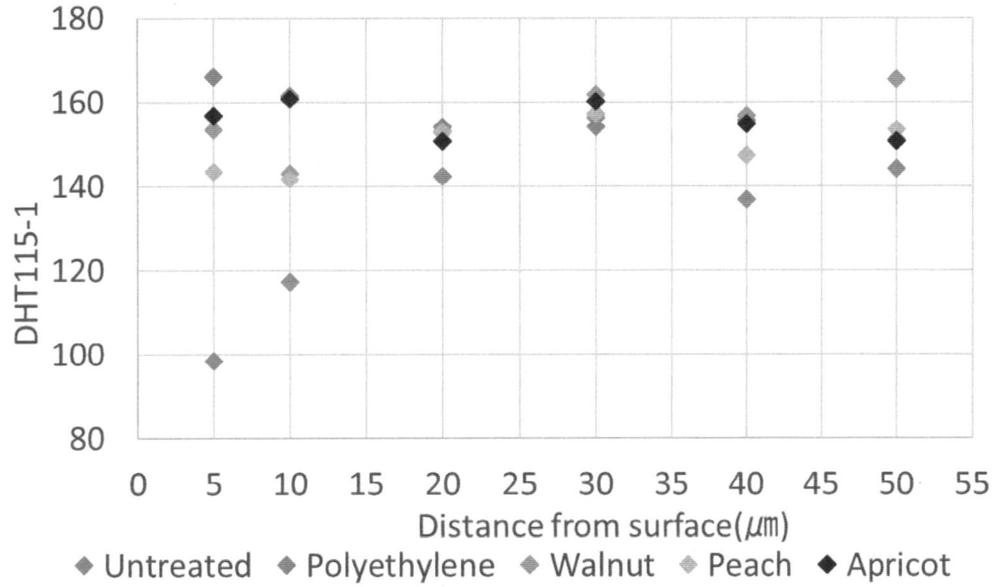

Figure 3 Result of hardness measurement

3.3 Surface Roughness Measurement Results

Figure 4 shows the results of surface roughness measurements. The surface roughness of the soft particle peening treated material increased compared to the untreated material. The surface roughness of polyethylene, which has the highest hardness of the projectile, was lower than that of the other projectile materials. On the other hand, the surface roughness of apricot, which has the lowest hardness of the projectile, also increased. It is not clear whether this is due to the properties of the copper alloy or the properties of the projectile. It is expected that the behavior of these projectile materials during particle impact differs from that of conventional metallic and ceramic particles. The collision behavior of these materials should also continue to be investigated.

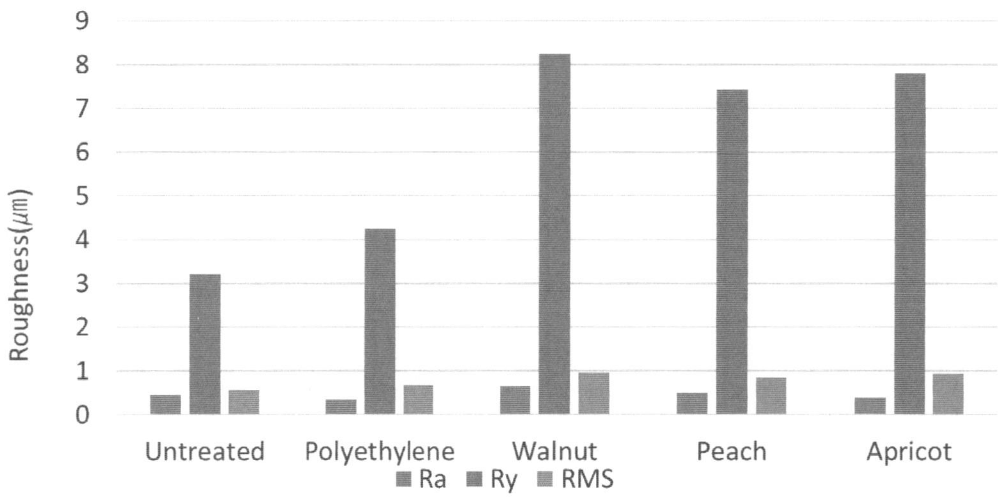

Figure 4 Result of surface roughness measurement

4. Conclusions

The effects of soft particle peening treatment on the fatigue strength properties of copper alloys were investigated and the following conclusions were obtained

1. Soft particle peening treatment is effective in improving the fatigue strength of copper alloys regardless of the type of projectile.
2. Even when softer particles than the work material are used, the hardness in the vicinity of the surface is improved.
3. Surface roughness increased regardless of the type of projectile.
4. Factors that increase hardness and surface roughness will continue to be investigated in the future.

References

[1] Kikuchi. S., et al., *Effect of shot peening using ultra-fine particles on fatigue properties of 5056 aluminum alloy under rotating bending* ,. Materials Science and Engineering: A 652, 279-286

[2] NAMBU. K., et al., *Influence of vacuum carburizing treatment on fatigue crack growth characteristic in DSG2* , Frattura ed Integrità Strutturale 9 (34)

[3] Nakamura. Y., et al., *Effect of Fine Particle Peening Using Hydroxyapatite Particles on Rotating Bending Fatigue Properties of β-Type Titanium Alloy*, Applied Sciences 11 (9), 4307

[4] NAMBU. K., et al., *Influence of soft particle peening treatment on fatigue strength of aluminum alloy a5052*, Fracture, Fatigue and Wear, 390-400

[5] Kameyama.Y., et al., *Effect of fine particle peening using carbon-black/steel hybridized particles on tribological properties of stainless steel,* Tribology International, Volume 78, October 2014, Pages 115-124

Application of Deep Learning to Predict Shot Peening Coverage

Y.H.A Chua[1], Z.B. Wang[1], H.C. Ang[1], and A. Shukri[1]

1. Advanced Remanufacturing and Technology Centre, 3 Cleantech Loop, #01/01 CleanTech Two, Singapore 637143

Abstract
Coverage is one of the key process control variables for the shot peening process. Currently, the method to quantify coverage relies heavily on human visual judgement, which is highly subjective. This study explored the use of deep learning to predict coverage using images captured from peened SS316 and Ti-6Al-4V materials. It was found that all deep learning models developed with the SuaKIT software could achieve a high prediction accuracy of at least 93% based on fixed intensity and material testing. However, the models performed poorly when tested using crossed datasets with different materials and intensities. Combining the data from different datasets can improve the prediction accuracy to at least 93.5%.

Keywords: Shot Peening, Coverage, Deep Learning

Introduction

Shot peening is a cold work process that is widely used in many industries such as aerospace and automotive industries. It is a cost-effective process that improves the fatigue life and stress corrosion cracking of a treated metallic component. Shot peening is conducted by bombarding spherical shots onto a component, causing the surface to be plastically deformed. This results in a layer of residual compressive stress on the surface of the component, which acts as a resistance to prevent premature failure or cracks propagation during cyclic loadings. [1]

Coverage is one of the key process control parameters for the shot peening process, and it is described as the percentage of a peened area over the total surface area of the component. It is important to monitor coverage as premature failure such as crack propagation will potentially occur on areas which have not been sufficiently peened during operation. [1] Coverage is typically obtained through operators' visual assessment with the aid of either handheld magnifier, or a microscope of at least 10 to 30 times magnification. However, such method of quantifying coverage is highly subjective and relies heavily on skilled and experienced operators. There are other recommended indirect methods in the specifications [2] such as the use of fluorescent tracers, dye markers or the use of replicas to help aid the coverage analysis process. However, such indirect methods not only require experienced and skilled operators to use them in order to achieve the best results, but also, they only aid the operator to visually differentiate the shot peened indentations on a peened surface better. As such, this has opened up opportunities for development to find better methods to quantify coverage, with the hope of automating it in the future and bringing shot peening into the right direction towards autonomy in Industry 4.0.

There have been several studies in the literature that has looked into understanding and improving the current method of assessing coverage. Some literature looked more into understanding the concept of coverage through the use of simulation or theoretical models, while others tried to tackle the problem directly by using image segmentation or surface roughness (both 3D and 2D), with the aid of visual or scanning systems, to automate coverage assessments. There are also 2 studies that explored the use of machine learning or deep learning to aid in the image segmentation process to improve the overall performance of the model. Qiu Jiyuan et al [3] was able to achieve 98.45% accuracy during the UNet++ model training, and 100% accuracy in classifying the coverage of 18 random images using the model. Lubna Shahid et al [4] explored the use of artificial neural network to aid in the image segmentation process to help better identify coverage on specimens with presence of interference such as machining streaks which resulted in better accuracy of the image segmentation model. However, there are no studies conducted so far that looked into the performance of deep learning models, which are trained with images directly captured from peened specimens to predict coverage.

Therefore, this study will explore the performance of deep learning models trained with images of different coverage percentage ranges, taken from two different material types (SS316 and Ti-6Al-4V). The main aim of this study is not only to understand how well deep learning performs for coverage quantification, but also to study the challenges and limitations of applying deep learning in the use case of industry application.

Experimental Methodology

Flat rectangular SS316 and Ti-6Al-4V specimens of surface roughness of 0.5μm in Ra were prepared to ensure good visibility of peening indentations upon peening. Each specimen was then split into 3 different sections on the top and bottom of the specimen, to allow shot peening of different coverage percentages on each section.

Shot peening was conducted using a compressed air system, and cast steel shots of 0.58mm in diameter with hardness range of 45-52 HRC was used for the experiments. Each material type (SS316 and Ti-6Al-4V) was peened at 3 different intensities, with 10 different coverage ranges per intensity target as shown in Table 1. It should be noted that despite difficulties in visually differentiating coverage above 98% (or 100%), coverage range of 100-200% was also tested to study if the deep learning model could pick up the differences for such coverage range. Nozzle impingement angle of 80 degrees was used for intensities of 0.2mmA and 0.3mmA, while 30 degrees was used for intensity of 0.1mmA. The reason for the lower nozzle impingement angle for intensity of 0.1mmA was due to the limitations to achieve a low intensity value at 80 degrees impingement angle.

Table 1 Peening Parameters for Each Material Type

Coverage Ranges	10-29%, 30-39%, 40-49%, 50-59%, 60-69%, 70-79%, 80-89%, 90-97%, 98%, 100-200%
Intensity	**Peening Parameters (Impingement Angle, Standoff)**
0.100mmA	30 degrees, 150mm
0.200mmA	80 degrees, 150mm
0.300mmA	80 degrees, 150mm

In order to train the deep learning model with a larger variety of data, each coverage range consists of specimens peened with at least 5 different coverage percentages. For example, shot peening was conducted at 30%, 32%, 34%, 36% and 38% for coverage range of 30-39%. Such small intervals of coverage percentages were obtained through the development of a coverage curve as shown in Figure 1. Each point of the coverage curve was evaluated by 2 experienced operators during the development, and the average coverage percentages were used to plot the final curve. The exposure times to peen each coverage percentage were then taken through the formula of the coverage curve obtained. After peening at the calculated exposure times, the specimens were re-evaluated again, and regrouped accordingly into the respective coverage ranges (if necessary) to ensure all data taught during the deep learning model were as accurate as possible. It should be noted that exposure times for coverage percentages above 100% was taken as a multiple of full coverage (98%) instead of through calculation from the coverage curve. In this study, the development of coverage curves was conducted for each intensity and material type.

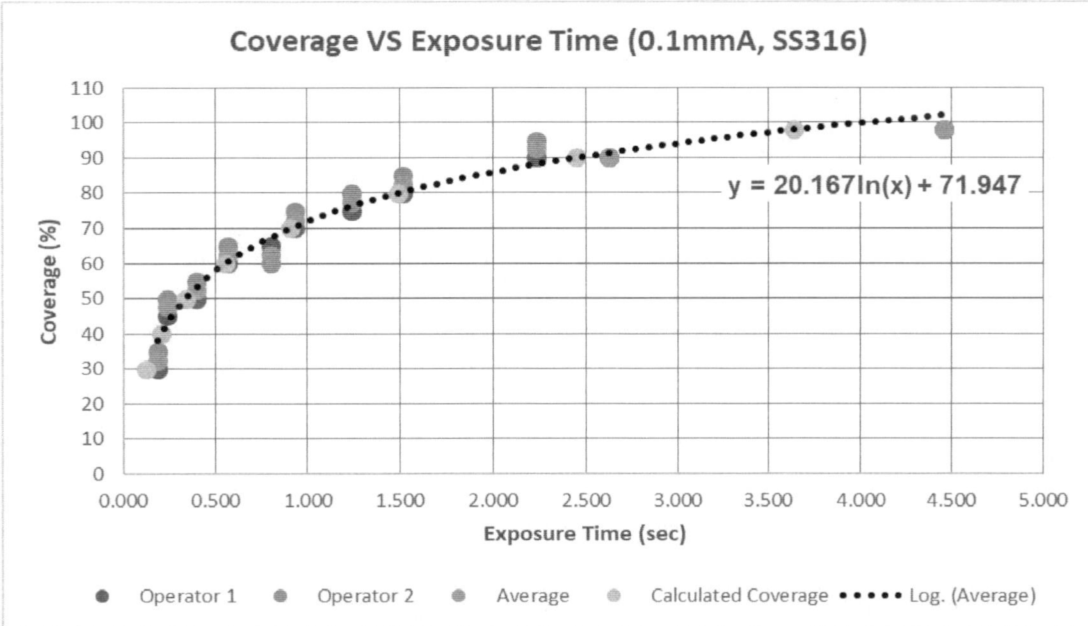

Figure 1 Coverage Development Curve for SS316, 0.1mmA

After all coverage percentages had been obtained from the shot peening process, a customized setup consisting of a telecentric lens with a field of view of 8mm by 7mm, a light source and a mono camera was used to capture images for the training of deep learning models. 30 images were captured and labelled for each coverage percentage in all 10 coverage ranges, resulting in a total of at least 1500 images (30 images x 5 coverage percentages x 10 coverage ranges) captured for each intensity per material type.

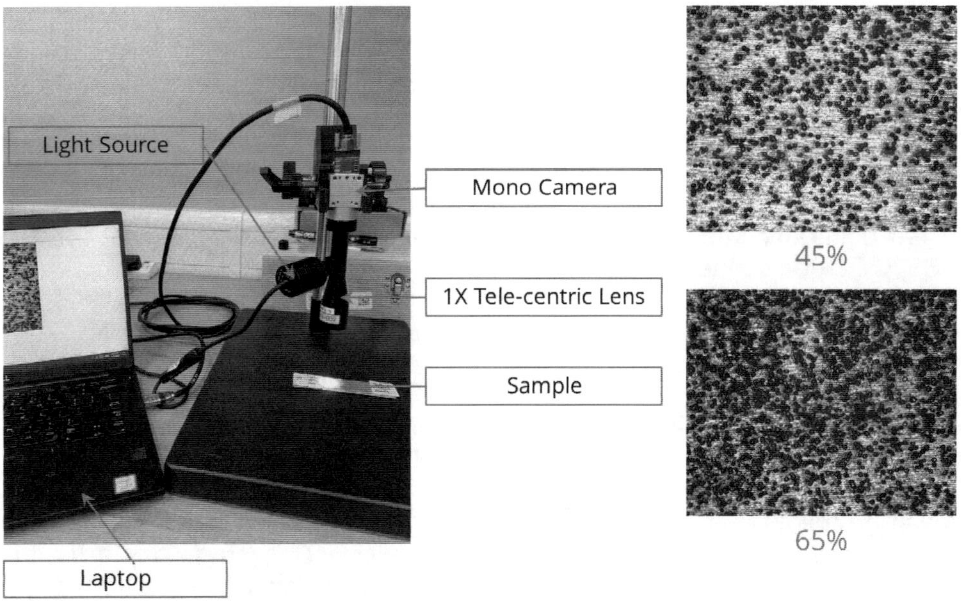

Figure 2 Customized Image Capturing Setup (left) and Sample Captured Images (right)

With all images captured and labelled according to coverage range, intensity and material type, a commercial deep learning vision software called SuaKIT [4] was used for the development and training of the deep learning models. Deep learning models were created for each intensity per material type through iterative sorting, validation and testing of image datasets. In general, images in a single dataset for each intensity and material of at least 1500 images were split into the following during the deep learning model development:

- **Training images** (60% of total image data collected): Images used to train the deep learning models

- **Validation images** (20% of total image data collected): Images used to test the performance of the models during the training phase
- **Testing images** (20% of total image data collected): Images used to test the performance of the models after the final model has been developed

In addition to studying the models' accuracy through the testing phase during the model development, the deep learning models were deployed into a separate software which was developed separately to simulate the actual performance of the models during an industrial deployment (Figure 3). This was done by capturing an additional set of 10 images from each coverage range from the peened specimens, followed by using the deployed model in the software to evaluate the overall model's accuracy. The output model accuracy from this simulated deployment test would then give an indication of the actual accuracy one would expect when the deep learning models are used in a production setting.

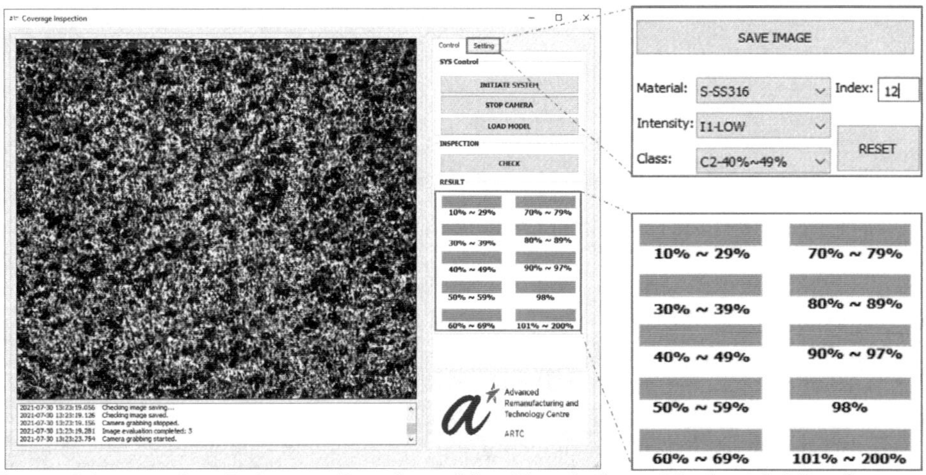

Figure 3 Inspection Software Developed for Simulation of Industrial Deployment

The deep learning models were also cross-tested against a separate dataset to study how robust the models were to identify coverage against another intensity value or material type. In this cross-testing study, the following was tested:

- **Cross Materials Study**: Ti-6Al-4V model trained with 0.200mmA images tested with SS316 images peened at 0.200mmA
- **Cross Intensities Study**: Both Ti-6Al-4V models trained with 0.200mmA and 0.300mmA images tested with Ti-6Al-4V images peened at 0.100mmA

Experimental Results

With the models developed, the performance of the models was evaluated during the model testing phase and the simulated deployment test. Model accuracy was the key metric used to evaluate the overall performance of the models, and this is defined as the percentage of correct coverage predictions over the total number of predictions made. In order to classify as a correct prediction, the model has to output the correct coverage range (i.e. 30-39%) upon evaluating an image data. Table 2 shows the summarized results from the models tested for a fixed dataset (same intensity and material).

Table 2 Summary of Model Prediction for a Fixed Dataset

Dataset	Model Accuracy	
	Model Testing Process (Out of Min. 300 Images)	Simulated Deployment Test (Out of 100 Images)
Ti-6Al-4V: 0.100mmA	100%	93%
Ti-6Al-4V: 0.200mmA	99.5%	95%
Ti-6Al-4V: 0.300mmA	99.7%	100%

SS316: 0.100mmA	100%	94%
SS316: 0.200mmA	100%	94%
SS316: 0.300mmA	100%	94%

It was observed that the developed models could achieve at least 99.5% and 93% model accuracy during the model testing phase and the simulated deployment test respectively. A deep dive on the false predictions showed that all models were mis-predicting images in the neighboring coverage range of the actual coverage range (i.e. 30-39% range was predicted instead of actual being in the 40-49% range). As such, the overall performance of the model showed very promising results for deep learning technologies for shot peening application.

In addition, several models were tested across different datasets to study how they performed against a dataset that the model was unfamiliar with, and it was observed that the models did not perform well for both cross material and cross intensity study. The cross-materials study, where the Ti-6Al-4V: 0.200mmA model was tested with 350 images from the SS316: 0.200mmA dataset, only achieved 16% accuracy, whereas the cross-intensities study, where both Ti-6Al-4V: 0.200mmA model and Ti-6Al-4V: 0.300mmA models were tested with at least 350 images from Ti-6Al-4V: 0.100mmA dataset, only achieved 23.4% and 30.2% respectively. The main possible reason for the poor model performances was due to the visual differences in images for different materials and intensities. Since the models were trained mainly on a single type of dataset, the model could not recognize images that looked vastly different from a different dataset, causing significant number of false detections. Figure 4 shows the differences between materials and intensities for 40% coverage. It was observed that not only the textures and light reflectivity were different between SS316 and Ti-6Al-6V, the size of peening indentations was also different between peening intensities.

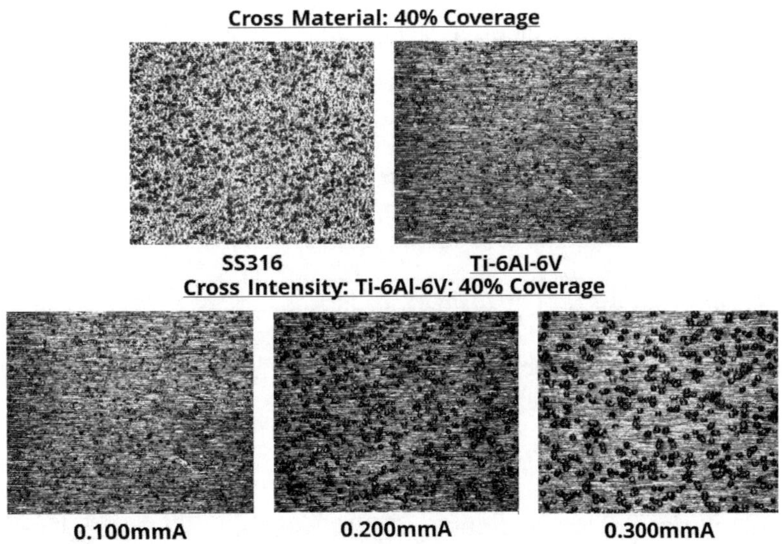

Figure 4 Images Captured for Different Materials and Intensities for 40% Coverage

Additional models were then built to study the performance of combined datasets, and this was done by combining both datasets for Ti-6Al-4V: 0.200mmA and SS316: 0.200mmA to form a combined materials model, as well as combining all datasets from Ti-6Al-4V (Ti-6Al-4V: 0.100mmA + Ti-6Al-4V: 0.200mmA + Ti-6Al-4V: 0.300mmA) to form a combined intensities model. It was observed that the models' accuracy significantly improved due to the model's ability to better recognize different images from different datasets. Table 3 shows the summarized results from the cross-materials and cross-intensities study. Similar to the fixed dataset study, the false predictions from the models were mainly mis-predicting images in the neighboring coverage range of the actual coverage range.

Table 3 Summarized Results for Cross Materials and Cross Intensities Study

Dataset	Model Accuracy	
Cross Materials (Ti-6Al-4V: 0.200mmA VS SS316: 0.200mmA)	16%	
Cross Intensities (Ti-6Al-4V: 0.200mmA VS Ti-6Al-4V: 0.100mmA)	23.4%	
Cross Intensities (Ti-6Al-4V: 0.300mmA VS Ti-6Al-4V: 0.100mmA)	30.2%	
Combined Materials (Ti-6Al-4V: 0.100mmA + SS316: 0.200mmA)	100% (Model Testing Process)	93.5 % (Simulated Deployment Test)
Combined Intensities (Ti-6Al-4V: 0.100mmA + Ti-6Al-4V: 0.200mmA + Ti-6Al-4V: 0.300mmA)	99.3% (Model Testing Process)	97.3% (Simulated Deployment Test)

Conclusions

This study shows the overall abilities and limitations of using deep learning inspection algorithm for coverage prediction. The deep learning models showed that they could perform well with high model accuracy of 93% and higher, but are limited to the type of data or images that the models were trained with. It is possible to improve the models' accuracy through introducing a larger dataset, but that would require re-training with new images.

In terms of application to industry, applying deep learning inspection algorithm can be particularly useful for production sites with 1 or 2 components, as the database could be easily built based on such limited component types. However, this could be challenging for factories or production sites that manufacture multiple types of components as all images or data type has to be collected and trained in the model for such robust applications.

There could be work-arounds towards implementing a quicker way of updating the deep learning models with additional data, such as ongoing research on online, transfer and continuous learning for quick teaching of the deep learning models. However, it would be most ideal to further study on how to teach a model to properly recognize images like how humans do in recognizing peening coverage templates and applying them on any material types and intensity values.

Acknowledgements

This research is supported by A*STAR under its Core Research Project (CRP) program in Advanced Remanufacturing and Technology Centre (ARTC). The authors would also like to Abhay Gopinath and Thomas Haubold from Rolls-Royce for their valuable support and advice throughout this study.

References

[1] Curtiss-Wright Surface Technologies. (2020). Surface Technologies Shot Peening Applications. https://www.cwst.co.uk/wp-content/uploads/2020/01/new-english-green-book.pdf.

[2] Standard, S. A. E. (2013). J2277 Shot Peening Coverage Determination. Developed by Surface Enhancement Committee, Publication Date: Apr, 1.

[3] Qiu, J., Li, W., & Yao, Q. (2021, May). Study on Shot peening Coverage of Metal Surface Based on Deep Learning. In Journal of Physics: Conference Series (Vol. 1920, No. 1, p. 012106). IOP Publishing.

[4] Shahid, L., & Janabi-Sharifi, F. (2019). A neural network-based method for coverage measurement of shot-peened panels. Neural Computing and Applications, 31(9), 4829-4836.

[5] Cognex Corporation (2019) *SuaKIT Vision Software* (V2.2.6) [Computer software]. https://www.cognex.com/products/deep-learning/suakit

Shot Peening on All-solid-state Lithium Metal Battery for High-speed Charging

M. Kodama[1], K. Takashima[1] and S. Hirai[1]

1 School of Engineering, Tokyo Institute of Technology, Japan

Abstract
All-solid-state lithium metal battery is a candidate of the next generation electrical vehicle's battery. However, lithium dendrite generated in the fast charging process penetrates the solid electrolyte layer and breaks the battery with an internal short circuit. With this problem, the charging speed of the all-solid-state lithium metal battery is limited. In this paper, the authors propose and demonstrate a shot peening processing on the solid electrolyte layer surface to increase the charging speed of the all-solid-state lithium metal battery. With the shot peening, crack generation on the solid electrolyte layer can be suppressed with the compressive residual stress, and the lithium dendrite penetration into the solid electrolyte layer should be suppressed. Moreover, the shot peening generates a roughness on the surface of the solid electrolyte layer, and the contact between the solid electrolyte layer to the lithium metal anode would be improved, and the electrochemical reaction should be enhanced. The experimental results show that the shot peening successfully increases the charging speed (critical current density). Moreover, there is a synergistic effect between shot peening and gold thin-film insertion, which is one of the electrochemical performance improvement methods. Shot peening with changing the shot size is also conducted. The results show that there is an optimum shot size that maximizes the charging speed.

Keywords All-solid-state lithium metal battery, shot peening, charging speed, shot size.

Introduction

To reduce the CO_2 emission, gasoline vehicles should be changed to electrical vehicles (EV). However, a conventional EV's cruising distance is only about 400 km and is much shorter than that of a gasoline vehicle (~1000 km) due to lithium-ion batteries' volumetric energy density limitation. Moreover, there is an ignition risk of organic electrolytes in the lithium-ion battery, and some cases are reported of fire accidents of EVs. An all-solid-state lithium-metal battery (ASSLiMB) is proposed to increase the battery volumetric energy density for long cruising distance and as high safety battery [1]. The ASSLiMB's anode is lithium metal, and the energy density is ten times higher than that of a conventional lithium-ion battery anode with graphite. Moreover, the inorganic solid electrolyte (SE) that is fire-resisting material is used as the electrolyte; thus, the ignition risk of the battery is negligibly small [2].

However, the charging speed of conventional ASSLiMB is limited. The reason for the limited charging speed is lithium dendrite. If the ASSLiMB is in high-speed charging (charging under high-current-density condition that exceed a critical current density (CCD)), the lithium dendrite grows into the SE layer that separates the anode and the cathode by developing a clack, and the lithium dendrite connects the anode and cathode, then the battery is in an internal short circuit [3]. For high-speed charging ASSLiMB, the lithium dendrite growth with clack propagation should be suppressed.

Electrochemical performance improvement methods are proposed to increase the charging speed of ASSLiMB [4]. In those methods, thin film (e.g., Au layer by sputtering) is inserted between the SE layer and Li metal anode. Overpotential is suppressed with the thin layer, and homogeneous lithium deposition is induced, and the growth of the lithium dendrite is suppressed, then the charging speed is increased. But, with this electrochemical performance improvement, the charging speed is not enough to use the ASSLiMB as EV's battery, and other methods to increase the charging speed are required.

Shot peening (SP) is one of the methods to suppress the clack generation by compressive residual stress [5-7]. As described above, the lithium dendrite grows in the SE layer by making a clack. If the clack to the SE layer is suppressed, the lithium dendrite growth would be suppressed. Therefore, there is a possibility to increase the charging speed of the ASSLiMB by the SP to the SE layer in ASSLiMB. Moreover, the SP and thin-film insertion method can have a synergetic effect because the dendrite growth suppression mechanism is different.
According to the background mentioned above, in this study, the characteristics of SP processed ASSLiMB is demonstrated with experiments to elucidate the influence of the SP on battery charging speed and electrochemical and mechanical characteristics of the ASSLiMB.

Experimental Methods
To evaluate the influence of SP on the battery performance and the characteristics of lithium batteries with oxide SE, measurements of charging speed, SE surface structure, and SE surface mechanical characteristics are conducted for an oxide SE of $Li_{6.25}Ga_{0.25}La_3Zr_2O_{12}$ [8]. Ga-stabilized oxide SE ($Li_{6.25}Ga_{0.25}La_3Zr_2O_{12}$), which is an oxide solid electrolyte with a high ionic conductivity, is used in this study. The starting materials are La_2O_3, ZrO_2, Ga_2O_3, and LiOH. The starting materials are weighed out in their intended stoichiometric proportions together with an excess of 10 wt.% LiOH to compensate for lithium loss during synthesis. The mixed powder of starting materials of 10 g is placed in a zirconia pot (50 ml) with 30 ml hexane and 20 zirconia balls (ϕ 10 mm × 10, ϕ 5 mm × 10). Mechanical milling is conducted using a planetary mill operating at 300 rpm for 2 h. The material is heat-treated at 800 °C in the air for 12 hours at an increase rate of 5 °C/min by an electric furnace with a platinum crucible. The resulting materials are ball-milled with hexane again, following the same procedure as the first ball milling. The resulting materials are dried and pressed into ϕ 5 mm pellets by 150 MPa uniaxial compression. The pellets are set into the platinum crucible and covered by the mother powders. The final sintering step is done in the air in an electric furnace at 1050 °C for 12 hours at an increase rate of 5 °C/min, followed by a decrease rate of 2 °C/min. After sintering, the surface of the sintered pellet is shaped with #1500 sandpaper. The crystal structure of the synthesized pellet is measured by XRD, confirming that the SE is in the cubic garnet phase, which can achieve high ionic conductivity.
The SP is processed on the SE pellet. Al_2O_3 powder is used as the shot, and three grain size (50 μm, 100 μm, and 300 μm) is used to evaluate the influence of the grain size on the SP and battery performance. The shot is propelled to the SE pellet from a gun with high-pressure air. The nozzle diameter of the gun is 2 mm, and the gun inlet air pressure is 0.4 MPa. The distance from the exit of the gun nozzle to the SE pellet is 10 cm. The SP processing time is 10 s on each side of the SE pellet.
To compare the effect of the SP to the electrochemical performance improvement method (thin gold film insertion method between the SE and Li metal) and to investigate the synergistic effect of the SP and the thin-film insertion method, gold thin-film insertion between the lithium metal and the SE was performed by DC sputtering on the SE pellet. The sputtering is processed on the lithium contact area of the SE pellet. The pellet is placed in the DC sputtering system and processed for 4 min in the air with an ion current of 5 mA.
To evaluate the electrode performance, the CCD of a lithium symmetric cell is measured. Lithium metal is attached to both sides of the SE pellet, and the cell was pressurized to 3 MPa with a jig and installed into an airtight container in an argon-filled glove box. The CCD is measured by connecting the symmetric cell to a battery evaluation system. A rectangular current is applied with a 2-hour cycle, and the current density is increased with each cycle to measure the CCD, that is, the current density at which the cell is short-circuited, at 60 °C.
The surface shape of the SE pellet is changed by SP. The surface of SE is an interface of the SE and lithium metal anode and is an electrochemical reaction plane. Therefore, the surface shape of SE has an important role in the interface reaction, so the SE surface 3D structure is observed by a topographic method with reflected-electron SEM images [9]. Four reflected-electron images are obtained with four detectors in the SEM. Because the four

detectors are set in different places in the SEM, different shadows can be observed on the reflected-electron images; the shadows contain information about the surface unevenness. The surface 3D structure is obtained with reconstruction software.

SP induces compressive residual stress on the surface of the processed material [5-7]. This compressive residual stress can increase fracture toughness and suppress cracking on the surface and can suppress the growth of the lithium dendrite. To study the change in the fracture toughness, the surface mechanical characteristics of the SE pellet were measured by the indentation system, in which a square indentation is made on the SE pellet with a diamond indenter. At the edge of the square indentation, a crack is generated. The length of the crack depends on the fracture toughness, which can be calculated by the following equation (1) [10].

$$K_c = 0.026\{(E^{1/2}P^{1/2}c)/a^{3/2}\} \quad (1)$$

In equation (1), K_c is the fracture toughness, E is Young's modulus, P is the indenter force, a is the average crack half-length, and c is the average indentation length. Young's modulus E of LLZO (130 GPa) is used for the calculation. The indenter force P is 1 kgf.

Experimental Results

Figure 1 illustrates the CCD of each sample. The CCD of the non-processed sample (non-proc.) is only 0.04mA/cm^2. With gold sputtering, which is an electrochemical performance improvement method, the CCD is increased to 0.38 mA/cm^2 (Au). With the SP with 50 μm grain, the CCD is increased to 0.27mA/cm^2 (S.P. 50 μm). By combining SP and gold sputtering, the CCD is increased to 0.73 mA/cm^2 (S.P. 50 μm + Au). With changing the grain size of the SP, the CCD is changed. With the increase in the grain size to 100 μm, the CCD is increased to 0.76mA/cm2 (S.P. 100 μm + Au). Nevertheless, the CCD is decreased to 0.32 mA/cm^2 with 300 μm grain (S.P. 300 μm + Au).

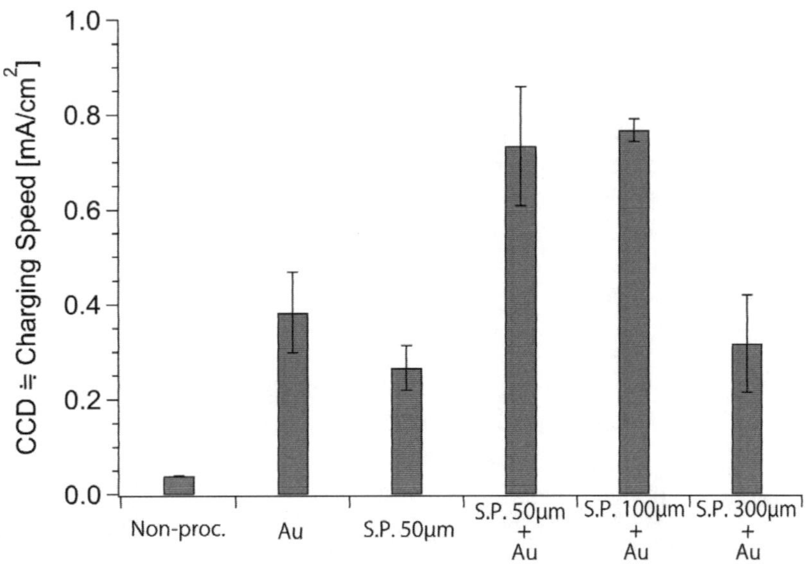

Figure 1. Critical Current Density of each processed samples.

Figure 2 illustrates the surface 3D SEM images of each processed sample. In the non-processed sample (fig.2 (a)) and a sample with gold sputtering (fig.2 (b)), the surface is almost flat. With the SP, surface roughness is generated, and the rough surface is not changed by the gold sputtering after the SP (fig2(c)-(f)). The distance between the bumps and height of the bumps are increased with the increase in the SP grain size.

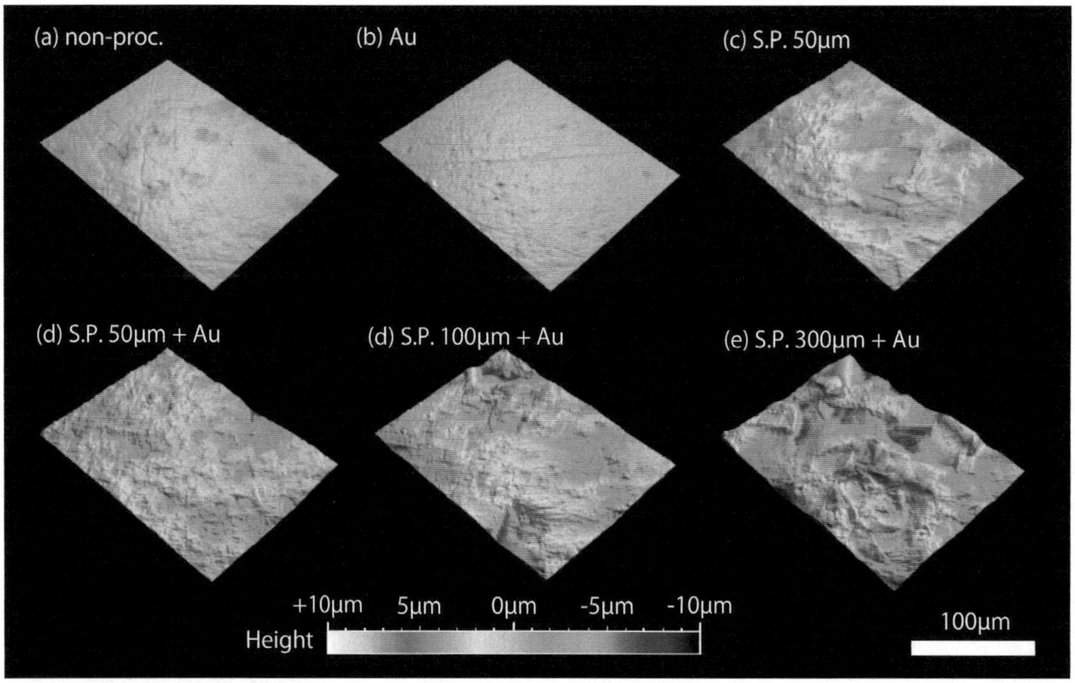

Figure 2. Surface 3D SEM Images of Each Processed Samples (a) non-processed sample, (b) Au sputtering sample, (c) S.P. 50 μm, (d) S.P. 50 μm + Au, (e) S.P. 100 μm + Au, (f) S.P. 300 μm + Au.

Figure 3 shows the fracture toughness of each processed sample. The fracture toughness of the non-SP processed samples (Non-proc. and Au) are around 2.0 MPa·m$^{1/2}$. The fracture toughness is increased by the SP, and 3.0 MPa·m$^{1/2}$ is achieved by the SP with 50 μm grain. The fracture toughness is increased with the increase in the grain size, and fracture toughness is 4.1 MPa·m$^{1/2}$ for 300 μm grain.

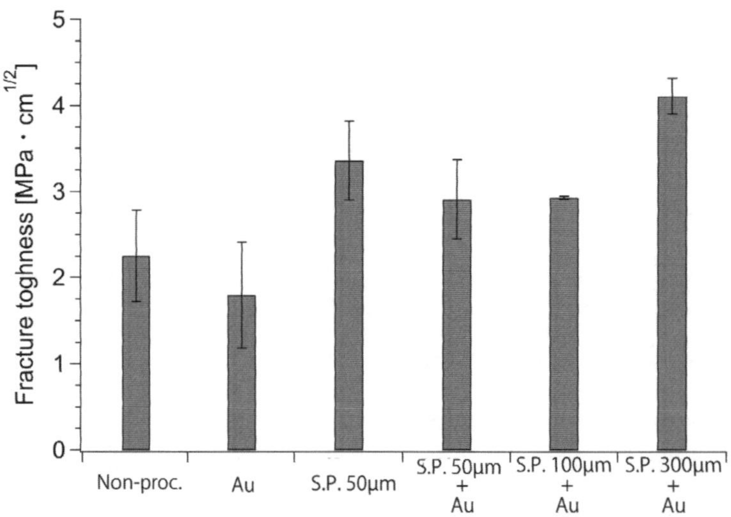

Figure 3. Fracture Toughness of Each Processed Samples.

Discussion

As shown in fig.1, the CCD of the non-processed sample was low. The CCD was increased to 0.38 mA/cm^2 by Au sputtering and to 0.27 mA/cm^2 by SP with 50 µm grain. Therefore, the performance improvement obtained by SP was comparable to conventional thin-film insertion. Moreover, by combining gold sputtering and SP with 50 µm grain, the CCD was increased to 0.73 mA/cm^2. Therefore, SP and gold sputtering can be used together to increase the CCD. With the increase in the grain size from 50 µm to 100 µm, the CCD was increased from 0.73 mA/cm^2 to 0.76 mA/cm^2, but with the increase from 100 µm to 300 µm, the CCD was significantly decreased from 0.76 mA/cm^2 to 0.32 mA/cm^2. Therefore, there is an optimum size of the grain to maximize the CCD.

The surfaces of the non SP processed samples were almost flat (Fig.2 (a),(b)), but surface roughness was made on the SE by SP (Fig2 (c)-(e)). With the increase in the grain size, the distance from top to top of the surface structure was increased, and the height of the structure was also increased. So, the increase of the grain size increases the surface roughness of the SP processed SE. This surface roughness allows mechanical penetration of the solid electrolyte into the lithium metal. This penetration increases the contact between SE and lithium metal anode and enhances the electrochemical reaction and increases the CCD. However, the interface roughness induces inhomogeneous electrochemical reactions and lithium deposition. This inhomogeneous lithium deposition can induce lithium dendrite. As a result of this balance of the enhancement of the electrochemical reaction and lithium dendrite generation, there would be the optimum grain size to maximize the CCD.

The fracture toughness was increased by the SP. This would be because the compressive residual stress is induced around the SE surface. This increase in fracture toughness also can be a reason for the increase of the CCD with the SP. The fracture toughness is slightly decreased by gold sputtering. In the sputtering process, nitrogen plasma is generated around the SE. The plasma is in non-equilibrium condition, and the gas temperature is around room temperature, but the electron temperature is higher than 5000 K. This high-temperature electron increases the processed SE temperature and releases the compressive residual stress, decreasing fracture toughness. The highest fracture toughness is achieved for SP with 300 µm grain. But the highest CCD was achieved for SP with 100 µm. Therefore, the CCD is not simply dependent on the fracture toughness, and the surface shape effect discussed above is also a factor of the CCD.

Conclusions

In this study, the influence of processing by SP to the SE pellet surface on the performance of the lithium metal electrode of an ASSLiB is elucidated by CCD, surface 3D structure, and mechanical strength measurements. The following conclusions can be drawn.

1. By attaching lithium metal to the SE pellet processed by shot peening to form the lithium metal electrode, higher CCD can be obtained than for the electrode formed when using an SE that has not been processed by shot peening. This improvement in performance is comparable to the results obtained with the conventional method of thin gold film insertion. Shot peening can be used in combination with conventional thin gold film insertion; higher electrode performance can be obtained by combining both shot peening and thin gold film insertion than by using either method alone.
2. Shot peening on the surface of the oxide SE pellets generates surface roughness on the pellets. This surface roughness can increase the contact area between the SE and the lithium metal because the surface roughness allows mechanical penetration of the solid electrolyte into the lithium metal.
3. The fracture toughness of the surface of the oxide SE pellet is increased by shot peening, thus suppressing the propagation of the lithium dendrite into the solid electrolyte. This increase in fracture toughness results in the increase in the CCD observed upon shot peening.
4. There is an optimum grain size for shot peening processing to maximize the CCD. In this study, the optimum grain size was 100 µm, and a maximum CCD of 0.76 mA/cm2 was achieved. It is estimated that the shot peening with optimum grain size generates the

optimum roughness on the SE that can enhance the interface electrochemical reaction and does not induce an inhomogeneous electrochemical reaction.

Acknowledgement
This study was supported by the Japan Keirin Autorace Foundation and JSPS KAKENHI Grant Number 22K14761.

References

[1] M.V. Reddy, C.M. Julien, A. Mauger, K. Zaghib, Sulfide and Oxide Inorganic Solid Electrolytes for All-Solid-State Li Batteries: A Review, Nanomaterials. 10 (2020) 1606.
[2] P. Bonnick, J. Muldoon, The quest for the holy grail of solid-state lithium batteries, Energy Environ. Sci. (2022) 10.1039.D2EE00842D.
[3] Y. Qi, C. Ban, S.J. Harris, A New General Paradigm for Understanding and Preventing Li Metal Penetration through Solid Electrolytes, Joule. 4 (2020) 2599–2608.
[4] J. Wakasugi, H. Munakata, K. Kanamura, Effect of Gold Layer on Interface Resistance between Lithium Metal Anode and $Li_{6.25}Al_{0.25}La_3Zr_2O_{12}$ Solid Electrolyte, J. Electrochem. Soc. 164 (2017) A1022–A1025.
[5] W. Pfeiffer, T. Frey, Strengthening of ceramics by shot peening, Journal of the European Ceramic Society. 26 (2006) 2639–2645.
[6] S. Bagheri, M. Guagliano, Review of shot peening processes to obtain nanocrystalline surfaces in metal alloys, Surface Engineering. 25 (2009) 3–14.
[7] K. Takahashi, Y. Nishio, Y. Kimura, K. Ando, Improvement of strength and reliability of ceramics by shot peening and crack healing, Journal of the European Ceramic Society. 30 (2010) 3047–3052.
[8] X. Huang, J. Su, Z. Song, T. Xiu, J. Jin, M.E. Badding, Z. Wen, Synthesis of Ga-doped $Li_7La_3Zr_2O_{12}$ solid electrolyte with high Li+ ion conductivity, Ceramics International. 47 (2021) 2123–2130.
[9] A. Sawaguchi, T. Kamimura, A. Yamashita, N. Takahashi, K. Ichikawa, F. Aoyama, Y. Asada, Informative three-dimensional survey of cell/tissue architectures in thick paraffin sections by simple low-vacuum scanning electron microscopy, Sci Rep. 8 (2018) 7479.
[10] Ponton. B., C., Rawlings. D., R., Vickers indentation fracture toughness test Part 1 Review of literature and formulation of standardized indentation toughness equations, Materials Sciences and technology, 5 (1989), 865-872.

Dry Ice Blasting to Increase Fatigue Strength

J. Hoffmeister[1], E. Schuch[1], M. Siedler[1] and T. Harter[1]

[1] Esslingen University, Faculty of Mechanical and Systems Engineering, Kanalstr. 33, 73728 Esslingen, Germany

Abstract
Dry ice blasting is a blasting process that is usually used to clean components. The main advantage of this process is the residue-free evaporation of the abrasive, so that the blasting material is not decontaminated. Another characteristic of the process is the low intensity. However, [1, 2] have already shown that the surface condition of components can be changed by dry ice blasting. In this paper, therefore, dry ice will be used specifically to harden material specimens. Two materials are used: a soft aluminum alloy EN AW-5754 and a soft annealed tool steel 90MnCrV8. Subsequently, the surface condition in terms of roughness, residual stresses and strain hardening was investigated and the cyclic properties of the specimens were determined.

Keywords: dry ice blasting, fatigue strength, residual stresses

Introduction

Dry ice blasting, also called dry ice cleaning, is a compressed air blasting process that uses dry ice at a temperature of -78.5 °C as blasting medium. Basically there are two advantages of the process: on the one hand, even sensitive workpieces can be treated with little damage due to the low hardness of the blasting medium and, on the other hand, dry ice evaporates after blasting, so that it is no longer necessary to clean the workpiece of the blasting medium. No other blasting process has this advantage [3].
Three respectively four effects play a role in the dry ice blasting process: mechanical, thermal and sublimation effects and additionally a solvent effect is reported in some cases [4]. In order to quantify the different effects on decoating processes, [5] has investigated and determined the effective proportions of the several effects with the aid of various experiments. Among others, in order to quantify the mechanical effect of dry ice blasting, [3] measured the Almen intensity [6], which is a parameter primarily used in consolidation blasting. However, preliminary investigations have shown that it is not possible to measure the arc height with the Almen-"N" [7], which is used for shot peening with low beam intensities. Therefore, so-called Aero Almen strips "AA" [8] were used, which consist of an aluminum alloy A2024-T3 instead of a spring steel SAE1070. The Aero Almen strips also enabled to measure deflections of the Almen strip during dry ice blasting. Comparing different blasting processes, comparable Almen intensities could be determined with nutshell granules at lower pressures. Tests concerning the removal rate of ceramic thermal barrier coatings showed similar removal rates for both blasting media at comparable blasting intensities. The author concluded that the mechanical effect of dry ice blasting is high. At temperatures of the workpiece of 20 °C, the mechanical percentage was quantified by 85 % and the thermal percentage by 15 %. However, the ratios depend on the temperature of the workpiece; the thermal effect increases as the temperature of the workpiece grows. The proportion of the sublimation effect was not detectable.
With regard to the effect on the surface layer condition, it has already been shown by [1, 2] that dry ice blasting can influence both the hardening depth profiles and the residual stress condition. Also other processes with a low Almen intensity, such as micropeening have an influence on the surface condition and can also increase the fatigue strength significantly [9–11]. Based on the findings that even low intensities have an influence on the fatigue strength, and that dry ice blasting with even lower intensities can change the surface layer positively, this work shows the influence of dry ice blasting on the fatigue strength.

Experimental Methods

In this project two materials were used: a tool steel 90MnCrV8 [12] in the soft annealed condition and an aluminum alloy EN AW-5754 [13]. The allowable and measured chemical composition of the two materials are shown in table 1: for the tool steel and in table 2 for the aluminum alloy. Both materials do not show any deviations from the standards.

table 1: Specification of the chemical analysis according to the standard [12] and the measured chemical analysis for 90MnCrV8

elements	C	Si	Mn	P	S	Cr	V	Mo	Fe
according standard in mass-%	0,85-0,95	0,1-0,4	1,8 - 2,1	max. 0,03	max. 0,03	0,2-0,5	0,05-0,2	-	rest
chemical analysis in mass-%	0,85	0,284	1,99	$<5\cdot10^{-4}$	$<5\cdot10^{-4}$	0,445	0,111	0,18	rest

table 2: Specification of the chemical analysis according to the standard [13] and the measured chemical analysis for EN AW-5754

elements	Si	Fe	Cu	Mn	Mg	Cr	Zn	others	Al
according standard in mass-%	max. 0,4	max. 0,4	max. 0,1	max. 0,5	2,6 - 3,6	max. 0,3	max. 0,2	max. 0,15	rest
chemical analysis in mass-%	0,173	0,304	0,028	0,267	2,92	0,034	0,039	0,15	rest

Bored flat specimens were manufactured from sheet metal by laser cutting. This results in a notch factor of K_t = 2.5. The area of the notches was manually treated by dry ice blasting. For comparison purposes, some of the aluminum specimens were also manually sand blasted (compare figure 1). None of both surface treatments was optimized.

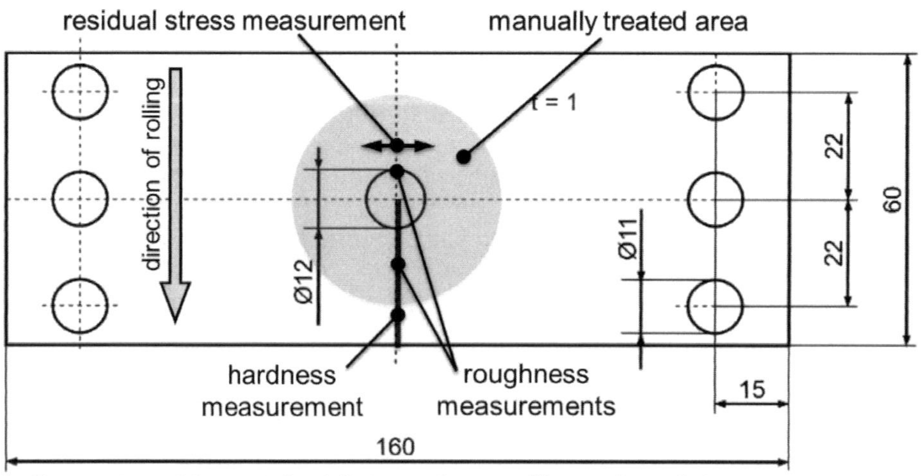

figure 1: Geometry of the notched flat specimens, illustration of the manually mechanical treated area and the measurement direction

The surface conditions of the specimens were characterized. All measurement point are illustrated in figure 1. The hardness distributions in the cross section trough the notch and the roughness in the base of the notch were measured. Additionally the residual stresses near the notch were measured by X-ray diffraction. The measurement conditions and basis for the calculation of the residual stresses are listed in table 3 for the investigated materials. For the evaluation of the residual stresses the cosα-method was used. The complete depth distributions of residual stresses were measured by alternating electro polishing and measuring steps. Because of the fact that the removed area was small, the stress relaxation was ignored.

table 3: basis for the calculation of the residual stresses

	90MnCrV8	EN AW-5754
Interference line	{211}	{311}
Radiation	Cr-Kα	Cr-Kα
Young's Modulus of the interference line $E^{\{\}}$	224.00 GPa	69.31 GPA
Poisson ratio of the interference line $\nu^{\{\}}$	0.28	0.348
Stress free diffraction angle $2\theta_0$	156.396 °	139,497 °

With both materials five tensile tests were performed. Finally, all specimen conditions and materials were cyclic loaded with a stress ratio $R = -1$ and the cyclic test results were statistically evaluated. In the long-life fatigue region, at least 12 specimens per surface condition were tested. The results were evaluated with the staircase method according Hück [14]. A test was finished as soon as the specimen was broken or the ultimate number of load cycles of 10^7 was achieved. In the high cycle fatigue region, the tests were evaluated according to the method of discrete load steps [15].

Experimental Results

The tool steel was changed significantly by the thermal influence of the laser cutting. This resulted in a martensitic zone in the laser-machined areas. This zone can be seen in the hardness distribution (see figure 2 left). The laser cutting changed the hardness from approximately 200 HV 0.1 of the soft annealed initial condition to approximately 600 HV 0.1 near the laser cut area. After dry ice blasting the hardness of the tool steel was not affected. The difference between the untreated and the dry ice blasted condition near the surface can be attributed to measurement inaccuracies or could be the result of the martensitic layer, which is not formed uniformly.

For the aluminum alloy (compare figure 2 right), the surface hardness increased slightly to bore from approximately 78 HV 0.1 to 88 HV 0.1. This can be related to laser cutting. There are no differences in hardness distribution between the three conditions.

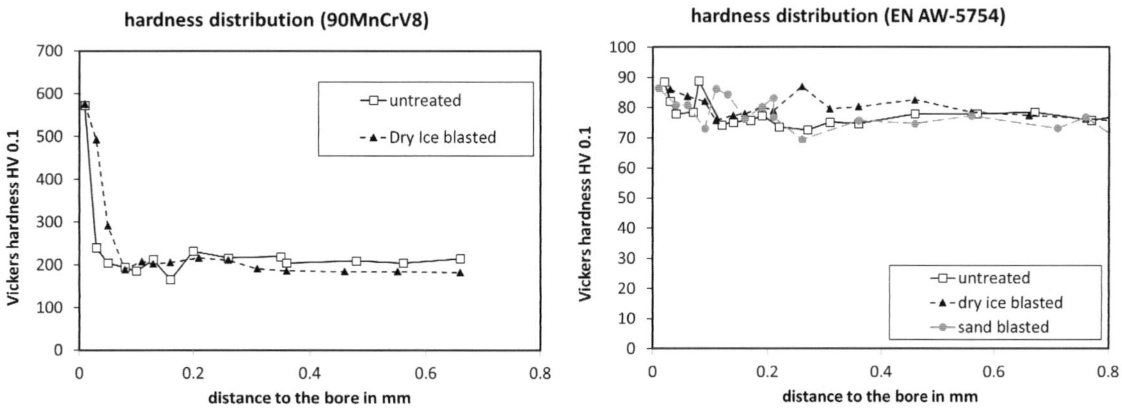

figure 2: Hardness distribution for 90MnCrV8 (left) and EN AW-5754 (right)

Due to the fabrication of the bored flat specimen by laser cutting a relatively high roughness is resulting in the Ø12 bore (notch). For the tool steel 90MnCrV8, the average roughness of the untreated condition in the bore was Rz = 11.7 µm. After dry ice blasting, the average roughness Rz = 13.4 µm was determined. When the scatter of the roughness measurement is taken into account, it can be stated that in the bore no changes were produced as a result of dry ice blasting. Additional roughness measurements in the area close to the bore on the plane showed an average roughness for the untreated condition of Rz = 8.6 µm and for the dry ice blasted condition of Rz = 2.7 µm. Here, an improvement is visible due to the dry ice blasting. For the aluminum alloy, an average roughness for the untreated condition of

Rz = 40.3 µm, for the dry ice blasted condition of *Rz* = 37.3 µm and for the sand blasted condition of *Rz* = 27.7 µm was measured in the bore. Once more, taking into account the scatter of the measurement, the roughness was not changed by dry ice blasting. Sand blasting slightly improved the roughness in the area of the bore. In addition, an edge rounding in the area of the bore could be detected due to sand blasting. Considering the roughness in the area of the bore on the planes, an average roughness *Rz* = 0.75 was measured for the untreated condition, *Rz* = 2.8 µm for the dry ice blasted condition and *Rz* = 37.5 µm for the sand blasted condition. The roughness of the aluminum alloy in the area of the bore was slightly degraded by dry ice blasting and massively degraded by sand blasting.
The residual stresses and full width at half maximum distribution (*FWHM*) for the two materials and the different surface layer conditions are shown in figure 3. For the tool steel 90MnCrV8 (figure 3 left), it can be seen that the untreated initial state already contains residual compressive stresses. These residual compressive stresses could be slightly increased by dry ice treatment. Unfortunately, it is not possible to make any clear statements about the depth, since the electrolytic ablation was no longer uniform starting from an ablation depth of approx. 40 µm on. This nonuniformity had an influence on the measurement results. The full width at half maximum distribution was not influenced by the dry ice blasting, so that probably no work hardening was induced.
The residual stresses of the aluminum alloy EN AW-5754 could be influenced more significantly by both mechanical surface treatments. The untreated condition shows low compressive residual stresses near the surface, which disappear relatively quickly with the depth. The dry ice blasted condition shows the highest surface residual compressive stresses of approximately σ^{rs} = -115 MPa at the surface. These compressive residual stresses decrease with the depth. The sand blasted condition also shows compressive residual stresses, but less compared to the dry ice blasted condition: σ^{rs} = -60 MPa. At depth, compressive residual stresses still build up. From a depth of approximately 20 µm, the residual stresses are similar for both treated conditions. However, the zero crossing of the residual stresses could not be measured, which is due to the electrolytic removal in interaction with the thin flat specimens. The removal causes the residual stresses to be redistributed, which leads to incorrect measurement results. Still, the depth effect of the blasting treatment can be estimated from the full width at half maximum distribution. The untreated condition shows approximately a constant leveling of the full width at half maximum values over the depth. The dry ice blasted condition shows slightly increased values, which reach the level of the untreated condition by about 60 µm. The sand blasted condition shows the highest full width at half maximum values at the surface. This initial value remain constant in the depth and drop from approximately 10 µm until they reach the values of the untreated condition at 160 µm. Sand blasting thus produces the greater work hardening effect than dry ice blasting, but dry ice blasting also work hardens the aluminum alloy.

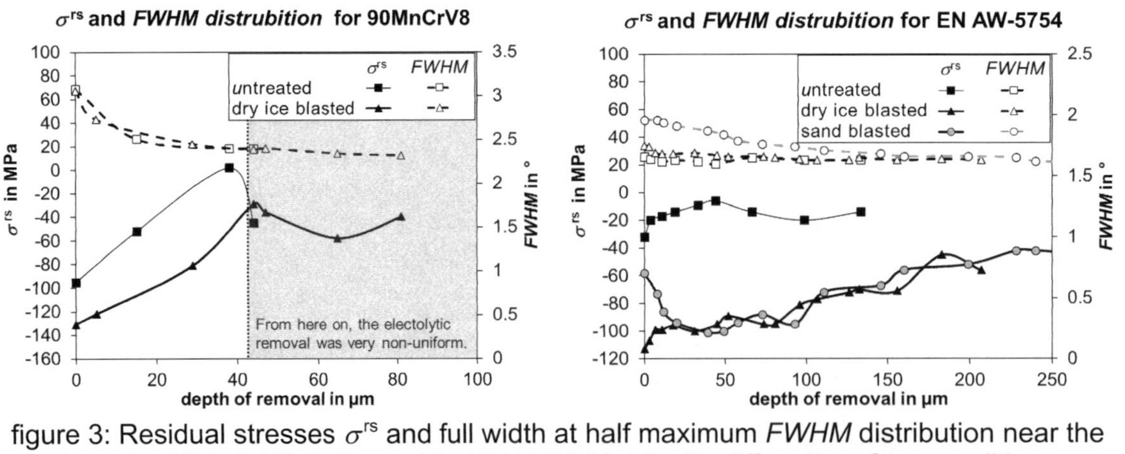

figure 3: Residual stresses σ^{rs} and full width at half maximum *FWHM* distribution near the bore for 90MnCrV8 (left) and EN AW-5754 (right) with different surface conditions

The results of the fatigue tests after the evaluation are illustrated in figure 4 and the resulting specific values are listed in table 4. Regarding the results of the tool steel 90MnCrV8 (see figure 4 left), the fatigue limit of the dry ice blasted tool steel could be increased slightly by 3 % for a failure probability of 50 % and by 6.5 % for a failure probability of 10 % compared to the untreated condition. The strength is therefore hardly influenced. Considering the scatter of cyclic test results, the increase can be neglected.
For the aluminum alloy EN-AW-5754 (figure 4 right), a significant increase of the fatigue limit was determined for both blasting processes. For the sand blasted condition, the fatigue limit was increased by 22.8 % for a failure probability of 50 % and 23.3 % for a failure probability of 10 % compared to the initial condition. The dry ice blasting process even achieved an increase of 28.9 % for a failure probability of 50 % and 30.5 % for a failure probability of 10 % compared to the initial condition. So it was possible to increase the fatigue limit by dry ice blasting more then by sand blasting, although the intensity of the dry ice blasting process is much lower.

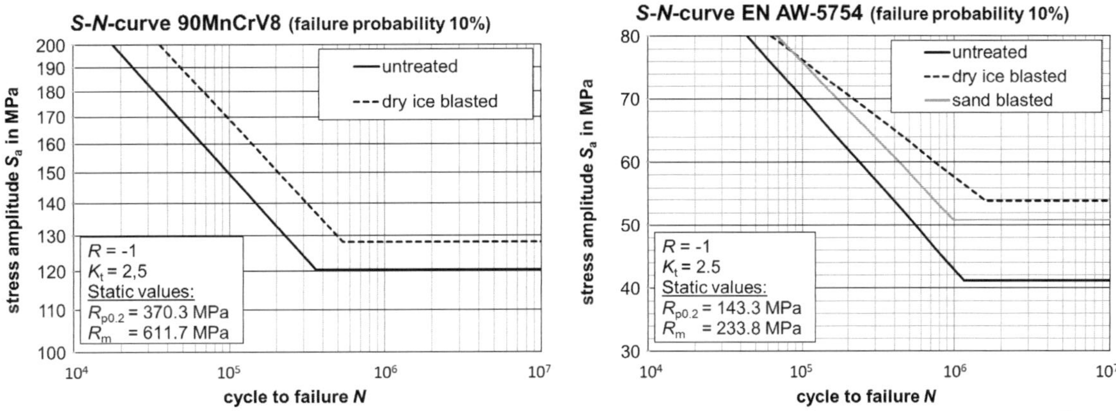

figure 4: *S-N*-curve with a failure probability of 10 % and a stress ratio R = -1 for bored flat specimens with a notch factor of K_t = 2.5 for 90MnCrV8 (left) and EN AW-5754 (right) with different surface conditions

table 4: summary of the statistic evaluation of the cyclic tests (*failure probability of 10 % or 50 %)

Material	**90MnCrV8**		**EN AW-5754**		
Surface condition	**un-treated**	**dry ice blasted**	**un-treated**	**dry ice blasted**	**sand blasted**
Fatigue limit $S_{F,10\%}$* in MPa	120.4	128.2	41.2	53.8	50.8
Fatigue limit $S_{F,50\%}$* in MPa	129.6	133.6	42.5	54.8	52.2
Edge load number of cycles N_D	361263	541212	115527	1592477	985305
***k*-factor (slope of the curve)**	5.93	6.11	4.92	8.13	5.86

Conclusion

It was shown that dry ice blasting can influence the surface condition of metallic materials positively. The roughness was not influenced by dry ice blasting for rough initial conditions. For low initial roughness, an improvement of the roughness was observed for the tool steel and a slight increase was measured for the aluminium alloy. The influence of dry ice blasting is also visible for the residual stresses. For the tool steel and for the aluminium alloy compressive residual stresses can be induced by the dry ice blasting. The hardening effect is clearly more obvious for the aluminium alloy than for the tool steel. The results of the cyclic investigations show that no change could be observed for the tool steel. However, this result is relativized when taking into account the increase in hardness in the peripheral area of the bore, which was caused by the laser cutting, and the high roughness. This combination

represents a difficult starting point for cyclic testing. For the aluminium alloy, a significant increase in fatigue strength was determined for the dry ice blasted condition. This is also illustrated by the additionally considered process of sand blasting, which achieves smaller increases in fatigue strength than the dry ice blasted condition.
In order to determine clearer limits of the dry blasting process in terms of the effect on different materials, even materials of higher strength, further investigations are necessary and in addition the process should be optimized for work hardening.

References

[1] Uhlmann, E.; Krieg, M. C.: *Shot peening with dry ice*. International Conference on Shot Peening 9th (ICSP9, 2005).

[2] Elbing, F.; Uhlmann, E.; Huehns, T.: *Verbesserung der Oberflächeneigenschaften von Metallen mit Trockeneisstrahlen*. In: Galvanotechnik (2003) 1, pp 75–82.

[3] Krieg, M. C.: *Analyse der Effekte beim Trockeneisstrahlen*, Dissertation. Berlin 2008.

[4] Haberland, J.: *Reinigen und Entschichten mit Trockeneisstrahlen*. Grundlegende Untersuchung des CO_2-Strahlwerkzeuges und der Verfahrensweise, Als Ms. gedr. Düsseldorf 1999.

[5] Motschmann, S.: *Einfluss von Herstellungs- und Lagerungsfaktoren auf die Eigenschaften von Trockeneispellets und das Strahlergebnis*, Dissertation. Stuttgart 2016.

[6] Surface Enhancement Committee: *Procedures for Using Standard Shot Peening Almen Test Strip*. 400 Commonwealth Drive, Warrendale, PA, United States.

[7] Surface Enhancement Committee: *Test Strip, Holder, and Gage for Shot Peening*. 400 Commonwealth Drive, Warrendale, PA, United States.

[8] US Air Force: Chapter 5: *Standard Almen and aero Almen strip tests* (DRAFT). URL: https://www.electronics-inc.com/wp-content/uploads/paintstripmanual.pdf. Abrufdatum 31.12.2020.

[9] Weingartner, R.; Hoffmeister, J.; Schulze, V.: *Investigation on the Mechanisms Improving Fatigue Strength in Surface Layers after Micropeening*. In: Wagner, L. (Hrsg.): ICSP12. Proceedings of the 12th International Conference on Shot Peening : Goslar, Germany, September 15-18, 2014. Clausthal-Zellerfeld 2014.

[10] Weingärtner, R.; Hoffmeister, J.; Schulze, V.: *Generation and Determination of Compressive Residual Stresses of Short Penetration Depths*. In: Materials Science Forum 768-769 (2013), S. 580–86.

[11] Weingärtner, R.; Hoffmeister, J.; Schulze, V.: M*echanische Oberflächenbearbeitung durch Mikrostrahlen*. In: HTM Journal of Heat Treatment and Materials 70 (2015) 2, S. 59–65.

[12] DIN EN ISO 4957:2018-11, *Werkzeugstähle* (ISO_4957:2018); Deutsche Fassung EN_ISO_4957:2018. Berlin.

[13] DIN EN 573-3:2019-10, *Aluminium und Aluminiumlegierungen- Chemische Zusammensetzung und Form von Halbzeug- Teil 3: Chemische Zusammensetzung und Erzeugnisformen*; Deutsche Fassung EN_573-3:2019. Berlin.

[14] Hück, M.: *Ein verbessertes Verfahren für die Auswertung von Treppenstufenversuchen*. In: Materialwissenschaft und Werkstofftechnik 14 (1983) 12, S. 406–17.

[15] DIN 50100:2016-12, *Schwingfestigkeitsversuch- Durchführung und Auswertung von zyklischen Versuchen mit konstanter Lastamplitude für metallische Werkstoffproben und Bauteile*. Berlin.

Definition and impact of overlap on the surface state after deep rolling of aluminum and steel alloys

J. Schubnell, M. Jung, A. Sarmast and E. Carl

Fraunhofer Institute for Mechanics of Materials (IWM), Freiburg, Germany

Abstract
The mechanical surface treatment by deep rolling is a widely used process in many industrial applications to increase the fatigue and wear resistance. Multiple process parameters (contact pressure, feed rate, tool diameter) can be varied to achieve an optimum treatment result. To reduce the optimization effort for finding adequate deep rolling parameters the overlapping of two subsequent deep rolled lines was calculated assuming linear-elastic (elastic overlap factor) and elastic-plastic material behavior by analytical calculations (Hertzian contact) and by Finite Element simulation. The impact of overlapping on the surface roughness and residual stress state after deep rolling was experimentally investigated for the materials EN AW 5083, S355J2+N and S690QL. A high (elastic) overlap factor of around 90% leads to surface defects for EN AW 5083 and S355J2+N. An optimum of around 60% was determined for these materials and 90% for S690QL.

Keywords Deep rolling, Overlap factor, Residual stress, Surface roughness

Introduction

Deep rolling is a non-cutting production process which is classified as a fine surface rolling method according to VDI 3177 [1] that is used since many years in multiple industrial applications to increase the fatigue and wear resistance of crank shafts, gear wheels, spindles, screws or threads, valves, wheels rims, turbine blades etc. [2]–[4]. The fatigue life improvement by deep rolling of specimens and parts was proved for a variety of material groups in numerous investigations [4]–[8]. Hydrostatic deep rolling has the advantage that multiple parameters (contact pressure, feed rate or line spacing and tool diameter) can be seperatly adjusted and used for process optimization. However, experimental optimization is time-consuming and costly. Numerical simulations [9] may decrease the time effort but there are still multiple parameters that have to be considered. Due to this reasons, the aim of the present study is to introduce a easily useable so-called overlap factor that unifies the process parameters and reduces in that way the optimization effort.

Definition of overlap factor

Main parameters for deep rolling kinematics are contact pressure *p*, feed rate *f* and tool radius *rt*, see Figure 1 (a). The newly introduced overlap factor *Oe* is defined by the overlapping area of the contact surfaces between the tool and the treated surface of two consequent passes, see Figure 1 (b). To define the contact ellipses Hertzian contact assuming elastic material behavior is utilized:

$$O_e = \frac{A_o}{A} = \frac{2}{\pi} \arccos\left(\frac{f}{2a}\right) - \frac{f}{\pi a}\sqrt{1 - \frac{f^2}{4a^2}} \tag{1}$$

with $a = c\left(\frac{3F\,r}{2E^*}\right)^{\frac{1}{3}}$ (2) and $r = 2/\left(\frac{2}{r_t} + \frac{1}{r_w}\right)$ (3) and $E^* = \frac{2E_tE_w}{(1-v_t^2)E_w + (1-v_w^2)E_t}$ (4)

Where *rt* is the radius of the deep rolling tool and *rw* is the radius of the workpiece (for flat specimen *rw*=∞,) E_1 and E_2 the young's modulus of deep rolling tool and workpiece and *F* the contact force. *F* can be simplified calculated by the following equation:

$$F = \pi\, r_t^2\, p \tag{5}$$

Where p is the operating pressure of the hydrostatically supported tool. Furthermore, the constants c_1 and β were calculated according to the following equations:

$$c = \frac{2452,92}{\beta^3} - \frac{560,021}{\beta^2} + \frac{97,5878}{\beta} \quad (6) \qquad \beta = \arccos\left(\frac{r}{2}\frac{1}{r_w}\right) \quad (7)$$

The overlap factor *Oe(f,p)* depends on the process parameter *f* and *p*, as well as the materials of tool and workpiece diameter and tool diameter. In the case of a flat workpiece the stepover distance *s* is used instead of *f*. However, the plastic deformation during the process is not covered by this simplified approach. To account for elastic-plastic material behavior an alternative definition of the overlap factor *Op* is introduced, that is identical with equation 1 but instead of utilizing Hertzian pressure to calculate the contact ellipse half axis *a* (eq. 2) the width of the contact surface is obtained by Finite Element (FE) simulation and is hereafter designated as *b*, see Figure 1(c) and equation 8.

$$O_p = \frac{2}{\pi} \arccos\left(\frac{f}{2b}\right) - \frac{f}{\pi b}\sqrt{1 - \frac{f^2}{4b^2}} \quad (8)$$

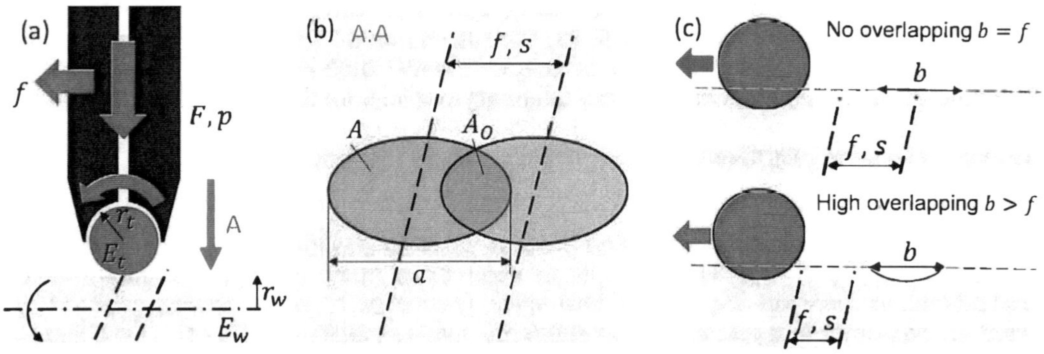

Figure 1. (a) Deep rolling process parameters, (b) overlap factor defined by contact ellipses, (c) overlap factor defined by plastic deformation of the surface after deep rolling

Material and Specimen

Three materials were investigated: The aluminum-alloy AlMg4.5Mn0.7 (EN AW 5083), the mild steel S355J2+N and the medium strength steel S690QL. The materials were chosen according to the investigations regarding deep rolling of welded joints [12,13]. The mechanical properties of this materials were determined by tensile tests according to DIN EN ISO 6892-1:2020-06 and are summarized in Table 1. The monotonic tensile curves are shown in Figure 2 (a). Deep rolling was applied with different contact pressure and feed rate at round bar specimens, shown in Figure 2 (b), at the laboratory of the Ecoroll AG. For this, a hydrostatic deep rolling tool type HG6-9 with an effective diameter of r_t=6.35 mm was used. Two rotational velocities were used, Vu=520 1/min and Vu=920 1/min, for S355J2+N and EN AW 5083. Altogether 33 parameter sets were investigated.

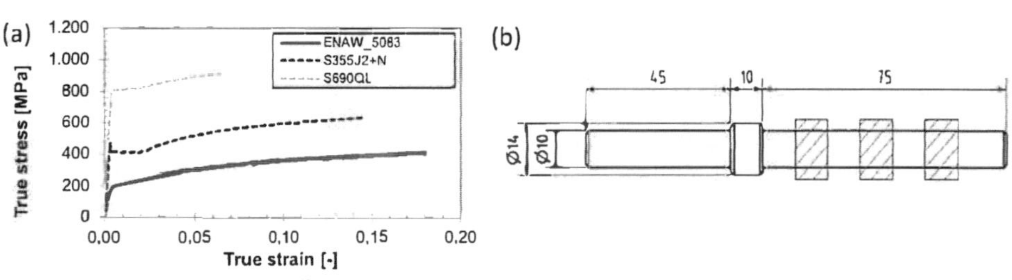

Figure 2. (a) Monotonic tensile curves of the investigated materials, (b) Specimen geometry

Table 1. Mechanical properties of the investigated materials.

	Yield strength [MPa]	Strength [MPa]	Young's modulus [GPa]	Elongation after fracture [%]	Hardness [HV10]
EN AW 5083	203	325	70	24.5	85
S355J2+N	402	505	210	32	165
S690QL	803	834	210	17	282

Determination of overlapping

The overlap factor *Op(f,p)* in dependence of feed rate *f* and contact pressure *p* was determined by analytical according to equation (1) to (7), as well as by FE-simulation, see Figure 3. An isotropic hardening model based on the stress-strain curves in Figure 2 (a) was used. The material behavior of the tool was assumed as linear elastic with E=210 GPa and v=0.3.

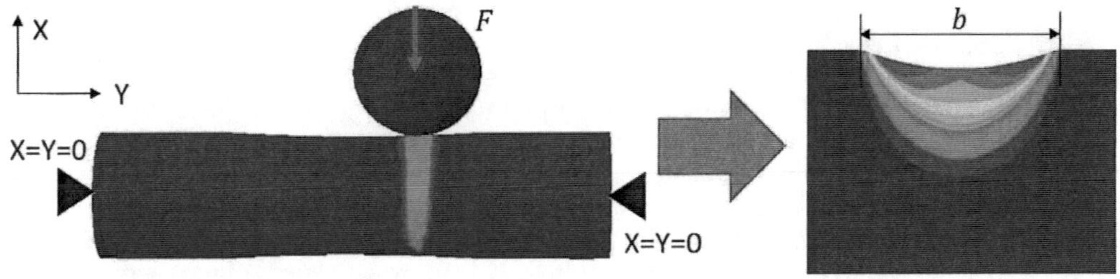

Figure 3. Finite Element model of the deep rolling process of a single line

The elastic overlap factor *Oe* and the plastic overlap factor *Op* in dependence of the contact pressure *p* and the feed rate *f* are displayed in Figure 4. As expected, the values of *Op* are significantly higher than for *Oe*. Also, it is shown that the overlap factors are much more influenced by feed rate than the contact pressure. High differences of *Oe* and *Op* are determined for a feed rate f > 0.1 mm/rot. For lower feed rates similar values of *Oe* and *Op* are calculated (*Oe*=0.89 and *Op=0.97* for f =0.05 mm/rot) for all materials.

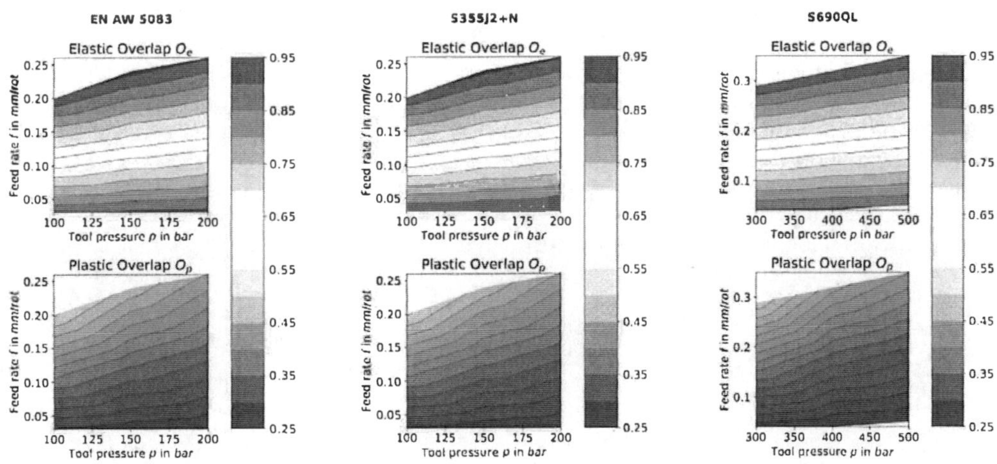

Figure 4. Influence of tool pressure and feed rate on the elastic and plastic overlap factor

Influence on the surface state

To investigate which overlap factor is needed for a proper treatment result the surface roughness and residual stress state were investigated. For all investigated parameter sets the overlap factor *Oe* for linear-elastic contact according to equation (1) to (7) and the overlap factor *Op* based on elastic-plastic material behavior was calculated.

The surface roughness Rz and maximum roughness Rt according to DIN 4768:1990-05 was measured within 4.8 mm transverse to the rolling direction by a tactile roughness measurement device type Hommel T8000 according to DIN EN ISO 3274:1998-04. In initial condition the specimens have a roughness of Rz=5,9 µm (EN AW 5083), Rz=7.3 µm (S355J2+N) and Rz=7.1 µm (S690QL). The results for Rz after deep rolling are displayed in Figure 5 in dependence of overlap factor and tool pressure. The results show that the achieved roughness is highly dependent on the tool pressure. While for EN AW 5083 and S355J2+N low pressures are favorable, S690QL shows best results with a high tool pressure. This finding points towards a dependency of the optimum tool pressure on the material's tensile strength. For all materials an elastic overlap factor of 60% and more was found to lead to the lowest surface roughness.
A high overlap factor is also not generally favorable for achieving low surface roughness. As the results for EN AW 5083 show, an overlap of 90% with a tool pressure of 150 bar leads to a very poor surface quality, see also Figure 6. Surface defects were also detected for *Oe*=90% for S355J2+N in combination with a high contact pressure of *p*=200 bar. In these cases, deep rolling leads to surface defects that may negatively affect the surface functions. For S690QL however, no surface defects were determined even for *Oe*=90% and a high contact pressure of *p*=500 bar. It is assumed that the yield strength of the material is high enough for a high intensity treatment (high overlap factor) in this case.
The surface-near residual stress state after deep rolling was determined by X-ray diffraction (XRD) analysis. The residual stresses were measured at the {422}-aluminum lattice plane with Cu-Kα radiation for EN AW 5083 and at the {211}-ferritic lattice plane with Cr-Kα radiation for S355J2+N and S690QL with a diffractometer type Bruker D8 Discover. The residual stresses in hoop- and axial-direction were measured with a collimator diameter of 1 mm. The evaluation was performed according to the $\sin\psi^2$-method assuming a planar stress state. The constants *E*=220 GPa and ν=0.29 (steel) and *E*=69.7 GPa ν=0.348 (Al) were used.
The results of the XRD-analysis are summarized in Figure 7. The residual stress values were normalized with the yield strength of the materials, see Table 1. As illustrated, the residual stresses in axial direction are less affected by the variation of overlapping and contact pressure for all investigated materials. In hoop direction, a tendency is shown that a higher contact pressure for S690QL leads to higher compressive residual stresses. Furthermore, the maximum compressive residual stresses in hoop direction were measured for a lower overlapping of *Oe*=30% (*Op*≈76%) for all materials. The highest compressive residual stresses in both directions were determined at *Oe*=60% (*Op*=76.8%) and p=100 bar for EN AW 5083, at *Oe*=60% (*Op*=89.3%) and p=200 bar for S355J2+N and at *Oe*=90% (*Op*=97.4%) and p=500 bar for S690QL.

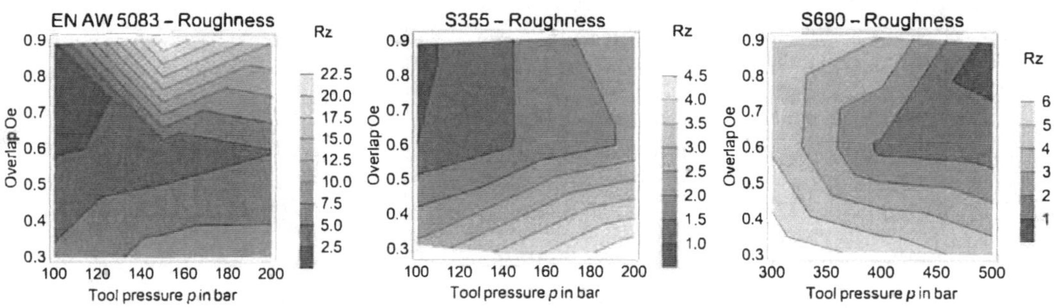

Figure 5. Impact of overlapping on the surface roughness after deep rolling

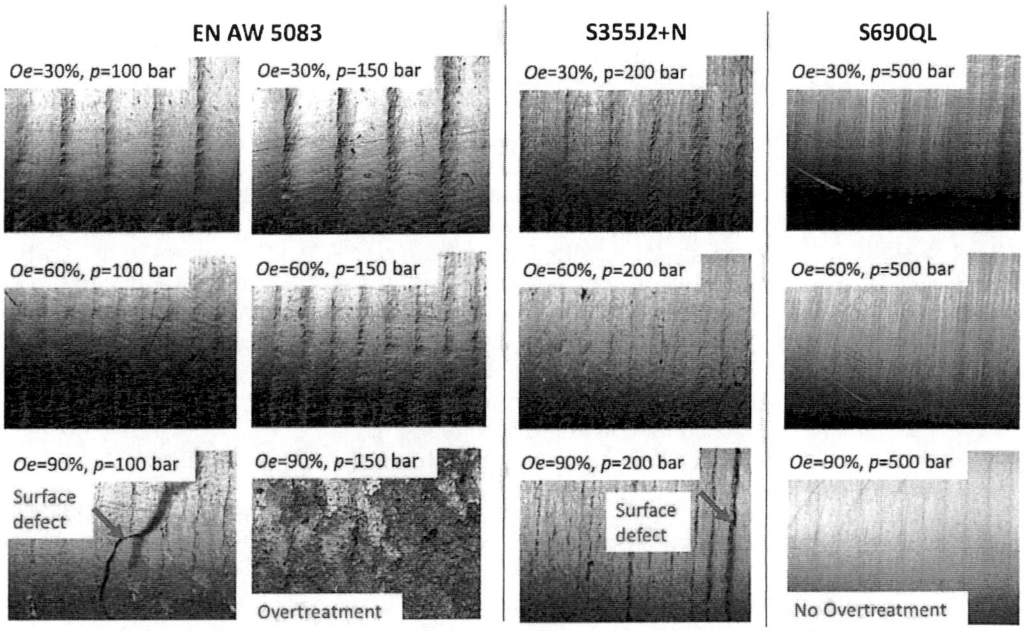

Figure 6. Impact of overlapping on the surface roughness after deep rolling

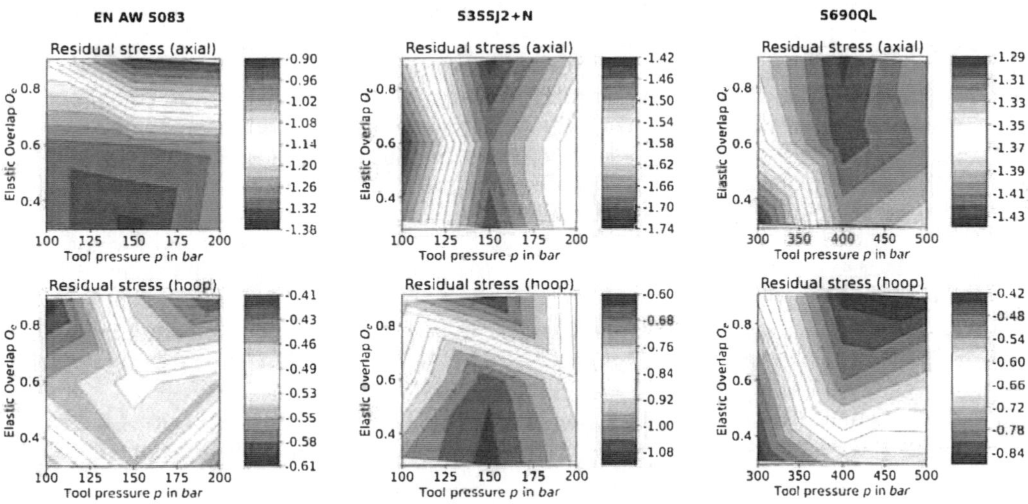

Figure 7. Impact of overlapping on the residual stress state after deep rolling

Conclusions

The overlap factor by deep rolling of subsequent treated line is defined assuming pure linear material behavior (*Oe*) and elastic-plastic material behavior (*Op*) to reduce the optimization effort and process parameters. Round bar specimen made of EN AW 5083, S355J2+N and S690QL were deep rolled with different overlap factors. The influence of the overlap factor on surface roughness and residual stress was studied. For EN AW 5083 and S355J2+N and optimum overlap factor of *Oe*=60% (*Op*≈87%) with a tool pressure of 100 bar was determined. For higher values of *Oe* or higher tool pressure surface defects were induced by deep rolling. For S690QL a high overlap factor of *Oe*=90% (*Op*≈97%) does not lead to surface defects. Surface roughness was optimized by applying a high overlap factor and a high contact pressure of 500 bar. The residual stress state at the surface after deep rolling was comparably unaffected by the variation of the overlap factor. However, a high influence of the overlap factor on the surface roughness was determined. For mild strength materials (EN AW 5083 and S355J2+N) increasing the overlap factor above the stated Oe=60% does not lead to lower surface roughness. Too high contact pressure leads to a destruction of the specimen's surface.

Funding

The IGF project „Festwalzen zur Schwingfestigkeitserhöhung bzw. Lebensdauerverlängerung zyklisch beanspruchter Schweißkonstruktionen aus Stahl und Aluminiumlegierungen" (No IGF-19537N) of the Research Association on Welding and Allied Processes of the DVS, Aachener Str. 172, 40223 Düsseldorf has been funded by the AiF within the program for sponsorship by Industrial Joint Research (IGF) of the German Federal Ministry for Economic Affairs and Climate Action based on an enactment of the German Parliament.

References

[1] VDI, "VDI 3177:1963-03 Surface finish-rolling," 1963.

[2] V. Schulze, *Modern Mechanical Surface Treatment: States, Stability, Effects*. New York: Wiley VCH, 2006.

[3] V. Schulze, F. Bleicher, P. Groche, Y. B. Guo, and Y. S. Pyun, "Surface modification by machine hammer peening and burnishing," *CIRP Ann.*, vol. 65, no. 2, pp. 809–832, Jan. 2016, doi: 10.1016/J.CIRP.2016.05.005.

[4] P. Delgado, I. I. Cuesta, J. M. Alegre, and A. Díaz, "State of the art of Deep Rolling," *Precis. Eng.*, vol. 46, pp. 1–10, Oct. 2016, doi: 10.1016/J.PRECISIONENG.2016.05.001.

[5] G. H. Majzoobi, K. Azadikhah, and J. Nemati, "The effects of deep rolling and shot peening on fretting fatigue resistance of Aluminum-7075-T6," *Mater. Sci. Eng. A*, vol. 516, no. 1–2, pp. 235–247, Aug. 2009, doi: 10.1016/J.MSEA.2009.03.020.

[6] I. Altenberger, R. K. Nalla, Y. Sano, L. Wagner, and R. O. Ritchie, "On the effect of deep-rolling and laser-peening on the stress-controlled low- and high-cycle fatigue behavior of Ti-6Al-4V at elevated temperatures up to 550 °c," *Int. J. Fatigue*, vol. 44, pp. 292–302, Nov. 2012, doi: 10.1016/j.ijfatigue.2012.03.008.

[7] N. Ben Moussa, K. Gharbi, I. Chaieb, and N. Ben Fredj, "Improvement of AISI 304 austenitic stainless steel low-cycle fatigue life by initial and intermittent deep rolling," *Int. J. Adv. Manuf. Technol. 2018 1011*, vol. 101, no. 1, pp. 435–449, Nov. 2018, doi: 10.1007/S00170-018-2955-0.

[8] S. Saalfeld, B. Scholtes, and T. Niendorf, "On the influence of overloads on the fatigue performance of deep rolled steel SAE 1045," *Int. J. Fatigue*, vol. 126, pp. 221–230, Sep. 2019, doi: 10.1016/J.IJFATIGUE.2019.05.001.

[9] D. Trauth, F. Klocke, P. Mattfeld, and A. Klink, "Time-efficient prediction of the surface layer state after deep rolling using similarity mechanics approach," in *Procedia CIRP*, Jan. 2013, vol. 9, pp. 29–34, doi: 10.1016/j.procir.2013.06.163.

[10] J. Schubnell, L. Mayer, E. Carl, and M. Farajian, "Deep rolling as an effective tool for the fatigue improvement of tubular welded joints," *IIW-doc. XIII-2858-2020*, 2020.

[11] J. Schubnell and M. Farajian, "Fatigue improvement of aluminium welds by means of deep rolling and diamond burnishing," *Weld. World*, vol. 1, pp. 1–10, Dec. 2021, doi: 10.1007/S40194-021-01212-1/TABLES/5.

Laser Shock Peening to enhance stress corrosion cracking and corrosion fatigue resistance in marine aluminium alloys

J. Kaufman[1,2], M. A. Steiner[3], S. R. Mannava[3], V. K. Vasudevan[4], J. Brajer[2], J. Pilař[2], L. Pína[1], T. Mocek[2]

[1] Faculty of Nuclear Sciences and Physical Engineering, CTU in Prague, Czechia
[2] HiLASE Centre, Institute of Physics of the Czech Academy of Sciences, Czechia
[3] Mechanical and Materials Engineering Department, University of Cincinnati, USA
[4] Department of Materials Science and Engineering, University of North Texas, USA

Abstract
This paper shows the effects of Laser Shock Peening (LSP), an alternative to Shot Peening, on marine aluminium alloys. LSP generated compressive residual stresses which varied depending on the peening strategy and method of confinement. Electrochemical tests indicated a region of increased passivity samples without protective coating as well as significantly reduced intergranular corrosion rate. 3-point bend tests showed significant improvement in fatigue strength. Slow strain rate testing showed strong work hardening of treated specimens although the influence on IGSCC was not clearly indicated.

Keywords marine aluminium alloys, laser shock peening, intergranular corrosion, sensitization

Introduction
5xxx series aluminium alloys are high strength-to-weight ratio weldeable materials known for their exceptional performance in extreme environments such as seawater. They are commonly used in the marine industry for the construction of hulls and other ship-related structures due to their excellent corrosion resistance. The 5xxx aluminium series can also be referred to as aluminium-magnesium alloys due to high magnesium content which is responsible for the enhanced strength. However, when exposed to elevated temperatures (<50 °C) for prolonged periods of time (years), the alloys become sensitized which makes them susceptible to intergranular corrosion (IGC) [1]. Combined with the tensile loading experienced during the ship operation, the IGC can lead to intergranular stress corrosion cracking (IGSCC) [2] and corrosion fatigue (CF) [3] which can result in catastrophic failures. For example, as of 2015, more than 6000 SCC and CF cracks have been documented across the whole Ticonderoga class fleet [4]. Several approaches were tried to address the problem, mainly through metallurgy and electrochemical inhibition but the results were unsatisfactory. In this work, we investigate the stress removal option where tensile stresses in critical areas are replaced with compressive stresses using LSP. While the positive effects of LSP on SCC [5] and CF [6] are well documented using other materials. Studies conducted both with [7] and without protective coating [8] indicate that LSP could also have a direct positive effect on corrosion behaviour.

Experimental methods
For residual stress (RS) measurements, a 6.35 mm thick plate of AA5083-H116 was sectioned using electric discharge machining into 40 mm x 40 mm square coupons. The chemical composition of the alloy is shown in Table 1. The coupons were peened in their received condition without any additional heat treatment. The corrosion fatigue samples were in the form of 60 mm x 12 mm x 4 mm blocks sectioned from a 25.4 mm thick rolled plate. And finally, the slow strain rate test (SSRT) were also sectioned from a 25.4 mm thick rolled plate with dimensions showed in Fig. 1. Both the CF and SSRT were sensitized at 100 °C for 60 days prior to the LSP treatment. The sensitized material was determined to be IGC susceptible following the ASTM G67 standard [9]

where the degree of sensitization (DoS) was determined to be 5 mg/cm^2 and 44.6 mg/cm^2 for the as-received material and the sensitized material, respectively.

Table 1 : Chemical composition of AA5083-H116.

Alloy element	Mg	Si	Mn	Fe	Other (Zn, Cr, Ti, Cu)	Al
Composition in wt%	4.0–4.9	0.4	0.4–1.0	0.4	0.65	balance

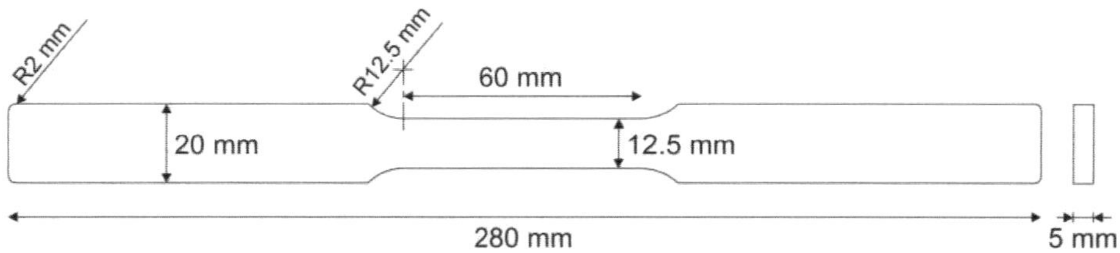

Figure 1 : SSRT sample dimensions.

RS measurements were conducted on an X-ray diffractometer Proto LXRD instrument using the sin$^2\psi$ method using Cr K_α (2.2897 nm) radiation at 2Θ = 139°. To measure residual stresses through depth, layers of the material were progressively removed using an electro polisher with 87.5:12.5 vol% CH_3OH:H_2SO_4 electrolytic solution. Corrosion fatigue tests were performed on Electromagnetic pulsator Testronic for high cycle fatigue testing. The 3-point bend test was load controlled and took place at room temperature at 45 Hz with R value 0.1. The supports were 20 mm apart and the specimen was submerged in 3.5% NaCl solution. The terminating condition was either when 10^7 cycles were reached, or when the frequency dropped by 20 Hz due to lowered toughness, signified by a crack that developed and started to propagate. The tensile testing was performed on Kappa 100 SS-CF electromagnetic creep testing machine with a uniaxial strain rate of 10^{-6}. The samples were placed into a specially designed cell which contained circulating 3.5% NaCl water solution kept at constant 40 °C. Prior to the test, the upper and lower wide parts of the specimen were coated with silicon to ensure that only the gauge was in contact with the electrolyte.

The LSP process uses high-energy nanosecond laser pulses to generate deep compressive residual stresses in metallic materials. The sample surface is usually covered with a protective coating in the form of black vinyl tape or paint, which absorbs the focused laser pulses (Fig. 2). The subsequent rapidly expanding plasma is confined against the sample surface by a thin water layer (1–2 mm thick), and a strong shock wave with a magnitude of several GPa [10], is formed. If the pressure exceeds the yield strength of the material, the surface is plastically deformed, and compressive stresses are generated. With the protective coating applied, LSP can be considered a cold working process since the heat associated with the laser absorption is screened from the sample surface and only plastic deformation takes place. In Laser Peening without Coating (LPwC), the surface is directly affected by heat, which can substantially affect the residual stresses in the near surface layer. All samples were peened using first power pre-amplifier of Yb:YAG laser system Bivoj operating at 1030 nm with 10 Hz repetition rate [11] and second harmonic generation option for 515 nm. The laser pulses were 14 ns long with a rectangular temporal and spatial profile with top-hat intensity distribution. Two peening conditions were investigated, one where protective coating was applied while using infrared laser (LSP IR) and second with no protective coating applied while using green laser wavelength (LPwC SHG). The LPwC SHG treatment took place underwater in a water tank. The LSP IR condition used pulses of 3 J focused to a 2.6 mm square laser spot with resulting power density of 3.17 GW/cm^2. The sample after peening is shown in Fig. 2b. When compared to the non-peened sample (Fig. 2a) The surface is clearly deformed but it retains its polished appearance. The LPwC SHG treatment used 1 J laser pulses with laser spot size of 1.5 mm, power density of 3.17 GW/cm^2 and pulse density of 1089 p/cm^2. The LPwC SHG sample after treatment is shown in Fig. 2c. As opposed to the LSP IR sample, a white recast layer was created on the surface.

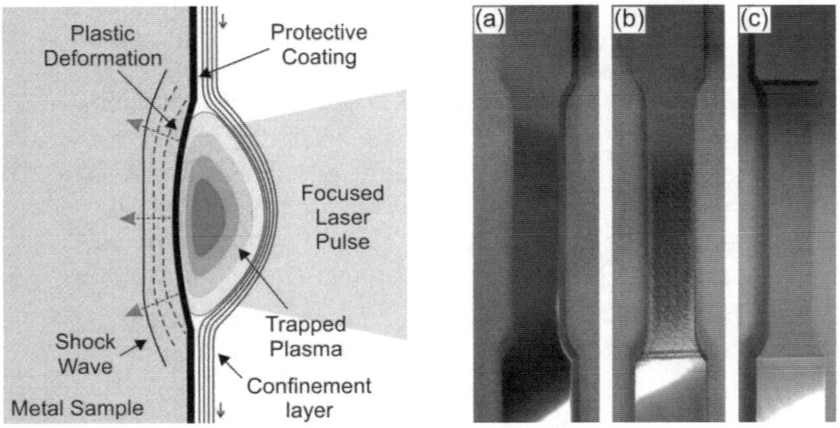

Figure 2: *(left) LSP mechanism and (right)* SSRT samples with (a) no laser peening, (b) LSP IR and (c) LPwC SHG.

Experimental results

Fig. 3 shows in-depth residual stress measurements. Both directions of measurement are shown, that means σ_S and σ_T which represent the scanning and transversal directions, respectively. In comparison with the baseline, the LSP IR treatment with protective tape generates significant compressive residual stresses, around -200 MPa on the surface, with maximum compressive stresses reached at a depth of about 200 µm. The stresses induced in both σ_S and σ_T directions are very similar and no significant variations are observed. The compression depth in both directions is about 1.5–1.6 mm, which is higher than in the LPwC case.

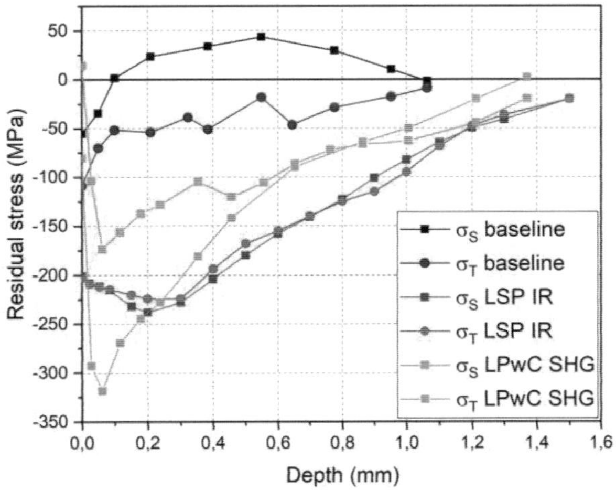

Figure 3: In-depth residual stress measurement where σ_S and σ_T denote stresses in scanning and transversal directions, respectively.

Unlike the LSP IR plots, the LPwC SHG treatment shows a clear distinction between σ_S and σ_T. In scanning direction, the maximum compressive stress reached is 180 MPa and the compression depth is about 1.4 mm. In the transversal direction, the stress curve changes with a maximum compression of about 325 MPa. Similar resultant stress anisotropy with LPwC treatment was observed in other studies. Correa et al. [12] explain the phenomenon in the context of interacting stress fields created by laser pulses overlapping in zigzag scanning patterns.

During SSRT, sensitized samples without any additional treatment were tested first both in air and NaCl solution to determine what effect the environment has on the tensile properties of the samples. The stress/strain curves are shown in Fig. 4a. We can therefore conclude that the decrease in maximum stress and strain to failure is a direct result of the environmental influence of the NaCl solution on the sensitized material. The results of SSRT of sensitized specimens after LSP and LPwC treatments are shown in Fig. 4b. It is immediately clear the material was cold worked by the laser treatment, resulting in its strengthening and raising of maximum stress before breaking. At the same time, the strain hardening lowered the ductility as the nominal strain at break decreased significantly. The maximum stress at failure was achieved with the LSP IR condition with ~8% increase when compared to the baseline although the strain at break dropped by 45%. LPwC SHG had a similar increase in maximum stress of 6% but the strain at break decreased by 18% only.

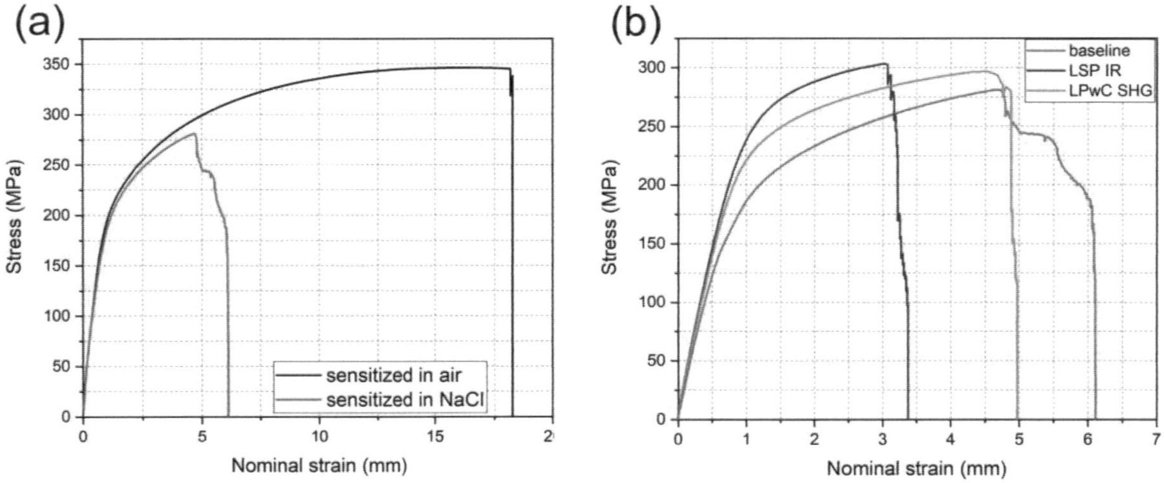

Figure 4: Stress/strain curves of (a) sensitized samples in air and 3.5% NaCl solution and (b) of sensitized samples in 3.5% NaCl solution after LSP and LPwC treatments.

The fracture analysis consists of SEM followed by metallography analysis. All the images are displayed in Fig. 5 to allow for direct comparisons between different sample conditions. The SEM show two different crack propagation mechanisms. Sensitized samples in air show transgranular crack features while all sensitized samples in NaCl show distinct intergranular cracking. The metallographic analysis further reveals that cracks in the NaCl solution were initiated at the specimen surface, specifically at the edges. At first, the intergranular cracks propagated perpendicular to the applied stress due to material weakening in the corrosive environment. As the cracks progressed, the fracture morphology changed to ductile until the specimens fractured completely. Furthermore, as the primary crack propagated, secondary intergranular corrosion cracks split off into the specimen volume. Moreover, next to the primary crack, smaller perpendicular cracks can be observed on the edge surfaces of specimens in NaCl solution which lead to weakening of specimen cross-section. The size and length of these smaller cracks varied among different testing conditions. The cracks were smaller in the LPwC SHG case. This assessment, however, is very rough since laser treated strain hardened specimens failed sooner than non-treated specimens and the surface cracks did not have the same amount of time to develop. Nevertheless, LSP IR specimens with about half the time to failure displayed side intergranular cracks about twice as short which implies linear propagation of the surface cracks in time. Based on this estimate and longer times to failure, we would expect longer surface cracks for the LPwC specimen compared to the LSP IR but the crack length was in fact shorter.

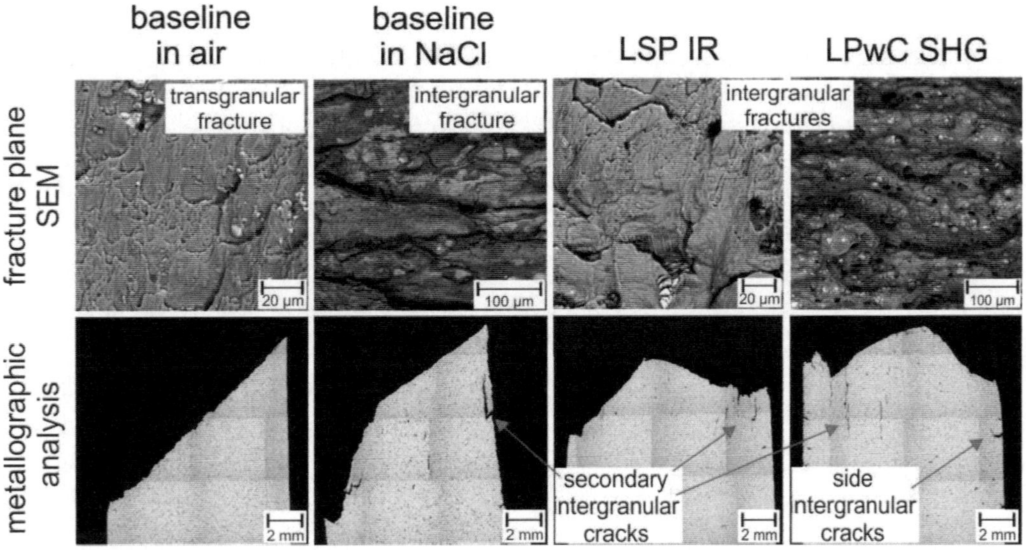

Figure 5: Fracture analysis results.

The results of the 3-point bend test are shown in Fig. 6. The sensitized sample displays the S–N curve shifted down compared to the as received sample which shows a clear negative effect of IGC on the CF of AA5083. The fatigue strength of the non-sensitized baseline material was 127 MPa which is 44 MPa higher than the sensitized baseline sample. Both the LSP and LPwC treatments had a positive impact on the fatigue, bringing the S–N curves back up and possibly improving the original non-sensitized fatigue strength. Better improvements were achieved with the LPwC SHG treatment where the fatigue strength improved by 69%. LSP IR showed lower 59% improvement, but still reached a slightly higher fatigue resistance than the non-sensitized sample.

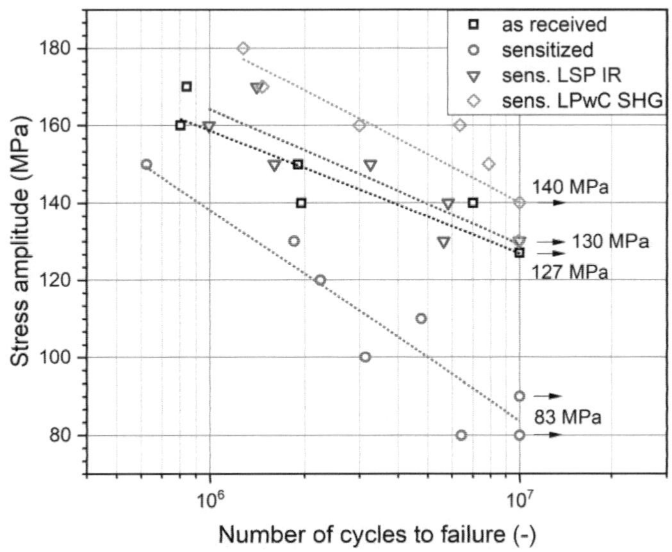

Figure 6: S-N curves of sensitized AA5083 after 3-point bend test in 3.5% NaCl solution

Prior to laser treatment, the effect of sensitization in terms of thermodynamic stability was evaluated by potentiostatic polarization measurement. In order to suppress the chloride driven attack, the samples were polarized at - 800 mV vs Ag/AgCl. The current density measured over time is shown in Fig. 7a. After some time, the current density of the sensitized samples starts rising exponentially, reaching a value of 330 μA/cm^2 over a period of 6 h. On the contrary, the current density of the as received sample stays close to zero and does not deviate over time. The as received and sensitized sample surface after the polarization is shown in Fig. 7b and Fig 7c, respectively. The IGC is clearly visible on the sensitized sample while no pitting is present. When it comes to the treated samples, LPwC SHG treatment shows significant improvement. The current density after 6 h is 5 μA/cm^2, which is nearly the same as that of the as received sample. LSP IR sample with protective coating showed only a slight improvement with final current density of 267 μA/ cm^2.

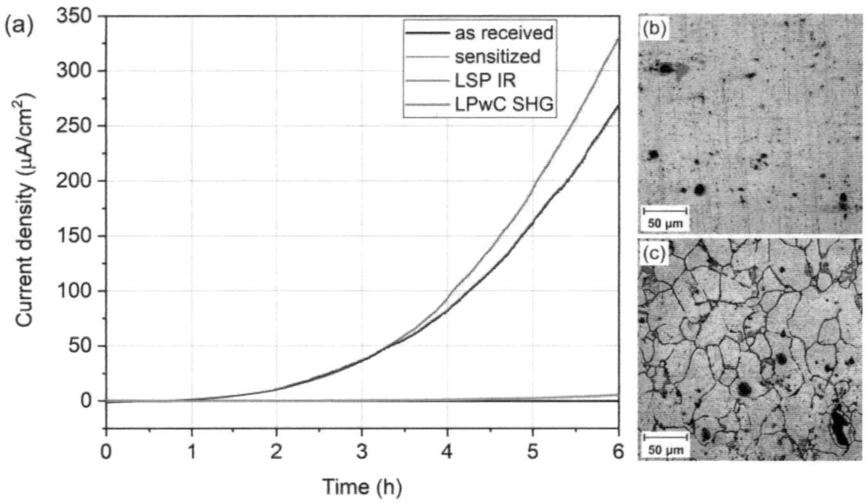

Figure 7: (a) Current density during potentiostatic polarization at -800 mV vs Ag/AgCl electrode of non-treated and treated AA5083 in 3.5% NaCl solution. (b) as received and (c) sample surface after 6 hours of potentiostatic polarization.

Conclusions

In this work, the effect of LSP and LPwC treatment on intergranular stress corrosion cracking and corrosion fatigue of sensitized AA5083-H116 was investigated. Two laser peening configurations were used, one with protective overlay and one without which were further separated by different water confinement. The following conclusions can be made:

1. Both LSP and LPwC imparts deep compressive residual stresses into the studied material. Residual stress anisotropy was observed in the case of LPwC treatment, where larger compressive stresses were measured in the laser advancing direction with respect to the peening pattern.
2. Laser plasma interaction with the aluminium surface associated with the LPwC process creates a recast layer with a modified oxide layer on top in the treated area. This leads to significantly lower intergranular corrosion rate.
3. Sensitization has a negative impact on corrosion fatigue of the material which manifests as a 35% drop in fatigue strength. LSP and LPwC treatment restored and even possibly improved the original fatigue resistance of the non-sensitized samples.
4. The SSRT measurements clearly showed the cold working effect of the LSP treatment. We have observed an increase in maximum stress but at the same time significant decrease in nominal strain at failure as the samples lost some ductility. The effect of LSP on SCC is not clear but some evidence suggests that LPwC treatment slows down the crack propagation.

References

[1] R. Zhang, S. P. Knight, R. L. Holtz, R. Goswami, C. H. Davies, and N. Birbilis, *A survey of sensitization in 5xxx series aluminum alloys*, Corrosion, vol. 72 (2016), pp. 144–159.

[2] J. L. Searles, I. G. Pelagia, and R. G. Buchheit, *Stress corrosion cracking of sensitized AA5083 (Al-4.5 Mg-1.0 Mn)*, Materials Science Forum. Vol. 396. (2002)

[3] R. L. Holtz, P. S. Pao, R. A. Bayles, T. M. Longazel, and R. Goswami, *Corrosion fatigue of al 5083-H131 sensitized at 70, 100, and 175 c relation to microstructure and degree of sensitization*, in DoD Corrosion Conference (2011).

[4] C. Cavas, *Cracks plague ticonderoga-class cruisers*, Navy Times, vol. 9 (2010).

[5] Y. Sano, M. Obata, T. Kubo, N. Mukai, M. Yoda, K. Masaki, and Y. Ochi, *Retardation of crack initiation and growth in austenitic stainless steels by laser peening without protective coating*, Materials Science and Engineering A, vol. 417 (2006), pp. 334–340, 2006.

[6] M. Abdulstaar, M. Mhaede, M. Wollmann, and L. Wagner, *Investigating the effects of bulk and surface severe plastic deformation on the fatigue, corrosion behaviour and corrosion fatigue of AA5083*, Surface and Coatings Technology, vol. 254 (2014), pp. 244–251.

[7] H. Wang, C. Ning, Y. Huang, Z. Cao, X. Chen, and W. Zhang, *Improvement of abrasion resistance in artificial seawater and corrosion resistance in NaCl solution of 7075 aluminum alloy processed by laser shock peening*, Optics and Lasers in Engineering, vol. 90 (2017).

[8] U. Trdan and J. Grum, *Evaluation of corrosion resistance of AA6082-T651 aluminium alloy after laser shock peening by means of cyclic polarisation and EIS methods*, Corrosion Science, vol. 59 (2012), pp. 324–333.

[9] *ASTM G67, Standard Test Method for Determining the Suscepti- bility to Intergranular Corrosion of 5XXX Series Aluminum Alloys by Mass Loss After Exposure to Nitric Acid (NAMLT)*, ASTM International (1999).

[10] R. Fabbro, et al, *Physics and applications of laser-shock processing*, Journal of laser applications vol. 10.6 (1998), pp. 265-279.

[11] P. Mason, M. Divoký, K. Ertel, J. Pilař, T. Butcher, M. Hanuš, S. Banerjee, J. Phillips, J. Smith, M. De Vido, *Kilowatt average power 100 J-level diode pumped solid state laser*, Optica 4 (2017), pp. 438–439.

[12] C. Correa, D. Peral, J. A. Porro, M. Díaz, L. R. de Lara, A. García-Beltrán, J. L. Ocaña, *Random-type scanning patterns in laser shock peening without absorbing coating in 2024-T351 Al alloy: A solution to reduce residual stress anisotropy*, Opt. Laser Technol. Vol 73 (2015), pp. 179–187.

Compressive residual stress relaxation behavior of shot peened AISI4140 steel during first cycle of repeated loading

Motoaki Hayama[1] and Jun Komotori[2]

1 Graduate School of Science and Technology, Keio University, Japan
2 Department of Mechanical Engineering, Keio University, Kanagawa, Japan

Abstract
This study aims to reveal the relaxation behavior of the compressive residual stress induced by shot peening during the first cycle of cyclic loading. An hourglass-type specimen was machined from AISI440 steel, and the surface was shot peened using high-speed steel. The relaxation behavior of the compressive residual stress was investigated using an *in-situ* X-ray stress measurement method. Results shows that the compressive residual stress relaxation occurred during the compression process of the first fatigue loading, and the relaxation occurred continuously after the first cycle.

Keywords Residual stress, Relaxation, in-situ X-ray measurement, fatigue.

Introduction

Shot peening is an effective method to improve the fatigue properties of steel. Compressive residual stress is one of the important factors affecting the fatigue properties of shot peened material as it prevents crack initiation and its propagation [1, 2]. However, the compressive residual stress is relaxed during the fatigue process [2–6] and reducing the benefit in improving fatigue properties. It is important to reveal the relaxation behavior of compressive residual stress to evaluate its effect on fatigue properties. It was reported that the relaxation of the compressive residual stress occurs primarily within the first cycle of fatigue [4–6]. However, the behavior of residual stress relaxation during the first cycle of repeated loading has not yet been clarified. Therefore, this study aims to examine the relaxation behavior of residual stress generated by shot peening during a cycle of repeated loading, using an in-situ X-ray stress measurement [6].

Experimental Methods

The material used in this study was the quenched-and-tempered AISI4140 steel with chemical composition (in wt%): 0.42C, 1.03Cr, 0.16Mo, 0.20Si, 0.75Mn, 0.016P, 0.013S, 0.12Cu, 0.05Ni, 0.0008O and balance Fe. The mechanical properties of AISI4140 steel are listed in Table 1. The material was machined into an hourglass-type fatigue test specimen with the smallest diameter of 5 mm and notch curvature of 24 mm. The smallest diameter section was polished using #100, #320 and #600 emery paper.
Shot peening was performed to generate the compressive residual stress for the smallest diameter section of the specimen. The peening conditions are listed in Table 2. The specimen was rotated on the latch during peening.

Table 1. Mechanical properties of AISI4140 steel

0.2 % proof stress	Tensile strength	Reduction in area	Vickers hardness (9.8 N)
1044 MPa	1133 MPa	54.8 %	335 HV

Table 2. Shot peening conditions.

Particle	High-speed tool steel (700 HV)
Particle diameter	125~150 μm
Projection method	Air blasting
Air pressure	0.6 MPa
Peening time	30 s

Hardness-depth profiles were obtained in cross section using a Vickers hardness tester at a load of 0.98 N. Residual stress was measured using portable X-ray diffraction stress tester. Measurements were performed in the longitudinal direction of the specimen and the stress value was calculated using the cosα method.
Fatigue tests were performed using the electrohydraulic-servo fatigue testing machine under axial loading with a stress ratio of −1 and test frequency of 20 Hz. The test was terminated when the specimen did not fail with a fatigue life of 10^7 cycles.
To investigate the change behavior of compressive residual stress during the fatigue process, *in-situ* X-ray stress measurement [6] was applied. A specimen was fixed to an axial loading fatigue test machine. The stress on the surface, i.e., the sum of applied stress and residual stress, was directly measured through X-ray diffraction without removing the specimen from the test machine. The applied stress was changed stepwise with a stress ratio R = −1, and the stress on the surface of the specimen were measured successively. The applied stress was maintained during the X-ray irradiation.

Experimental Results and Discussion
Fig. 1 shows the distribution of the Vickers hardness of the peened specimen; no increase in hardness was observed on the surface of the specimen owing to peening. The distribution of residual stress on the peened specimen is shown in Fig. 2; approximately 500 MPa compressive residual stress was generated on the peened AISI440 surface, and the total depth of the compressive residual stress region was approximately 90 μm.
The S-N diagram of the shot peened specimen obtained from the fatigue tests with axial loading is shown in Fig. 3; the fatigue life tended to increase as the stress amplitude decreased, and the specimens did not fail at stress amplitude of 500 MPa. The plots with '+' symbol represent the result of the interrupted fatigue test for residual stress measurement, and the ones with the '→' symbol represent the specimens, which did not fail until 10^7 cycles.

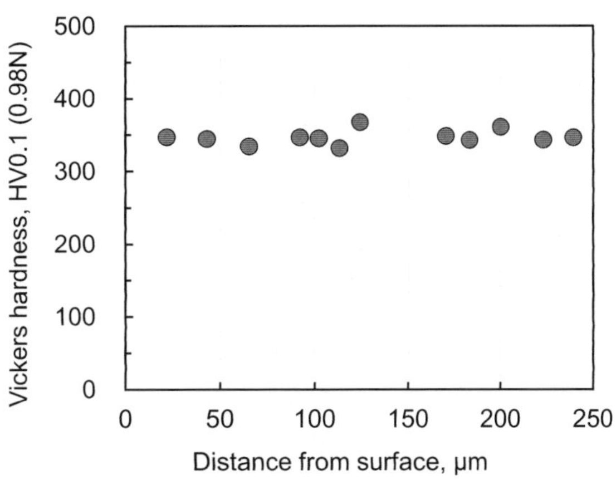

Fig. 1 In-depth Vickers hardness distributions of the peened surface

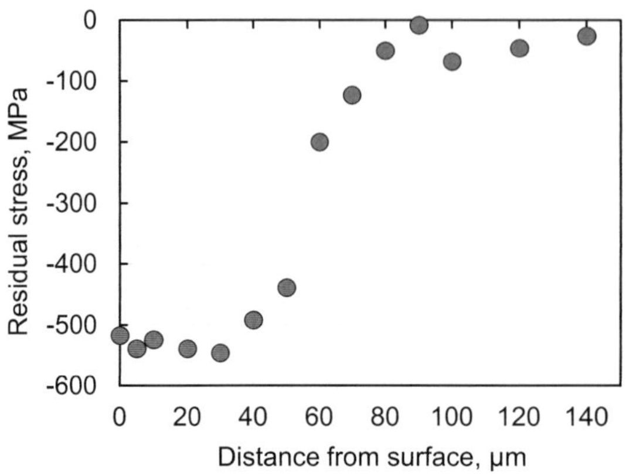

Fig. 2 Distribution of residual stress on the peened surface

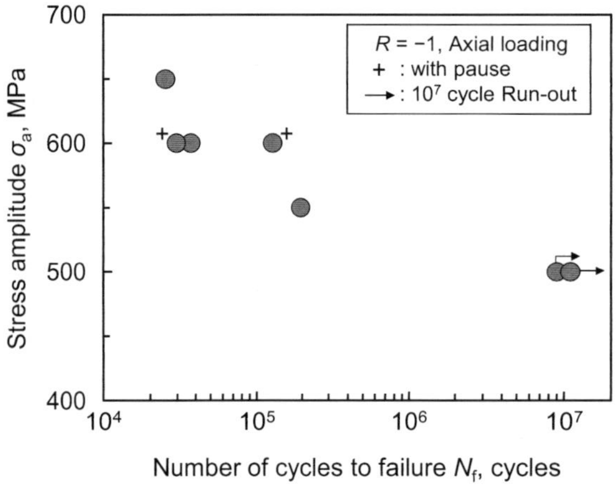

Fig. 3 S-N diagram of the peened AISI440 steel

Fig. 4 shows the surface residual stress of the specimen before and after the fatigue test at stress amplitude of 500 MPa. Before the fatigue test, approximately −500 MPa residual stress existed on the surface of the specimen. However, approximately −300 MPa residual stress remained after the fatigue test as the compressive residual stress was relaxed during the fatigue process.

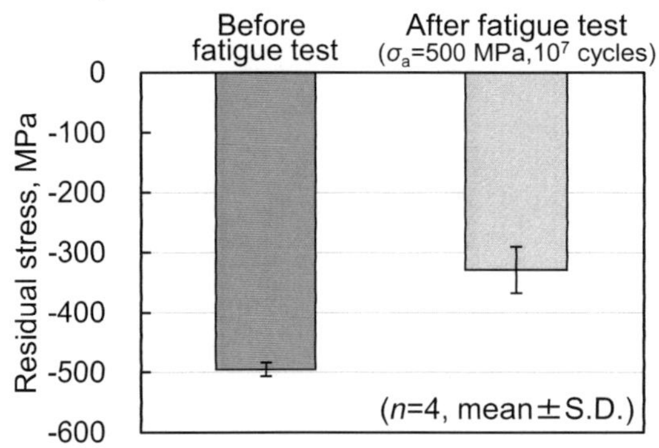

Fig. 4 Surface residual stress of the peened specimen before and after the fatigue test at stress amplitude of 500 MPa

To investigate the relaxation behavior of the compressive residual stress during the on-cycle of fatigue loading, in-situ X-ray stress measurement was applied. Figs. 5(a)~(c) show the relations between the applied stress and surface stress of the specimen during the fatigue test at stress amplitude of 600 MPa. The horizontal axis of the figure represents the applied stress, and vertical axis represents the surface stress measured through X-ray. Fig. 5(a) shows the results for the first cycle. The initial residual stress was approximately −450 MPa. The surface stress decreased linearly with the increase in the compressive applied stress until the applied stress reached approximately −300 MPa. The result indicates that the compressive residual stress did not change. However, non-linear relation between the applied stress and surface stress was observed to determine if the compressive applied stress exceeded −300 MPa. The result indicates that the compressive residual stress was relaxed during the compressive loading. This is owing to local yielding of the compressive residual stress layer. When the compressive stress was unloaded and tensile stress was applied, the surface stress varied linearly with the applied stress. Therefore, these results indicate that the compressive stress loading process is a significant phase in the compressive residual stress relaxation.

Figs. 5(b) and (c) show the results for the second and tenth cycles, respectively. In the second cycle, a nonlinear relationship, i.e., slight release of compressive residual stress can be observed in the compressive loading process. This result indicates that cyclic deformation occurred on the surface compressive layer. In the tenth cycle, relations between the applied stress and surface stress are linear, which indicate low relaxation of compressive residual stress.

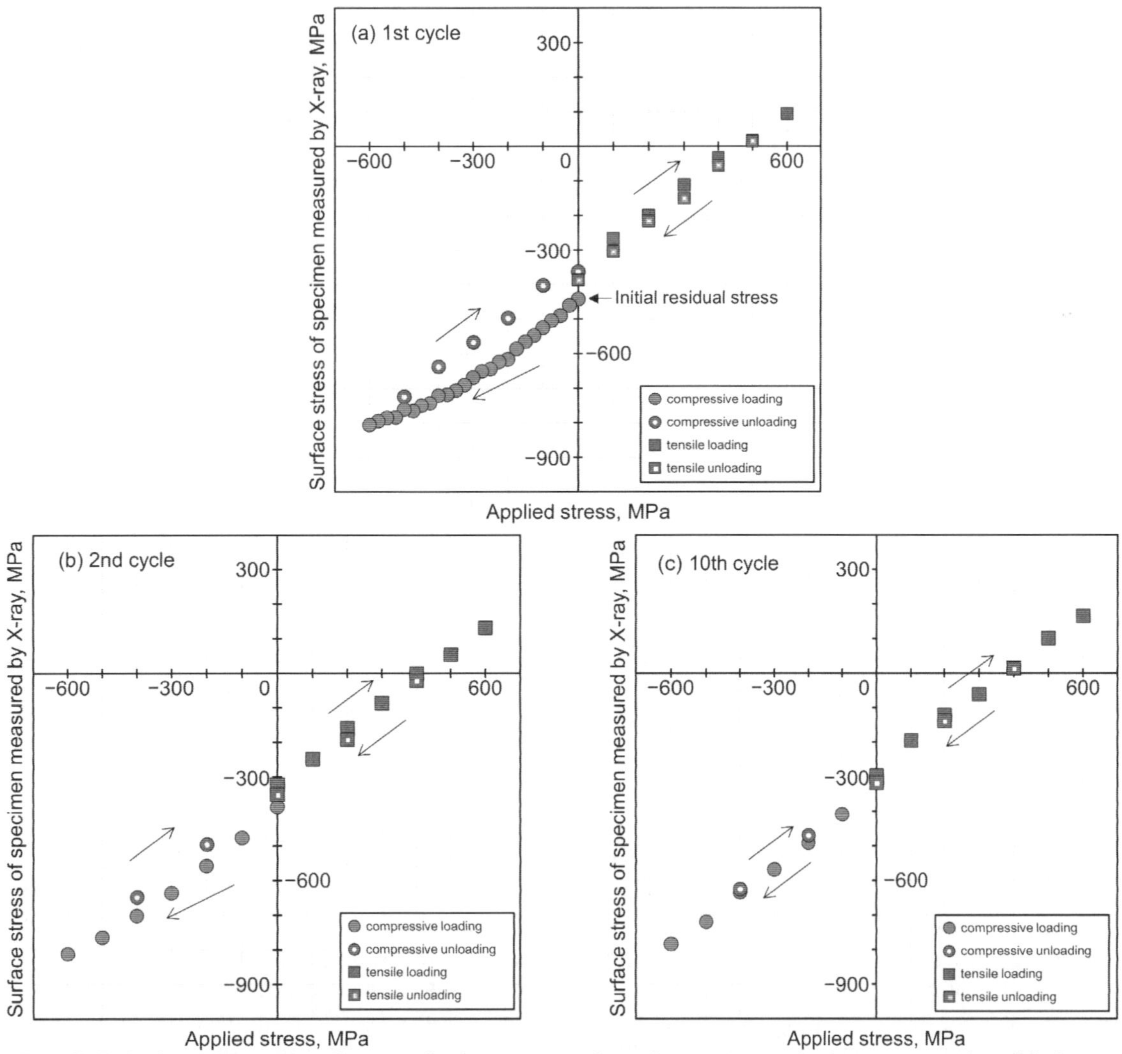

Fig. 5 Relations between the applied stress and surface stress during one cycle of fatigue loading at stress amplitude of 600 MPa; (a) first cycle, (b) second cycle, (c) tenth cycle

The relations between the number of loading cycles and residual stress are shown in Fig. 6. It can be seen that the compressive residual stress, which was initially approximately−450 MPa, was relaxed to approximately −400 MPa during the first cycle. The compressive residual stress was continuously relaxed after the first cycle; however, the amount of release per cycle was found to be decreasing. This is because the cyclic deformation of the surface compressive residual stress layer decreases as the number of cycles increases, as show in Fig. 5.

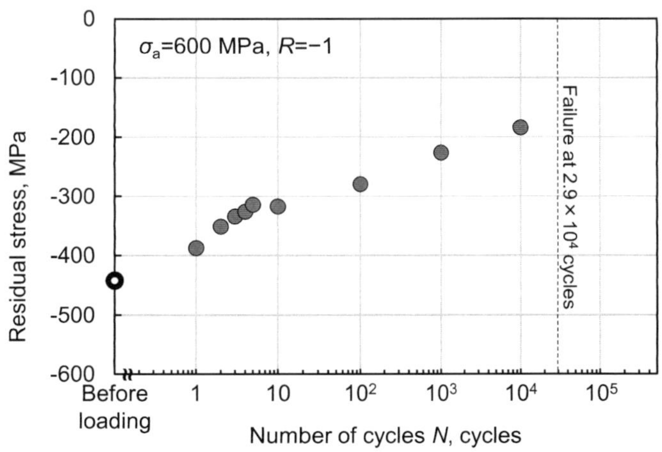

Fig. 6 Change in the residual stress during the fatigue test (stress amplitude = 600 MPa)

Conclusions

In this study, the relaxation behavior of shot peening induced compressive residual stress during a cycle of fatigue loading was investigated using *in-situ* X-ray stress measurement. The following conclusions were drawn:

- Compressive residual stress induced by shot peening is relaxed by the compressive stress in the first cycle of repeated loading. This is owing to the compressive yield of the residual stress layer.
- After the first cycle of loading, the compressive residual is gradually released. The cause of relaxation after the initial cycle is considered to be the cyclic deformation of the compressive residual stress layer.

References

[1] C. Wang, G. Wu, T. He, Y. Zhou and Z. Zhou, *Numerical Study of Fatigue Crack Propagation in a Residual Stress Field Induced by Shot Peening*, Journal of Materials Engineering and Performance, Vol. 29 (2020) pp. 5525–5539.

[2] S. Kikuchi, Y. Hirota and J. Kotmori, *Effect of Specimen Hardness and Shot Particle Hardness on Residual Stress and Fatigue Properties of SCM435H Steel Performed by Fine Particle Peening*, Journal of the Society of Materials Science, Vol. 60 (2011) pp. 547–553.

[3] A. Avilés, R. Avilés, J. Albizuri, L. Pallarés-Santasmartas and A. Rodríguez, *Effect of shot-peening and low-plasticity burnishing on the high-cycle T fatigue strength of DIN 34CrNiMo6 alloy steel*, International Journal of Fatigue, Vol. 119 (2019), pp. 338–354.

[4] J. Zhang, J. Zhang, B. Yang and X. Li, *Very High-Cycle Fatigue Properties and Residual Stress Relaxation of Micro-shot-Peened EA4T Axle Steel*, Journal of Material Engineering and performance, Vol.28 (2019) pp. 6407–6417.

[5] S. Kodama, *Residual Stress and Fatigue Strength of Metals*, Journal of the Society of Mechanical Engineers, Vol.75 (1972) pp.1026–1033.

[6] M. Hayama, S. Kikuchi and J. Komotori, *Evaluation of the compressive residual stress relaxation behavior by in situ X-ray stress measurement*, ISIJ International, Vol. 62, No. 4 (2022), pp. 758–765.

Fatigue properties of high strength steel treated by laser peening with handheld microchip lasers

Y. Mizuta[1], S. Tamaki[2], T. Kato[3], Y. Sakino[3], T. Hosokai[1] and Y. Sano[1]

1 SANKEN, Osaka University, Ibaraki, Japan
2 LAcubed Co. Ltd., Yokohama, Japan
3 Kindai University, Higashi-Hirochima, Japan
4 Institute for Molecular Science, National Institutes of National Sciences, Okazaki, Japan

Abstract

Laser peening (LP) introduces compressive residual stresses (RSs) on the surface of metallic components covered with water by irradiating them with high intensity laser pulses [1-7]. The advantage of LP is the possibility of fine execution management and the capability to introduce deep compressive RS on the material surfaces. It is well known to be highly effective in inhibiting stress corrosion cracking and fatigue cracking on material surfaces [7]. In addition, LP has an excellent effect on improving the fatigue strength of welds, which compensates for the disadvantage of high strength steels when welded. LP has a high potential for enhancing material surface, but the high-power laser used requires clean room facilities, large equipment and severe operating conditions. Therefore, the application of the LP has been limited to the countermeasure against high cycle fatigue of jet engine fan blades and stress corrosion cracking of nuclear reactor structures. If microchip lasers, which are small and easy to handle, could be used as a light source for LP, it would be possible to apply them not only to production processes in factories but also to existing steel structures such as bridges, to which conventional lasers have been difficult to apply for the above reasons. Based on this idea, we have developed a compact mobile LP device and verified its performance. [8-9]

Keywords Laser peening, microchip laser, handheld laser, residual stress, fatigue

Introduction

In LP, peening intensity can be instantaneously controlled by adjusting process parameters such as laser pulse energy and irradiation density corresponding to "coverage" of shot peening. Furthermore, LP can be applied to objects with complicated surface structure by real-time control of the focal position of irradiated laser pulses. In this study, we prepared LP experiments with a pulse energy of about 8 mJ attainable by microchip lasers [10-13] and peened the welded joint samples of an HT780 high-strength steel to find a way to compensate for the drawback of high-strength steels when welded: First, HT780 base metal was laser-peened to confirm the conditions for introducing favorable compressive RSs. Next, LP was applied to fatigue samples cut out from HT780 butt-welded joints prepared by CO2 gas shield arc welding, and then followed by uniaxial fatigue experiments with a stress ratio of 0.1. The results showed that LP with a pulse energy of about 8 mJ imparted compressive RSs in a near-surface layer of HT780 to a depth of about 0.2 mm and improved fatigue properties to the same level attained by LP with a pulse energy of 200 mJ from a conventional laser.

Preparation of Fatigue samples (HT780)

Fatigue samples were cut out from a butt-welded joint of 9 mm thick HT780 high-strength steel plates. The V-groove was filled with solid wire for 780 MPa class steel by carbon dioxide gas shielded arc welding. The shape and dimensions of the fatigue samples are shown in Fig. 1.

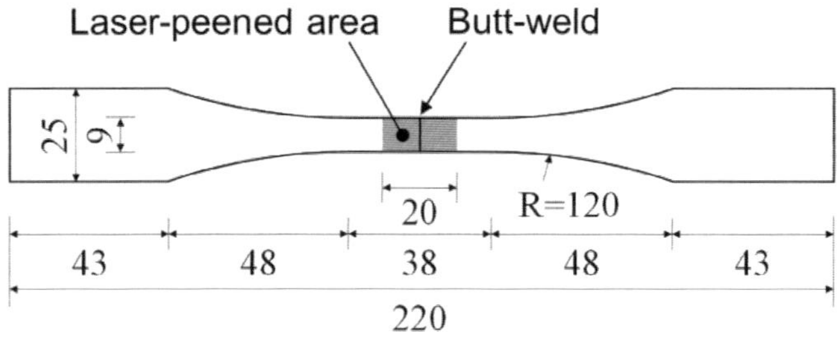

Fig. 1 HT780 butt-welded joint sample for fatigue testing.

LP was applied to both the front and back sides of the central part of the samples (shown in red in Fig. 1). A newly-developed compact LP device was used for the treatment. The LP conditions were 7.7 mJ pulse energy, 0.49 mm spot diameter, and 800 pulse/mm^2 irradiation density.

Effects on Surface Residual Stress of HT780

The effects on RSs were confirmed using an X-ray diffraction (XRD) device. The depth distribution of RSs was estimated by alternating electropolishing and XRD. The unpeened material was also subjected to the same procedure and the results were compared to evaluate the effect of LP. Fig. 2 shows the RS depth profile in the laser-peened and unpeened HT780 base metal. The depth of compression reaches about 0.2 mm from the surface. On the top surface of the peened material, the absolute value of RS is quite small compared to those just below the surface, which is probably due to the mill scale covering the base metal.

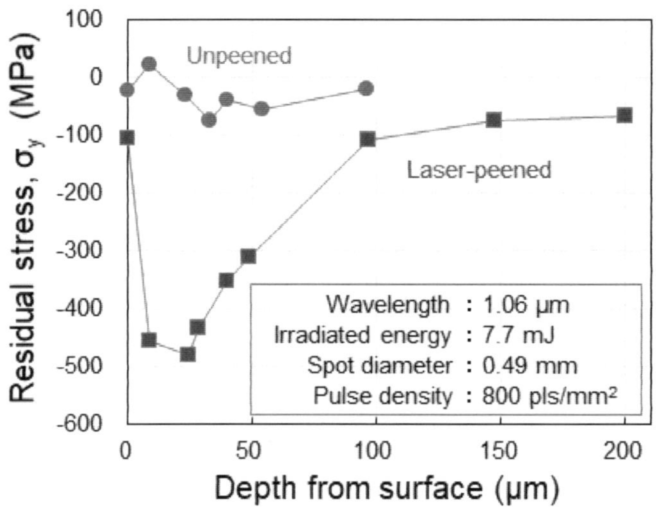

Fig. 2 Residual stress depth profiles of HT780

Effect on Fatigue Properties

A 100 kN uniaxial fatigue testing machine was used for fatigue loading with stress ranges (Δσ) of 200, 250, and 300 MPa. The stress ratio (R) was 0.1. The runout of the experiment was set to 10^7 cycles. For comparison, samples without LP were also subjected to fatigue testing. Fig 3 plots the fatigue test results in the form of the S-N curve.

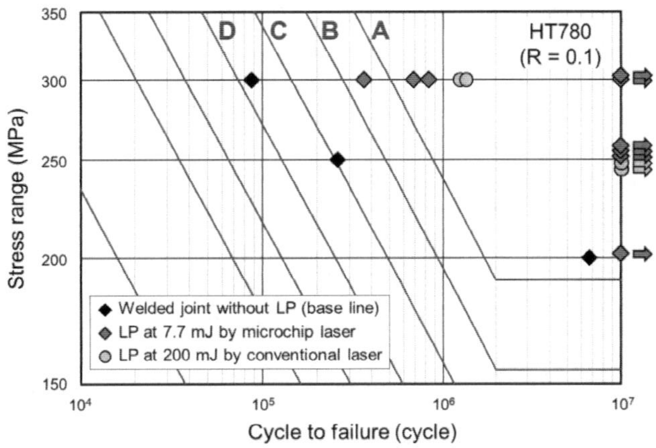

Fig. 3 Fatigue test results of HT780 butt-welded joint samples.

The blue lines indicated by A to D are the design curves for welded structural members specified by the Japan Society of Steel Construction. The design curve of the present samples is D. All the data are on the higher strength side than D, satisfying the standard.

The sample with Δσ = 200 MPa, all three samples with Δσ = 250 MPa, and two of the five samples with Δσ = 300 MPa reached runout. Comparing with the results of the unpeened reference sample, it can be seen that the fatigue strength is significantly improved by LP with a pulse energy of 7.7 mJ, which may improve the category of the welded joint by two grade from D to B.

The fatigue data of the welded joint samples after LP using a conventional laser is also plotted in the figure. It seems that the average fatigue life extension effect is almost the same between the present results with a pulse energy of 7.7 mJ and the previous study using a 200 mJ pulse energy.

Conclusions

HT780 high-strength steel was subjected to LP using a newly-prototyped device with a compact robot arm and a thumb-sized microchip laser with a pulse energy of about 8 mJ. The results cleared that RSs and fatigue properties were significantly improved by LP despite that the pulse energy was orders of magnitude smaller than those of the conventional LP devices.

Acknowledgments

This work was partially supported by the JST-MIRAI Program (grant number JPMJMI17A1), the Supporting Industry Program of the Small and Medium Enterprise Agency (grant number 20334933), the Amada Foundation (grant number AF-2020239-C2), and the Impulsing Paradigm Change through Disruptive Technologies Program (ImPACT) of the Council for Science, Technology and Innovation.

References

[1] Y. Sano, N. Mukai, K. Okazaki and M. Obata, *Residual stress improvement in metal surface by underwater laser irradiation*, Nucl. Instrum. Methods Phys. Res. B, Vol. 121 (1997), pp 432–436.

[2] Y. Sano, M. Obata, T. Kubo, N. Mukai, M. Yoda, K. Masaki and Y. Ochi, *Retardation of crack initiation and growth in austenitic stainless steels by laser peening without protective coating*, Mater. Sci. Eng. A, Vol. 417 (2006), pp 334–340.

[3] Y. Sakino, Y. Sano and Y.-C. Kim, *Application of laser peening without coating on steel welded joints*, Int. J. Struct. Integr., Vol. 2 (2011), pp 332–344.

[4] Y. Sakino and Y. Sano, *Investigations for lowering pulse energy of laser-peening for improving fatigue strength*, Q. J. Jpn. Weld. Soc., Vol. 36 (2018), pp 153–159.

[5] T. Kato, Y. Sakino and Y. Sano, *Effect of laser peening with a microchip laser on fatigue life in butt-welded high-strength steel*, Appl. Mech., Vol. 2 (2021), pp 878–890.

[6] A.H. Clauer, *Laser shock peening, the path to production*, Metals, Vol. 9 (2019), p 626.

[7] Y. Sano, *Quarter century development of laser peening without coating*, Metals, Vol. 10 (2020), p 152.

[8] Y. Sano, T. Kato, Y. Mizuta, S. Tamaki, K. Yokofujita, T. Taira, T. Hosokai and Y. Sakino, *Development of a portable laser peening device and its effect on the fatigue properties of HT780 butt-welded joints*, Forces in Mech., Vol 7 (2022), p 100080.

[9] Y. Sano, K. Masaki, Y. Mizuta, S. Tamaki, T. Hosokai and T. Taira, *Effects of laser peening with a pulse energy of 1.7 mJ on the residual stress and fatigue properties of A7075 aluminum alloy*, Metals, Vol. 11 (2021) p. 1716.

[10] T. Taira, *Domain-controlled laser ceramics toward giant micro-photonics*, Opt. Mater. Express, Vol. 1 (2011), pp 1040–1050.

[11] Y. Sato, J. Akiyama and T. Taira, *Process design of microdomains with quantum mechanics for giant pulse lasers*, Sci. Rep., Vol. 7 (2017), p 10732.

[12] L. Zheng, A. Kausas and T. Taira, *Drastic thermal effects reduction through distributed face cooling in a high power giant-pulse tiny laser*, Opt. Mater. Express, Vol. 7 (2017), pp 3214–3221.

[13] L. Zheng, A. Kausas and T. Taira, *>30 MW peak power from distributed face cooling tiny integrated laser*, Opt. Express, Vol. 27 (2019), pp 30219–30224.

Fatigue property of Shot peened Maraging Steel fabricated by Additive manufacturing

Toshiya Tsujia[1], Yuji Kobayashia[1], Kiyotaka Masaki [2]

1 SINTOKOGIO, LTD., Development group, 180-1 Komaki Ohgi-cho Toyokawa, Aichi, Japan
2 Okinawa College, National Institute of Technology, 905 Henoko Nago, Okinawa, Japan

Abstract
Shot peened maraging steel fabricated by additive manufacturing were evaluated the fatigue strength by rotating bending fatigue test. It is to investigate the influence of shot peening for the fatigue property of additive manufactured parts. The specimen was employed the surface which was as-built and the surface which was polished. As a result, shot peened specimens increased 2.5 times higher fatigue strength than As-built specimens. Fatigue fracture origin changed form surface to internal by shot peening. Furthermore, shot peened specimens were higher fatigue strength and longer fatigue life than Polished specimen. Fatigue fracture origin changed from surface to internal by shot peening. Therefore, shot peening drastically affects to increase the fatigue strength of additive manufactured material due to prevention of surface crack propagation by the inducing compressive residual stress.

Keywords Shot peening, Additive manufacturing, Fatigue strength, Maraging steel.

Introduction

Additive manufacturing has been researched by many researchers as new manufacturing method to fabricate complex parts which cannot fabricate conventional manufacturing methods such as grinding, forging, casting etc. [1-4].

Most of metal additive manufacturing are manufactured by melting material powder by thermal source such as laser beam, electron beam and stacking the molten metal. The surface roughness on additive manufactured parts is Ra > 4 [μm] [5]. This surface roughness affects the fatigue strength of additive manufactured parts. Consequently, additive manufactured parts were lower fatigue strength than parts which made by conventional manufacturing methods [6-7]. The surface roughness is generally removed by machining. However, it is difficult to access the machining the tool to parts due to complex geometry by the taking advantage of additive manufacturing. Therefore, surface enhancement method to be easy to process for the complex geometry is required.

Shot peening is one of the surface enhancement method by inducing the work hardening, compressive residual stress. This process is carried out the impact of peening media which has media diameter from 0.05[mm] to 2.0 [mm]. Hence, it is relatively easy to access the complex geometry and to enhance the parts. Many researchers studied the fatigue strength of shot peened specimen fabricated by additive manufacturing. However, the surface roughness of additive manufactured specimen in most of researches was machined [8-9].

In this study, we investigated the influence of shot peening on fatigue strength against the additive manufactured maraging specimen which retained the surface roughness of as-built condition.

Experimental Methods

Specimen

Maraging powder was employed. The powder was stacked in the direction of longitudinal direction of specimen by selective laser melting method. Polished surface and As-built surface were employed to investigate the influence of surface roughness on fatigue strength.

These specimens were carried out solution and aging treatment. Shapes and geometry of finished specimen is shown in Fig.1. Manufacturing procedure of both specimens were shown in Fig.2.

Three shot peening conditions which changed the mainly media diameter was employed to vary the residual stress distribution induced by shot peening. Kinds of specimens were shown in Table1.

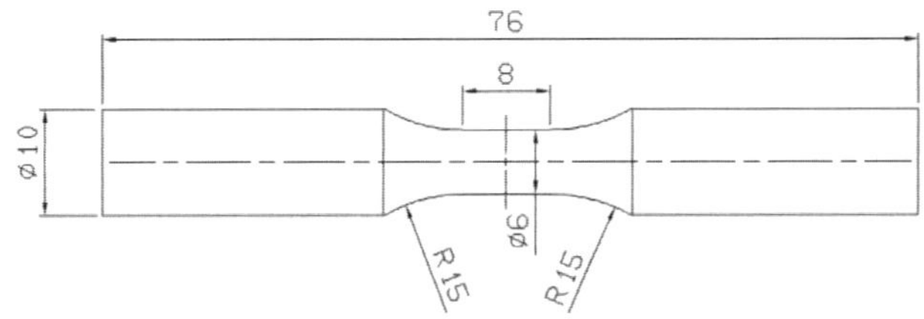

Fig.1 Shapes and geometry of specimen.

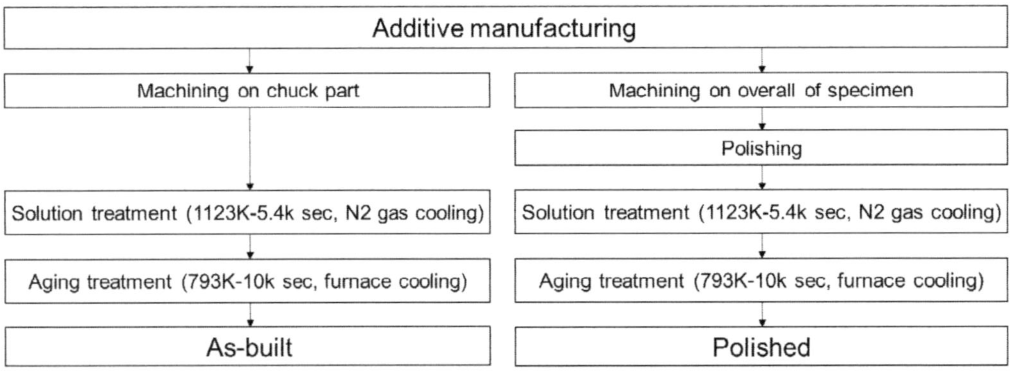

Fig.2 Manufacturing procedure of specimen.

Table1 Kinds of Specimens.

Symbol	Surface condition	Shot Peening Conditions					
		Media	Media diameter [mm]	Media Hardness [HV]	Peening Method	Air Pressure [MPa]	Arc height
As-built	As-built	-	-	-	-	-	-
Polished	Polished	-	-	-	-	-	-
FSP	As-built + Shot Peening	Cast steel	0.1	800	Direct Pressure	0.2	0.286 [mmN]
SP		CCW	0.6	600		0.3	0.430 [mmA]
HSP		CCW	1.2	600		0.3	0.901 [mmA]
Polished + FSP	Polished + Shot Peening	Cast steel	0.1	800	Direct Pressure	0.2	0.286 [mmN]
Polished + HSP		CCW	0.6	600		0.3	0.430 [mmA]

Evaluation Methods

Microstructure surface roughness, hardness distribution and residual stress measurement were evaluated to investigate the characteristics of each specimens. Furthermore, cross section of minimum diameter part in specimen was carried out to investigate the defect due to additive manufacturing in specimen.

Cross section observation was carried out by laser microscope. The specimen was mirror polished sample which was cut at minimum diameter part in transverse direction. This sample was mounted by hot set resin before mirror polishing.

Microstructure was observed by laser microscope after etching against sample which used in cross section observation.

Surface roughness was measured by stylus-based profiler. Roughness measurement standard was JIS B 0601-2001. Roughness measurement direction was longitudinal direction of specimen.

Hardness distribution was measured by micro Vickers hardness tester. Measurement load was 2.94[N].

Residual stress was measured by X-ray diffraction method. Analyzing method of residual stress by X-ray diffraction was $\sin^2\psi$ method. X-ray constant was -318[MPa/deg.]. Measurement angle was selected the 5 angles that $\sin^2\psi$ value of measurement angles was equal interval from 0 to 0.5. Oscillation angle was 5[deg.]. Depth direction measurement of residual stress was carried out by removing the layer to optional depth by electropolishing.

Fatigue test Method

Fatigue test was carried out under room temperature and atmosphere by Ono-type rotating bending fatigue testing machine. Number of run-out was 2.0 X 10^7 [cycles]. Stress ratio was R=-1. Fractured surface was observed by optical microscope and scanning electron microscope to evaluate the crack initiation point.

Experimental Results and Discussions

Cross-section and Microstructure observations

Representative cross-section and microstructure observation results are shown in Fig.3.

In the cross-section observation, many defects were observed from surface to center of specimen. These defects are uniformly distributed in specimen. This trend was observed in all conditions.

Microstructure of all specimens was martensite by solution and aging treatment. Unmelted particles were observed on As-built surface. On the other hands, unmelted particles were not observed on surface of shot peened specimens which had As-built surface due to removing it by the impact of peening media.

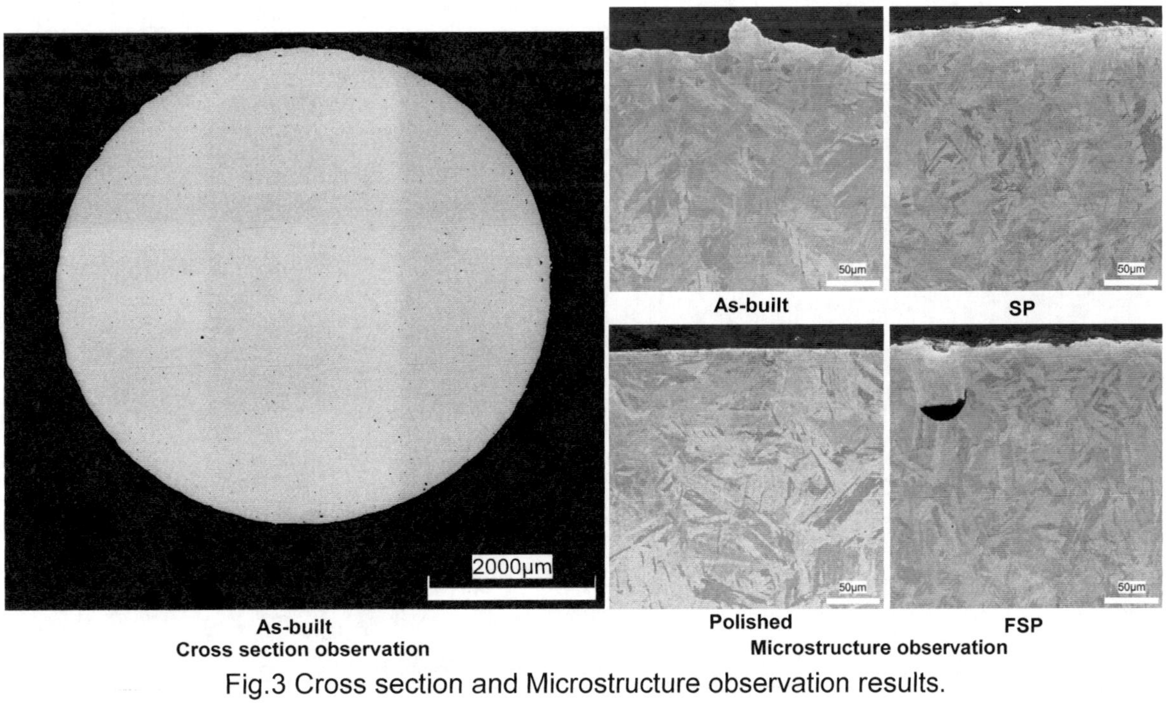

Fig.3 Cross section and Microstructure observation results.

Surface roughness

Average surface roughness measurement results are shown in Fig.4.

Average surface roughness of As-built was approximately 7.25 [μm]. All shot peened specimen which had As-built surface decreased the surface roughness due to smoothing the surface roughness and removing the unmelted particles by the impact of peening media. In the case of same hardness of peening media, surface roughness decreased with increasing the media diameter. The larger media diameter is, the higher media impact energy obtains.

Therefore, HSP condition can smooth the high surface roughness of As-built surface due to higher impact energy of peening media. In this study, even though media diameter of FSP was the smallest, FSP condition was the most effective condition to reduce the surface roughness of all shot peening conditions. The result indicates that the ablation generated in FSP condition due to higher media hardness than that of SP and HSP conditions.

Average surface roughness of Polished was approximately 0.27 [μm]. All shot peened specimen which had Polished surface increased the surface roughness. The surface roughness increased with increasing media diameter in these conditions.

Hardness distributions

Hardness distributions are shown in Fig.5.

Hardness of this specimen was approximately 630 [HV] from surface to internal.

Work-hardening of all shot peened specimen were not obtain in these conditions. These results imply that the work-hardening coefficient of this material was low.

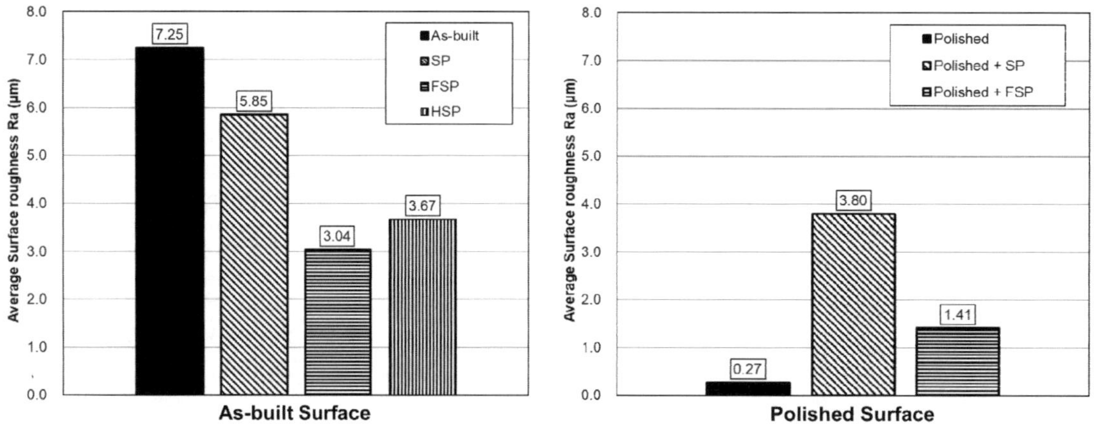

Fig.4 Surface roughness results.

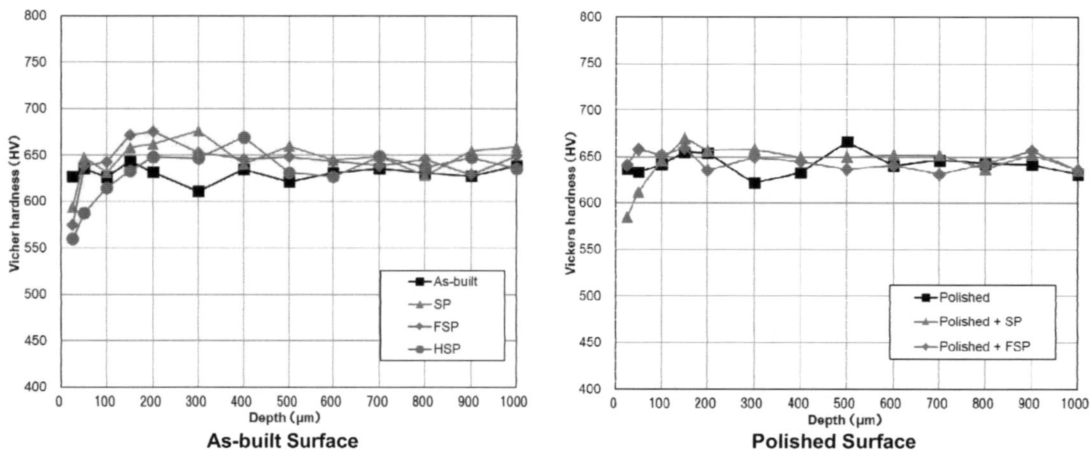

Fig.5 Hardness distributions.

Residual stress distributions

Residual stress distributions are shown in Fig.6.

Polished and As-built had tensile residual stress from surface to internal. In our past research [8], the additive manufactured maraging steel which was carried out solution treatment was lower tensile residual stress than the specimen in this study. Hence, this comparison indicates that the precepted particles due to aging treatment affects the residual stress.

Shot peened specimens were induced compressive residual stress up to approximately -1400 [MPa] by shot peening. Surface compressive residual stress was difference in each shot peening conditions. As a tendency, the smaller media diameter was, the higher surface compressive residual stress increased. The crossing point was approximately 60[μm] in FSP and Polished + FSP, approximately 200[μm] in SP and Polished + SP, approximately 300[μm] in HSP. In addition, the depth of maximum residual stress point was approximately 20[μm] in FSP and Polished + FSP, approximately 60[μm] in SP and Polished + SP, approximately 100[μm] in HSP. The depth of maximum residual stress point and the crossing point increased with increasing the arc height.

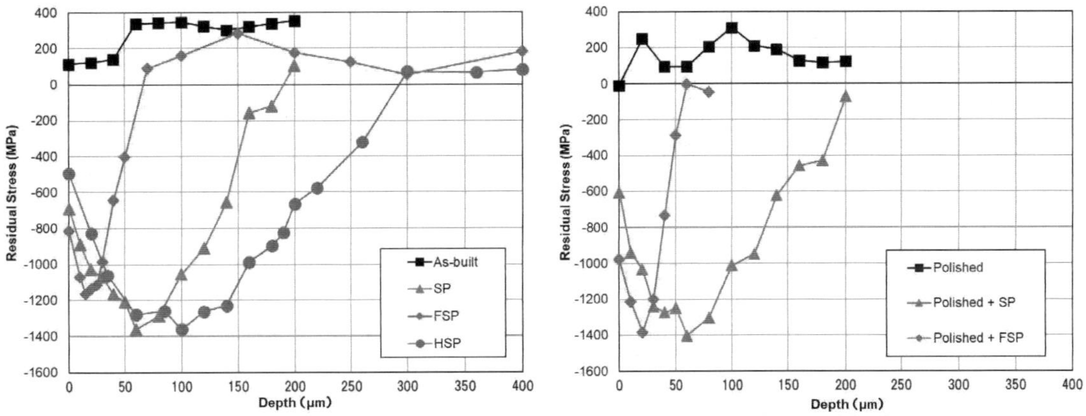

As-built Surface Polished Surface

Fig.6 Residual stress distributions.

Fatigue test results

1.As-built and Polished

Fatigue test results are shown in Fig.7. Fracture surface observation results are shown in Fig.8.

Fatigue fracture was generated surface and internal. Surface fatigue fracture was generated the valley of surface roughness, matrix crack of microstructure, and surface defect. Internal fatigue fracture was generated matrix crack of microstructure and internal defect.

Fatigue strength of As-built was 200[MPa], this result was the lowest fatigue strength of all conditions. On the other hand, it of Polished was 450[MPa]. As a result, Polished was 2.25 times higher fatigue strength than As-built. The fatigue life of Polished extended comparison with As-built. The residual stress and hardness distributions between As-built and Polished were almost same. Furthermore, fatigue fracture of As-built was generated from the valley of surface roughness, whereas it of Polished was generated from surface defect or matrix crack of microstructure. Because, stress concentration due to surface irregularity decreased to remove surface irregularity by shot peening. Therefore, surface roughness strongly affects fatigue fracture in this comparison.

2. As-built and FSP, SP, HSP

Fatigue strength of FSP, SP and HSP were 500[MPa]. There was not the difference of fatigue property in FSP, SP, and HSP. The fatigue strength of theses increased 2.5 times against As-built. The fatigue life of its extended in comparison with As-built. Surface roughness of all shot peened specimens decreased in comparison with As-built. From the comparison with As-built and Polished, the reduction of surface roughness affects to increase the fatigue strength. Therefore, the one of influence factor to increase the fatigue strength is to reduce the surface roughness by shot peening. Fatigue fracture origin of shot peened specimens changed from surface to internal. SP was the highest surface roughness of all shot peened conditions, however fatigue fracture generated from the internal. On the other hand, Polished which was the lowest surface roughness in this study failed from the surface.

Furthermore, the surface hardness has no relation to prevent the surface fatigue crack initiation, because the surface hardness of shot peened specimens was almost equivalent to Polished. Therefore, compressive residual stress strongly affected to increase the fatigue strength due to prevention of surface crack initiation and propagation.

3. Polished and Polished + FSP, Polished + SP

Fatigue strength of Polished + FSP, Polished + SP were 500[MPa]. The fatigue strength of these increased 1.1 times against Polished, however there were not the difference of fatigue property between shot peened specimens for As-built surface and shot peened specimens for Polished surface. The fatigue life of these extended in comparison with Polished. These fatigue fracture origins changed from surface to internal, even though these were higher surface roughness than Polished which was surface fatigue fracture from surface defect or the matrix crack of microstructure. In this comparison, the surface hardness of shot peened specimens slightly decreased, furthermore surface roughness of these increased in comparison with Polished. Therefore, compressive residual stress due to shot peening strongly affected to increase the fatigue strength due to prevention of surface crack initiation and propagation. In addition, the reduction of surface defect due to shot peening, which is employed large shot media occurred, because there were not fatigue fracture from the surface defect in SP condition.

4.Relashitonship between Residual stress distribution and Internal fracture

Fig.9 shows relationship between Stress amplitude and Depth of internal fatigue fracture origin from the surface. The internal fatigue fracture origin was generated at location more than 70[μm] in FSP and Polished +FSP, at location more than 240[μm] in SP and Polished +SP, at location more than 400[μm] in HSP. These depths were deeper than crossing point of each shot peening conditions. Therefore, compressive residual stress by shot peening affects the internal fatigue crack initiation. Furthermore, it has possibility to achieve internal defect harmless due to additive manufacturing when the size of defect is not large enough to reach the threshold stress intensity factor including the residual stress.

Conclusions

(1) Additive manufactured specimens were induced compressive residual stress up to -1400[MPa] by shot peening. Furthermore, the surface roughness of As-built surface was decreased from 7.25[μm] to 3.04[μm] at maximum by shot peening.

(2) In comparison with As-built and Polished, Polished was higher fatigue strength than As-built due to the reduction of stress concentration to remove the surface irregularity.

(3) In comparison with As-built and SP, FSP, HSP, fatigue strength of shot peened specimens increased 2.5 times against As-built due to the prevention of the fatigue crack initiation and propagation by the inducing compressive residual stress and the reducing stress concentration to remove the surface irregularity.

(4) In comparison with Polished and Polished + FSP, Polished + SP, the fatigue strength of shot peened specimens increased 1.1 times against Polished due to the prevention of the fatigue crack propagation by the inducing compressive residual stress the reducing the surface defect size.

(5) Compressive residual stress distribution affects the depth of internal fatigue fracture origin. It has possibility to achieve internal defect harmless due to additive manufacturing when the size of defect is not large enough to reach the threshold stress intensity factor including the residual stress.

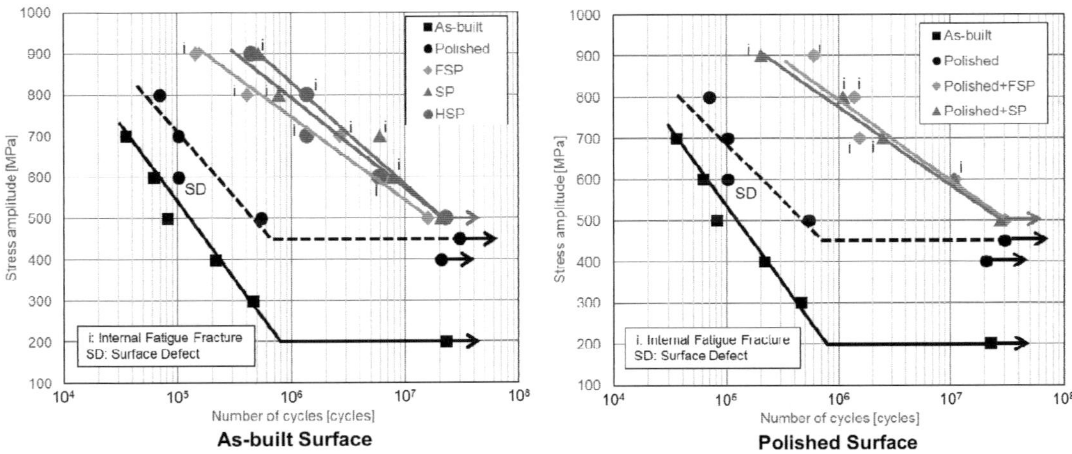

Fig.7 S-N curves.

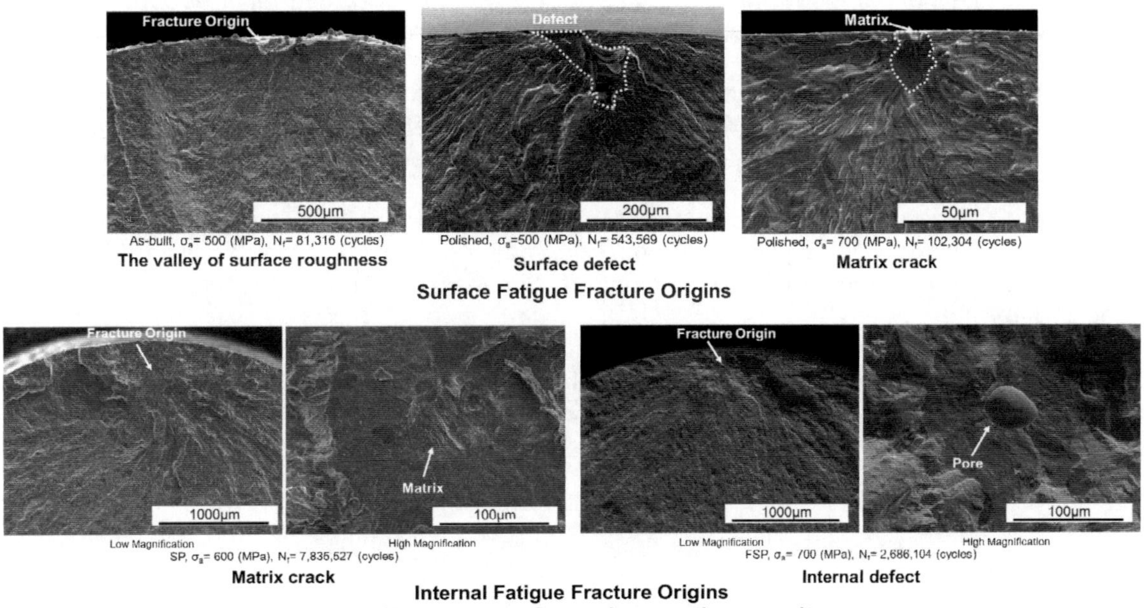

Fig.8 Fracture surface observation results.

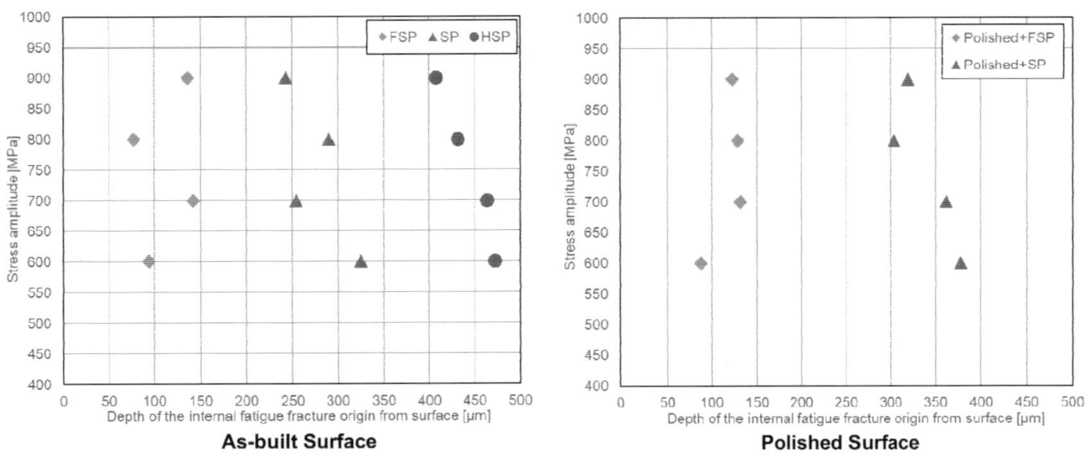

Fig.9. relationship between Stress amplitude and
Depth of internal fatigue fracture origin from the surface.

References

[1] Yuji Kobayashi, Kiyotaka Masaki, *Shot peening effect on pieces manufactured by 3D printer*, The proceedings of 13th International Conference on Shot Peening(2017), pp.141-144

[2] Hideki Kyogoku, *The Current Status and Development of Metal Additive Manufacturing Technology,* Materia Japan, Vol.57(2018), Issue 4, pp 140-144.

[3] Akihiko Chiba, *Microstructure of Alloys Fabricated by Additive Manufacturing Using Electron Beam Melting,* Journal of The Society of Instrument and Control Engineers, Vol.54 (2015), Issue 6, pp 399-404

[4] Takayuki Nakamoto, Takahiro Kimura, Naruaki Shinomiya, *Development of High-Performance Parts Using Metal Additive Manufacturing,* Materia Japan, Vol.56(2017), Issue 12, pp 704-707.

[5] Valmik Bhavar, Prakash Kattire, Vinaykumar Patil, Shreyans Khot, Kiran Gujar, Rajkumar Singh, *A Review on Powder Bed Fusion Technology of Metal Additive Manufacturing,* Additive Manufacturing Handbook (2017), pp.252-261

[6] Bastien Vayssettea, Nicolas Saintiera, Charles Bruggera, Mohamed El Maya, Etienne Pessardb, *Numerical modelling of surface roughness effect on the fatiguebehavior of Ti-6Al-4V obtained by additive manufacturing*, International Journal of Fatigue, Vol.123(2019), pp.180-195.

[7] Andrew H.Cherna, PeeyushNandwanab, TaoYuancMichael, M.KirkabRyan, R.Dehoffb, Peter K.Liawd, Chad E.Dutya, *A review on the fatigue behavior of Ti-6Al-4V fabricated by electron beam melting additive manufacturing,* International Journal of Fatigue, Vol.119(2019), pp.173-185.

[8] Kiyotaka Masaki, Yuji Kobayashi, Yuta Mizuno, *Influence of Shot Peening Treatment on Rotating Bending High-Cycle Fatigue Properties of Additive Manufactured Maraging Steels Material*, Journal of the Society of Materials Science, Vol.67(208), Issue10, pp.891-897

[9] Shiori Tsuchiya, Koji Takahashi, *Influence of Laser Peening on the Fatigue Strength of 3D Additive Manufactured Maraging Steel*, Transactions of Japan Society of Spring Engineers, Vol.2021(2021), Isuue66, pp.7-12

INFLUENCE OF SHOT PEENING MEDIA AND PARAMETERS ON CARBURIZED STEEL

M.R. Chini[a], F.Wagner[a], S. Benali[a], C. Dides[a], D. Bouttes[b], L. Besnault[c], M. Courteaux[d], C. Grente[e], M.Millot[f]. C. Berthod[g], G. Bal-Fontaine[g], P. Picard[h]

[a] IRT M2P, 4 rue Augustin Fresnel, 57070 Metz, France - maria-rita.chini@irt-m2p.fr,
[b] Saint-Gobain Research Provence, 550 avenue Alphonse Jauffret, 84300 Cavaillon, France, david.bouttes@saint-gobain.com
[c] MELIAD, 6 Rue des Orfèvres, 44840 Les Sorinières
[d]STELLANTIS, Centre Technique de Belchamp 25420 Voujeaucourt
[e]RENAULT Technocentre, API TCR LAB 00A, 1 avenue du Golf, 78084 GUYANCOURT, France.
[f] SAFRAN TECH Rue des jeunes Bois, 78117 Châteaufort
[g] VALEO ELECTRICAL SYSTEMS 10, rue du Revolay -38070 Saint Quentin-Fallavier – France
[h]SCHAEFFLER, 93, route de Bitche 67506 Haguenau-France

Abstract

The aim of this study is to determine the shot peening media and parameters which can introduce the strongest residual stress profiles while keeping a limited surface roughness on carburized steels. Another objective is to evaluate the ceramic shots Zirshot HDC and compare them to traditional metallic media. The influence of initial roughness and hardness of samples before shot peening was also investigated.

Keywords: *shot peening, ceramic shot, DOE, roughness, residual stress*

Introduction

In industry, shot peening is widely used to improve the fatigue properties through the introduction of compressive residual stress on a superficial layer [1]. Most studies are focused on only two of the parameters of this process, Almen intensity and coverage, but few studies show the effect of different types of media. Moreover, carburized steels are often difficult to shot peen due to their high hardness, and it is known that the media selection is of great importance [2], whereas the shot peening process can bring significant performance improvement [3]. There is hence a need for multi-parameter investigation of the shot peening process for this type of material. This development was carried out during TRANSFUGE project.

Experimental methods

Four kinds of specimens of 16MnCr5 steel were studied. Specimens were carburized, and then annealed at two different temperatures to obtain two different surface hardness (700HV and 800HV). Two different surface finishing were investigated, one whose surface roughness (R_a) is 0,3 µm and another with 1,6 µm of R_a. The geometry chosen is the same as an Almen A strip but 10 mm thick.

For studying simultaneously, the influence of shot peening parameters and the previously mentioned material parameters (surface hardness and initial roughness) a wide Design of Experiment (DOE) has been set up. The 3 process parameters that are studied are: media, Almen intensity and incidence angle. First a complete fractional DOE was stablished and then optimized using the software Minitab to reduce the number of configurations to study from 96 to 36. The table 1 shows the different parameters which were studied and their values.

Residual stress profiles were measured by X-ray diffraction using Cr-Kα radiation. Surface roughness was characterized by two methods: standard 2D measurements using the Mitutoyo SJ-210 roughness meter and 3D surface analysis using a GT contour microscope from Bruker. Statistical analyses of the results were carried out using Minitab software.

Table 1 : Process and material parameters

Parameter	Value
Almen Intensity	low; high
Media	CW400, CW600, ASH170, ASH230, ZC400, ZC600
Incidence angle	85°;45°
Coverage	200%
Projection	Nozzle – 10mm diameter
Initial roughness (Ra)	0,3μm ;1,6 μm
Surface hardness	700HV; 800HV

Experimental results and discussion
Influence of Shot Peening on Roughness

All parameters with a normalized effect greater than a certain value are considered to have a significant impact on roughness. This corresponds to the P value of the experiment plan using an alpha coefficient of 5%. The Pareto charts in Figure 1 show the normalized effect of each parameter and their combinations for R_a, R_z, S_a and S_z. Statistical analysis showed that all shot peening parameter have a direct effect on final roughness R_a, except for the incidence angle. The four most influential parameters are initial roughness, Almen Intensity, media, and surface hardness, in that order.

On the other hand, statistical analysis on other surface parameters R_z, R_p, R_v, S_a, S_z and S_m(0,8μm) showed that after initial roughness, media has the greatest influence, and not almen intensity.

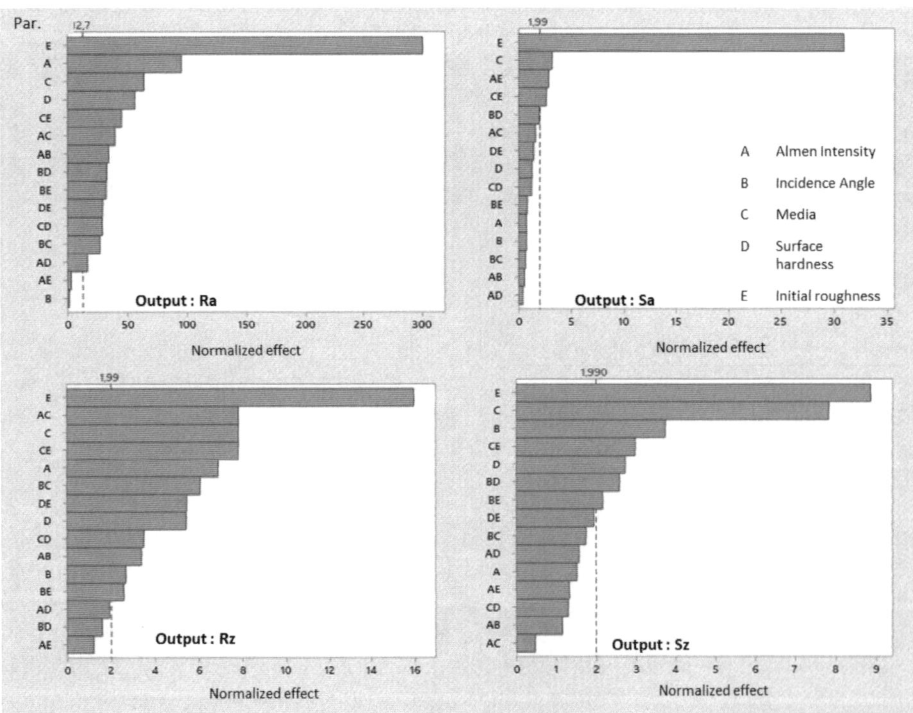

Figure 1 : Pareto chart for normalized effects on average roughness Ra, Rz, Sa and Sz for α=0.05.

Figure 2 shows the combined effect of parameters. This diagram reveals that initial roughness defines two different populations, and therefore has a strong interaction with other parameters. Final roughness is usually close to the initial one.

In this diagram the interaction between Almen intensity and media, the most influential parameters, is made explicit. The greater slope of the ceramic shots implies that these medias are more sensitive to Almen intensity than metallic ones for this kind of applications.

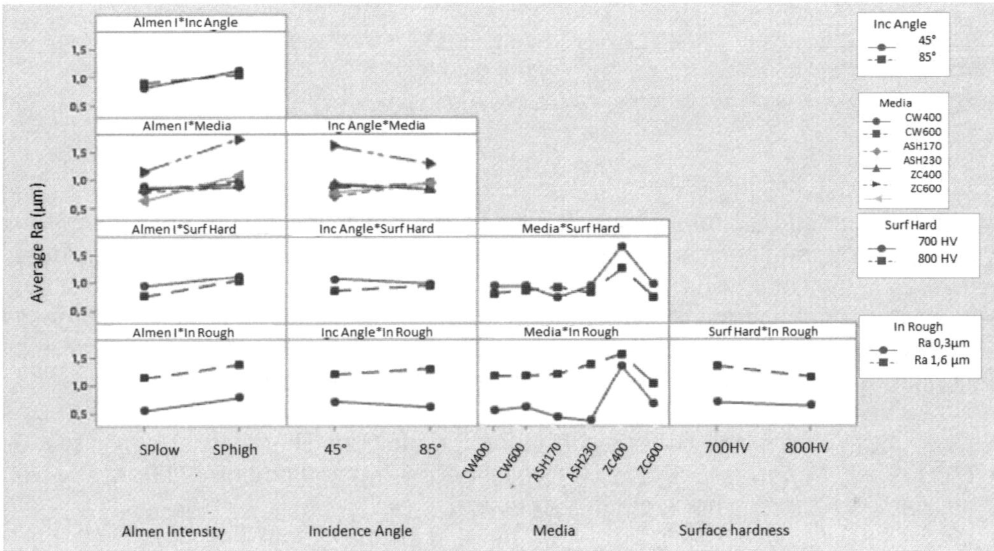

Figure 2 : Interaction diagram for Roughness parameter Ra (Adjusted averages)

Influence of Shot Peening on residual stress

To be able to analyze statically the effect of process and material parameters on residual stress, each one of the 36 residual stress profiles were characterized by 4 parameters: surface residual stress, maximum residual stress, peak position and affected depth. The Pareto charts in Figure 3 show the normalized effect of each DOE parameter and their combinations on Maximum Residual Stress. This diagram show that Media is the only parameter that has an influence. Statical analysis showed the same result for the surface residual stress and peak position, however affected depth is only determined by Almen intensity.

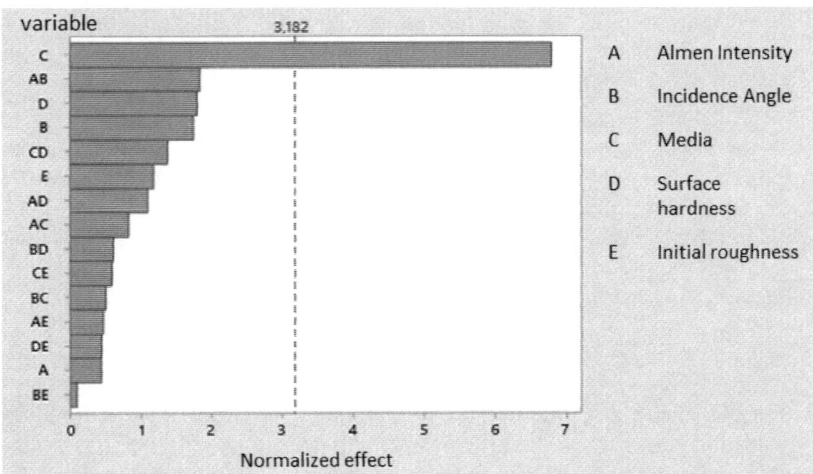

Figure 3 : Pareto chart for normalized effects calculated for maximal residual stress. α=0.05.

The interaction diagram for maximum residual stress in Figure 4 show clearly that each media produces a specific Maximum residual stress, independently of the Almen intensity or the incidence angle

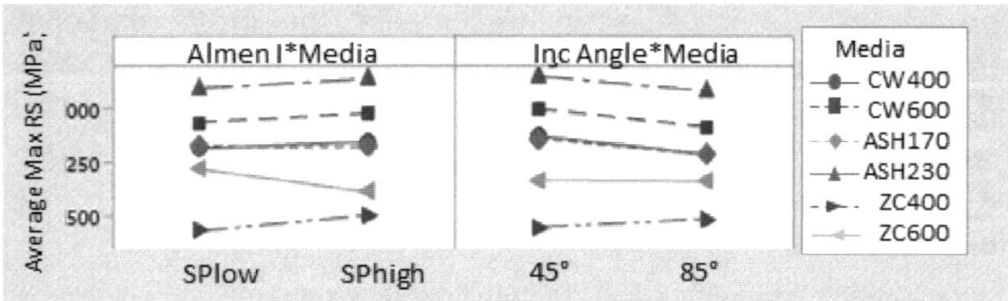

Figure 4 : Interaction diagram for Maximal Residual Stress (Adjusted averages)

Optimization

Different studies [4-8] associate a stress concentration factor K_t with the roughness parameter R_z. Therefore, certain shot peening conditions neutralize the beneficial effects of the residual compressive stresses by creating surface defects which increase the stress concentration factor. Consequently, it is not enough to evaluate the influence of shot peening on roughness and residual stress independently to choose the most suitable parameters. For this, diagrams which display simultaneously the influence on both for each configuration must be used.

Figure 5 shows maximum stress and R_z for each studied condition. Theses graphics show a distribution which is highly dependent on media for grinded samples (initial Ra 0,3µm), while in case of rougher samples there is not a clear distribution.
For grinded samples, media is a key parameter, being ceramic shots the ones that can introduce the strongest residual stresses but also strongly damage the surface. This suggests that ceramic shots should not be used at iso parameters regarding metallic media. Lower Almen intensity must be used for ZC shots to achieve suitable RS profiles and lower surface roughness.

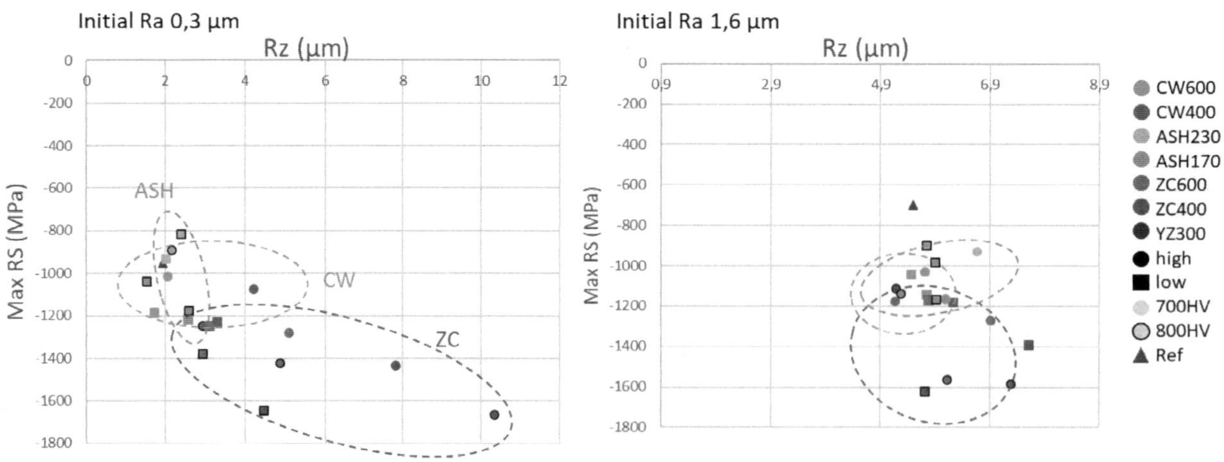

Figure 5 : Maximum residual stress and Rz after shot peening for each tested configuration

Conclusion

Analyses showed that media selection is a key parameter to consider for optimizing the ratio between the roughness generated by the shot peening and the intensity of the residual stresses. In particular, the ratio of hardness of the media to the target controls different behaviors: increasing this ratio allows to increase the residual stresses at the surface, while increasing the roughness sensitivity to the process parameters. The complete analysis allows to determine the influence of each factor tested. ZC shots should not be used at iso parameters regarding metallic media. Lower Almen intensity must be used for ZC shots to achieve suitable RS profiles and lower surface roughness

Acknowledgements

References (optional)

[1] Withers, P. J., *Residual stress and its role in failure. Reports on progress in physics*, (2007). 70(12), 2211

[2] Wohlfahrt H., The Influence of Peening Conditions on the Resulting Distribution of Residual Stress. ICSP 2, 1984, p. 316-331.

[3] Benedetti M. et al., Influence of shot peening on bending tooth fatigue limit of case hardened gears, International Journal of Fatigue, 2002, vol. 24, 11, p.1127-1136.

[4] Koster, W. P., & Field, M., *Effects of machining variables on the surface and structural metals.* PROCEEDINGS OF THE NORTH AMERICAN MANUFACTURING RESEARCH CONFERENCE, SME. (2001, January)

[5] Siebel, E., *Influence of surface roughness on the fatigue strength of steels and non-ferrous alloys*. Engineers Digest, 18, (1957) pp 109-112.

[6] Javidi, A., *Influence of machining on the surface integrity and fatigue strength of 34CrNiMo6 steel.* (2008).

[7] Neuber, H., *Theory of notch stresses: Principles for exact stress calculation* (Vol. 74). JW Edwards. (1946).

[8] D. Arola, C.L. Williams, *Estimating the fatigue stress concentration factor of machined surfaces,* International Journal of fatigue 24 (2002) pp 923-930.

Bending Fatigue Resistance of Duplex Shot Peened Austempered Ductile Iron

W. Abela[1], G. Cassar[1], A. Steuwer[2], A. Zammit[1]

1 Department of Metallurgy and Materials Engineering, University of Malta, Msida MSD 2080, Malta
2 Research Support Services Directorate, University of Malta, Msida MSD 2080, Malta

Abstract
This study investigates conventional and duplex shot peening as a means to further improve the fatigue behaviour of austempered ductile iron. This involves characterisation of surface regions through microstructural analysis, phase analysis, and synchrotron radiation stress measurement, amongst other techniques. Additionally, rotating bending fatigue testing revealed that both shot peening variants enhanced the fatigue strength of ADI by 65%. Moreover, duplex shot peening brought about an improvement in fatigue life relative to conventionally shot peened specimens. This was attributed to a reduction in surface roughness of approximately 30% along with amplified magnitudes of compressive residual stresses.

Keywords Austempered ductile iron, duplex shot peening, fatigue strength, fractography, strain induced transformation.

Introduction
Austempered ductile iron (ADI) has a remarkable combination of mechanical properties and tribological characteristics, such as high tensile and fatigue strength, good wear and rolling-contact fatigue resistance, and appreciable ductility [1, 2]. The potential of ADI as a viable material in the automotive industry may be further enhanced through shot peening (SP). The residual compressive stresses, together with the increased dislocation density and the strain induced transformation of austenite into martensite at the surface, significantly improve the fatigue life of the peened components along with the hardness of the surface layer. Zammit et al. [3] investigated the effect of shot peening on Cu-Ni ADI, observing that the fatigue strength was increased by 60%. Similarly, Benam et al. [1] observed improvements ranging from 27% to 50% for Cu-Ni ADI austempered at temperatures ranging from 320°C to 400°C respectively. The former study [3] also reported consistent improvements in fatigue life of approximately 35% at all stress levels, with peak stresses of around 560 MPa. On the other hand, Benam et al. [1] observed that improvements in fatigue life were much less pronounced at stress levels exceeding 500 MPa. Overall, several studies corroborate the positive impact of shot peening on fatigue performance, results ranging from an improvement of 23% [4] to 115% [5], with other authors reporting values falling anywhere within the range. A variant of the conventional shot peening process is duplex shot peening, which employs a secondary peening operation, typically using smaller shots and a lower peening intensity [6]. While this mechanical surface engineering treatment has never been applied to ADI, promising results have been reported for other ferrous materials, such as medium carbon steel [7] and duplex stainless steel [8].
To investigate the potential of shot peening and duplex shot peening of ADI, this study employs a holistic approach involving the characterisation of as-treated and shot peened ADI specimens. This allowed for the analysis of various aspects, such as compositional and microstructural changes, brought about by the heat treatment and surface engineering techniques implemented in this study. The accompanying improvements in mechanical properties, specifically the fatigue behaviour of ADI, are evaluated through rotating bending fatigue testing at different applied stress levels.

Experimental Methods

Hourglass-type specimens, having dimensions shown in Figure 1, were machined from standard Y-shaped ductile iron keel blocks. The as-cast material had a nodule count of 213.6/mm^2 and a nodularity exceeding 85%. The main elements making up the ductile iron are 90.7 wt% Fe, 3.3 wt% C, 2.4 wt% Si, 1.7 wt% Cu, 1.6 wt% Ni. The specimens were first austenitised in an electric furnace at 900°C for 120 minutes, then austempered in a salt bath at 360°C for 90 minutes, followed by cooling in air to room temperature.

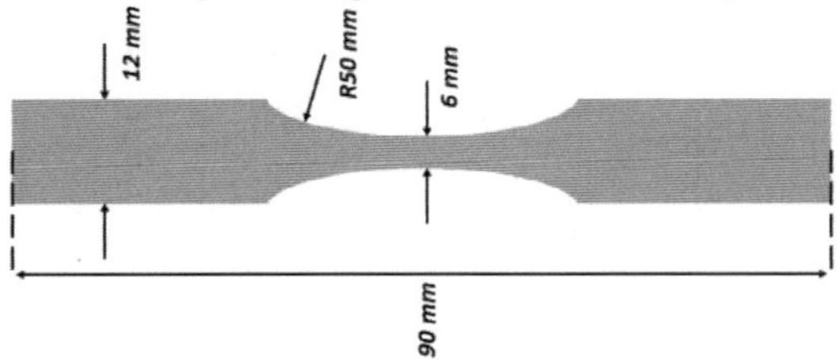

Figure 1: Hourglass-type fatigue specimen

Shot peening was carried out using S330 steel shots having a diameter of 0.84 mm at a peening intensity of 0.47 mmA (SP330) and S110 shots having a diameter of 0.28 mm at an intensity of 0.32 mmA (SP110). Duplex peening (SP3-110) was done by first peening with the larger S330 shots, followed by the smaller S110 shots. Benchmark unpeened specimens (UT-ADI) were mechanically polished to a mirror finish.

Microhardness measurements were done using a Mitutoyo MVK-H2 microhardness testing machine equipped with a Vickers pyramidal diamond indenter, 500 gf load and a dwell time of 10 s. Reported values are the arithmetic mean of a minimum of five measurements.

Two dimensional surface roughness analysis was carried out using a NanoMap-500LS contact profilometer. Fifteen measurements were recorded at a scanning speed of 10 µm/s over a distance of 2500 µm under a contact load of 25 mg.

Rotating bending fatigue tests were carried out on an Italsigma 2TM831 four-point loading rotating-bending fatigue test machine, using a loading ratio R of -1 and a frequency of 60 Hz, in accordance with standard ISO 1143:2010. Each test was stopped when a specimen failed or upon reaching 10^7 cycles, at which point it was considered to be a run-out.

Residual stress measurements were conducted through energy-dispersive synchrotron radiation, on the high-energy beamline ID15A at the European Synchrotron Research Facility (ESRF) in Grenoble, France. A germanium single-crystal detector cooled with liquid nitrogen was utilised as part of the setup. The 2θ value was set to 5°, with the variable being the wavelength of the X-ray source, providing the full X-ray spectra at the selected diffraction angle. The maximum energy throughout these measurements was set at approximately 300 keV.

Results and Discussion

Microhardness and Profilometry

Microhardness measurements show that both SP330 and SP3-110 experienced an increase in surface microhardness of approximately 50% (Figure 2).

Additionally, due to its nature, which involves plastic deformation of the surface, shot peening is expected to bring about an increase in surface roughness. This is also consistent with the results observed in this study (Figure 3), where shot peening using S330 shots brought about an increase in R_a of 675% relative to the ground condition (condition of coupons prior to peening). An increase in surface roughness relative to ground ADI was also observed for duplex shot peened coupons. Duplex shot peening resulted in smaller increase in surface roughness, bringing about a reduction of 26% in R_a relative to SP330 coupons. This reduction in surface roughness can be attributed to peening by the smaller shot, which has a tendency to 'hammer in' the peaks generated by the first coarser peening operation, thus levelling the surface. The results are in agreement with previous work by Scuracchio et al. [9], where the lowest roughness was recorded for a secondary peening stage using shots having a diameter of 0.3 mm, which is similar to the nominal diameter of S110 shots utilised in this study.

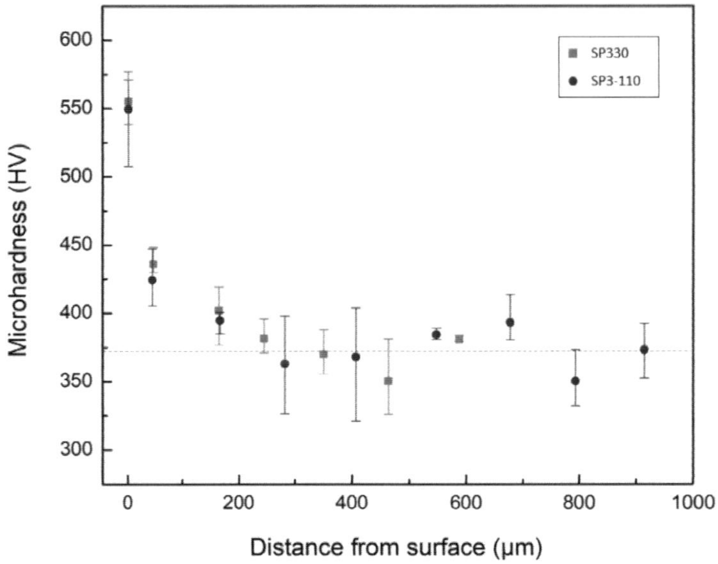

Figure 2: Microhardness profiles for SP330 and SP3-110 coupons. The dotted line represents the microhardness of the ADI microstructure. The error bars represent the maximum and minimum microhardness values recorded.

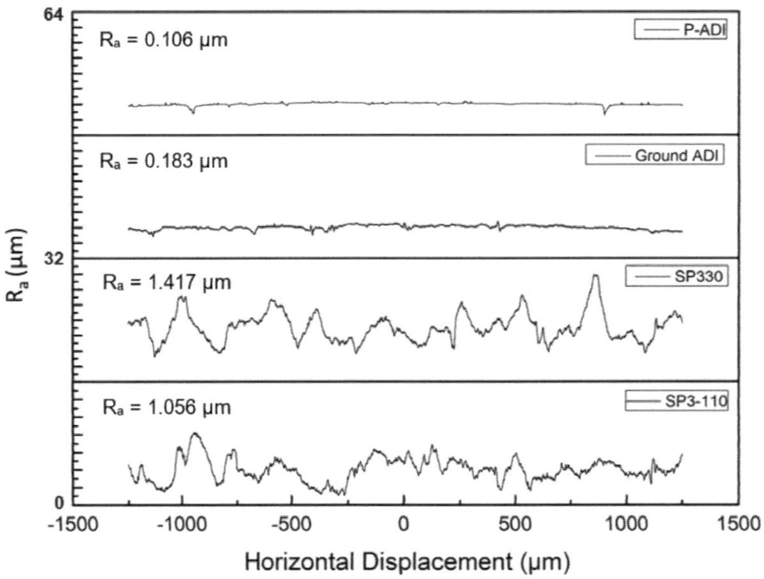

Figure 3: Axial roughness profiles for P-ADI, ground ADI, SP330, and SP3-110. Each profile is plotted on the same scale.

Figure 4(a) shows that ADI in the polished condition reveals a significant degree of residual stresses, which varies from compressive to tensile particularly in the ferritic phase. On the other hand, the surface layers of the austenite making up the ausferrite structure is under tensile stress up to a depth of around 250 μm. Residual stresses in the as-polished specimens can be attributed to machining, heat treatment, or the following grinding and polishing operations. Figure 4(b) displays the residual stress profiles for the austenitic and ferritic phases in SP330. The residual stresses take on the characteristic profile of shot peened materials, comprising of a compressively stressed layer of maximum intensity just beneath the surface, and a tensile stressed layer deeper towards the material core. Overall, shot peening of SP330 coupons brought about maximum compressive stresses of approximately 985 MPa and 785 MPa for austenite and ferrite, peaking in amplitude at depths of around 145 µm and 160 µm below the surface, respectively. This is a result of the plastic deformation induced on the surface, coupled with the strain-assisted transformation of austenite into martensite [3, 10].

Duplex shot peening produced a stress profile similar to that observed for SP330, varying only in terms of the maximum stress amplitude, as illustrated in Figure 4(c). SP3-110 coupons exhibit a maximum compressive stress of 1.17 GPa in the austenitic phase, an increase of approximately 19% over the single shot peened counterpart. Similarly, but to a lesser extent, stress in the ferritic phase increased to a magnitude of around 810 MPa, an increase of roughly 4% over the SP330 condition.

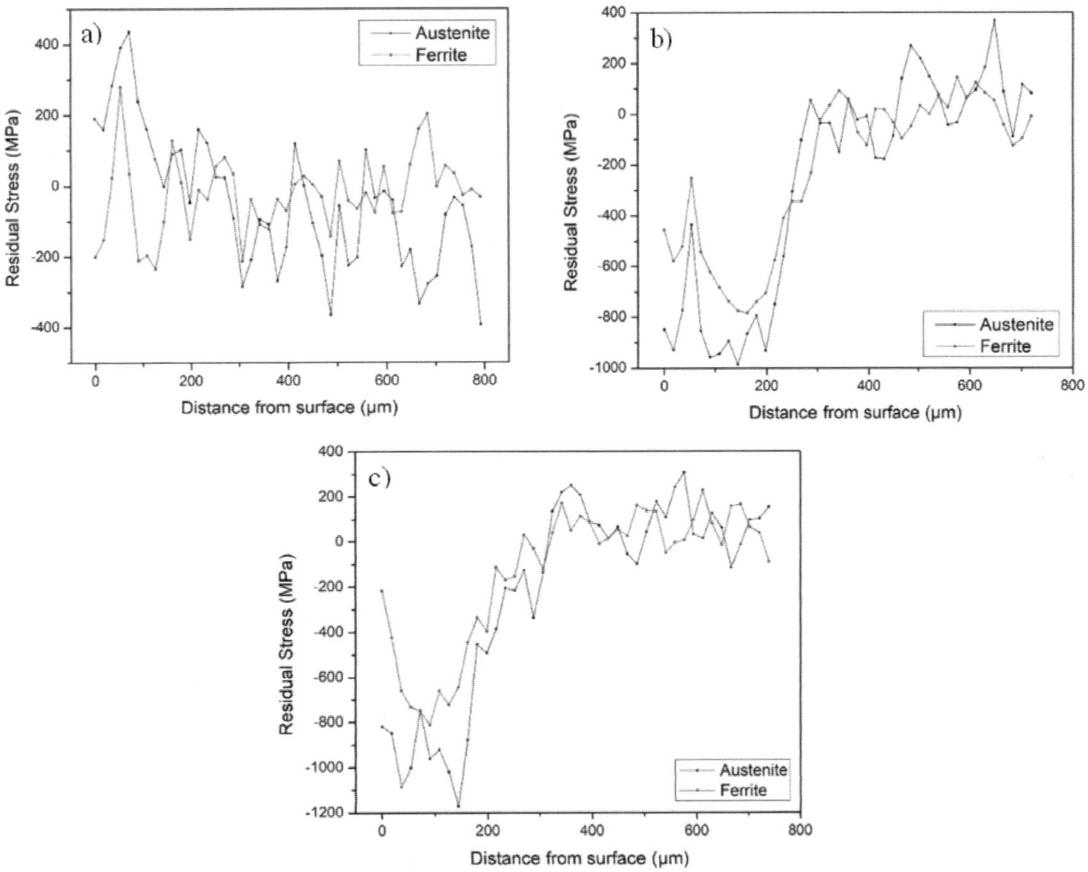

Figure 4: Residual stress-depth profiles for a) P-ADI, b) SP330, and c) SP3-110

Rotating-Bending Fatigue Testing

As shown in Figure 5, the fatigue strength of polished ADI was found to be of 363 MPa. Overall, results reported in literature [3, 5, 11] identify the fatigue limit of ADI as ranging between 200 MPa and 500 MPa, depending on chemical composition, heat treatment, fatigue testing conditions, and numbers of cycles defining the fatigue limit. The fatigue strength is increased to 600 MPa for SP330 coupons, an improvement of approximately 65% over the polished condition. This improvement is a result of the compressive stresses generated through shot peening and the aforementioned solid state transformation, along with work hardening of the ausferrite microsctructure. The detrimental effect brought about by an increase in surface roughness as a result of shot peening, is greatly overshadowed by the positive effects. Reported improvements in fatigue strength following shot peening of ADI fall in the range between 23% [4] to 115% [5], depending on the chemical composition of the as-cast material, the heat treatment employed, the resultant microstructure and phase composition and the shot peening parameters applied.

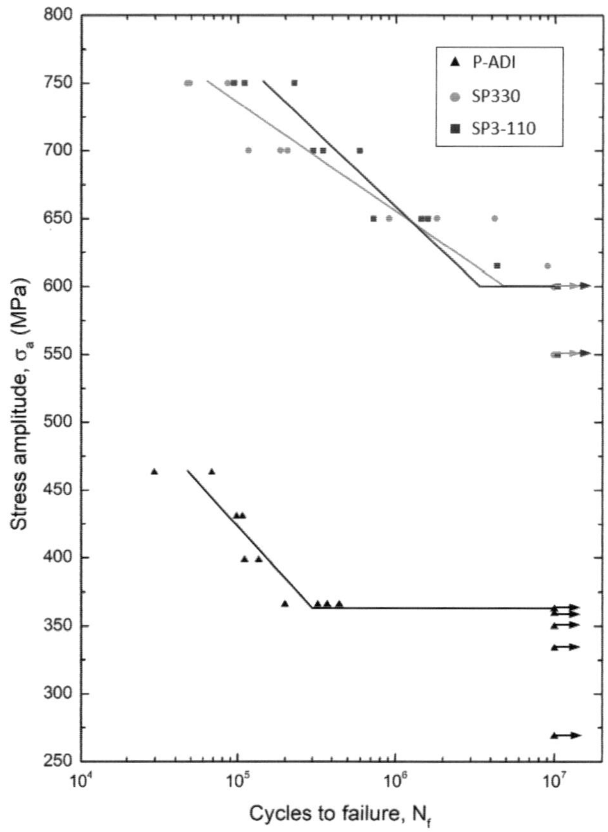

Figure 5: S-N semi-logarithmic curves for P-ADI, SP330 and SP3-110 hourglass-type specimens.

Figure 5 also shows the comparison between shot peened and duplex shot peened specimens. Based on the same principles discussed for SP330, duplex shot peening also brought about significant improvements over P-ADI specimens, with a fatigue strength of 600 MPa. As illustrated in Figure 5, duplex shot peening brought about significant improvements in fatigue life at high stresses. At stress levels of 700 MPa and 750 MPa, the average fatigue life of SP3-110 coupons is extended by approximately 140% over the SP330 condition. This trend was reversed at lower stresses, with SP330 specimens proving to be superior to their duplex counter parts. For a fatigue life of 10^7 cycles, both conditions displayed a fatigue strength of 600 MPa. Due to a reduction in R_a equivalent to around 30%, it was expected that the secondary peening stage would improve the fatigue performance of SP3-110 when compared to SP330 specimens. Despite this, since the surface roughness of SP330 coupons was still significantly low when compared to other results reported in literature, it is postulated that any such improvement may not have been very significant. As outlined in the previous section, duplex shot peening was also effective in obtaining enhanced compressive. Despite this, SP3-110 was not found to be superior throughout.

Conclusions

This study aimed at determining the effect of conventional and duplex shot peening on the fatigue performance of Cu-Ni ADI.

1. Following shot peening, the surface microhardness increased by approximately 50% relative to the as-austempered condition.
3. While shot peening brought about significant roughening, duplex shot peening was successful in reducing this roughness by approximately 30%.
4. Both conventionally and duplex shot peened specimens displayed high magnitudes of compressive residual stresses. This led to an improvement of 65% in fatigue strength over as-treated ADI.
5. The bending fatigue strengths of both groups of shot peened specimens are within the ranges specified in the 'AGMA – Austempered Ductile Iron for Gears' technical sheet [12], verifying their potential as viable gear materials for automotive applications.
6. At high stresses, duplex shot peening brought about significant improvements in fatigue life relative to conventionally shot peened specimens.

Acknowledgements

The authors would like to thank the European Synchrotron Radiation Facility (ESRF) in Grenoble for carrying out the stress measurements. The authors would also like to acknowledge the positive impact of ERDF (Project ref. no. 012: "Developing an Interdisciplinary Material Testing and Rapid Prototyping R&D Facility".

References

[1] A. S. Benam, S. Yazdani, and B. Avishan, "Effect of shot peening process on fatigue behavior of an alloyed austempered ductile iron," *China Foundry,* vol. 8, pp. 325-330, 2011.

[2] A. Zammit, M.Bonnici, M. Mhaede, R. Wan, and L. Wagner, "Shot peening of austempered ductile iron gears," *Surface Engineering,* vol. 33, pp. 679-686, 2017.

[3] A. Zammit, M. Mhaede, M. Grech, S. Abela, and L. Wagner, "Influence of shot peening on the fatigue life of Cu-Ni austempered ductile iron," *Materials Science and Engineering: A,* vol. 545, pp. 78-85, 2012.

[4] S. Ezoe, M. Hashimoto, I. Asano, and Y. Tanno, "Effects of shot peening and heat treatment on endurance limits of austempered ductile cast iron gears," in *ICSP-6*, San Francisco, California, USA, 1996, pp. 14-23.

[5] A. Ebenau, D. Lohe, O. Vohringer, and E. Macherauch, "Influence of shot peening on the microstructure and the bending fatigue strength of bainitic-austenitic nodular cast iron," in *ICSP-4*, Tokyo, Japan, 1990, pp. 389-398.

[6] D. Breuer, "Dual peening," *The Shot Peener,* vol. 27, pp. 18-20, 2013.

[7] S. Matsumara and N. Hamasaka, "High strength and compactness of gears by WHSP (Double hard shot peening) Technology," *Komatsu Technical Report,* vol. 52, pp. 1-5, 2006.

[8] A. V. Jeberaj, A. Bhagade, D. Vijapuri, and E. Gupta, "Influence of dual shot peening on the enhancement of surface characteristics of duplex stainless steel," *Materials Today: Proceedings,* vol. 24, pp. 921-929, 2020.

[9] B. G. Scuracchio, N. B. d. Lima, and C. G. Schon, "Role of residual stresses induced by double peening on fatigue durability of automotive leaf springs," *Materials and Design,* vol. 47, pp. 672-676, 2013.

[10] K. H. S. Silva, J. R. Carneiro, R. S. Coelho, H. Pinto, and P. Brito, "Influence of shot peening on residual stresses and tribological bhavior of cast and austempered ductile iron," *Wear,* vol. 440-441, 2019.

[11] C.-K. Lin, P.-K. Lai, and T.-S. Shih, "Influence of microstructure on the fatigue properties of austempered ductile irons - I. High-cycle fatigue," *International Journal of Fatigue,* vol. 18, pp. 297-307, 1996.

[12] AGMA 939-A07, "Austempered Ductile Iron for Gears," ed: AGMA, 2007, pp. 1-10.

Using Laser Shock Peening to Enhance Metallic Additive Manufacturing

Sanin Zulić[a], Sunil Pathak[a], Michael Fitzmire[b], Ebrahim Asadi[b], Jan Kaufman[a], Vojislav Petrović[c], Fernando Gustavo Warchomicka[d], Marek Boehm[a], Ondrej Stransky[a], Benjamin, Meier[c], Jan Brajer[a]

[a] HiLASE Centre, Institute of Physics of the Czech Academy of Sciences, Na Slovance 1999/2, Prague, Czech Republic – sanin.zulic@hilase.cz

[b]The University of Memphis, Herff College of Engineering, 3720 Alumni Ave, Memphis, TN 38152, The United States of America; easadi@memphis.edu

[c]JOANNEUM RESEARCH Forschungsgesellschaft mbH, MATERIALS – Institute of Surface Technologies and Photonics Leobner Straße 94, 8712 Niklasdorf, Austria; vojislav.petrovic@joanneum.at

[d]Graz University of Technology, Kopernikusgasse 24/I, 8010 Graz, Austria; fernadno.warchomicka@tugraz.at

Keywords: *Additive Manufacturing, Laser Shock Peening, Fatigue Life Improvement, Surface Integrity Improvement*

Abstract: In the present work, the effects of laser shock peening on mechanical properties of metal additive manufacturing parts made of SS 304L, SS 316L, AlMg10Si, and Ti6Al4V were discussed. The samples were evaluated for residual stress, microstructure, computed tomography scanning, and/or fatigue life testing. All the measurements/testings were carried out on the As-Built and LSP treated samples and the results were compared.

1. Introduction

Additive Manufacturing (AM) technology has brought numerous benefits to mankind and considering the benefits of the technology such as manufacturing of complex shapes and lightweight parts, easy customization, design freedom, etc.[1]. It is expected that this technology will be a more vital part of the manufacturing processes in the coming years. Even though this technology has found its place in many industries, and it's regularly used nowadays, there are still areas where additive manufacturing cannot be applied because of certain limitations [4]. These limitations are not allowing additive manufacturing to be used for critical parts, either because of technical or safety reasons. By employing Laser Shock Peening (LSP), many of these limitations can be overcome, mitigated, or even completely eliminated. Laser Shock Peening can significantly improve the mechanical properties of the AM parts and bring AM technology into the most demanding applications and industries. In the present work laser shock peening on SS 304L, SS 316L, AlMg10Si, Ti6Al4V. The residual stress measurements were presented for SS 304L and SS316L samples, while the porosity in AlMg10Si was discussed and for Ti6Al4V fatigue life is presented in this paper.

2. Laser Shock Peening process

For the results presented in this paper, LSP experimental work has been carried out at HiLASE' premises in the laboratory that is dedicated to the LSP research. Figure 1 presents the scheme of the LSP arrangement. For all the experiments a BIVOJ laser system was used with all the characteristics important for the LSP processing. The laser beam was specially fixed, and the samples were moved by the robotic arm with a predefined path.

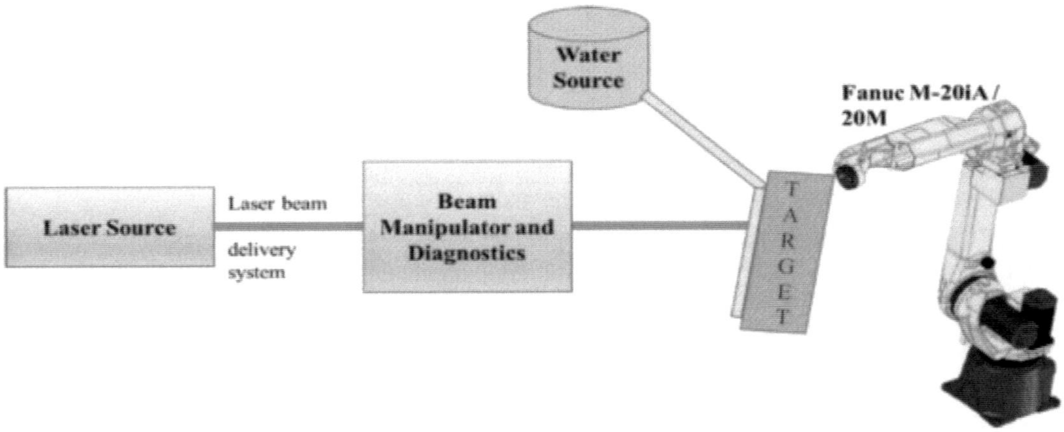

Figure 1: Scheme of the laboratory for LSP at HiLASE Center [3]

3. Experimental results

3.1 Residual stress change

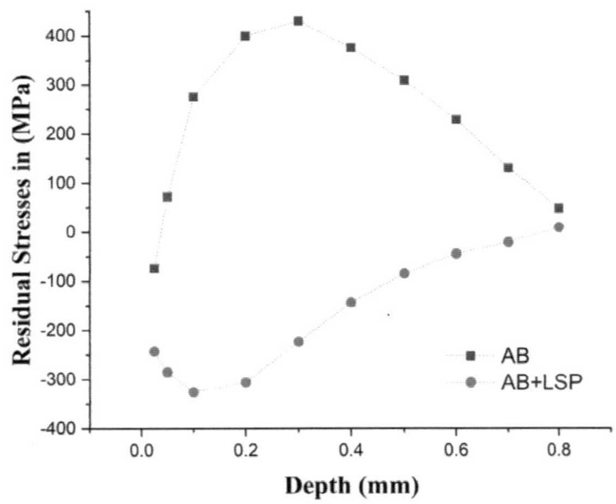

Figure 2: Change of the residual stress state from tensile to compressive for 304L additively manufactured part

Figure 2 presents how the LSP is changing the state of residual stresses from the tensile regime to the compressive regime for additively manufactured samples made of stainless steel 304L. In this particular case, samples have not been heat-treated for stress relaxation and this is the reason for high tensile stresses. The laser energy per pulse for this processing was 5J and we have used the sport size 2 mm X 2 mm. The pulse width was 14 ns, and the corresponding power density was 4.9 GW/cm^2. Figure 3 presents what the residual stress profile looks like when the samples are heat-treated for the stress release purpose. In both cases, a significant amount of compressive residual stresses has been achieved deep into the material. In the end, this typically brings improvement in fatigue properties of the part. The corresponding power density, in this case, was 2.5 GW/cm^2, achieved with similar laser parameters as in the first case.

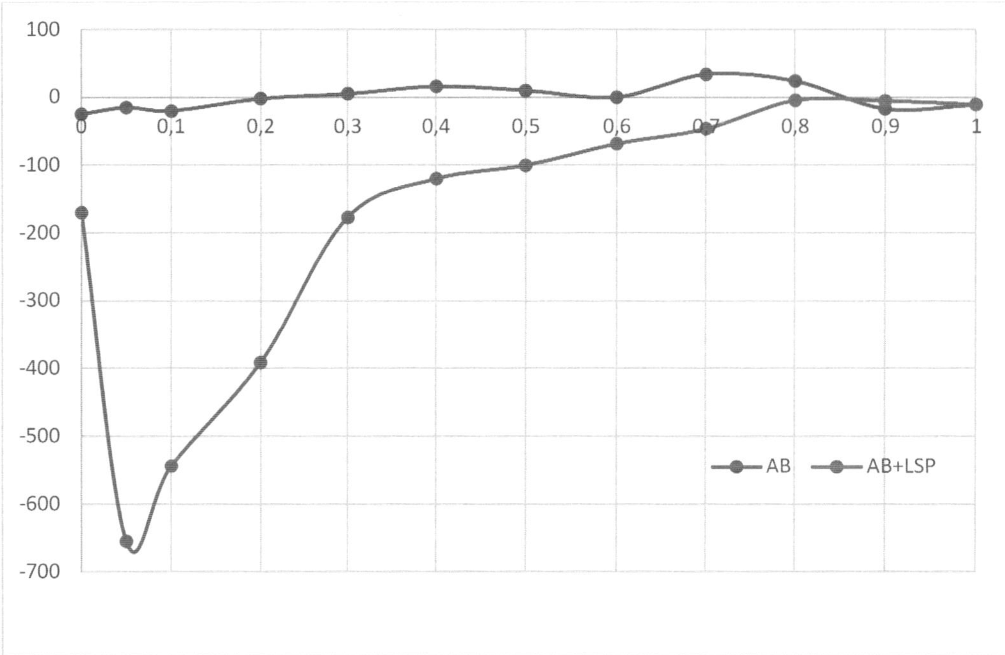

Figure 3 Residual stress distribution for additively manufactured 316L parts before LSP (AB) and after LSP Microstructure evaluation

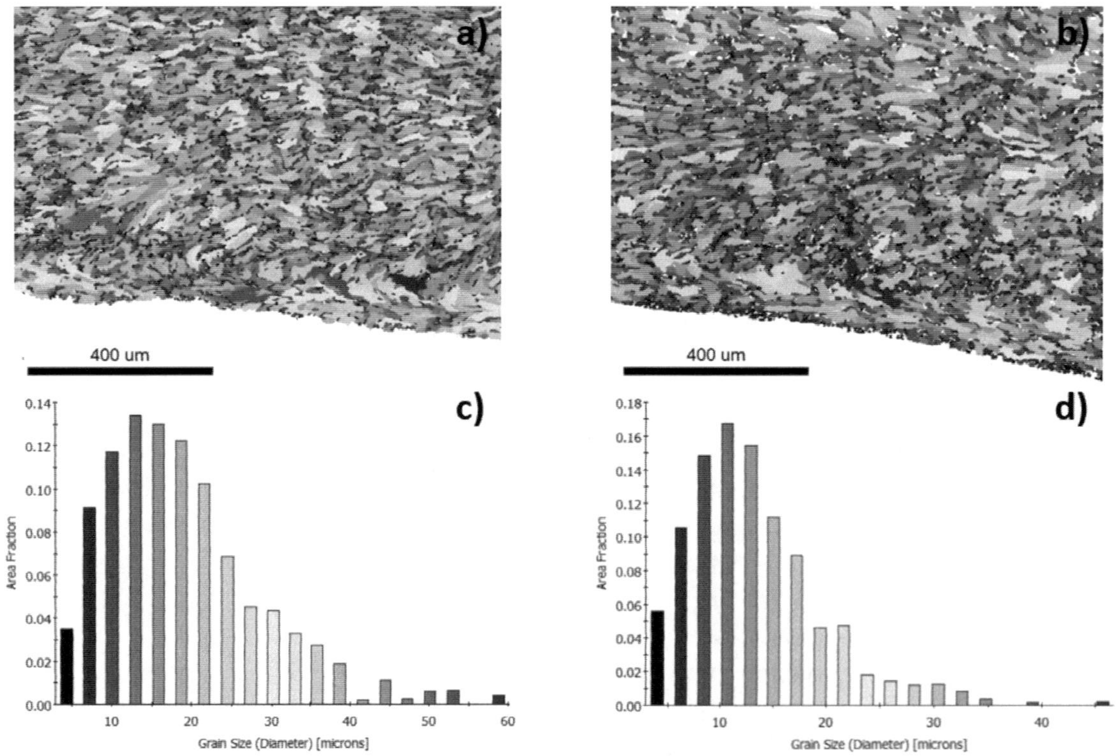

Figure 4 SEM of additively manufactured stainless steel 304L parts before (a) and after LSP (b) and grain size distribution before (c) and after (d) LSP

Figure 4 presents the scanning electron microscopy (SEM) of the cross-sectional cut unpeened sample (Fig. 4a) and LSPeened sample (Fig. 4b). Whereas Fig. 4c and 4d depict the grain size distribution over the area fraction within the unpeened and LSPeened samples respectively. It can be seen by comparing Fig. 4b with Fig. 4a that the grain sizes have been reduced due to the transfer of shock waves causing plastic deformation within the material. The observation can be further verified with the grain size distribution as presented in Fig. 4c and 4d. It is visible from Fig. 4d that grains of more than 40 microns were totally diminished from the measured range and smaller grains of 2 µm to 20 µm have gathered the larger area. Such improvement in grain structures has resulted in higher values of compressive residual stresses as presented in Fig. 3 also.

3.2 Porosity mitigation

During printing, the strategies for the outer skin and inner infill differ. Where these two strategies meet, we are observing a systematical drawback which is the formation of porosity. This porosity was successfully mitigated using laser shock peening up to a certain depth as can be seen from the Computed Tomography scan results shown in figure 5. The power density used during LSP processing was 3.6 GW/cm^2.

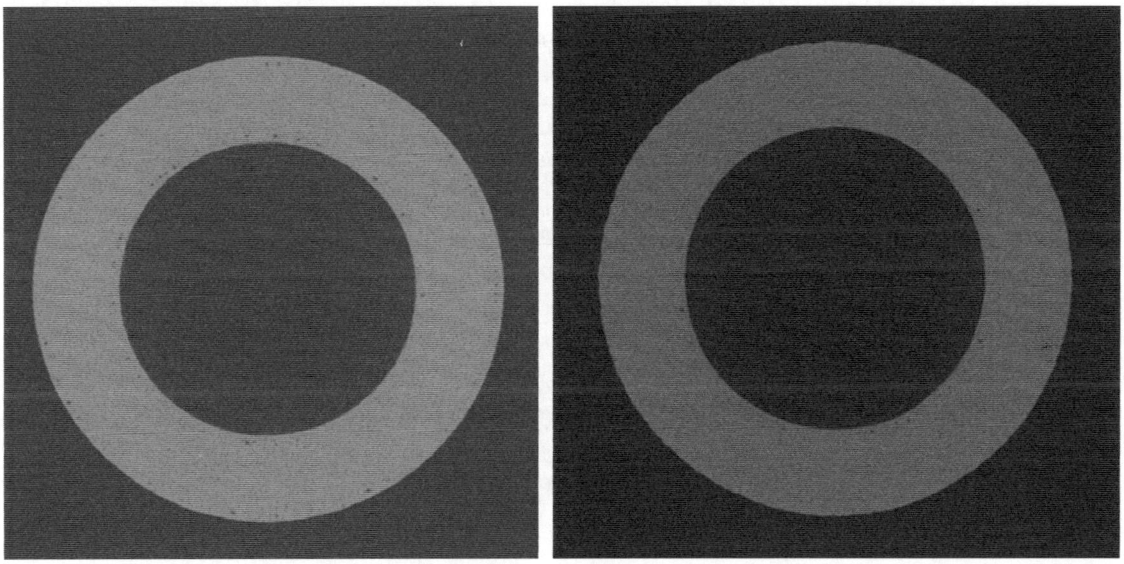

Figure 5 Porosity mitigation for AlSi10Mg; left before LSP, right after LSP

3.3 Fatigue life improvement

Fatigue life improvement is the major benefit of LSP. In the case of additive manufacturing, fatigue life improvement can be longer than for the parts that are traditionally produced. The fatigue life of the AM parts can be prolonged several dozen times. The reason for this is that AM parts are quite often of the r quality after the production, compering to the parts traditionally produced. At HiLASE Centre, we have achieved an improvement in fatigue life of Ti6Al4V 100 times. In the case of the stainless steel 316L samples with a stress riser, the improvement was significantly less, but still, it was a valuable improvement (up to 4 times) [3].

4. Conclusions

Laser Shock peening is a promising method for the post-processing treatments of metal additively manufactured parts. It can completely change the state of residual stresses in the additively manufactured part from tensile to a compressive state of residual stresses. By imparting high compressive residual stresses, deep into the part. LSP significantly can prolong the lifetime of the component that is additively produced and make it available for the applications that require the highest quality. The refinement of the microstructure is another benefit that LSP brings to additive manufacturing. Also, the mitigation, or complete elimination of the porosity, which is often a problem for additive manufactured parts, is possible with LSP. Certainly, LSP is a superior technology for certain improvements in additive manufacturing.

References:

[1] DebRoy, H.L. Wei, J.S. Zuback, T. Mukherjee, J.W. Elmer, J.O. Milewski, A.M. Beese, A. Wilson-Heid, A. De, W. Zhang, Additive manufacturing of metallic components – Process, structure and properties, Progress in Materials Science, 92, 2018, 112-224.
[2] Vrancken et. al. Heat treatment of Ti6Al4V produced by Selective Laser Melting: Microstructure and mechanical properties, Journal of Alloys and Compounds, Volume 541, 2012, Pages 177-185
[3] Zulić, D. Rostohar, J. Kaufman, S. Pathak, J. Kopeček, M. Böhm, J. Brajer, T. Mocek, Fatigue life enhancement of additive manufactured 316l stainless steel by LSP using a DPSS laser system, Surface Engineering, 2022. DOI: 10.1080/02670844.2022.2060463
[4] Hackel, J. R. Rankin, A. Rubenchik, W. E. King, M. Matthews, Laser peening: A tool for additive manufacturing post-processing, Additive Manufacturing, 24, 2018, 67-75,
[5] Jafari, W. W. Wits, The utilization of selective laser melting technology on heat transfer devices for thermal energy conversion applications: A review, Renewable and Sustainable Energy Reviews, 91, 2018, 420-442.

The Effect of Shot Peening on Additively Manufactured 316L Stainless Steel for the Marine Transportation Industry

L. Bonnici[1], J. Buhagiar[1], G. Cassar[1], K. A. Vella[1], J. Chen[2], X.Y. Zhang[2], M.Y. Liu[2], Z.Q. Huang[2], A. Zammit[1]

1 Department of Metallurgy and Materials Engineering, University of Malta, Msida, Malta
2 School of Materials Science and Engineering, Southeast University, Nanjing, Jiangsu Province, China

Abstract
In the marine transportation industry, the harsh environment during voyage leads to severe corrosion failure or mechanical breakdowns. Hence, continuous and preventive maintenance is required. Additive manufacturing (AM) can potentially be used to reduce the downtime associated with replacement of components by printing the required parts directly from the design itself. In this study, shot peening (SP) was applied as a posterior treatment on AMed 316L stainless steel (SS) with the aim to increase the mechanical properties at the surface and the surface characteristics without creating a detriment to the alloy's corrosion resistance. A 44% increase in surface hardness from 231 $\pm$ 38 $HV_{0.2}$ to 334 $\pm$ 61 $HV_{0.2}$ was measured after SP, while the affected depth was of around 200 μm. SP also resulted in R_a values of 10 $\pm$ 1 μm for as-printed samples, which was reduced to 5 $\pm$ 1 μm after SP. Additionally, the SP treatment induced a surface compressive residual stress of 589 MPa. Corrosion results were not favourable for the SP-ed samples. However, the overall results suggest that the use of AM SP parts may indeed be a suitable contender for use in selected maritime components.

Keywords
Additive manufacturing, shot peening, residual stress measurement, corrosion resistance

Introduction
Maritime activities are the pillar of global economy and trade. Primary necessities such as food, petroleum and goods are imported by means of water transportation. Recent studies show that 70% [1] of the global goods are carried by sea.
The marine environment gives rise to corrosion, erosion and detrimental effects. Due to this, parts such as propellers and shafts must be frequently replaced [2]. The issue of high replacement demands in this industry can be solved with AM. Selective laser melting (SLM) is one of the efficient processes of AM. Parts which are manufactured through SLM are used in various industries such as the medical, dental and aerospace industries, lightweight structures and heat exchangers [3].
However, parts produced via AM have notoriously poor surface finish and high porosity, posing deterioration of such critical parts, weakening their performance, particularly resistance to cyclic loading [4]. To address this challenge, SP was applied to AM 316L SS. Studies performed on polished surfaces, such as that of Sugavaneswaran *et al.* [3], show significant improvements in hardness and roughness following SP.
During this study, samples were left unpolished to deduct a preparation step and also make use of the 3D printer's potential, that which it can manufacture parts rapidly to solve emergent issues. This advantage was explored by Bagherifard *et al.* [5] where a 9% increase in hardness and an 18% reduction in roughness were identified after performing SP on 316L SS. However, in the authors best knowledge, studies on corrosion tests on as-printed and printed and shot peened samples are novel and the aim of this study was to analyse the corrosion properties of unpolished samples amongst other material characterisation studies.

Experimental Methods

316L SS samples were printed with an AmPro SP100 3D Printer, China, using a SLM process. A laser power of 200 W, laser speed of 950 mm/s, beam offset of 10 µm, layer thickness of 30 µm and a hatch distance of 80 µm were used. Table 1 shows the used powder's chemical composition.

Table 1: Chemical composition for powder 316L SS[6]

Element	Fe	Cr	Ni	Mo	Si	Mn	P	C	S
wt %	68.64	16.59	10.77	2.48	0.76	0.72	0.025	0.0078	0.0037

SP was carried out using a modified Industrial Surface Treatments sand blaster. S230 steel shots were used, at a pressure of 6 bar, shot flow rate of 50 %, nozzle to specimen distance set to 100 mm, peening time of 21 seconds and intensity of 0.21 mmA. The designation of the samples is AP for the as-printed samples and PSP for the printed and shot peened samples.
Electrochemical tests were carried out with a Gamry Interface 1000™ potentiostat / galvanostat / ZRA. A 3-electrode setup was used. A 1 cm^2 surface area from the sample to be tested was exposed to 300 ml of substitute oceanwater. This was prepared according to ASTM D 1141 – 98: *Standard practice for the preparation of substitute ocean water*. The procedure was comprised of an initial OCP measurement for 2 hours. This was followed by cyclic polarisation sweeps at a rate of 0.167 mV/s between the following ranges: −100 mV versus OCP and -500 mV versus reference. The polarisation scan was instructed to reverse when an apex current of 0.5 mA/cm^2 was reached at a scan rate of 0.167 mV/s. The test was stopped after a few steps once the protection potential was reached. Three repeats were done for each set.
Subsequent material characterisation included optical and scanning electron microscopy, microhardness measurements and topographic analysis. X-Ray Diffractometry (XRD) phase analysis was performed using a Rigaku Ultima IV diffractometer. This was set for Glancing Incidence Asymmetric Bragg (GIAB) with an angle of incidence of 3° and Cross Beam Optics (CBO). The source of radiation used was that of Cu$K\alpha$. The same diffractometer was used to carry out the surface residual stress measurements, set in θ / 2θ configuration. Stresses were measured according to the standard BS EN 15305:2008 - Non-destructive testing - Test method for residual stress analysis by X-ray diffraction. A Rigaku Multipurpose attachment IV (MPA-U4) was used for Ψ tilting and stress measurements were carried out using the parallel beam method. The peak was measured for the [3 1 1] fcc at a theoretical nominal of 2θ = 90.68°. The 2θ range was set from 87° to 93°, with seven Ψ tilts from 0° to 60°.

Experimental Results

Figure 1 shows the microstructure along the building direction of the AP 316L SS to be composed of an austenitic matrix, which was verified by XRD (Figure 3). At higher magnifications, shown in Figure 2, columnar and cellular dendritic structures were observed. Such structures are formed following molten metal solidification.

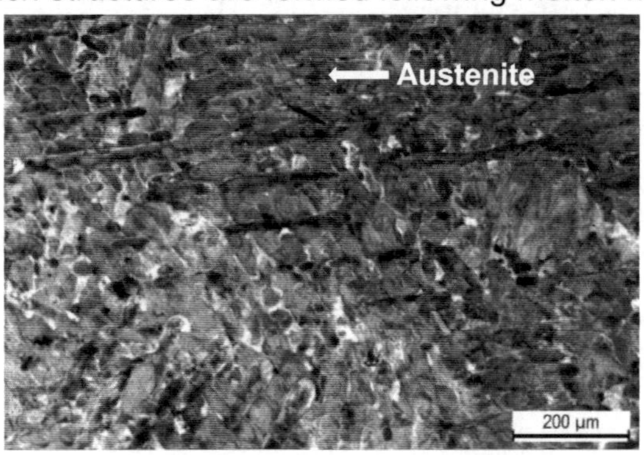

Figure 1: Etched as-printed microstructure highlighting the austenite matrix

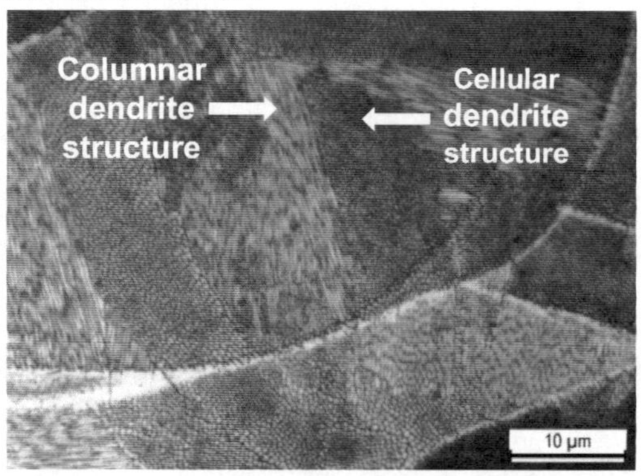

Figure 2: Columnar and cellular dendritic structures in the as-printed microstructure

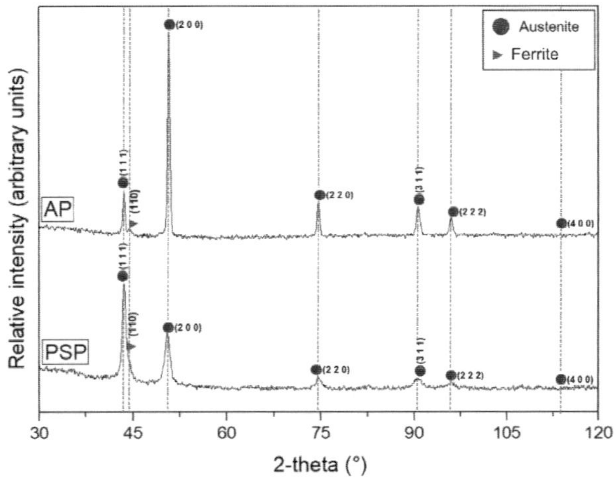

Figure 3: XRD patterns for as-printed and printed and shot peened samples with austenite (PDF number: 00-023-0298) and ferrite (PDF number: 00-003-0411) as the main peaks

Table 1 shows the results obtained in this work. Shot peening resulted in a reduction of 50% for the R_a and 77% for the R_z roughness. This result can also be observed from Figure 4, where the shot peened surface (Figure 4b), showing the individual dimple characteristic, shows to be less rough than the AP surface (Figure 4a) which shows the printing striations. In addition, SP generated a compressive residual stress of 589 MPa at the surface and a 45% increase in surface micro-hardness. The micro-hardness depth profile in Figure 5 shows a maximum hardness value at the surface reaching 334±16 $HV_{0.2}$ decreasing gradually into the bulk, reaching the substrate average hardness of 230±10 $HV_{0.2}$ at around 200 μm.

Table 1: Summary of results for as-printed and printed and shot peened specimens

	n, population for standard error	As-Printed (AP)	Printed and Shot Peened (PSP)
Surface stress (MPa)	2	60 ± 18	-589 ± 9
R_a (μm)	20	10 ± 1	5 ± 1
R_z (μm)	20	95 ± 9	20 ± 5
Surface micro-hardness (HV_{200})	15	231 ± 10	334 ± 16
E_{corr} (mV)	3	-2 ± 14	-198 ± 15
Breakdown potential (mV)	3	500 ± 36	334 ± 92
Protection potential (mV)	3	-133 ± 17	-53 ± 17

Figure 4: a) As-printed surface, highlighting the printing striations b) Printed and shot peened surface, highlighting the dimples

Figure 6 shows the representative cyclic polarisation curves for AP and PSP, with the marked E_{corr}, breakdown voltage (E_{break}) and protection voltage (E_{prot}). Upon inspecting the respective values in Table 1, it can be seen that AP performed better during the corrosion test than PSP.

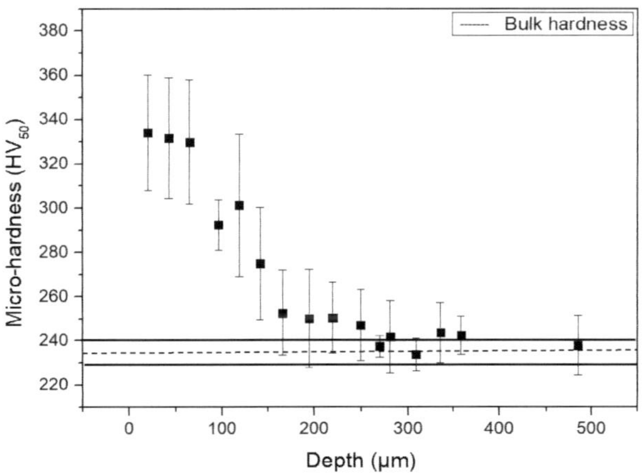

Figure 5: Micro-hardness depth profile for printed and shot-peened

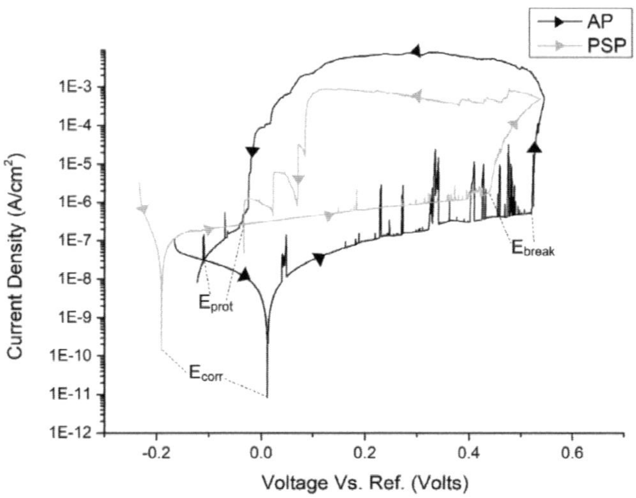

Figure 6: Representative cyclic polarization curves for as-printed and printed and shot peened samples

Discussion and Conclusion

The microstructure of AM 316L SS is made from austenite. Figure 3 shows a peak of ferrite at an angle of 44.5° [1 1 0]. Similar observations were reported for studies performed on 316L SS manufactured by SLM by Sun *et al.* [7], Kurzynowsk*i et al.* [8] and Kamariah *et al.* [9] . Columnar and cellular dendritic structures were evident, which are attributed to the effect of laser melting during the AM process. These structures were also evident in a study by Jin Oh *et al.* [11] while studying AMed 316L SS.

Following SP, no peak shifts occurred. However, the peak at 44.5° [1 1 0] could be either ferrite or martensite or a mixture of both, as found by Krawczyk *et al.* [12]. This could be a result of the strain induced by the SP. The main differences identified between the AP and PSP diffractographs (Figure 3) were the broadening of XRD peaks in the SP-ed sample and change in relative intensity at the [1 1 1] and [2 0 0] peaks. The broadening of the peaks is attributed to the residual stresses generated. In a study carried out by AlMangour and Yang [13], it was explained that this broadening is formed by plastic deformation leading to macro and micro residual stresses induced by peening. Additionally, the decrease in peak intensity is attributed to the structural change at crystalline level which is caused by severe surface deformation via peening. Prabhakaran *et al.* [14] argue that the reason for such phenomenon is that residual stresses affect the crystal lattice structure and therefore, the lattice spacing. Moreover, the original surface stresses of AP were in the tension state, which were generated during characteristic fast melting and solidification processes. However, SP has resulted in the generation of compressive stresses reflecting the transfer of shot kinetic energy to retained elastic stress, as already explained.

The surface characteristics were also significantly changed after SP. The surface peaks that were present before peening, were flattened, leaving individual and dense dimple characteristics of SP. The features show approximately constant dimensions, confirming the uniformity in the shot sizes and impact velocities. Peak flattening was confirmed by the surface topography results, where a decrease of 50% and 77% were discovered for R_a and R_z respectively. As the shots impinge the surface, the crest is crushed, diminishing the surfaces' roughness. Similar results were obtained in a study performed by Bagherifard *et al.* [5] where unpolished 316L SS was shot peened with S230 shots and an 18% decrease in R_a was found.

Even though there was no full phase transformation, an increase in hardness following SP was still evident. This is due to the cold work taking place as the shots impact the surface. As this is done, plastic deformation takes place, inhibiting any crack formation or propagation, strengthening and hardening the material. These results are comparable to those found by AlMangour *et al.* [13] where a 52% increase in hardness was discovered following SP 17-4 stainless steel. The relationship of the hardness along the depth shows that SP effects a 200 μm depth, which decreases steadily towards the substrate's bulk, proving that there is still subsurface hardening.

When analysing the corrosion results, AP shows more favorable conditions than those for PSP. This was confirmed from the results obtained in Table 1. The higher E_{corr} for AP shows a more corrosion resistant material, while a higher breakdown potential shows a more pitting resistant material. The detrimental effects of PSP could be influenced by the change in surface morphology by creating a larger surface area by the dimples, possibility of multiple phases of ferrite and martensite [15] and also, due to high dislocation densities [16] induced by the SP process [17].

From this study, it can be concluded that SP can be adopted on additively manufactured 316L SS, resulting in an increase in hardness and a decrease in surface roughness. This suggests that SP can be considered as an appropriate surface engineering technique for a number of applications relevant to the marine transportation industry. However, the adverse behavior noted for its corrosion performance should be the subject to further surface engineering treatment, in addition to SP.

Acknowledgments

This study was part of the SEAM project, funded by the Malta Council for Science and Technology (MCST) as part of the Sino-Malta fund 2019-02.

References

[1] J. Hoffmann, "Review of Maritime Transport," in *United Nations Conference on Trade and Development*, New Yord and Genova, 2018.

[2] J. Sienkiewicz and T. J. Cuff, "Extreme Maritime Weather: Improving Safety of Life at Sea," World Meteorological Organization, [Online]. Available: https://public.wmo.int/en/resources/bulletin/Products_and_services/Extereme_maritime_weather. [Accessed 22 May 2022].

[3] M. Sugavaneswaran, A. Vinoth Jebaraj, B. Kumar, K. Lokesh and A. J. Rajan, "Enhancement of surface characteristics of direct metal laser sintered stainless steel 316L by shot peening," *Surfaces and Interfaces,* vol. 12, pp. 31-40, 2018.

[4] H. Bikas, P. Stavropoulos and G. Cryssaolouris, "Additive manufacturing methods and modelling approches: a criticl review," *International Journal of Additive Manufacturing Technology,* vol. 83, pp. 389-405, 2016.

[5] S. Bagherifard, S. Slawik, I. Fernandez-Pariente, C. Pauly, F. Muclich and M. Guagliano, "Nanoscale surface modification of AISI 316L stainless steel by severe shot peening," *Materials and Design,* vol. 102, pp. 68-77, 2016.

[6] J. V. A. M. T. C. Ltd., *Certificate of Quality 316L Stainless Steel,* Suqian, Jiangsu: Jiangsu Vilory Advanced Materials Technology Co. Ltd., 2021.

[7] Z. Sun, X. Tan, S. B. Tor and W. Y. Yeong, "Selective laser melting of stainless steel 316L with low porosity and high build rates," *Materials and Design,* vol. 104, pp. 197-204, 2016.

[8] T. Kurzynowski, K. Gruber, W. Stopyra, B. Kuznicka and E. Chlebus, "Correlation between process parameters, microstructure and properties of 316 L stainless steel processed by selective laser melting," *Materials Science and Engineering A,* vol. 718, pp. 64-73, 2018.

[9] M. Kamariah, W. Harun, N. Khalil, F. Ahmad, M. Ismail and S. Sharif, "Effect of heat treatment on mechanical properties and microstructure of selective laser melting 316L stainless steel," in *4th International Conference on Mechanical Engineering Research* , Malang, Indonesia, 2017.

[10] L. E. Murr, E. Martinez, J. Hernandez, S. Collins, K. N. Amato, S. M. Gaytan and P. W. Shindo, "Microstructures and Properties of 17-4 PH Stainless Steel Fabricated by Selective Laser Melting," *Journal of Materials Research and Technology,* vol. 1, no. 3, pp. 167 - 177, 2012.

[11] W. Jin Oh, W. Jin Lee, M. Seon Kim and J. S. S. D. Bae Jeon, "Repairing additive-manufactured 316L stainless steel using direct energy deposition," *Optics and Laser Technology,* vol. 117, pp. 6-17, 2019.

[12] B. Krawczyk, P. Cook, J. Hobbs and D. L. Engelberg, "Corrosion behaviour of cold-rolled type 316L stainless steel in HCl containg environments," in *Corrosion*, Houston, 2017.

[13] B. AlMangour and J.-M. Yang, "Improving the surface quality and mechanical properties by shot-peening of 17-4 stainless steel fabricated by additive manufacturing," *Materials and Design,* vol. 110, pp. 914-924, 2016.

[14] S. Prabhakaran, A. Kulkarni, G. Vasanth, S. Kalainathan, P. Shukla and V. Vasudevan, "Laser shock peening without coating induced residual stress distribution, wettability characteristics and enhanced pitting corrosion resistance of austenitic stainless steel," *Applied Surface Science,* vol. 428, pp. 17-30, 2018.

[15] L. Peguet, B. Malki and B. Baroux, "Effect of austenite stability on the pitting corrosion resistance of cold worked stainless steel," *Corrosion Science,* vol. 51, pp. 493-498, 2009.

[16] M. Godec, C. Donik, A. Kocijan, B. Podgornik and D. Skobir Balantic, "Effect of post-treated low-temperature plasma nitriding on the wear and corrosion resistance of 316L stainless steel manufactured by laser powder bed fusion," *Additive Manufacturing,* vol. 32, 2020.

[17] S. Geng, J. Sun and L. Guo, "Effect of sandblasting and subsequent acid pickling and passivation on the microstructure and corrosion behavior of 316L stainless steel," *Materials and Design,* vol. 88, pp. 1-7, 2015.

Application of Laser Shock Peening on Nitrided Steel for Power Transmission Components

M. Guerbois[1,2], C. Dides[2], A. Bonnin[3], L. Berthe[1], and V. Favier[1],

1: Laboratoire PIMM, Arts et Metiers Institute of Technology, ENSAM, CNRS, CNAM, HESAM Université, 151 Boulevard de l'Hôpital, 75013 Paris, France,

2 : Institut de Recherche Technologique Matériaux, Métallurgie et Procédés, 4 Rue Augustin Fresnel, 57070 Metz, France

3: Airbus Helicopters

Abstract
Laser Shock Peening (LSP) has been performed on 32CrMoV13 nitrided steel. Surface properties, residual stresses (RS) and hardness had been investigated to study the effect of LSP on these kinds of materials. High level of compressive RS had been measured after LSP through the depth. No changes had been noticed in terms of hardness and microstructure. Degradation of roughness had been observed after treatment. This paper presents the first results about LSP on nitrided steel.

Keywords
Laser shock peening, Deep nitrided steel, Compressive residual stresses, Roughness, Gear tooth root.

Introduction
Gear teeth root load-carrying capacity is one of the key factors in gear design for transmission components. For fatigue life enhancement, thermochemical treatments on such material are a promising method. In order to improve the performances even more, the shot peening can be performed afterwards. Because of the high surface hardness (700HV to 800Hv) of nitrided parts, introducing additional compressive residual stresses with conventional shot peening requests specific conditions and shots. In addition, the perfect control and repeatability of the peening operation is very difficult. In the opposite, Laser Shock Peening (LSP) technology involves a better control and therefore repeatable loading. It generates very high pressure on material surface and so introduce higher and deeper compressive residual stress than conventional shot peening. Today, most of the study about laser peening are performed on aluminum or titanium alloys. In this study, LSP is applied on samples made of 32-CrMoV-13 deep nitrided steel. The present study aims to identify material response for laser peening parameters. Its main goal consists of finding a set of LSP parameters which allows the surface and sub-surface compression residual stress enhancement with a minimum impact on surface morphology and roughness. Moreover, no change in nitrided material's metallurgic features (hardness, microstructure in particular) is desired. This study is performed as part of IRT-M2P TRANSFUGE project, which aims to study innovative material and treatment for gear parts. In the following, process parameters and experimental measurements have been normalized by reference values to compare the results obtained without LSP and with LSP while guaranteeing the confidentiality of the values obtained.

Experimental Methods

1. 32-CrMoV-13 deep nitrided steel characteristics

The 32-CrMoV-13 deep nitrided steel studied has high surface hardness (~800HV), and compressive residual stresses deep into the treated material.

2. Mechanical behavior of the 32-CrMoV-13 under high pressure (give a range of pressure) laser induced shock wave.

Thin specimens of nitrided 32-CrMoV-13 steel were shocked with a high-intensity laser and water confinement configuration. Specific foils of 400 µm, 600 µm and 800µm thickness had been prepared to conduct these experiments. Two sets of targets had been prepared: a nitrided set (representative of the surface composition) and a raw set (representative of the bulk).
The laser irradiation produces a high-pressure shock wave at the metal surface which propagates through the thickness of the foil. This propagation results in a displacement of the rear free surface of the target. The velocity of this displacement is measured using VISAR (Velocity Interferometer System for Any Reflector) diagnostic. The velocity associated with the elastic precursor provides the Hugoniot limit value and thus the minimal pressure of the shock wave required to produce plastic deformation within the material at a strain rate representative of LSP conditions (around 10^6 s^{-1}). Seddik et al's paper give details about this method [1].

3. Laser shock peening process (LSP)

A flashlamp-pumped Nd:YAG with a Gaussian temporal profile has been used for all the LSP treatments presented here. This laser can provide up to 14 J of energy with 7 ns pulse, at 2 Hz of repetition rate.
LSP treatments had been performed on 35x35x20mm^3 specimens, using the pattern introduced on Figure 1.

Figure 1: Pattern used for LSP treatment

Pulse energy had been chosen according to the VISAR results to provide a high enough power density to plastify the material. Only one pulse density is used in this paper and is defined by PD1. Two overlappings conditions have been tested and their defined by %OVA and %OVA x2 in this paper. A higher overlapping, %OVA x4, has been also used to assess the material's response under high intensity LSP treatment.

4. Residual stresses measurement

Residual stress (RS) measurements have been carried out on X-Raybot (Chromium detector, λ = 2.2909 Å, beam diameter 2 mm) at the PIMM Lab. Displacement of the 211 peak had been studied. Electropolishing using salt-saturated solution has been used to perform in depth RS measurement. Correction due to the several polishing has been applied on the RS value measured [2].

5. Profilometry and micro-hardness measurement

The profilometry of the surface has been determined using contact Dektak 150 Surface Profiler. Step and Average Roughness (Ra) values have been measured with this tool.

Hardness measurements have been carried out using 200g load for 12 seconds on a Clemex CMT micro hardness tester.

Experimental Results

1. Profilometry

Figure 2: Step and roughness variation caused by LSP shows the variation of roughness and the step caused by LSP at power density PD1 for % OVA and % OVA x2 overlapping.

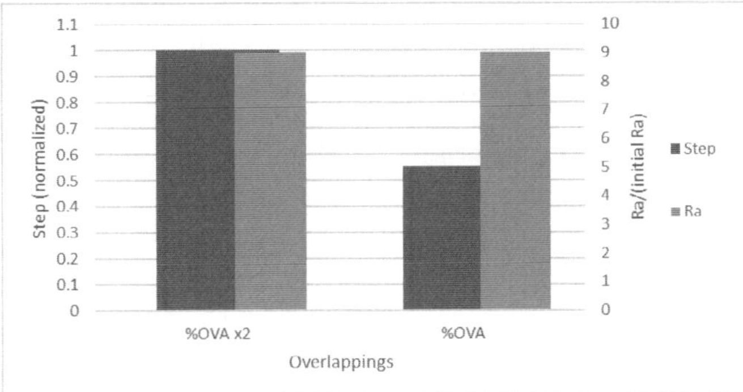

Figure 2: Step and roughness variation caused by LSP

The step measured after LSP varies from 0.55 step unit to 1 step unit while the overlapping goes from % OVA to % OVA x2.

In terms of roughness, it appears that Ra after LSP is 9 times higher than the initial one. Switching from % OVA to % OVA x2 does not make any difference in terms of roughness.

2. Micro hardness

Figure 3 presents the micro hardness profile observed on 3 different positions of an LSP treated sample (PD1 and % OVA x4).

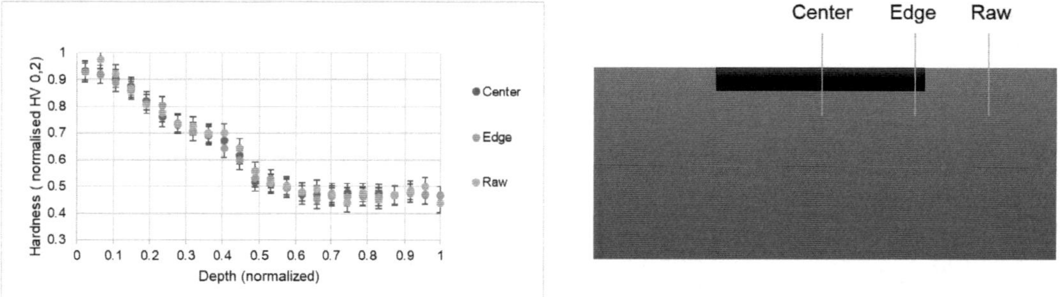

Figure 3 : Micro hardness profile measured on 3 positions (center edge and outside a LSP patch)

The 3 curves follow the same trend. The maximum hardness is reached on the surface and remains constant for 0.1 depth unit. This value of hardness decreases from 0.1 depth unit to 0.6 depth unit where it remains constant at 50% of the maximum hardness.

3. Residual stresses:

Figure 4 shows the typical RS profiles obtained after LSP and compare them to the initial RS profile.

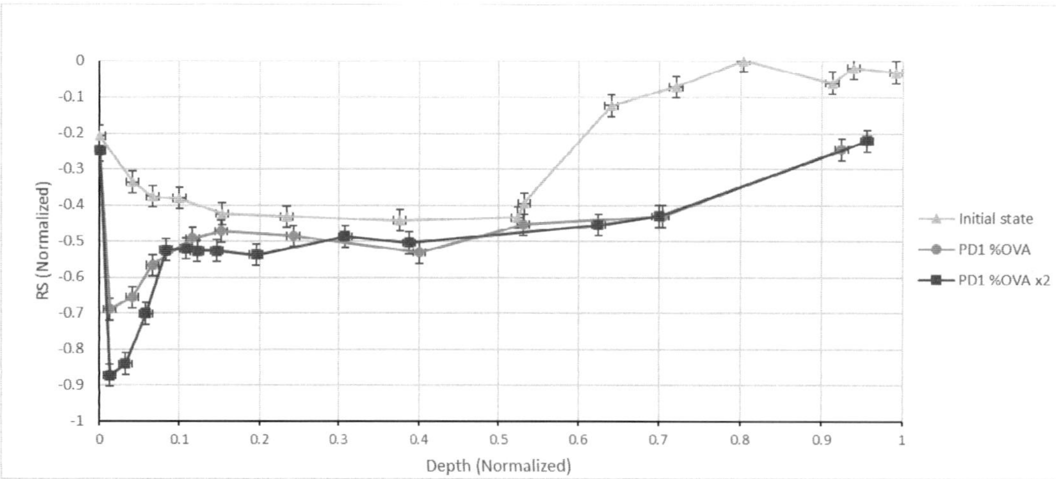

Figure 4 : RS profiles obtained for % OVA and % OVA x2 overlapping values at PD1 GW/cm²

Those RS profiles have been obtained for a power density PD1. Overlapping % OVA and % OVA x2 have been investigated. The initial RS profile start on the surface with a -0.2 RS unit, reach -0.43 RS unit from 0.15 depth unit to 0.52 depth unit. From 0.52 depth unit to 0.8 depth unit, the compression level decrease to reach 0 RS unit, and then remain at zero.

RS profiles post LSP shows some additional compression in comparison with the initial state profile. Compression peaks are observed at 0.02 Depth unit. The maximum of compressive stresses reaches -0.9 RS unit for the % OVA x2 sample versus -0.7 RS unit for the % OVA sample. From 0.1 depth unit to 0.52 depth unit, the stress level remains constant for LSP profiles, at -0.5 RS unit.

After 0.52 depth unit, the stress level decreases until it reaches 0 RS unit at 1.2 depth unit. At 0.8 depth unit, the RS level is still at -0.35 RS unit for the LSP treated samples, while it was 0 RS unit for the initial state.

Discussion and Conclusions

Results presented above gives an overview of the effect of LSP on 32CrMoV13 nitriding steel. Profilometry and roughness measurements show a 9 times higher roughness than the initial state, regardless of the overlapping. However, doubling the overlapping induces a two times higher step. These results allow us to conclude that the surface roughness does not follow the variation of overlapping for these overlapping rates, while the step variation tends to follow the overlapping variations. These results also call into question the position of LSP step in a full manufacturing process. The consequences of finishing processes on such surfaces should be investigated, even though Petan and al shows that the negative effect on roughness can be overcomed by the amount of RS introduced by LSP [3].

It appears that the initial hardness profile is not affected by the LSP treatment at PD1 and %OVA x 4 overlapping. These results are in adequation with the preservation of the microstructure observed after EBSD observation that is not shown in this paper.

Variations of overlapping influences compressive residual stresses level. The intensity of the RS peak of the % OVA x2 overlapping is nearly 30% higher than the % OVA peak. The RS peak width is not affected by the variation of overlapping. It has been showed by Halilovič and al. that the width of the RS profile increases with the pulse duration [4]. It is also interesting to notice that LSP provides compressive RS up to 1 depth unit. Increasing the compression depth should provide a benefic effect on fatigue life.

All these results show the ability of LSP to introduce high level of compressive RS in 32CrMoV13 nitriding steel without impacting the hardness and the microstructure. The roughness degradation may have some negative impact on fatigue. However, this effect is competing with the positive impact of RS level and depth

References

[1] R. Seddik *et al.*, “Identification of constitutive equations at very high strain rates using shock wave produced by laser,” *Eur. J. Mech. - A/Solids*, vol. 92, p. 104432, 2022.

[2] M. . Fitzpatrick, A. . Fry, P. Holdway, F. . Kandil, J. Shackleton, and L. Suominen, “Determination of residual stresses by X-ray diffraction - Issue 2,” *Meas. Good Pract. Guid.*, no. 52, 2005.

[3] L. Petan, J. L. Ocaña, and J. Grum, “Influence of laser shock peening pulse density and spot size on the surface integrity of X2NiCoMo18-9-5 maraging steel,” *Surf. Coatings Technol.*, vol. 307, pp. 262–270, 2016.

[4] M. Halilovič, S. Issa, M. Wallin, H. Hallberg, and M. Ristinmaa, “Prediction of the residual state in 304 austenitic steel after laser shock peening - Effects of plastic deformation and martensitic phase transformation,” *Int. J. Mech. Sci.*, vol. 111–112, pp. 24–34, 2016.

Fine, Hard, and High Density Ceramic Beads for Shot Peening

D. Bouttes[1], L. Zhou[2], F. Wiss[3] and A.-L. Beaudonnet[3]

[1] Saint-Gobain Research Provence, Cavaillon, France
[2] Saint-Gobain Research Shanghai, Shanghai, China
[2] Saint-Gobain SEPR, Le Pontet, France

Abstract
We investigate the use of new high density, fine, and hard ceramic beads for shot peening. By studying the effect of such media on a series of representative alloys, we show that they can impart very efficiently intense residual stresses, especially at the surface. Due the higher hardness of the media compared to the treated alloys – even carburized steel – this efficient modification of the sub-surface is accompanied by a roughness modification that is strongly dependent on the intensity of the treatment. The combination of the intense residual stresses and controlled surface topography modification brings some new opportunities for improving the shot peening process of different alloys.

Keywords Fine shot peening, ceramic shot, roughness, residual stress.

Introduction

Shot peening with hard media on a metallic substrate is a standard method to obtain a controlled surface roughness, as well as to induce compressive residual stresses. The media size is typically from 10µm to 1000µm, and they can be metallic shots, glass or ceramic beads [1]. Ceramic beads are selected when high hardness or no metallic contamination is needed, and provides significantly longer lifetime than glass media. Standard ceramic media have a density of 3.9g/cc, and a Vickers hardness of 700HV. As the hardness ratio of media to substrate is a key parameter for shot peening [2], it can be advantageous to use even harder media. Hence, higher hardness and higher density ceramic media with a size of 400 – 1200µm were introduced a few years ago, with density of 6.2g/cc – close to metallic media – and hardness of nearly 1200HV [3].

The use of fine media for shot peening has been studied for many years, and confirmed the interest to increase the lifetime of the treated parts. For instance, corrosion resistance was reported to be increased on TA6V [4]; fatigue resistance was also improved an aluminum alloy [5,6], or steel [7,8]. Yet these studies considered either standard ceramic shots with moderate density (3.9g/cc) and hardness (700HV), or metallic shots. The use of finer, denser, and harder ceramic media is still to be explored.

In the following, we evaluate a new ceramic media with high density (6g/cc), hardness (1250HV) and sizes of 300µm and below. In order to estimate the interest for a broad range of application, a selection of alloys was considered, with very wide hardness levels, from aluminum to carburized steel.

Experimental Methods

High density, fine ceramic shots Microshot YZ (Saint-Gobain) were tested for shot peening application. The media size and main properties are reported on Table 1. The selected Almen intensities correspond to low to moderate pressure: from 1 to 2.5 bar.

Media	Type	Density	Hardness	Size	Almen tested
YZ50	Ceramic	6 g/cc	1250 HV	50 µm	0.07 and 0.10mm N
YZ100	Ceramic	6 g/cc	1250 HV	100 µm	0.10 and 0.15mm N
YZ300	Ceramic	6 g/cc	1250 HV	300 µm	0.06 ; 0.10 and 0.15mm A
ZC600	Ceramic	6.2 g/cc	1180 HV	600 µm	0.25mm A
S110	Cast steel	7.8 g/cc	600 HV	300 µm	0.10 and 0.15mm A

Table 1: Media, main characteristics, and Almen intensities used for the shot peening tests

Shot peening tests were performed on different flat samples: aluminum 7075-T6, titanium TA6V, AISI1070 (45-48HRC) steel, and 16MnCr5 carburized steel. We used a suction blasting machine ARENA DF1000 for 100µm and 50µm media, and a direct pressure ARENA PF1500 for larger media. The Almen intensity was controlled with standard N and A Almen strips. For all tests, the coverage rate was 200%.

On the carburized steel sample, a double shot peening peening was performed by first peening with Zirshot HDC ZC600 at 0.25 mmA, then YZ100 at 0.10mmN.

After shot peening, the surface topography and defects were assessed. Roughness were measured by a linear profilometer Mitutoyo Sj210). The surfaces were also observed by SEM Hitachi TM3030, with back-scattered electron imaging, as well as a "topographic mode" that allows to visualize the surface roughness.

X-Ray residual stress analyses were carried out with an X-Ray diffractometer X3000 (Cr Kα radiation, sin2θ method).

Experimental Results

Roughness and surface

The surfaces shot peened with the high density ceramic media showed well defined craters, with depths and width depending on the media size, substrate type, and intensity. We show some SEM pictures on **Figure 1** and **Figure 2**. The typical width was about one quarter to one third of the media diameter, increasing as the target hardness decreased. The surfaces inside the craters were smoother than the ones obtained after shot peening with metallic shots, and the crater shapes better defined. As can be seen on figure **Figure 3**, the roughness increased with media size, Almen intensity, and decreasing hardness of the substrate. Even for the harder substrate (carburized steel, 800HV) and the 100µm beads, we observed a significant modification of the surface roughness: e.g. the Ra increased from 0.30µm to 0.64µm when shot peening at 0.15mm N.

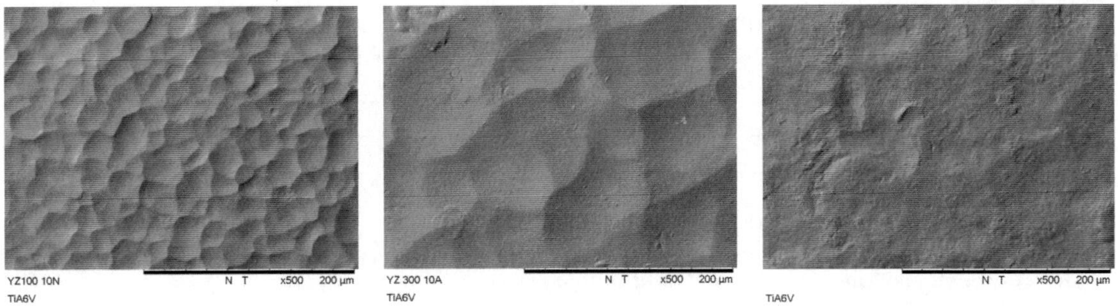

YZ100 (0.10mm N) YZ300 (0.10mm A) S110 (0.10mm A)

Figure 1: Surface of TA6V samples after shot peening – SEM images in topographic mode.

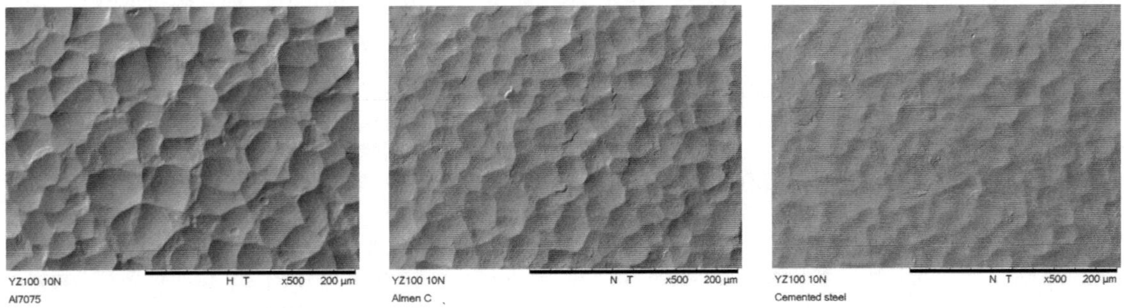

Al 7075 (0.10mm N) AISI 1070 (0.10mm N) Carburized steel (0.10mm N)

Figure 2: Surface of different samples after shot peening with YZ100 at – SEM images in topographic mode.

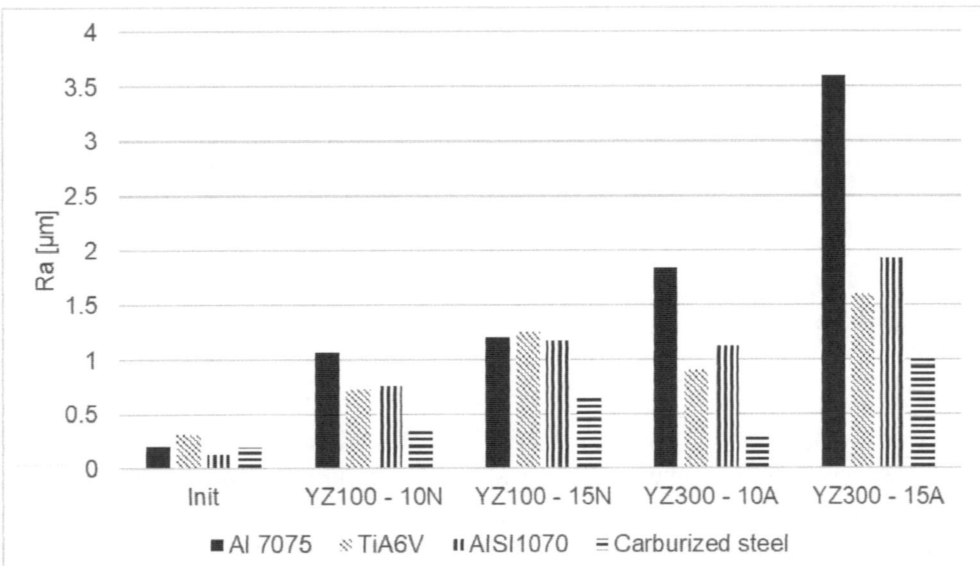

Figure 3: Ra measured before and after shot peening with the high density ceramic beads of size 100µm and 300µm.

Regarding the double shot peening test, after shot peening with ZC600 at 0.25mm A, we observed a roughness reduction after the second step, see Table 2.

	Ra [µm]	Rz [µm]
Initial surface	0.20	1.5
ZC600	0.90	4.8
ZC600 + YZ100	0.75	4.45

Table 2: Roughness measured on the carburized steel sample after shot peening with ZC600 at 0.25mmA, then after a second shot peening with YZ100 at 0.10mm N.

Residual stresses

The residual stresses obtained on the various substrate of course depend strongly on the material properties. We report the measurements on aluminum and titanium samples on Figure 4.

For aluminum, with the 50µm and 100µm ceramic beads, we recover stress profiles that are similar to shot peening with fine metallic particles [9] at similar Almen intensities. The depth of the residual stress is quite shallow, from 20 to 30µm, however the maximum stress and the stress at the surface exceed 300MPa.

On titanium, the media of 300µm, either cast steel shot or ceramic, impart similar residual stress profiles. The depth of the stress profile is reduced below 40µm by using the smaller 100µm ceramic beads, while the surface stress is kept similarly at about 700MPa.

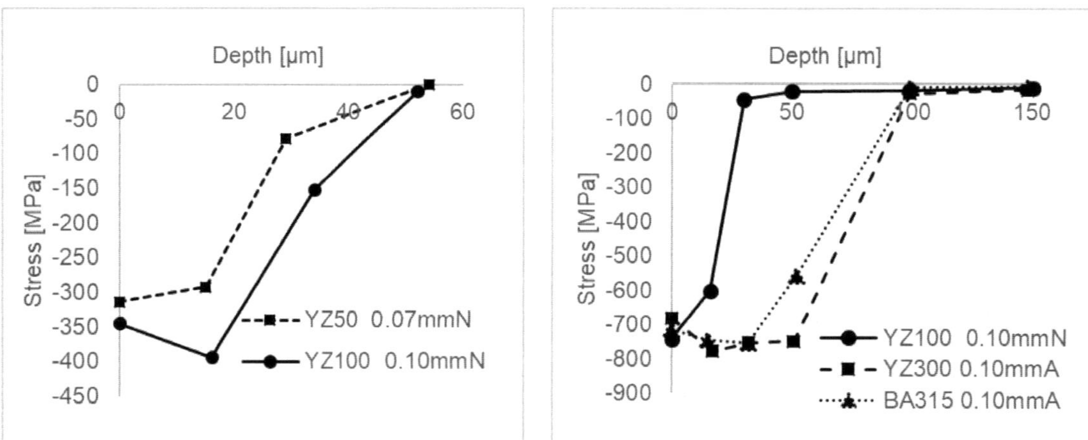

Al 7075 TA6V

Figure 4: Residual stress profiles obtained with several shot peening conditions on Aluminum 7075, and Titanium TA6V.

The residual stresses obtained on the 16MnCr5 carburized steel are reported on Figure 5. With the tested ceramic beads, the maximum stresses reach 1400MPa, which compares favorably to what can be obtained with metallic shots, but at generally at higher Almen intensity [1]. Particularly striking are the surface stresses, that are close to 1200MPa, and even 1400MPa when using double shot peening with the YZ100 as second shot peening media.

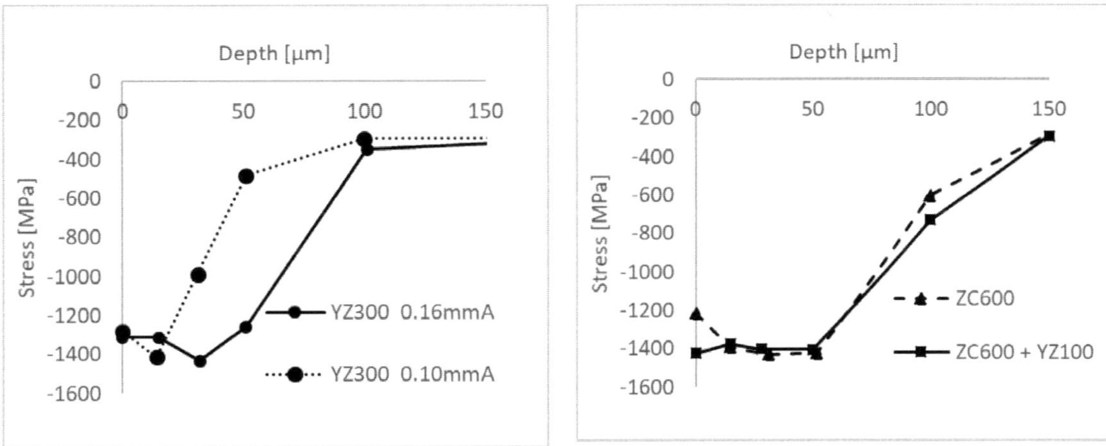

Figure 5: Residual stress profiles obtained on carburized steel.

Discussion and Conclusions

We performed a series of shot peening tests with fine and hard ceramic beads on several alloys from light soft alloys to carburized steel. Compared to all the considered alloys, the ceramic shots were significantly harder. We recovered some results that were observed in similar cases where this ratio is pronounced:

- A roughness modification that is pronounced;
- Intense residual stresses, especially at the surface.

These fine and hard ceramic media hence offer some opportunity for new treatments:

- Efficiently induce shallow but very intense residual stresses or microstructural changes – from 20 to 50µm: to improve corrosion resistance properties, treat thin parts, or use as second peening step.
- By selecting carefully, the conditions, obtain intense compressive residual stresses on very hard alloys like carburized steels, while controlling the roughness level.
- Replacing fine metallic shots to provide similar residual stresses, but with smoother surface and no prejudicial contamination e.g. on titanium alloys.

References

[1] Wohlfahrt H., *The Influence of Peening Conditions on the Resulting Distribution of Residual Stress*. ICSP-2 Proceedings (1984), pp.316-331.

[2] Cabrero, J., et al.: *High density ceramic shot for peening application*. ICSP-17 Proceedings, (2017), pp. 62-67.

[3] Baiker, S., et al.: *Shot peening: a dynamic application and its future* (2006).

[4] Ahmed A. A. et al.: *Effect of micro shot peening on the mechanical properties and corrosion behavior of two microstructure Ti–6Al–4V alloy*, Applied Surface Science, Vol. 363 (2016), pp.50-58.

[5] Kikuchi S. et al.: *Effects of Ultrafine Particle Peening on Fatigue Properties of ASTM 5056 Aluminum Alloy*, ICSP (2014).

[6] Benedetti et al.: *High- and very high-cycle plain fatigue resistance of shot peened high-strength aluminum alloys: The role of surface morphology*, International Journal of Fatigue, Vol 70 (2015), pp.451-462.

[7] Sugimoto K. et al.: *Effects of fine particle peening on fatigue strength of a TRIP-aided martensitic steel*, International Journal of Fatigue, Vol. 100 (2017), pp.206-214.

[8] Bagherifard et al.: *Fatigue Strength of a Low-Alloy Steel Shot Peened with Ultra Fine Hard Steel Media*, ICSP (2014).
[9] Oguri, K: *Fatigue life enhancement of aluminum alloy for aircraft by Fine Particle Shot Peening (FPSP)*, Journal of Materials Processing Technology, Vol. 211 (2006), pp.1395-1399.

Overview of Laser Shock Processing Development in South Africa: Industrial Applications

D. Glaser[1,2], M. Newby[2], C. Polese[3], R. Scheepers[4]

1 Council for Scientific and Industrial Research (CSIR), Photonics Centre, Meiring Naudé Road Brummeria, Pretoria, South Africa
2 Nelson Mandela University, Mechanical Engineering, University Way, Summerstrand, Gqeberha, South Africa
3 University of the Witwatersrand, School of Mechanical, Industrial and Aeronautical Engineering & DSI-NRF Centre of Excellence in Strong Materials, Jan Smuts Avenue, Braamfontein, Johannesburg, South Africa
4 Eskom Holdings SOC Ltd, Johannesburg, South Africa

Keywords: Laser Shock Peening, Low Pressure Steam Turbine Blade, Raised Bore Drilling, Distortion Correction

Abstract

The Laser Shock Peening (LSP) process can be utilised to enhance the life of critical metal components or to provide precision engineered distortions for forming applications. In South Africa, the CSIR, in collaboration with industry and academia, is developing LSP technology to address local requirements. In the power generation industry, Low Pressure (LP) steam turbine blades are typically susceptible to stress corrosion cracking and fatigue. The resulting LSP performance on FV566 material shows deeper levels of beneficial compressive residual stress with minimal surface roughness compared to the mechanical Shot Peening process already implemented. In the mining sector, large steel components may require refurbishment and life extension solutions due to their harsh operating environment. The physical size of these components has pushed for the development of a mobile LSP capability. Three-point bending fatigue testing on AISI4330V reveals over 20x lifetime enhancement compared to the base material. In the aerospace sector, distortion correction of a machined AA7050-T7451 component has successfully been demonstrated.

Introduction

Laser Shock Peening (LSP), also referred to as Laser Shock Processing, is a residual stress engineering technology specifically used to introduce beneficial compressive residual stresses into critical components to enhance fatigue and/or stress corrosion cracking (SCC) resistance [1]. The precise introduction of surface strains can also be implemented for distortion engineering applications.

The schematic in Figure 1 below illustrates the mechanism of LSP technology. A laser pulse with a short duration (typically 5-25 ns) is used to produce laser intensities in the order of GW/cm^2. The high intensity irradiation results in the near instantaneous vaporization of the target into a plasma. The surface of the target region is covered by a medium that is transparent to the laser wavelength to achieve inertial confinement of the plasma generated [2]. A photograph in Figure 1 illustrates the thin water layer after a subsequent laser shot. The plasma reaches a few thousand degrees Celsius within nano-seconds, and expands at a rapid rate. The confined plasma generates peak pressures with a magnitude in the range of Gigapascals acting over a short timeframe, which results in the formation of a shockwave within the target material. The shock wave propagates with sufficient strength to exceed the material's dynamic yield strength, and therefore introduces plastic strains within the surface region. The peening action is generated by the elastic response of the surrounding material to form beneficial compressive residual stresses through the target surface. Due to the unique shock-wave mechanism, the LSP process can achieve deep levels of compressive residual stress (typically >1 mm), with minimal surface deterioration, therefore enabling the achievement of significant performance enhancements.

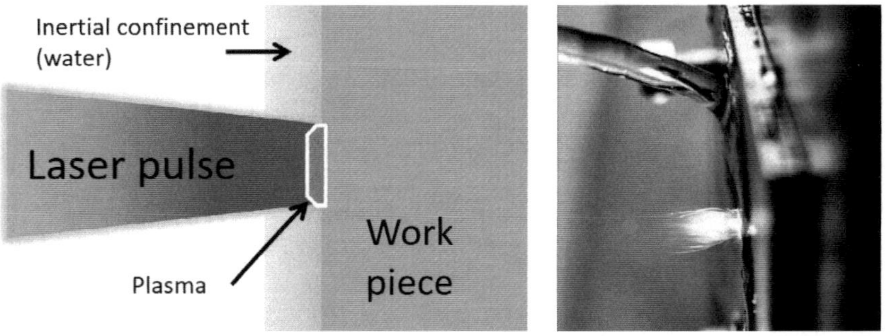

Figure 1: A schematic of the LSP process mechanism (left), and a photograph after the LSP process (right)

The LSP process was successfully commercialized in the late 1990's in the USA and Japan. In the USA, LSP was initially used to enhance the life of critical aircraft engine fan blades that were susceptible to Foreign Object Damage (FOD). The process developed in the USA typically includes a sacrificial ablative overlay to prevent thermal effects at the surface due to laser irradiation, as well has high energy levels (>10 J) [3]. In Japan, the initial commercial applications were developed to mitigate Stress Corrosion Cracking (SCC) on nuclear reactor components, where the process involves direct irradiation of the target with specifically tuned parameters to address direct laser-material interaction effects (typically low energy levels of < 0.2 J) [4].

Although LSP technology has matured significantly since early adoption with many innovative applications, the technology is still regarded as fairly niche, and is only applied industrially in a limited number of countries globally. In South Africa, LSP R&D activities initially commenced via academic studies with local Higher Education Institutions (HEIs). Due to the significant value offering of LSP as an advanced engineering solution, a technology pull has resulted in the effort to develop the LSP process for industrial applications in South Africa. The current paper provides a brief overview of some of the key local applications under development for sectors such as power generation, mining, and aerospace.

In the power generation industry, Low Pressure (LP) steam turbine blades are typically susceptible to stress corrosion cracking and fatigue. The photograph in Figure 2 illustrates typical LP blades in a turbine assembly, where blade lengths may approach or exceed 1 m. Conventional mechanical Shot Peening (SP) is already applied to the highly stressed blade attachment regions at service intervals as a mitigation measure to failure phenomena [5]. The potentially deeper levels of compressive residual stress possible with a lower surface roughness presents scope for superior performance offered by LSP.

Figure 2: Photographs of LP steam turbine blades (left), a CAD model of a single blade (centre), and a close up of a fir attachment region after coupon removal (right) [5]

In the mining sector, large machines may require refurbishment and life extension solutions for critical metal components. An example of current applications investigated includes life enhancement for Raise Bore Drilling (RBD) components which experience significant alternating loads within a harsh operating environment and are therefore susceptible to both fatigue and stress corrosion cracking. A schematic of a generic RBD operation is depicted in Figure 3 (A). The RBD process involves first drilling a pilot hole from the drill rig at the surface to an underground tunnel. A large reamer head with cutters is attached to the drill string which applies tension and torsion causing cutting of the rock from the tunnel to the surface. The cuttings fall into the tunnel which are then removed by loading machinery. The entire load chain is exposed to significant loads which may vary, depending on rock conditions.

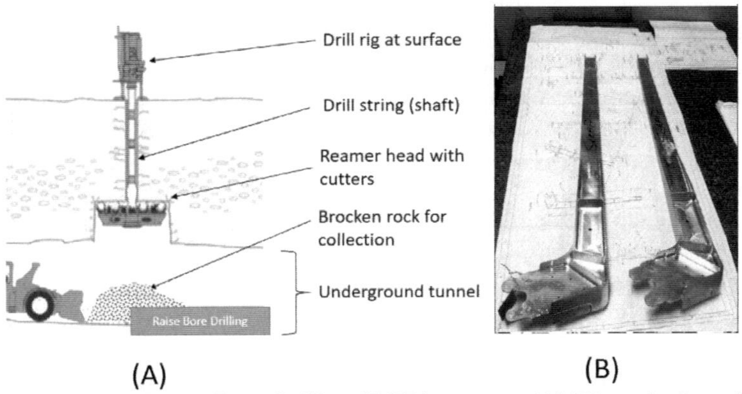

Figure 3: A schematic of the Raise Bore Drilling (RBD) process (A) [6], and of machined aerospace components that are scrapped due to distortion (B)

In the local aerospace sector, South African companies manufacture several precision aluminum components for both domestic and international markets. Manufacturing of precision thin-walled aluminum components often encounters distortion challenges, due to pre-existing residual stresses within the billet as well as machining induced residual stresses. Since LSP can apply precision residual strains to a surface, the laser process may potentially be applied as a distortion correction technology. The current project identified a typical component that has historically presented a number of challenges with distortions, resulting in undesirable scrapping of components. The identified component is depicted in Figure 3 (B), where the current project aimed to demonstrate the technology capability on a previously scrapped component. The challenge for this project is that the resulting distortion of each component is different, therefore the capability to precisely tailor the resulting engineered strain is required.

Experimental Methods

There are currently two LSP work-cells based at the Council for Scientific and Industrial Research (CSIR) in South Africa. A system operating at a wavelength of 1064 nm is used for Proof-of-Concept (POC) work in collaboration with local academia, and a 532 nm system is being devised for industrial applications. Due to the nature of the application in the mining sector, there is also a need for a mobile system for refurbishment applications where components are too large and heavy to handle at the current LSP processing facility. The LSP process in SA is applied in what is considered a "mid-range" energy level (compared to the initial high energy systems in the USA, or low energy systems in Japan). The LSP process is typically performed in the direct ablation mode [4], also referred to as Laser Peening without Coating (LPwC), which therefore requires refined process parameters to achieve suitable performance.

The material for the LP steam turbine blades is a 12% chromium martensitic stainless steel (FV566), which has been extracted from ex-service turbine blades. Process development reported here is on samples of 20x20x15 mm where results for residual stress are obtained by Synchrotron X-Ray Diffraction (SXRD) conducted at the ESRF facility (Grenoble, France), detailed further in a prior description by the authors for experiments [7]. Scale slices with a thickness of 15 mm have also been extracted from ex-service blades (as depicted in Figure 4), and processing has been implemented by motion of the target using a 4 degree of freedom precision motion system.

The material for the mining application for RBD is a AISI 4330V material with a microhardness in the range of 355-361 HV_{20}, which is typically employed due to its material property combination of high strength and impact toughness. The results presented are with respect to three-point bending fatigue testing conducted on samples with dimensions of 150x20x15 mm (R ratio of 0.1). The purpose of the current batch of fatigue samples was to first validate selected process parameters for the "mid-range" LSP approach. Processing has been performed at a laser power intensity of 10 GW/cm^2 and a wavelength of 1064 nm.

The component for the aerospace application reported is from an AA7050-T7451. The component was previously scrapped due to excessive allowable distortion. Process development occurred on off-cut material from the aluminium billets (80 mm thickness) used to machine new components. Mechanical testing for static and dynamic properties was required to demonstrate that the LSP process does not adversely affect material performance. An Almen strip approach adapted for the LSP process was devised to identify suitable parameter combinations that would allow for the required deformation to be generated [8]. Distortion results are also presented for an AA6082-T651 application for a non-aerospace application for components used in a precision machine.

Experimental Results and Discussion

The results from the LSP process applied to the steam turbine blade material are shown in Figure 4. Synchrotron XRD profiles reveal the current conventional shot peening process achieves a depth of Compressive Residual Stress (CRS) of around 0.25 mm, while LSP reveals a depth of CRS beyond 1 mm. The resulting surface roughness reveals a lower Ra of 1 μm for LSP compared to 3.2 μm for conventional mechanical SP. The LSP process has therefore been shown to produce deeper levels of compressive residual stresses combined with a lower surface roughness compared to mechanical SP. For the fir tree root configuration of LP blade attachments, the geometry is somewhat restricting to obtain line of sight at a near normal angle of incidence. For the conventional SP process, the effectiveness of the shot impact is reduced as the angle of incidence from the surface normal is reduced. However, for the LSP process, the plasma exerts a hydrodynamic pressure which always acts normal to the surface. The application of the LSP process to complex fir tree root sections is therefore a highly attractive technology option for LP blades that require further life extension solutions.

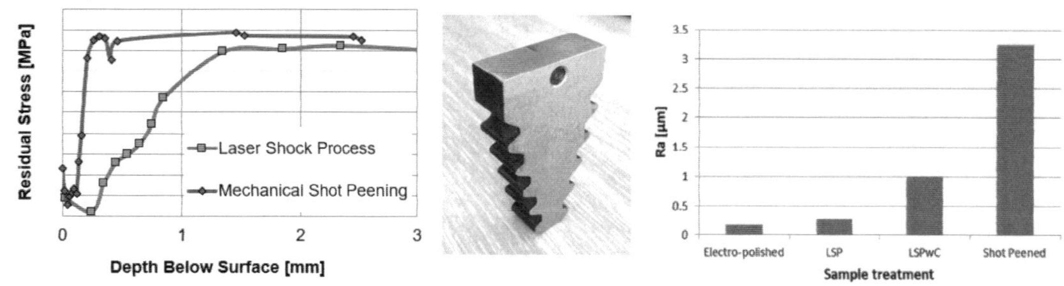

Figure 4: Synchrotron XRD results (left), A scale fir tree root section processed (centre), and resulting surface roughness (right) [7]

Three-point bending fatigue testing on the AISI 4330V material for the mining application reveal the effect of high magnitude and deep levels of compressive residual stress with minimal surface roughness. Figure 5 illustrates the fatigue results obtained at three different maximum stress levels from 1100 to 1186 MPa. The three base samples all failed in the range of 96 022 and 241 685 cycles, where all LSP samples achieved run-out defined at 5 million cycles (which reveals at least 20x life enhancement). This promising result illustrates the feasibility of "mid-range" energy LSP processing, where further testing will focus on technology validation to represent the component specific geometry and in-service loading conditions.

The components of interest for the RBD application have dimensions on the order of several meters, with a mass in the range of tons. Such large and heavy components cannot currently be accommodated at the CSIR LSP processing facility. To effectively cater to the needs for mining applications, a mobile LSP capability is currently under development. The photograph in Figure 5 below illustrates the mobile laser module after transportation testing from the CSIR (in Pretoria) to a customer site in Johannesburg (≈50 km). The mobile module ensures that the sensitive laser can be transported safely and operated effectively at a customer site. During transport, the sensitive laser system is protected from shock loading using passive isolation. The laser unit is housed within a sealed system to protect sensitive optical components from any possible contamination by dust. The module also provides active vibration isolation for the sensitive laser system to cater for possible workshop machinery in operation. The unit also includes a number of system diagnostics of the laser to ensure that all systems are within the required operational envelope.

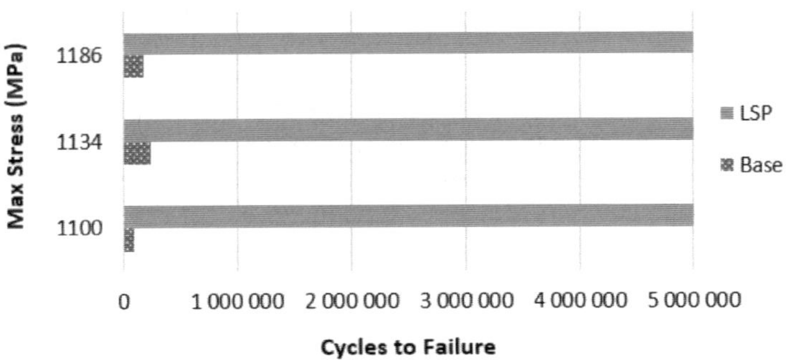

Figure 5: Three-point bending fatigue test results for the AISI 4330V material (left), and a photograph of the mobile laser unit under development during on-site testing (right)

The measured distortion of the aerospace component before and after application of the LSP process is shown in Figure 6 (A). Different parameter strategies of applying either a large single patch, or multiple smaller LSP stripes were explored. Both approaches were found to effectively achieve the required result, however the stripe approach enabled finer control of the process. The LSP process has several process parameters that can be varied to control the resulting effects. Typical parameters include power intensity, spot size, and spot coverage. This application however required proof that the LSP process does not adversely affect the material properties, and component structural integrity. The approach adopted was to therefore keep all LSP process parameters constant, and only vary the size of the LSP region to control resulting distortion.

The results in Figure 6 (B) illustrate the effectiveness of LSP to correct for the distortion on AA6082-T651 components which a local engineering company required for a precision machine. Four components were provided, each with a different level of distortion. The approach devised for the aerospace application was applied to four components to successfully correct for the varying levels of distortion. The AA6082-T6 components are the first LSP components to be processed locally to serve in an industrial application, which represents an important milestone for LSP technology development in South Africa.

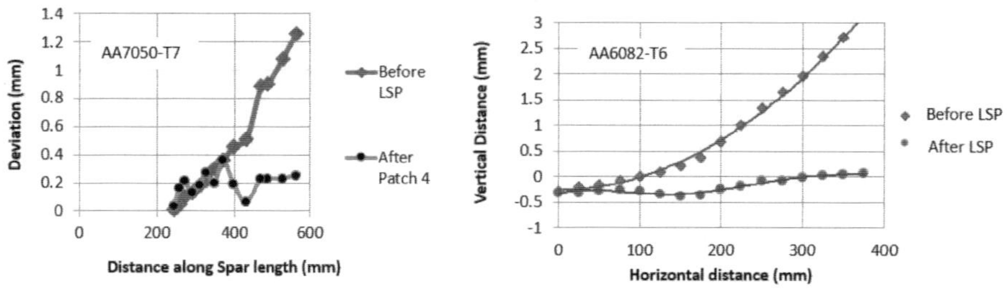

(A) (B)

Figure 6: Distortion correction results before and after LSP for the AA7050-T7451 (A) and AA6082 material (B).

Conclusions

LSP technology may offer numerous benefits, however the technology is still highly niche with only a limited number of groups globally able to provide industrial and commercial services. While LSP activities in South Africa were initially academically orientated, local industry has demonstrated specific technological needs for the process. Local industrial development of LSP applications in SA is currently focused primarily on applications in the power generation sector, mining sector, and aerospace sector. The application for LP turbine blades and RBD components offers significant life extension for components susceptible to stress corrosion cracking and fatigue due to deep levels of compressive residual stress obtained with low surface roughness. The aerospace application developed for an AA7050-T7451 component distortion correction has subsequently successfully been implemented for AA6082-T651 components, which represents the first in-service LSP components locally. Work is on-going to enable industrial uptake of LSP in partnership with local stakeholders. Due to the large and heavy nature of components in the local mining industry, a mobile LSP capability is under development to enable on-site applications.

References

[1] L. Petan, J. Grum, J.A. Porro, J.L. Ocaña, R. Šturm, *Fatigue Properties of Maraging Steel after Laser Peening,* Metals *9.12 (2019): 1271.*

[2] R. Fabbro, P. Peyre, L. Berthe, and X. Scherpereel*, Physics and applications of laser-shock processing*, Journal of laser applications, (1998) 10(6), pp 265-279.

[3] A.H. Clauer, Laser shock peening, the path to production, Metals (2019) Vol. 9.6, pp 626

[4] Y. Sano, *Quarter century development of laser peening without coating*, Metals (2020) Vol. 10.1, pp. 152.

[5] M.N. James, M. Newby, D.G. Hattingh, and A. Steuwer, *Shot-peening of steam turbine blades: Residual stresses and their modification by fatigue cycling.* Procedia Engineering, (2010), Vol. 2(1), pp. 441-451.

[6] A. Lashgari, M. Majid Fouladgar, A. Yazdani-Chamzini and M. J. Skibniewski, Using an integrated model for shaft sinking method selection, Journal of Civil Engineering and Management, (2011), Vol. 17(4)

[7] M. Newby, A. Steuwer, D. Glaser, C. Polese, D. G. Hattingh, and C. Gorny, *Synchrotron XRD evaluation of residual stresses introduced by laser shock peening for steam turbine blade applications*, Mechanical Stress Evaluation by Neutron and Synchrotron Radiation 4 (2018), pp. 97-102.

[8] D. Glaser, C. Polese, A. M. Venter, D. Marais, and J. R. Plaisier, *Evaluation of laser shock peening process parameters incorporating Almen strip deflections*, Surface and Coatings Technology, (2022) Vol. 434, pp. 128158

Effect of surface treatment on the fatigue strength of additive manufactured Ti6Al4V alloy

Carlos Navarro[a], Jesús Vázquez[a], Jaime Domínguez[a], Antonio Periñán[b], Marta Herrera García[b], Fernando Lasagni[b], Simon Bernarding[c], Sebastian Slawik[c], Frank Mücklich[c], Francisco Boby[d], Lloyd Hackel[d]

,

[a]University of Seville, Spain
[b]Advanced Center for Aerospace Technologies (CATEC), Seville, Spain
[c]University of Saarlandes
[d]Curtiss Wright Surface Technologies, Metal Improvement Company, Seville, Spain
Francisco.Boby@cwst.com

Abstract

Different alloys can be used for Additive Manufacturing (AM) with good structural strength. Among the titanium alloys, Ti6Al4V is the most used, especially for aerospace applications. There have been many analyses of the mechanical properties of additive manufactured Ti-6Al-4V with very good static strength results in general. However, there are still some difficulties to get fatigue properties close enough to the ones of specimens manufactured using traditional processes

Considering the high effect of surface roughness on the fatigue strength of AM specimens, this work deals with the effect produced by some surface treatments on the fatigue properties. Five treatments have been used for comparison. All specimens were annealed previously to reduce residual stresses, as well as sand blasted to reduce the roughness. The treatments considered are: 1) no treatment after annealing and sand blasting; 2) shot peening; 3) shot peening plus Chemical Assisted Surface Enhancement (CASE); 4) laser shock peening, and 5) HIP. After fatigue testing, a comparison of the results has been carried out. It was found that laser peening produced the best results, followed by shot peening plus CASE and shot peening, with the lowest strength produced by HIP as well as just sand blasting after thermal treatment.

Keywords: Additive manufacturing; fatigue strength of AM elements; shoot peening; laser shock peening; superfinishing; fatigue strength improvement.

Introduction

Additive manufacturing (AM) of metal parts is a relatively new technology, which is developing very fast. Different metal alloys are used for AM with relatively good strength. Stainless steels, cobalt-chrome, aluminium or titanium alloys are some of those used for structural proposes. Among the titanium alloys, T16Al4V is probably the most used, especially for aerospace applications. Specimens manufactured by AM are used since some years ago in satellites and launchers. They allow reducing the weight, not only because the low density material used, but also because the possibility of taking advantages of topological optimization, presenting monotonic mechanical properties similar to specimens manufactured through traditional technologies like casting. However, in the case of fatigue, AM elements still have problems of strength and reliability [1-3] and that prevents their application in the aeronautical industry, where the reliability is one of the main concerns. The porosity, residual stresses and anisotropy of the additive manufactured specimens as well as the surface roughness and microstructure, which frequently is not well controlled, make the fatigue strength of the elements an issue where there is a lot of work to be done in order to improve this mechanical property and the reliability.

Selective Laser Melting (SLM) is one of the most used AM techniques for Ti6Al4V alloy. The main features regarding fatigue of the work specimens made by this procedure are, as afore said, residual stresses, porosity, roughness, non-uniform microstructure and anisotropy [4-7], which reduce the fatigue strength and increase the scatter of the fatigue lives produced for the same cyclic load applied to different specimens taken from the same manufacturing batch. In addition to select the optimum manufacturing parameters to reduce defects, anisotropy, residual stresses, etc., several mechanical, thermo-mechanical and thermal treatments have been proposed to improve the strength and reduce the scatter [8-20]. These range from annealing or sand blasting to machining [9-13], Hot Isostatic Pressing (HIP) [13-16] or shot peening [17-19], or any combination of them [20]. Many experimental analysis have been published about the effect of these treatments on the fatigue strength, but mainly dedicated to the effect of thermal treatments, machining or HIP or combination of them [8-16]] and only a few about the effect of other surface treatment like shot or laser peening [17-20]. Actually, two of these works only consider the effect of shot peening on roughness, and on residual stresses and microhardness profiles close to the surface, but not directly on the fatigue strength, although they consider that the effect on those parameters will have also a beneficial effect on the fatigue strength. On the other side, Wycisk et al. [19] and Bagherifard et al. [20], analyse directly the effect of shot peening on the fatigue strength of the specimens, Wycisk did it on Ti6Al4V and Bagherifard on AlSi10Mg, with much better results in the case of AlSi10Mg

The main objective of this work is to analyse the effect of some surface treatments on the fatigue strength of additively manufactured specimens. The analysis was made by comparing the fatigue lives obtained with the different treatments selected. The treatments considered are mainly shot peening and laser shock peening. Considering the deleterious effect of roughness, anisotropy and residual stresses existing in the as built elements, before applying any surface treatment, all specimens were sand blasted and annealed. In the case of shot peening, two different surface treatments were considered, one was just shot peening and the other was shot peening followed by a surface treatment called Chemical Assisted Surface Enhancement (CASE), produced by Curtiss-Wright®. This treatment improves the surface finish produced by shot peening. In order to have a reference for the fatigue strength of the surface treated specimens two other treatments were considered. One was just sand blasting and annealing and the other was HIP after sand blasting and annealing. Real as built specimens instead of sandblasted plus annealing could have been used as reference. However, considering that it is well known that the real as buildt conditions produce very poor fatigue strength results and that most applications of SLM manufactured specimens use previous soft surface treatments, to at least clean the specimen surface, as well as heat treatment, it was decided using sandblasted and annealed conditions as reference. The HIP treatment is not actually a surface treatment, but it was included thinking in its use as reference for future analysis of the effect of the combination of HIP with other surface treatments.

This document is organized as follows. First, the manufacturing procedure, treatments considered and test procedure are shown. Following the testing results are described. Those results are discussed trying to understand the effect of the treatments and, finally, some conclusions about the treatments are obtained.

Methods and results

Methods

Considering that most fatigue failures are initiated at stress concentrations, where there is a stress gradient close to the surface, a series of four point bending tests was designed. 30 prismatic specimens (80x22x10 mm with a radius of 2 mm on corners) were produced by AM with the largest dimension oriented in Z (vertical) direction in the manufacturing chamber. This direction was selected because it is usually the most unfavourable to resist fatigue due to the orientation of the main defects produced during manufacturing [6,13]. The specimens were manufactured in a Renishaw AM250 SLM system with the following parameters: powder size, 15-45 μm; layer thickness, 60 μm; laser power output 200 W; scan rate, 0.86m/s. All specimens were sand blasted and annealed (730°C-2h; slow cooling in the furnace; high vacuum argon atmosphere) before doing any other treatment or testing directly. The sandblasting treatment was carried out with glass microspheres and the following parameters: direction, 90° to the surface; exposure time 15 s; work pressure, 5 bar; sandblasting distance; 30 mm. In order to analyse the effect of the treatments, the 30 specimens were divided into five groups of six specimens for each group. Each group was subjected to a different treatment before testing:

Group 1. Specimens were tested directly after sand blasting and annealing, in order to have a reference. This group will also be called "AS BUILT".

Group 2. All specimens were shot peened with the following parameters: 0.6 mm steel balls; 55-62 HRC balls hardness; intensity 14A and 500% coverage. This group will also be named "SP".

Group 3. n this group, specimens were shot peened before subjected to the treatment called CASE, by Curtis Wright®. In this treatment, parts are placed in an acid solution with non-abrasive ceramic media and the ceramic media is ‘excited’ using a vibratory bowl. Following, parts are reprocessed in a burnishing solution that restores the chemical stability and polishes the surface. The main objective of this treatment is reducing the roughness produced by shot peening. Actually, the CASE treatment is a kind of what is usually known as REM finishing. This group will also be named “SP+CASE”.

Group 4.The specimens were laser shock peened before testing. (wavelength: 1064 nm; pulse length: 10 ns; spot diameter: 2.6 mm; 6 shots at each point; power density 6 GW/cm²). This group will also be named “LP”. Figure 1 shows the residual stresses measured in this group.

Group 5.In order to have another reference for fatigue strength to compare the lives obtained with the proposed surface treatments, specimens of this group were treated by HIP (920ºC-2 h; 100 MPa; inert gas atmosphere). This group will also be named “HIP”.

Residual stresses produced by the four treatments were measured by the blind hole drilling technique.In all cases, Vishay Micro-Measurement residual stress rosettes type EA-031RE-120 were used .The drilling and measurement process were carried out using a Sint MTS300 equipment, following the Integral Method [21,22]. A residual stress profile was obtained for each specimen. Figure 1 shows the average residual stress profile in axial direction for each group of specimens and the scatter of the 6 measurement at each point represented by ± the standard deviation.

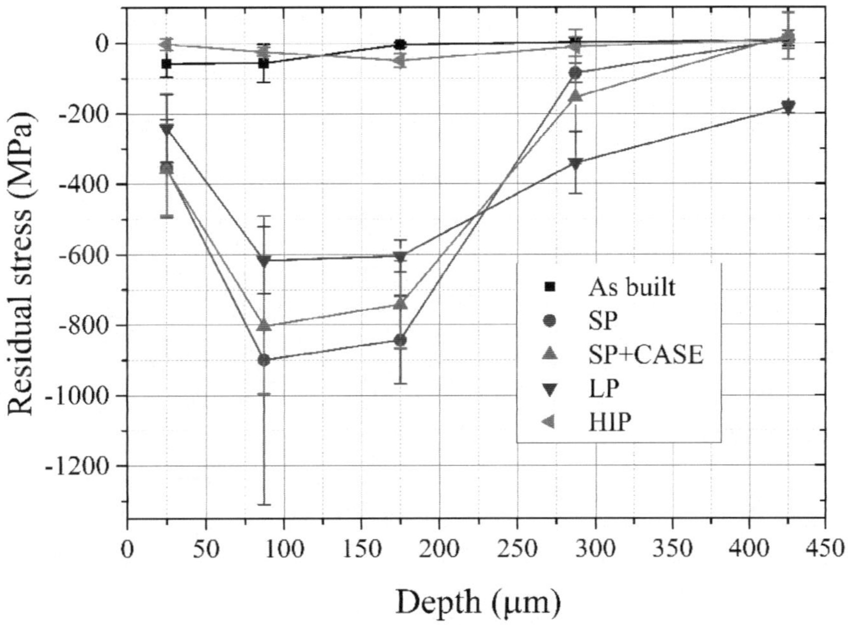

Figure 1. Residual stress produced by sand blasting, shot and laser shock peening, measured by the blind hole drilling technique.

Considering the large effect of roughness on fatigue, it was measured in all specimens produced. The surface roughness of all specimens was measured by the contact stylus arm with a RC filter technique, using a Mitutoyo Surftest 501 series 178 equipment. The probe tip radius was 5 microns and the evaluation length used to obtain R_t was 12.5 mm. Table 1 shows the average roughness measured in axial direction (Z) for each group as well as the standard deviation of the measurements carried out in the six specimens of each group. It can be seen that shot peening reduces to almost one third the original roughness, while CASE divides by two the roughness produced by the previous shot peening. Actually, the objective of CASE is to eliminate almost all peaks while maintain troughs, and that is why reduces R_a and R_t to 50%, approximately. It can be also seen that laser peening, producing a more uniform overpressure on a larger area of the surface for each laser pulse, barely changes the original roughness.

Table 1: Average surface roughness of specimens groups after treatments and standard deviation.

	AS BUILT	SP	SP+CASE	LP	HIP
R_a (μm)	11.7±1.0	4.6±2.4	2.7±0.3	10.2±1.9	9.8±1.1
R_t (μm)	97.5±11.8	38.6±18.6	19.7±3.4	89.2±15.3	77.9±12.3

Shot peening and laser peening treatments improve the fatigue strength close to the surface but not in the interior [23,24]. Thus, they are useful especially in case of stress gradients close to the surface, like bending, notches or fretting fatigue. Having this into account, all tests in this work have been carried out in four point bending. Figure 2 shows the specimen and loading system in one of these tests, where h = 10 mm, L = 70 mm and t = 15 mm. Test frequency was 8 Hz, and stress ratio, R = 0.1.

Results and Discussion

Figure 3 shows test fatigue live obtained for all specimens for different stress levels with the lines fitted for every group. Each group is identified with a different symbol, but for each test, the symbol is shown as solid if the failure initiated from the interior and hollow if the failure initiated from the surface. During the initial tests appeared some problems that invalidated two tests. That is why there are only four test results for the AS BUILT treatment. To facilitate the comparison with other data from the literature, the stress amplitude is represented by σ_{eff}, which is the equivalent maximum stress for a stress ratio R = - 1. To obtain σ_{eff}, the expression proposed by Walker (equation 1) [25,26] was used to transform the results of tests carried out with R = 0,1 to the equivalent maximum stress with R = -1. In equation 1, the exponent should be fitted for each material and

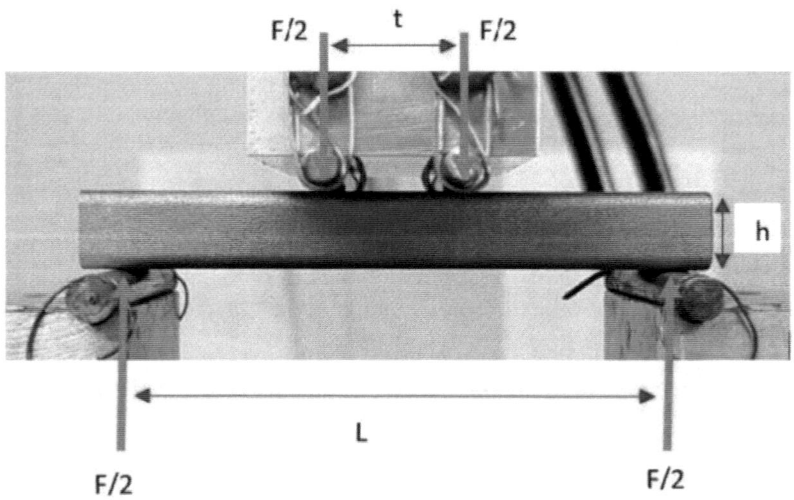

Figure 2. Four-point bending tests setup.

$$\sigma_{eff} = \sigma_{max}\left(\frac{1-R}{2}\right)^{0.28} \tag{1}$$

and σ_{max} is the maximum stress applied in tests with R = 0.1. In the present case a value of 0.28 was used for the exponent, according to experimental results obtained for conventionally fabricated T16Al4V specimens and uniaxially tested with stress ratios ranging from R = -0.5 to R = 0.5 [27]

From an initial view, it can be seen that HIP and AS BUILT groups are clearly less resistant than other groups. The SP group shows a higher strength, especially for high cycle fatigue; CASE increases this strength with near ten times the lives presented by SP. Finally, LP group shows the best results among all groups. Considering the initiation point of the specimens, all specimens of AS BUILT group initiated the crack on the surface, as well as the HIPed specimens. In the case of shot peened and LP, all specimens failed from the interior, but one, which failed from a defect on the surface, for SP, and on the rounded corner, for LP. All SP+CASE specimens failed with cracks initiated in the interior.

The results for HIPed specimens are very similar to those obtained by Kasperovich and Hausmann [28] with as built surface but HIPed specimens. This low strength can be understood considering that HIP does not eliminate or reduce the surface defects [11,15,29,30]. So the crack initiates similarly than in the case of AS BUILT specimens. Tests with HIPed and surface finished specimens are planned for the future, where a high improvement of the fatigue strength is expected, as occurred to Kasperovich and Hausmann with HIPed and machined specimens.

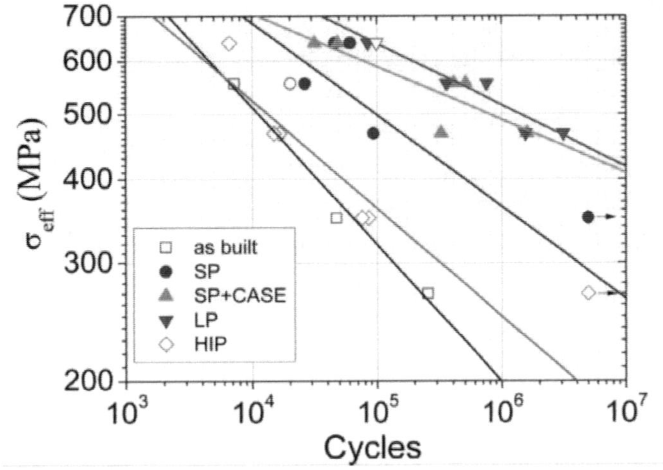

Figure 3. Fatigue test results. Each group is identified with a different symbol. Symbols are shown as solid for failures initiated from the interior and hollow for failures initiated from the surface

Regarding the shot peened specimens, results are in the same line that those obtained by Bagherifard et al. [20] for Al10SiMg, with a clear improvement of the strength. In Ti6Al4V specimens the results are also similar to those obtained by Wycisk et al [19] under axial loads. As far as authors know there are not results in the literature for the other twe surface treatment considered here.

Figure 4 depicts two fracture surfaces with the initiation points marked by arrows. Figure 5 shows SEM images of the initial defects of two of the specimens tested, one produced by lack of fusion and the other a gas pore. Figure 6 represents the position of each initiation point for all specimens that failed with an initiation point at the interior. It can be seen that the initiation point for SP is the surface or at any defect beneath it, while for SP+CASE all cracks initiate beneath the surface. The improvement of the strength produced by SP+CASE compared to SP could be explained because the roughness and the deleterious effect of shot peening on the surface, which also produces small micro cracks, while the CASE treatment reduces the roughness and eliminates those micro cracks. The improvement of the surface finish together with the residual compressive stress field close to the surface eliminates the crack initiation produced by the surface defects. Also, the elimination or reduction of the microcracks due to the CASE treatment, as well as the reduction of the effect of roughness on the stresses very near to the surface make the cracks to initiate deeper, where the stresses are lower because the stress gradient produced by bending and the residual compressive stresses in that zone, which are higher.

Figure 4. Details of initiation points on fracture surface of specimens: a) B1S4 Sup (as built, N_f = 16214 cycles); b) B4S31 (laser shock peened, N_f = 3099304 cycles)

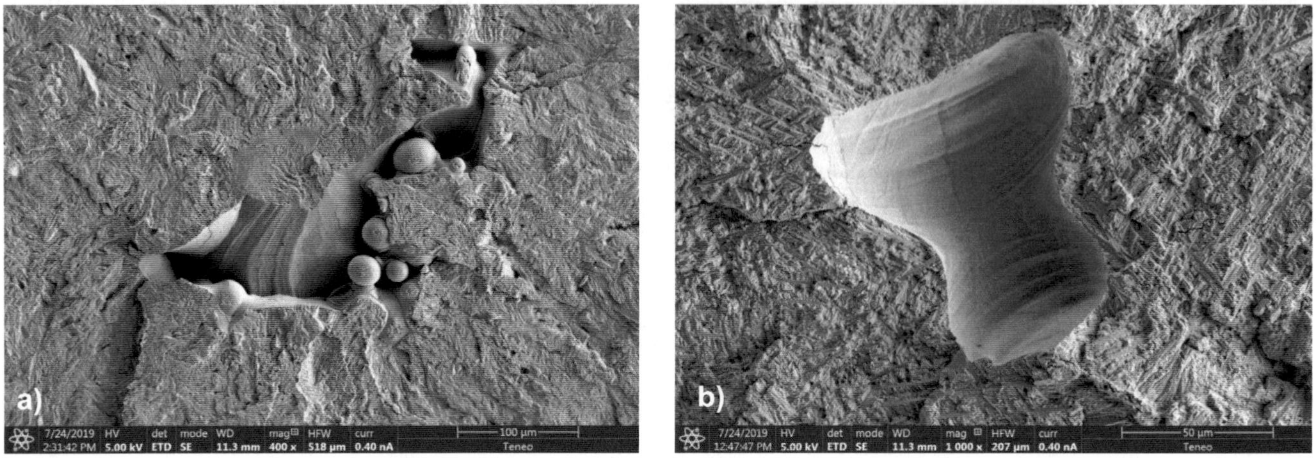

Figure 5. SEM images of two initial defects on fracture surface of specimens: a) B1S9Inf (shot peened, N_f = 45992 cycles); b) B4S32 (laser shock peened, N_f = 753603 cycles)

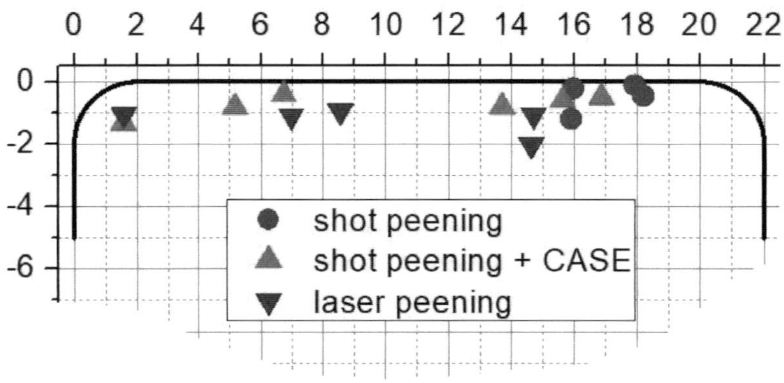

Figure 6. Location of the crack initiation in each test where the fail initiated from an interior defect. Distances in mm.

Regarding laser peening, the increase of the fatigue strength relative to the other two treatments could be explained by two reasons. On one side, the increase of the depth and width of the residual compressive stress field make the surface initial cracks to stop, allowing growing only cracks initiated below the residual stress field. On the other side, the deeper initiation points means lower stresses in the initiation zone and, so, higher fatigue lives.

Conclusions

This paper shows the improvement of fatigue strength obtained with different treatments (shot peening, shot peening+CASE, laser peening, HIP) in Ti6Al4V specimens made using additive manufacturing. These treatments have mainly three effects: Surface roughness modification, compressive residual stresses and decreasing pore size (only HIP) [14-16]. The last treatment has little effect on fatigue life. The reason is that due to the stress gradient the cracks initiate from the surface and the surface roughness in the as built specimens and in the HIP specimens is similar. The shot peened specimens experiment a ten fold increase in fatigue life and most initiation points from the interior due to the residual stresses, but due to the high roughness and possible damage due to the treatment there is one crack starting from the surface. The specimens with shot peening+CASE have the same residual stresses and a smoother surface, improving then fatigue life and having all the initiation points at the interior. Finally, the laser peening specimens have a high surface roughness but a different residual stress distribution which produces the highest fatigue strength. Other parameters that may influence fatigue behavior as the microestructure have not been studied at this point.

References

[1] Daniewicz, S.R., Shamsaei, N. (2017). An Introduction to the fracture and Fatigue Behavior of Additive Manufactured Parts, International Journal of Fatigue, 94, pp. 167.

[2] Daniewicz, S.R., Johnson, A., Thompson, S. M. and Shamsaei, N. (2018). Structural Integrity of Additive Manufactured Parts, in Laser-Based Additive Manufacturing of Metal Parts, L. Bian, N. Shamsaei y J. M. Usher (eds.), CRC Press, pp. 111-140

[3] Gorelik, M. (2017). Additive Manufacturing in the Context of Structural Integrity, International Journal of Fatigue, 94, pp. 168-177.

[4] Galarraga, H., Lados, D. A., Dehoff, R. R., Kirka, M. M. and Nandwana, P. (2016). Effects of the microstructure and porosity on properties of Ti-6Al-4VELI alloy fabricated by electron beam melting (EBM), Additive Manufacturing 10, pp. 47–57.

[5] Nicoletto, G. (2017). Anisotropic high cycle fatigue behavior of Ti–6Al–4V obtained by powder bed laser fusion, International Journal of Fatigue 94, pp. 255–262

[6] Yadollahi, A and Shamsaei, N. (2017). Additive manufacturing of fatigue resistant materials: Challenges and opportunities, International Journal of Fatigue 98, pp. 14–31.

[7] Mercelis, P. and Kruth, J. P. (2006). Residual stresses in selective laser sintering and selective laser melting, Rapid Prototyping Journal, 12, pp. 254-265.

[8] Seifi, M., Salem, A., Satko, D., Shaffer, J. and Lewandowski, J. J. (2017) Defect Distribution and Microstructure heterogeneity Effects on Fracture Resistance and Fatigue Behavior of EBM Ti-6Al-4V, International Journal of Fatigue, 94, pp. 263-287.

[9] Leuders, S., Thone, M., Riemer, A., Niendorf, T., Troster, T., Richard, H. A. and Maier, H. J. (2013). On the Mechanical Behaviour of Titanium Alloy TiAl6V4 Manufactured by Selective Laser Melting: Fatigue Resistance and Crack Growth Performance, International Journal of Fatigue, 48, pp. 300-307.

[10] Gong, H., Rafi, K., Starr, T.and Stucker, B. (2012) Effect of Deffects on Fatigue of As Built Ti6Al4V Parts Fabricated by Selective Lasser Melting, Solid Freeform Fabrication Symposium, pp. 499-506.

[11] Li, P., Warner, D. H., Fatemi, A. and Phan, N. (2016). Critical assessment of the fatigue performance of additively manufactured Ti–6Al–4V and perspective for future research, International Journal of Fatigue, 85, pp. 130–143.

[12] Wycisk, E., Solbach, A., Siddique, S., Herzog, D., Walther, F. and Emmelmann, C. (2014). Effects of defects in Laser Additive Manufactured Ti-6Al-4V on Fatigue Properties, Physics Procedia, 56, pp. 371-378.

[13] Shamsaei, N. and Simsiriwong, J. (2017). Fatigue Behaviour of Additively-Manufactured Metallic Parts, Procedia Structural Integrity, 7, pp. 3-10.

[14] Greitemeier, D., Palm, F., Syassen, F. and Melz, T. (2019). Fatigue Performance of Additive Manufactured TiAl6V4 Using Electron and Laser Beam Melting, International Journal of Fatigue, 94, pp. 211-217.

[15] Atkinson, H.V. and Davies, S. (2000). Fundamental Aspects of Hot Isostatic Pressing: an Overview, Metallurgical and Materials Transactions A, 31A, pp. 2981-3000.

[16] Massuo, H., Tanaka, Y., Morokoshi, S., Yagura, H., Uchida, T, Yamamoto, Y and Murakami, Y. (2018). Influence of Defects, surface roughness and HIP on the Fatigue Strength of Ti-6Al-4 Manufactured by Additive Manufacturing, International Journal of Fatigue,117, pp.163-179.

[17] Maamoun, A.H., Elbestawi, M.A. and Veldhuis, S.C. (2018). Influence of Shot Peening on AlSi10Mg Parts Fabricated by Additive Manufacturing, J. Manuf. Mater. Process. 2, 40, doi:10.3390/jmmp2030040.

[18] Lesyk, D.A., Martinez, S., Mordyuk, B.N., Dzhemeinsky, V.V., Lamikiz, A. and Prokopenko, G.I. (2020). Post-processing of the Inconel 718 alloy parts fabricated by selective laser melting: Effects of mechanical surface treatments on surface topography, porosity, hardness and residual stress, Surface and Coatings Technology, 381, 125136. Doi: 10.1016/j.surfcoat.2019.125136.

[19] Wycisk, E., Emmelmann, C., Siddique, S. and Walther, F. (2013). High Cycle Fatigue (HCF) Performance of Ti-6Al-4V Alloy Processed by Selective Laser Melting, Advanced Materials Research, 816-817, pp. 134-139.

[20] Bagherifard, S., Berette, N., Monti, S., Riccio, M., Bandini, M. and Guagliano, M. (2018). On the fatigue strength enhancement of additive manufactured AlSi10Mg parts by mechanical and thermal post-processing, Materials & Design, 145, pp. 28-41.

[21] Schajer, GS. (1988). Measurement of Non-Uniform Residual Stresses Using the Hole-Drilling Method. Part I—Stress Calculation Procedures. J Eng Mater Technol, 110, pp. 338-343. doi:10.1115/1.3226059.
[22] Schajer, GS. (1988). Measurement of Non-Uniform Residual Stresses Using the Hole-Drilling Method. Part II—Practical Application of the Integral Method. J Eng Mater Technol, 110, pp. 344-349. doi:10.1115/1.3226060.
[23] Vázquez, J., Navarro, C. and Domínguez, J. (2012). Experimental Results in Fretting Fatigue with Shot and Laser Peened Al 7075-T651 Specimens, International Journal of Fatigue, 40, pp. 143-153.
[24] Martín, V., Vázquez, J, Navarro, C and Domínguez, J. (2020). Effect of shot peening residual stresses and surface roughness on fretting fatigue strength of Al 7075-T651, Tribology International, 142, 106004.
[25] Walker K. (1970). The effect of stress ratio during crack propagation and fatigue for 2024-T3 and 7075-T6 aluminum. Effects Environ Complex Load History on Fatigue Life, ASTM STP 462, pp. 1–14.
[26] Dowling NE, Calhoun CA, Arcari A. (2009). Mean stress effects in stress-life fatigue and the Walker equation. Fatigue Fract Eng Mater Struct., 32, pp. 163-179.
[27] Metallic Materials Properties Development and Standardization (MMPDS-05), Battelle Memorial Institute; 2010. <http://app.knovel.com/hotlink/toc/id:kpMMPDSM11/metallic-materials-properties/metallic-materials-properties>.
[28] Kasperovich, G. and Hausmann, J. (2015). Improvement of Fatigue Resistance and Ductility of Ti6Al4V processed by Selective Laser Melting, Journal of Materials Procesing Tecnology , 220, pp.202-214.
[29] Tammas-Williams, S., Withers, P.J., Todd, I and Prangnell, P.B. (2016). The Effectiveness of Hot Isostatic Pressing for Closing Porosity in Titanium Parts Manufactured by Selective Electron Beam melting, *Metall and Mat Trans A,* 47A, pp. 1939-1946, doi:10.1007/s11661-016-3429-3.
[30] Shunyu Liu, Yung C. Shin. (2019). Additive manufacturing of Ti6Al4V alloy: A review, Materials & Design, 164, 107552, doi: 10.1016/j.matdes.2018.107552.

Laser Peening Effects on High-Cycle Fatigue Properties of Al-Li Friction Stir Welded Joints: Crystal Plasticity Modeling & Quantitative Evaluation

Maziar Toursangsaraki and Yongxiang Hu*

State Key Laboratory of Mechanical System and Vibration, School of Mechanical Engineering, Shanghai Jiao Tong University, Shanghai, China

Abstract

To optimize surface treatment procedures, their strengthening mechanisms need to be evaluated on the material properties. In this study, physics-based relations were employed with crystal plasticity constitutive equations to quantify the strengthening effects of microstructural attributes on the high-cycle fatigue properties of the AA2195-T6 friction stir welded joints after applying multiple laser peening layers. The developed model quantified the contributions of laser-peening-induced near-surface compressive residual stresses, dislocation density increase, and crystal morphology variations on the evolutions in fatigue indicator parameters. Grain-average fatigue indicator parameters were utilized to evaluate the fatigue property modification trends with depth, while their extreme values estimated the corresponding improvements at grain boundaries in near-surface regions after laser peening. As the crystal plasticity model predicted, the enhancement in joint fatigue properties mainly stemmed from the alleviation of cyclic mean stress under near-surface compressive residual stresses. This effect was followed by the increased resistance to cyclic plastic deformation due to the rise in dislocation density and texture intensity, and the decrease in crystal morphology heterogeneity under the homogenization of crystal morphology after laser peening. The decrease in extreme-value and grain-average fatigue indicator parameters in depth were the mechanisms behind enhancements in joint surface fatigue properties and the relocation of fatigue failure regions from surface to subsurface areas. Moreover, experimental high-cycle fatigue observations revealed the improvement in joint fatigue properties and validated the numerical approaches.

Keywords: Crystal plasticity, Laser peening, Friction stir welding.

Introduction

Friction stir welding (FSW) has been proven beneficial for joining Al-Li alloys owing to the superior joint properties compared to conventional fusion welding processes. However, its major limitation is the relative strength reduction at the joint center under the dissolution of precipitate phases and crystal texture alleviation caused by the thermo-mechanical procedures during welding. Hence, modification processes are necessary to regain the degraded joint mechanical properties. Laser peening (LP) has been extensively utilized to improve the fatigue properties of Al FSW joints. This process induces deeper near-surface plastic deformation, compressive residual stresses, and lower surface damage among mechanical surface treatment methods [1]. The LP-induced enhancements in material fatigue properties have been attributed to compressive residual stresses, work-hardening, and texture variations in near-surface regions under LP [2]. Microstructure-sensitive modeling techniques are necessary to quantify the strengthening contributions of LP effects on improving the joint fatigue properties.

The crystal plasticity (CP) modeling method simulates the crystal lattice and slip formation as the source of elastic-plastic deformation behavior. CP finite element method (CPFEM) includes the microstructural attributes of polycrystals in the CP constitutive modeling. Toursangsaraki et al. conducted the Eigenstrain technique on the CPFEM framework to predict the LP-induced residual stress distribution under texture anisotropy of A2024-T31 sheets [3]. Stopka et al. employed the fatigue indicator parameter (FIP) developed by

Fatemi-Socie to include the slip plastic shear strain and slip normal stress as the mechanisms behind high-cycle fatigue (HCF) crack initiation for AA7075-T6 [4].
Physics-based CPFEM models have been developed to quantify the strengthening effects of microstructural attributes, including grain size, crystal morphologies, dislocation density, deformation twinning, and precipitate phases, in material deformation mechanisms. Toursangsaraki et al. developed local physics-based CPRVE LP models at central regions of Al-Li FSW joints to quantify the LP strengthening effects on enhancing the joint tensile properties, including compressive residual stresses, and dislocation density elevation, and crystallographic texture intensification [5]. There has yet not been a material modeling approach to quantify the LP strengthening contributions to modifying HCF properties.
This study developed the physics-based CPFEM fatigue model of AA2195-T6 FSW joints to investigate the modification mechanisms in their fatigue properties under the application of multiple LP layers. This model quantified the contributions of LP effects on the variations in the distribution of FIPs, including compressive residual stresses, dislocation density increase, and crystal morphology alterations in near-surface regions. Experimental HCF tests were conducted on FSW joints at different LP conditions to observe the improvements in the joint fatigue properties under LP and validate the modeling outputs.

Experimental Methods

Rolled sheets of AA2195-T6 were FSWed in a butt-joint application. FSW joints generally show three relatively discrete zones, including the stirring zone (SZ), thermo-mechanically affected zone (TMAZ), and heat-affected zone (HAZ). An H13 tool with an 18 mm shoulder diameter and 6 mm threaded-pin diameter was employed using a normal force of 6 kN and tilt angle of 1°. The tool rotational and welding speeds were set at 270 mm/min and 500 RPM, respectively.
Double-sided LP was applied at the joint center to modify the joint fatigue properties. Specimens were fixed on an industrial robot arm that moved it in the LP direction with16 mm/s constant velocity to impose a uniform pattern of LP with the overlap ratio of 20% at each two successive shock impacts. An Nd: YAG laser beam with the wavelength of 1064 nm, power density of 10 J, the repetition frequency of 5 Hz, and duration of 15 ns was applied, which induced circular 4mm diameter impact spots on the surface.
Stress-controlled HCF tests were employed with a maximum nominal stress of 200 MPa, a ratio of 0.1, and a frequency of 15 Hz to reveal the LP effects on the joint fatigue life and failure mechanisms.

Modeling process

Because fatigue failure happened at SZ in all LP conditions, CPRVEs were only developed at this zone to evaluate LP modifications on the HCF properties of FSW joints. LP effects were measured by experiments after each LP layer and inserted into the CPRVE of SZ to quantify their contributions to the strengthening mechanisms under LP. This study utilized the previously measured LP-induced near-surface residual stresses, dislocation density, and crystal texture variations for similar FSW joints in the reference [5]. This study used the Eigenstrain distribution data calculated in the reference [5] to match the simulated LP-induced residual stress distribution with depth in the SZ CPRVE with the experimentally measured values by the X-ray diffraction (XRD) application. Then, the XRD measured distribution of dislocation density values was inserted to its corresponding CPRVE depth. Near-surface crystal orientation distributions were obtained by the electron backscattered diffraction (EBSD) process after each LP layer and synthesized into the CPRVE at SZ in its corresponding depth after each LP layer.
CP constitutive equations resolve the Cauchy stress to the crystal slip system α by multiplying its components with the dyadic product of the slip direction and normal plane unit vectors to calculate the slip shear stress. The cumulative deformation gradient is described by the multiplicative decomposition of rigid body rotation of crystal lattice and crystal slip deformation to represent the microscopic elastic/plastic deformation [6]. The physics-based CP model considers the slip strengthening effects of the various material microstructural

features, including grain size, crystal morphologies, and dislocation density. The critical resolved shear stress τ_{cr}^{α} of each slip system has been obtained as the cumulative effects of intrinsic crystal slip strength of Al alloys (τ_0^{α}), dislocation density effects ($\Delta\tau_{dis}^{\alpha}$), and grain size strengthening ($\Delta\tau_{HP}^{\alpha}$) by [5]

$$\tau_{cr}^{\alpha} = \tau_0^{\alpha} + \Delta\tau_{HP}^{\alpha} + \Delta\tau_{dis}^{\alpha} \tag{1}$$

The Hall-Petch relation has empirically predicted the strengthening effects of Grain-size on the crystal slip strength as

$$\Delta\tau_{HP}^{\alpha} = \frac{K_0}{\sqrt{D_G}} \tag{2}$$

with $K_0 = 0.06\,\text{MPa}\,\sqrt{\text{m}}$ being the Hall-Petch constant for Al alloys, and D_G defining the grain size average value. The dislocation density effects on the slip strength have been obtained by

$$\Delta\tau_{dis}^{\alpha} = \mu G b\sqrt{\rho} \tag{3}$$

with *G* being the Al matrix shear modulus, *b* describing the Burger's vector of the gliding dislocations, μ=0.2 defining the calibrated slip interactions for the Al matrix, and ρ as the total dislocation density, which is the summation of dislocation density values at different slip systems. The dislocation density evolutions under plastic deformation have been related to the grain size and current value of dislocation density value by

$$\dot{\rho}^{\alpha} = \left(\frac{C}{D_G} + K_a\sqrt{\sum_{\beta=1}^{n}\rho^{\beta}} - k_b\rho^{\alpha}\right)\left|\dot{\gamma}^{\alpha}\right| \tag{4}$$

where *C* is a grain size effect constant, and k_a and k_b include the generation and annihilation phenomenon of dislocation density, respectively. The UMAT subroutine is employed to insert the physics-based CP constitutive equations into the ABAQUS solver.
Fatemi and Socie FIP considers the critical plane where the highest maximum normal stress and cyclic plastic shear strain range occur. By considering the crystal slip as the source of plastic deformation, this FIP relation has been presented for slip planes by [4]

$$P_{FS} = \frac{\Delta\gamma_{p,max}^{\alpha}}{2}\left(1 + k_{FS}\frac{\sigma_{n,max}^{\alpha}}{\sigma_0}\right) \tag{5}$$

with $\Delta\gamma_{p,max}^{\alpha}$ describing the maximum cyclic plastic slip shear strain range on the slip plane, $\sigma_{n,max}^{\alpha}$ as its normal stress, and σ_0 defining the material yield strength. The constant value of $k_{FS} = 1$ characterizes the intensity of normal stress effects for the Al matrix.
The slip misorientations among neighboring grains cause the discontinuity of dislocation motions at the grain boundaries and the concentration of stress-strain values at these interfaces during material deformation. Moreover, crystallographic orientations define the grain strength by the material Taylor factor. The FCC crystal grains with the dominance of $\{110\}_{Al}$ orientations are softer. On the other hand, the relative dominance of $\{111\}_{Al}$ and $\{100\}_{Al}$ texture components are induced under plastic deformation and increase the grain strength [7]. The localization of plastic deformation at the softer side of hard/soft grain boundaries is caused by the combined effect of crystallographic misorientation and mechanical property heterogeneity. Thus, the grain boundaries with relatively higher misorientation angles and the mechanical property differences are more sensitive for extreme-value FIPs and fatigue crack formation under HCF loading. FIP extreme value distributions have been fit to the Gumbel probability distribution function to statistically analyze their distributions by

$$p = \exp(-e^{-\alpha_n(y_n - u_n)}) \quad (6)$$

where $p = F_{Y_N}(y_n)$ defines the possibility of Y_N value being lower than or equal to y_n which is a member of the population. The factor of u_n describes the characteristic highest members of the extreme-value population. Hence, higher u_n amounts indicate larger maximum values occurring within the extreme-value population. The α_n value denotes an inverse measure of occurring dispersions around the most significant characteristic extreme values. Higher α_n shows Lower dispersion in maximum extreme values.
Joint mechanical properties were considered homogeneous in the welding direction. Thus, joint cross-section 2D plane-strain CPRVEs were sufficient to model its HCF properties. DREAM.3D instantiated the EBSD-obtained crystal texture into the ABAQUS software input file to generate joint local CPRVEs with CPE4R elements. Two types of CPRVE were developed for the HCF modeling at SZ. The surface CPRVE illustrated extreme-value FIP variations at near-surface grain boundaries. On the other hand, half-depth CPRVEs simulated LP-induced variation in grain-average FIPs with depth. The simulated HCF tests consisted of a displacement-controlled cyclic loading with the maximum mean stress value of 200 MPa and a ratio of 0.1 for both surface and half-depth CPRVEs corresponding to the HCF experiments. The current CPFEM fatigue model utilized the previously obtained physics-based CP parameters for similar FSW joints presented in the reference [5].

Results and discussions

This study utilized previously measured microstructural evolutions of similar FSW joints under LP of similar FSW joints to evaluate the joint HCF enhancement under multiple-LP [5]. The texture recrystallization at SZ under FSW thermo-mechanical effects generated the refined grain texture with high-angle grain boundary distribution. LP-induced surface plastic deformation relatively reduced the misorientation angles through homogenizing the crystal texture towards $\{111\}_{Al}$ component. Fig. 1 illustrates the contributions of multiple-LP effects on the Gumbel distributions of near-surface FIP extreme values at SZ. Each LP effect decreased near-surface FIPs contributing to the improvement in surface HCF properties.

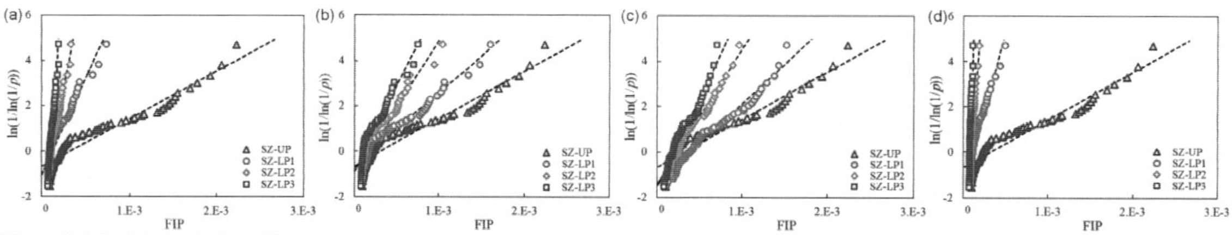

Fig. 1 Multiple-LP effects on the Gumbel distributions of surface extreme-value FIPs at SZ: (a) residual stresses; (b) dislocation density; (c) crystal morphologies; (d) combined effects.

Table 1 compares various contributions of LP effects on the FIP factors after the third layer of LP. α_n was most significantly affected by the compressive residual stresses at 74% of the combined LP effects, followed by crystal morphology variations at 15% and dislocation density evolutions at 11%. Corresponding LP contributions on u_n values was 59%, 21%, and 20%, respectively. This also showed a more noticeable improvement in the peak FIP extreme-value variations under texture modification than dislocation density.

Table 1 LP-induced variations in Gumbel distribution parameters of extreme-value FIPs.

LP effects	α_n /contribution %	u_n /contribution %
Unpeened	2e3	6.9e-4
Residual stresses	4.3e4 / 74%	1.2e-4 / 59%
Crystal morphology	8.3e3 / 15%	2.6e-4 / 21%
Dislocation density	7.2e3 / 11%	2.7e-4 / 20%
Total effects	7e4(270%↑)	9e-5(91%↓)

Table 2 compares various contributions of LP effects on the variations of grain-average FIPs with depth at SZ. Compressive residual stresses were the main contributors to the decrease in FIP values at about 62% of all contributions under all LP conditions, followed by dislocation density elevation at about 25% and crystal morphology variations at about 12%. Thus, compressive residual stresses led to the main HCF enhancement mechanism by reducing slip normal stress and cyclic plastic shear strain as crack formation mechanisms. Dislocation density elevation and Crystal morphology evolutions ranked second and third, respectively, in the joint fatigue property improvement in depth by increasing the material yield strength and resistance against slip cyclic plastic shear strain.

Table 2 Contributions of LP impacts in the variations of grain-average FIPs.

LP layers	RS / %	Dislocation / %	Texture / %	Total /Decrease %
Unpeened	5.9e-4/0	5.9e-4/0	5.9e-4/0	5.9e-4/0
LP1	1.5e-4 / 64%	3.2e-4 / 24%	4.1e-4 / 12%	9.7e-5/84%
LP2	1.02e-4/ 62%	2.1e-4 / 25%	3.1e-4/ 13%	4.2e-5/93%
LP3	8e-5 / 61%	1.6e-4 / 25%	2.6e-4/ 14%	3e-5/95%

Fig. 2 shows the grain-average FIP distributions at the slip planes for different LP conditions at SZ. FIPs kept reasonably constant values at around 5.9e-4 with depth before LP. Near-surface compressive residual stresses led to the highest effects on FIP values with depth, followed by dislocation density increase and crystal morphology variations. The highest FIP value alterations occurred at the first LP layer for all LP effects, and the corresponding evolutions decreased after each successive LP layer due to the saturation of near-surface plastic deformation and work-hardening under LP.

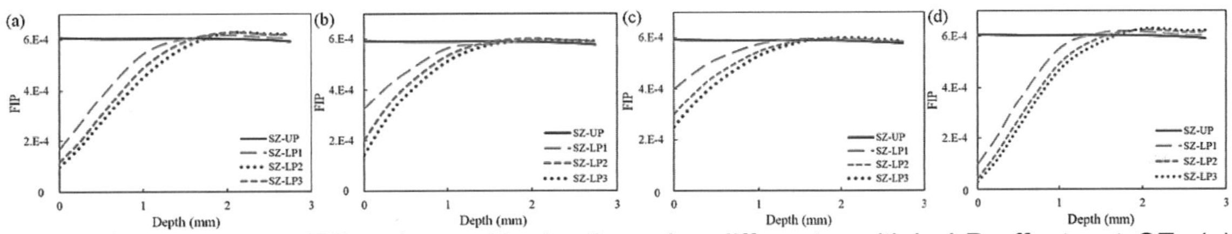

Fig. 2 Grain-average FIP values with depth under different multiple-LP effects at SZ: (a) Residual stresses; (b) Dislocation density; (c) Crystal morphology; (d) Combined effects.

Fig. 3a shows the experimental HCF results for the enhancement in HCF properties of AA2195-T6 and its FSW joints under multiple LP effects. Fig. 3b&c compare the joint fatigue fracture surfaces for the unpeened condition and third LP layer. The FCI for the unpeened condition occurred at the SZ surface because of lower fatigue resistance in the material surface and moved to the depth after LP, second, and third layers of LP because of the HCF resistance modification at surface-surface regions after LP. The in-depth FCI areas presented slightly less brittle failure mechanisms owing to the higher requirement for local accumulation of plastic deformation at these locations before failure [8].

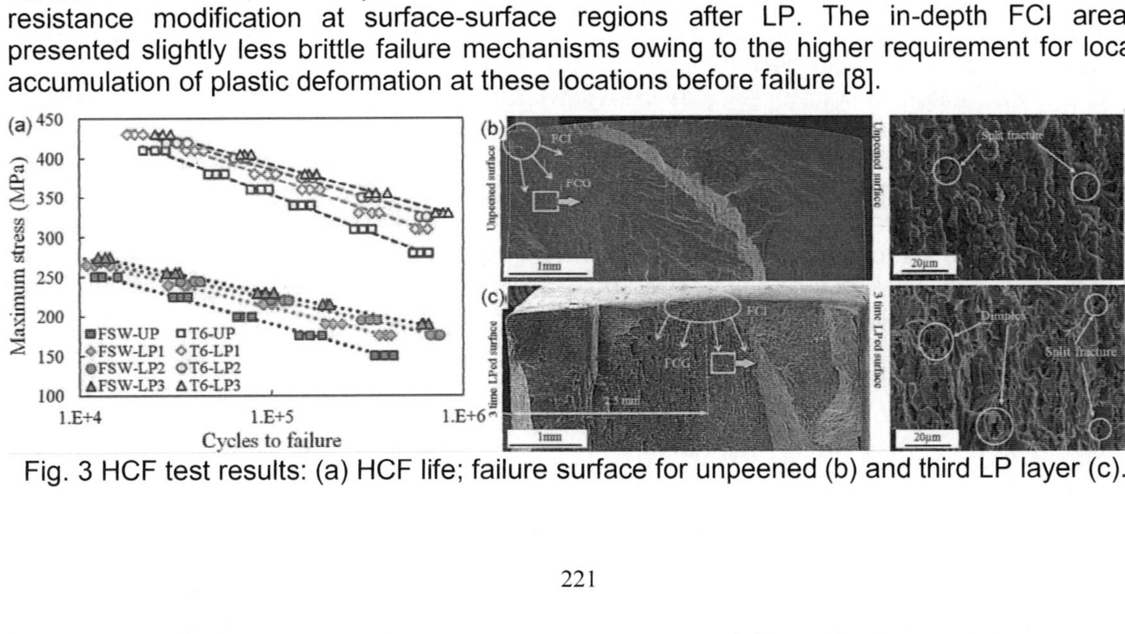

Fig. 3 HCF test results: (a) HCF life; failure surface for unpeened (b) and third LP layer (c).

Discussion and Conclusions

This study evaluated the mechanisms behind multiple-LP-induced modification in the HCF life of the AA2195-T6 FSW joints by developing the joint local physics-based CPFEM fatigue model at SZ. Overall conclusions were sorted as follows:

(1) The CPRVE fatigue model reasonably analyzed the LP modification mechanisms on the joint fatigue properties at SZ by quantifying various contributions of LP effects in reducing FIP values in this region.

(2) HCF strengthening features under LP were sorted as 1- cyclic mean stress reduction because of compressive residual stresses; 2- increase in resistance against cyclic microplastic deformation under the elevation of dislocation density and texture intensity; 3- the decrease in misorientations and heterogeneity of mechanical properties at grain boundaries under the texture homogenization effects.

(3) The most significant LP contributions in decreasing joint grain-average FIPs at SZ stemmed from the near-surface compressive residual stresses, followed by dislocation density evolutions and crystal texture variations after LP. Under the grain-boundary misorientation reduction, the texture variations effects in reducing surface extreme-value FIPs were higher than dislocation density effects.

(4) HCF test results indicated the improvements in the joint fatigue life and failure mechanisms by relocating FCI regions from the surface areas to the depth under multiple-LP effects and validated the modeling results.

Acknowledgments

This work was supported by the National Natural Science Foundation of China (Grant number U21A20135), "Shuguang Program" of Shanghai Education Development Foundation, Shanghai Municipal Education Commission (Grant number 20SG12), and Shanghai R&D public service platform project (Grant number 19DZ2291400).

References

[1] R. Yang, Y. Hu, M. Luo, H. Cheng, Z. Yao, *Distortion control of thin sections by single-sided laser peening*, Optics and Lasers in Engineering, 115 (2019) 90-99.

[2] Y. Sano, K. Masaki, T. Gushi, T. Sano, *Improvement in fatigue performance of friction stir welded A6061-T6 aluminum alloy by laser peening without coating*, Materials & Design (1980-2015), 36 (2012) 809-814.

[3] M. Toursangsaraki, H. Wang, Y. Hu, D. Karthik, *Crystal Plasticity Modeling of Laser Peening Effects on Tensile and High Cycle Fatigue Properties of 2024-T351 Aluminum Alloy*, Journal of Manufacturing Science and Engineering, 143 (2021).

[4] K.S. Stopka, M. Yaghoobi, J.E. Allison, D.L. McDowell, *Effects of boundary conditions on microstructure-sensitive fatigue crystal plasticity analysis*, Integrating Materials and Manufacturing Innovation, 10 (2021) 393-412.

[5] M. Toursangsaraki, Y. Hu, T. Zhang, *Crystal plasticity quantification of laser peening strengthening effects on AA2195-T6 friction stir welded joints*, The International Journal of Advanced Manufacturing Technology, (2022) 1-21.

[6] W.D. Musinski, D.L. McDowell, *On the eigenstrain application of shot-peened residual stresses within a crystal plasticity framework: Application to Ni-base superalloy specimens*, International Journal of Mechanical Sciences, 100 (2015) 195-208.

[7] W. Liu, H. Huang, J. Tang, *FEM simulation of void coalescence in FCC crystals*, Computational Materials Science, 50 (2010) 411-418.

[8] S. Farfan, C. Rubio-Gonzalez, T. Cervantes-Hernandez, G. *Mesmacque, High cycle fatigue, low cycle fatigue and failure modes of a carburized steel*, International Journal of Fatigue, 26 (2004) 673-678.

Localized indentation and its relationship with overall deformation of thin metal sheets induced by laser peen forming

Ruoyu Yang and Yongxiang Hu*

State Key Laboratory of Mechanical System and Vibration, School of Mechanical Engineering, Shanghai Jiao Tong University, Shanghai, China

Abstract

Laser peen forming (LPF) is a non-contact incremental forming method that introduces plastic deformation by laser shocks to modify the curvature of components. This study focuses on the localized indentation profiles of thin metal sheets induced by a single shock, and explores their relationship with overall plastic deformation after multiple shocks. LPF experiments with different spot sizes and laser power densities are conducted on thin metal sheets with a thickness of 0.5 mm. Two types of indentation profiles are produced, which are crater and conical profiles, after LPF with spot sizes of 0.8 mm and 4.0 mm. It is found that the indentation profile type remains the same except for the depth value after LPF with different power densities. Then, overall deformation behaviors are also investigated with different process parameters. Downward and upward deformation directions are generated after LPF with two spot sizes, respectively. Two deformation directions are also unchanged via adjusting laser power densities only. The overall deformation behavior is found to have a close relationship with the localized indentation profile. Therefore, an effective method is proposed to predict the overall deformation direction by localized indentation after a single shock.
Keywords: Laser peen forming, thin metal sheet, indentation profile, overall deformation.

Introduction

Laser peen forming (LPF) is a novel forming method to bend thin metal sheets. Laser-induced plasma expansion produces stress waves transmitting into the target material, which causes plastic deformation without extra thermal effect. Much attention has been paid to LPF treatment of numerous metallic specimens, including stainless steel [1], aluminum alloys [2], and titanium alloys [3].

Apart from the conventional convex curvature of components induced by LPF [4], Ocaña et al. obtained the concave curvature on 50 μm-thick metal sheets [1]. Hu et al. observed two kinds of deformation directions by changing material thicknesses and laser power densities [5]. Similar experimental results were also discovered by LPF with ultra-short laser pulses [6]. Further investigations showed that laser spot size also took a critical role in deformation behaviors of thin metal sheets with a thickness of 0.5 mm [7]. The section thickness, laser power density, and spot size significantly affect deformation behaviors. Therefore, more attention should be paid to the spot size effect on deformation behaviors as well as laser power densities.

Apart from the overall deformation behavior of thin metal sheets after multiple laser shocks, the localized indentation profile induced by a single shock is also a remarkable feature. Most related work shows a crater profile under the laser peening treatment [8, 9]. However, Clauer et al. pointed out that if the material is thin enough, the laser shock loading will produce a dimple on the topside and a bulge on the backside at the same time [10]. Considering that the accumulation of localized indentations results in the overall deformation behavior of thin metal sheets, the localized indentation profile after each laser shock could be closely related to the overall deformation behaviors. It would provide an efficient method to identify the deformation behaviors induced by LPF. However, limited studies have been conducted on the process parameter effects on indentations, and the relationship between localized indentation profiles and overall deformation behaviors has not been illustrated.

This study investigates process parameter effects on localized indentation profiles and the overall deformation behaviors of thin metal sheets induced by LPF with different spot sizes and laser power densities. The relationship between indentation profile types and overall deformation directions is established. An effective method is proposed to predict the overall deformation direction after multiple shocks by localized indentation after a single shock.

Experimental Methods

LPF experiments were conducted on Ti6Al4V thin sheets with the size of 40×30×0.5 mm^3. Half of the specimens were treated by a single laser shock to analyze the localized indentation profile. The others were treated by multiple laser shocks for overall deformation analysis, and the size of the treated region was 13.6×13.6 mm^2. The designed specimen geometry was machined by wire cutting and polishing. The laser source was a Q-switched Nd: YAG pulsed laser with a wavelength of 1064 nm and a pulse duration of 15 ns in FWHM. The laser beam was fixed during the LPF experiment, and an industrial robot constantly moved specimens to ensure the laser spots covering the specified surface region. The black tape and flow water were used as the absorbent and confinement layers, respectively. Laser pulse energies were 2, 4, 6, 8, 10, and 12 J. The spot diameters were 4.0 and 0.8 mm. The 4.0-mm spots were directly obtained by focusing the beam through a convex lens. The laser power densities were 1.06, 2.12, 3.18, 4.25, 5.31, and 6.37 GW/cm^2, corresponding to pulse energies of 2, 4, 6, 8, 10, and 12 J. Besides, 0.8-mm spots were achieved by adding a suitable aperture to ensure that the laser power density is consistent with that of 4.0-mm spots. After LPF treatments, the indentation profiles were measured using the KS-1100 optical surface profilometer with an accuracy of 0.5 μm.

Results and Discussions

Fig. 1 shows both topside and backside of indentation profiles on thin sheets after single-shock laser peening with the power density of 6.37 GW/cm^2. A crater profile is observed on the topside after a single shock with a spot size of 0.8 mm, as shown in Fig. 1(a). That is, pile-up is generated around the dent. The indentation diameter is 0.8 mm, which corresponds to the spot size. The depth of the profile is about 3.0 μm, and the measured displacement on the backside is only within ±0.2 μm, which could be considered negligible. Conversely, the indentation profile results are transformed into another situation when the spot size is increased to 4.0 mm, as shown in Fig. 1(b). A conical profile is generated on both sides of the thin sheet, which agrees with the phenomenon observed by Clauer et al [10]. Besides, the indentation diameter is much larger than the spot diameter, which is up to 12 mm. The laser spot diameter could be recognized from the change of the profile slope. The indentation depth on the topside is about 61.1 μm, and it is 46.6 μm on the backside. Both of them are over ten times the depth of the previous indentation with the crater profile.

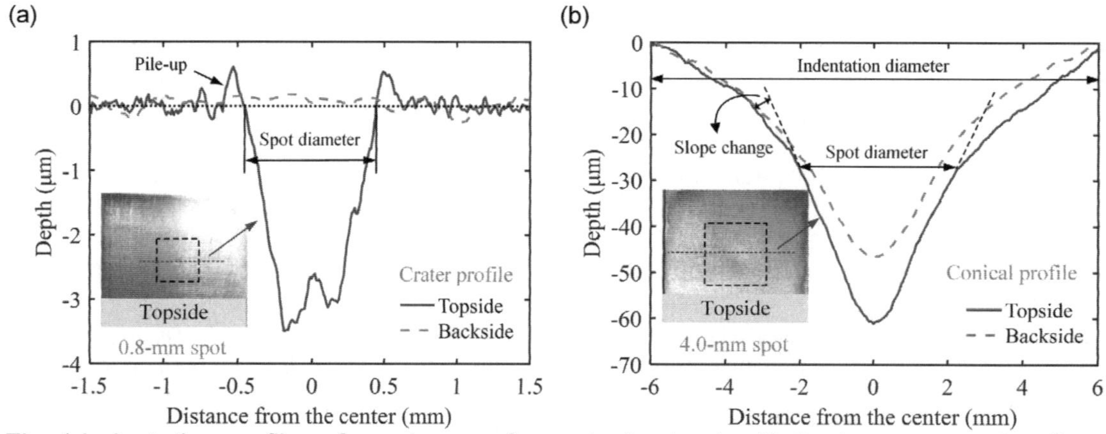

Fig. 1 Indentation profiles of specimens after a single shock with the laser power density of 6.37 GW/cm^2 and spot sizes of (a) 0.8 mm and (b) 4.0 mm.

Decreasing the laser power density from 6.37 to 1.06 GW/cm^2, two indentation types do not change except for the decrease in depth value, as shown in Fig. 2. For the thin metal sheets laser shocked with a spot size of 0.8 mm, all the indentations are crater profiles treated with different laser power densities. The depth value on the topside decreases with the power density, which are 3.0, 2.2, 1.7, 1.3, and 1.0 µm. Besides, all the indentation depths on the backside are negligible. For the thin sheets shocked with a spot size of 4.0 mm, all the indentations are conical profiles treated with different laser power densities. The indentation depths on the topsides also decrease with the laser power density, which are 61.1, 31.8, 21.3, 13.9, 8.1, and 2.0 µm, respectively. The indentation displacements on the backsides show a similar tendency, which are 46.6, 24.9, 14.7, 10.2, 5.9, and 1.5 µm. The experimental results demonstrate that the LPF with distinct spot sizes can change the indentation type of thin metal sheets, while the laser power density only affects the depth value.

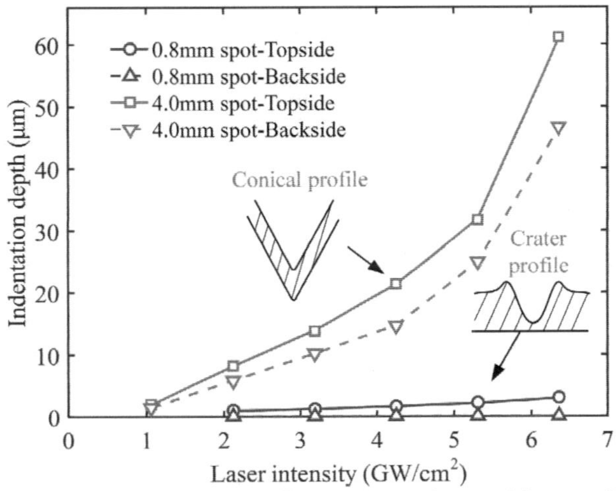

Fig. 2 Indentation depths with different laser intensities and spot sizes.

Fig. 3 provides the overall deformation behaviors of thin metal sheets after multiple laser shocks with different process parameters. It can be observed that the deformation profiles of two sheets are totally different after shocked with different spot sizes, as shown in Fig. 3(a). The thin sheet treated with 0.8-mm spots deforms upward and then downward from the clamping end, leaving a convex curvature with a maximum displacement of -410.5 µm at the free end. For the thin sheet treated with 4.0-mm spots, it deforms downward and then upward, leaving a concave curvature with the maximum displacement of 356.4 µm. The phenomenon of two bending directions induced by LPF has been demonstrated previously, but the study of localized deformation within the impacted region is limited. Essentially, the plastic deformation produced by multiple laser shocks is the accumulation of a single shock. Therefore, the overall deformation in the impacted region after multiple shocks can be regarded as the amplification of the localized indentation induced by a single shock. After multiple laser shocks with 4.0-mm spots, the downward deformation is induced in the impacted region, and the value is -112.8 µm. It is consistent with the conical profile after a single shock. For the thin sheet treated with 0.8-mm spots, the upward deformation is induced in the impacted region after multiple shocks, and the value is 94.5 µm. Dividing the displacement by the number of shocks (about 400), the displacement of the impacted region after a single shock is supposed to be 0.25 µm. Therefore, the slight deformation cannot be characterized in experiments because it is overwhelmed by the crater profile generated on the topside, and no upward displacement can be observed on the backside due to the machining and measurement error.

Fig. 3(b) provides the maximum displacements at the free end with different laser power densities and spot sizes. For the thin sheets treated with 0.8-mm spots, they all bend downward. The displacements are -410.5, -335.1, -211.7, -152.7, and -25.5 µm, varying the laser power density from 6.37 to 2.12 GW/cm^2. For the thin sheets treated with 4.0-mm spots, they all appear to bend upward. The displacements are 356.4, 209.2, 136.2, 97.1, 37.9, and 13.9 µm, corresponding to the power density from 6.37 to 1.06 GW/cm^2. The experimental results show that the spot size can change the bending direction induced by LP, while the laser power density only affects the displacement value. Considering the process parameter effects on the overall deformation behaviors and indentation profiles together, it can be easily observed that the overall deformation behavior after multiple laser shocks is significantly correlated with the localized indentation profile produced by a single shock. All the localized crater profiles lead to downward bending directions, while conical profiles result in upward bending directions. Therefore, it is feasible to predict the overall deformation direction by localized indentation profile after a single shock.

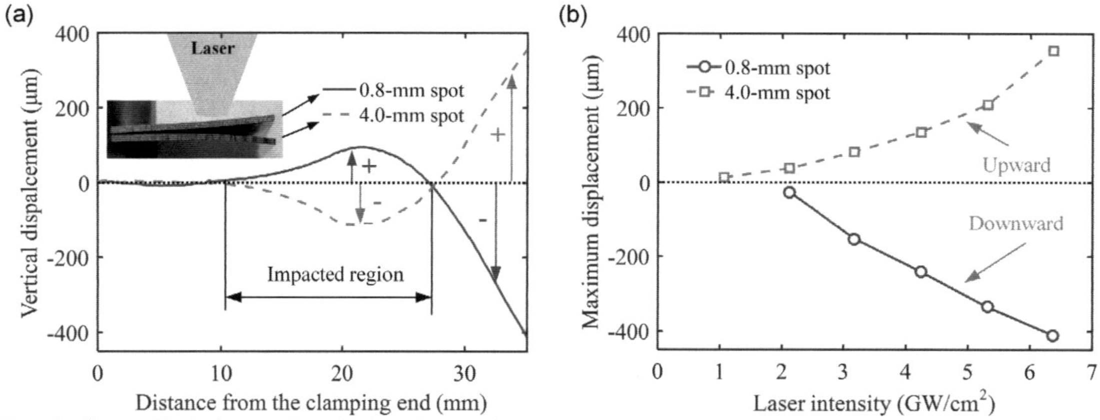

Fig. 3 Overall deformation behaviors after multiple laser shocks with different process parameters: (a) deformation profiles along the central axis under the laser power density of 6.37 GW/cm^2; (b) maximum displacements with different power densities.

Conclusions

This study analyzes the localized indentation profiles and overall deformation behaviors of thin metal sheets in LPF. Ti6Al4V thin sheets with a thickness of 0.5 mm are studied under laser shocks with different laser power densities and spot sizes. The main conclusions are as follows:

(1) Two types of indentations are observed, which are crater and conical profiles. The indentation depth of the crater profile is only a few microns on the topside with negligible deformation on the backside, while the depth of the conical profile can reach tens of microns on both sides.

(2) The indentation type is basically not affected by the laser power density but is mainly determined by the spot size under experimental conditions.

(3) The overall deformation behaviors are found to be closely related to the indentation type, and it is possible to predict the overall deformation direction by localized indentation. A crater profile causes a downward bending direction after multiple laser shocks, while conical profiles lead to upward bending.

Acknowledgments

This work was supported by the National Natural Science Foundation of China (Grant number U21A20135) and "Shuguang Program" of Shanghai Education Development Foundation and Shanghai Municipal Education Commission (Grant number 20SG12).

References

[1] J.L. Ocaña, M. Morales, C. Molpeceres, O, García, J.A. Porro, J.J. García-Ballesteros, *Short pulse laser microforming of thin metal sheets for MEMS manufacturing*, Applied Surface Science, vol. 254(2007), pp 997-1001.

[2] Z. Zhang, Y. Hu, Z. Yao, *Shape Prediction for Laser Peen Forming of Fiber Metal Laminates by Experimentally Determined Eigenstrain*, Journal of Manufacturing Science and Engineering, vol. 139(2017), pp 041004.

[3] Y. Hu, M. Luo, Z. Yao, *Increasing the capability of laser peen forming to bend titanium alloy sheets with laser-assisted local heating*, Materials and Design, vol. 90(2016), pp 364-372.

[4] H. Wielage, F. Vollertsen, *Classification of laser shock forming within the field of high speed forming processes*, Journal of Materials Processing Technology, vol. 211(2011), pp 953-957.

[5] Y. Hu, X. Xu, Z. Yao, J. Hu, *Laser peen forming induced two way bending of thin sheet metals and its mechanisms*, Journal of Applied Physics, vol. 108(2010), pp 073117.

[6] Y. Sagisaka, K. Yamashita, W. Yanagihara, H. Ueta, *Microparts processing using laser cutting and ultra-short-pulse laser peen forming*, Journal of Materials Processing Technology, vol. 219(2015), pp 230-236.

[7] R. Yang, Y. Hu, M. Luo, H. Cheng, Z. Yao, *Distortion control of thin sections by single-sided laser peening*, Optics and Lasers in Engineering, vol. 115(2019), pp 90-99.

[8] Y. Zhu, J. Fu, C. Zheng, Z. Ji, *Influence of laser shock peening on morphology and mechanical property of Zr-based bulk metallic glass*, Optics and Lasers in Engineering, vol. 74(2015), pp 75-79.

[9] Y.B. Guo, R. Caslaru, *Fabrication and characterization of micro dent arrays produced by laser shock peening on titanium Ti–6Al–4V surfaces*, Journal of Materials Processing Technology, vol. 211(2011), pp 729-736.

[10] A.H. Clauer, D.F. Lahrman, *Laser Shock Processing as a Surface Enhancement Process*, Key Engineering Materials, vol. 197(2001), pp 121-144.

Bonding of Magnesium Alloy with Ceramics by Shot Peening

Yasunori Harada[1] and Kenta Sugihara[2]

1 Graduate School of Engineering, University of Hyogo, Japan
2 Graduate Student, University of Hyogo, Japan

Abstract

Light metal materials such as aluminum and magnesium alloys are widely used in the field of transportation equipment where weight reduction of mechanical parts is required. However, it is known that wear resistance of light metals is lower than that of steel materials. Therefore, surface modification is being attempted by various coating techniques. In the present study, to improve the surface characteristics of corrosion resistance and wear resistance of light metals, surface modification was performed by bonding laminated sheet with ceramic powders and resin sheet to magnesium alloy using shot peening. The bondability of the laminated sheet was examined. Wear resistance was also evaluated. The laminated sheet could be bonded without peeling from the substrate. Bonding of hard powders was effective in improving wear resistance.

Keywords Bonding, magnesium alloy, ceramic, wear resistance, corrosion resistance.

Introduction

In recent years, light metals such as magnesium and aluminum alloys have been attracting attention in the field of transportation equipment such as aircraft and automobiles [1], [2]. Aluminum alloys have a high corrosion resistance and excellent recyclability. Magnesium alloys have a high specific strength, and are excellent in vibration absorption and dent resistance. However, corrosion resistance and wear resistance of light metals are lower than those of steel materials [3]. [4]. Therefore, many light metals are surface treated. The Yttrium-incorporated aluminizing of magnesium alloy for improved tribological and corrosion properties was carried out [5]. The properties of composite coatings electro co-deposited on magnesium alloy using triangular waveform pulse current were examined [6]. The effect of heat treatment on properties of Ni-P coatings deposited on magnesium alloy was investigated [7]. Coatings of functional materials have been used for magnesium alloys. Evaporation is performed by evaporating metals and oxides. The characterization and properties of hybrid coatings deposited onto magnesium alloys was examined [8]. Oxidation behavior of PVD processed magnesium alloy was carried out [9]. On the other hand, surface modification by plastic deformation of the surface has been performed for a long time. Typical techniques include burnishing and shot peening. It is known that surface hardness and fatigue strength increase in shot peening. It is widely used for automobile parts such as springs and gears, aircraft-related products such as jet engines and wings, and pressure vessels for chemical plants. Much research has been done to enhance the peening effect. The effect of warm shot peening treatments on surface properties and corrosion behavior of magnesium alloy was carried out [10]. The influence of shot peening on high cycle fatigue properties of the high-strength wrought magnesium alloy was examined [11]. The authors have been studying the bonding of dissimilar materials to metal surfaces by shot peening. Pure titanium foil and pure nickel foil were joined to improve corrosion resistance and wear resistance of the metal materials [12], [13]. In the present study, dissimilar materials were joined to magnesium alloy by shot peening. To improve corrosion resistance and wear resistance, attempts were made to join metal laminated sheets with resin or ceramic. The bondability of dissimilar materials was evaluated by bending test and microstructure observation.

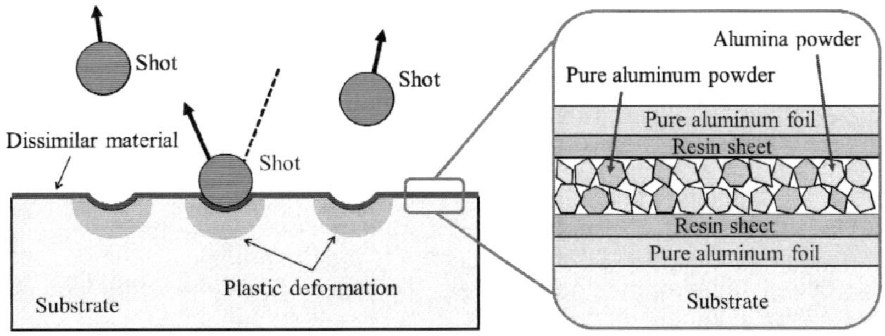

(a) Bonding of dissimilar material (b) Composition of laminated sheet

Figure 1 Bonding method of laminated sheet by shot peening.

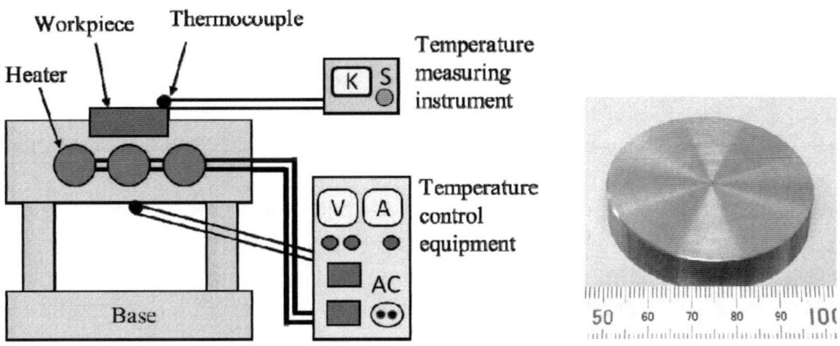

(a) Heating device (b) Workpiece AZ31

Figure 2 Equipment for bonding dissimilar material.

Table 1 Working conditions for bonding laminated sheet.

Equipment	Centrifugal peening machine
Shot material	High carbon cast steel, 450 HV
Shot diameter	1.0 mm
Impact speed	60 m/s
Heating temperature	573 K (300 °C)
Workpiece	Magnesium alloy AZ31, ϕ50x10mm
Hard powder	Almina Al_2O_3 (ceramic), 0.06-0.3 mm
Insert powder	Pure aluminum, 0.1 mm
Resin sheet	Thermoplastic polyimide, 0.025 mm
Insert foil	Pure aluminum, 0.02-0.04 mm
Surface finish	#120 (grain size 0.12 mm)
Atmosphere	In air, 298 K (25 °C)

Experimental Procedures

In shot peening, a metal workpiece undergoes plastic deformation near the surface due to the hit of a large number of shots at a high speed. Plastic deformation is utilized for bonding workpieces with dissimilar materials. A method of bonding dissimilar materials by shot peening is shown in Figure 1. Dissimilar material set on the substrate is hit with the shot. It is firmly bonded to the substrate (a). Dissimilar material is a laminated sheet composed of ceramic powders and the resin sheets (b). Ceramic was a hard alumina with a powder size of 0.06 to 0.3 mm. To examine the effect of powder size on bondability, powder size was divided into five groups. The average powder size of each group was 0.06, 0.12, 0.18, 0.25, 0.30 mm. The resin sheet was thermoplastic polyimide and had a thickness of 0.025 mm. Pure aluminum powders were mixed as an insert material because the bonding with the resin sheet was insufficient with only ceramic powders. The mixing ratio of ceramic particles to pure aluminum powder ranged from 10: 0 to 8: 2. Furthermore, pure aluminum foil was used as an insert material to improve the bondability with the substrate. Since the shot directly collides with the dissimilar material, plastic deformation of the substrate is small. Since the bondability of magnesium alloy is low, the use of pure aluminum foil is effective in improving the bondability [14].

A heating device for bonding dissimilar materials is shown in Figure 2. To promote plastic deformation of magnesium alloy, the dissimilar materials were joined in a warm state (a). The heating temperature was 573 K. The workpiece was a commercially available magnesium alloy AZ31 (b). An impeller type machine was used for shot peening. The shot material was made of cast steel with an average diameter of 1.0 mm. The peening speed was 60 m/s. The projection distance was approximately 270 mm. The surface of the workpiece was cleaned with emery papers prior to the shot peening. The dimensions of the materials used for the experiment are summarized in Table 1.

The structure of the cross section was observed by an optical microscopic observation near the surface where the laminated sheet was bonded. The bondability of dissimilar materials was tested by a three-point bending test. Wear resistance was evaluated by a wear test. The testing machine was the Suga type wear test machine (NUS-ISO3). On the surface after wear test, the depth from the surface was measured. It was a machine flat abrasion wear test using a rotating wheel. The force of 26.8 N was applied from the abrading wheel contact to the workpiece.

Experimental Results and Discussion

When the alumina powders were 100 %, the laminated sheet was peeled off regardless of the particle size. To improve the bondability, the pure aluminum powders were mixed as an insert material. Figure. 3 shows the bondability in the relationship between the mixing ratio of pure aluminum powder and the ceramic powder size. As the diameter of ceramic powder increased, the bondability improved. When the diameter of ceramic powder was 0.25 mm or less, however the bondability was not improve. Therefore, the mixing ratio of the pure aluminum powders was set to 20 %. As a result, the bondability of the laminated sheet was improved regardless of the diameter of ceramic powders.

Figure 4 shows the appearances of the shot-peened workpieces. When the powder size was small, the resin sheet was observed in the peeled workpiece. The peeled surface was broken with a layer of ceramic powder. Since the powder was hard, the laminated sheet was not easily deformed during the shot peening process.

In the workpiece showing good bondability, a cross section near the surface was observed by an optical microscope. Figure 5 shows an enlargement of the surface and a cross section near the surface. The mixing ratio of pure aluminum was 20 %, and the powder diameter of alumina was 0.18 mm. It was observed that no cracking or peeling occurred on the surface of the laminated sheet after shot peening (a). No voids were observed at the boundary between the laminate and the substrate in the cross section of the surface (b).

A three-point bending test was conducted to examine the bondability of the laminated sheet. The workpiece was bent until the substrate broke. Figure 6 shows the appearance of the workpiece and surface condition after bending test. The mixing ratio of pure aluminum was 20 %, and the powder diameter of alumina was 0.18 mm (a) and 0.3 mm (b). Although the laminate sheet broke with the substrate, no peeling was observed from the substrate. It was found that the bondability of the laminated sheet was good.

Wear test was performed on the workpieces showing good bondability. Figure 7 shows the appearances of the workpieces after wear test and its enlargement. The laminated sheet containing ceramic powders with a diameter of 0.18 mm were joined. The mixing ratio of the pure aluminum powders used as the insert material was 20 %. For comparison, the workpiece joined with laminated sheet without ceramic powders was also shown. In the workpiece without ceramic powder (a), metallic luster was observed in the worn area. In the case of the workpiece containing ceramic powders, however, the worn area was black (b). Ceramic powders were observed on the enlarged surface.

Wear test was performed on the workpieces covered with ceramic powders. Figure 8 shows the relationship between the powder size and the amount of wear. For comparison, the results of workpiece with shot peening, workpiece bonded with pure aluminum foil, and workpiece bonded with pure aluminum foil and resin sheet were also shown. In workpieces without ceramic powders, the wear depth ranged from 0.42 to 0.52 mm. For workpieces containing ceramic powders, the wear depth was approximately 0.3 mm. There was little effect of the size of ceramic powders on the depth of wear.

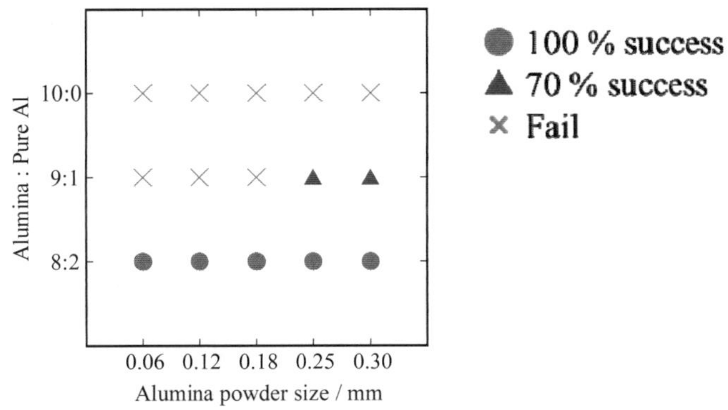

Figure 3 Bondability of laminated sheet.

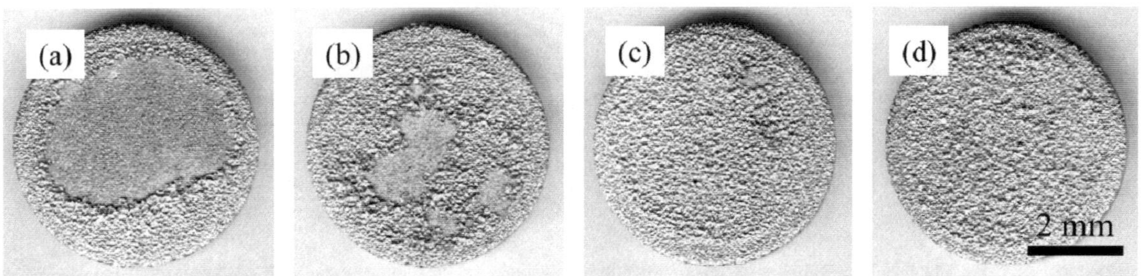

10 % Al powders, 0.12 mm | 10 % Al powders 0.18 mm | 10 % Al powders, 0.25 mm | 20 % Al powders 0.3 mm

Figure 4 Appearances of shot-peened workpieces.

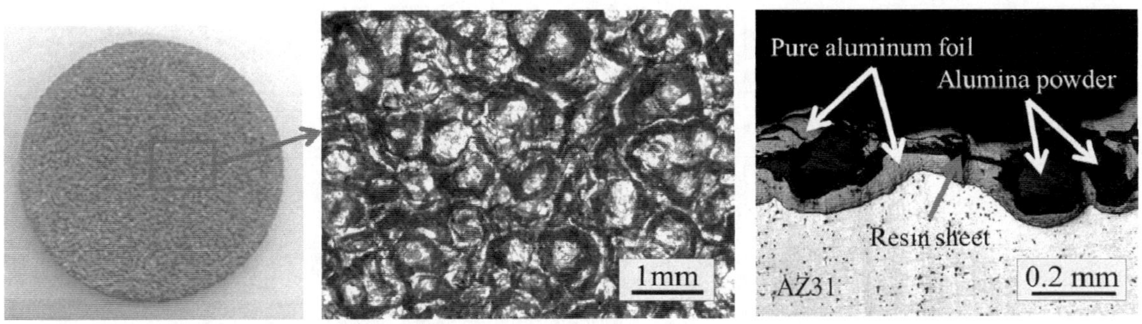

(a) Surface condition after shot peening (b) cross section near surface

Figure 5 Surface condition and cross-section near surface.

(a) 20% Al powders, 0.18 mm

(b) 20% Al powders, 0.3 mm

Figure 6 Appearances of workpieces after bend test.

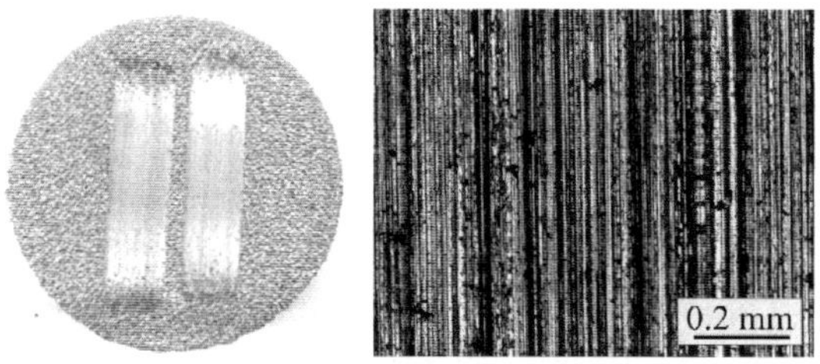

(a) Laminated sheet without ceramic powders

(b) 20 % Al powders, 0.18 mm

Figure 7 Surface conditions of workpieces after wear test.

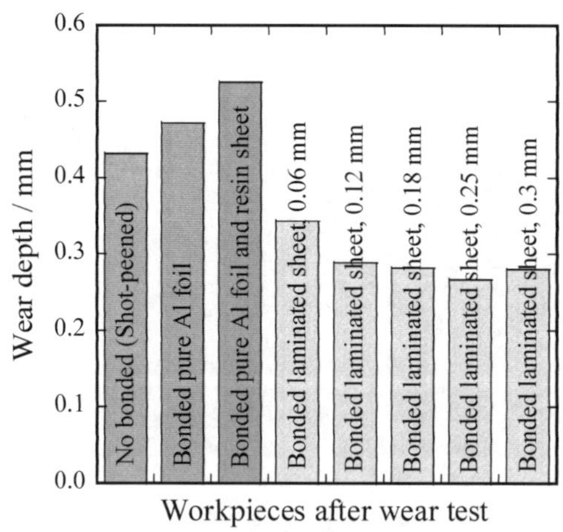

Figure 8 Relationship between powder size and wear depth.

Conclusions

The bonding method with the laminated sheet using shot peening was carried out. The laminated sheet with ceramic powders were successfully bonded to the surface of magnesium alloy by the hit of many shots. Pure aluminum powders used as the insert material to improve the bondability to the substrate. Bondability was evaluated by bending. Although the laminated sheet was fractured with the substrate, the laminated sheet peeled off from the substrate. The present method using shot peening was effective in wear resistance of the light metal products.

Acknowledgment

The authors would like to thank Mr. R. Takahashi, the former undergraduate student at University of Hyogo, for his help in the experimental work. This research was supported in part by a grant from the Light Metal Educational Foundation, Inc.

References

[1] D.N. Kuritsyn, M.V. Siluyanova, V.V. Kuritsyna, *Friction Stir Welding in the Aircraft Production*, Materials Science Forum, Vol. 992 (2020), pp 447-452.
[2] A.D. Evstifeev, I.V. Smirnov, Yu.V. Petrov, *Strain Rate Dependences of the Critical Stresses in Aluminum Al-Mg Alloys upon Impact Loads*, Physics of the Solid State, Vol. 62 (2020), pp 1967-1972.
[3] O.T. Olumide, O.B. Olamide, O.O. Timothy, *Corrosion, Corrosion Fatigue, and Protection of Magnesium Alloys: Mechanisms, Measurements, and Mitigation*, Journal of Materials Engineering and Performance, Vol. 31 (2022), pp 1707-1727.
[4] J. Chen, S. Hui, L. Guang, Z. Ziyun, T. Jiayun, C. Lang, *Improving Wear and Corrosion Properties of Magnesium Alloy by Cold Sprayed Nano WC-17Co Coating*, Journal of Superhard Materials, Vol. 43 (2021), pp 213-221.
[5] M. Nouri, D.Y. Li, *The Yttrium-Incorporated Aluminizing of Mg-3%Al Alloy for Improved Tribological and Corrosion Properties*, Journal of Materials Engineering and Performance, Vol. 31 (2022), pp 3218-3227.
[6] D.P. Pushpanathan, N. Alagumurthi, S.P. Devaneyan, *Investigations on the properties of composite coatings electro co-deposited on AZ80 Mg alloy using triangular waveform pulse current*, Surface Topography: Metrology and Properties, Vol. 9 (2021), pp 045078.
[7] M. Buchtik, M. Krystynova, J. Masilko, J. Wasserbauer, *The effect of heat treatment on properties of Ni-P coatings deposited on a AZ91 magnesium alloy*, Coatings (Web), Vol.9 (2019), pp 461.
[8] T. Tanski, K. Labisz, K. Lukaszkowicz, A. Sliwa, K. Golombek, *Characterisation and properties of hybrid coatings deposited onto magnesium alloys*, Surface Engineering, Vol. 30 (2014), pp 927-932.
[9] P. Perez, G. Garces, P. Adeva, Oxidation behaviour of a PVD processed Mg-14Ti alloy, Materials and Corrosion, Vol. 54 (2003), pp 145-151.
[10] L.B. Peral, A. Zafra, S. Bagherifard, M. Guagliano, I. Fernandez-Pariente I., *Effect of warm shot peening treatments on surface properties and corrosion behavior of AZ31 magnesium alloy*, Surface & Coatings Technology, Vol. 401 (2020).
[11] P. Zhang, J. Lindemann, *Influence of shot peening on high cycle fatigue properties of the high-strength wrought magnesium alloy AZ80*, Scripta Materialia, Vol. 52 (2005), pp. 485-490.
[12] Y. Harada, K. Mori, S. Maki, *Lining of metal plates with foils using hot shot peening*, Journal of Materials Processing Technology, Vol. 80-81 (1998), pp. 309-314.
[13] Y. Harada, N. Tsuchida, K. Fukaura, *Joining and shaping fit of dissimilar materials by shot peening*, Journal of Materials Processing Technology, Vol. 177 (2006), pp. 356-359.
[14] C.W. Tan, B. Chen, X.G. Song, L. Zhou, S.H. Meng, L.Q. Li, J.C. Feng, *Influence of Al Interlayer Thickness on Laser Welding of Mg/Steel*, Welding Journal, Vol. 95 (2016), pp. 384.s-394.s.

Analysis of the residual stress stability of high frequency mechanical impact (HFMI)-treated transverse stiffeners of mild steel (S355) and high strength steels (S700) under p (1/3) and straight-line shaped spectrum loading

Paul Diekhoff[1], Th. Nitschke-Pagel[1], K. Dilger[1]

D.Löschner[2], R. Schiller[2],I. Engelhardt[2]

[1] *Institut für Füge- und Schweißtechnik, Technische Universität Braunschweig*

[2] *Laboratory for Steel and Lightweight Structures, Institute for Material and Building Research, University of Applied Science Munich, Munich, Germany*

ABSTRACT

The effectiveness of high-frequency mechanical impact treatments (HFMI) and the associated increase in fatigue resistance is still being discussed, because it is generally assumed that the generated residual compressive stresses probably may be reduced by variable amplitude loading (VAL) and may therefore lose their beneficial effect. This study analyses further results for the sequence effect of VAL of a p (1/3) and straight-line shaped spectrum loading on the fatigue strength as well as the residual stress behavior of HFMI-treated transverse stiffeners (TS) of mild steel (S355) and high strength steels (S700). Fatigue test results with Random and High-Low loading for the two states as-welded (AW) and HFMI-treated joints together with in-situ residual stress measurements will be presented. The effect of VA loading on the residual stress behavior of as-welded (AW) and HFMI-treated joints and on the resulting fatigue improvement will be discussed.

Keywords: High strength steel, Welded joints, spectrum loading strength, residual stresses

1. Introduction

Hammer peening of welds is a well known simple technique to improve the fatigue strength of welds. Improved air driven or ultrasonic excited tools use higher hammer frequencies to enable a localized high treatment intensity. The target is to combine a simple tool-handling even under rough service conditions with an improved near surface material condition which enables a high benefit in applicable stress levels or endurance. The improvement of the fatigue strength is related to the generated compressive residual stresses, the plastic strain induced hardening of thin surface layers and a smoothening of the weld toe radii depending on the shape of the used tool. Within the scope of the common design rules [2,12] the effectiveness of such treatments is mainly related to the generated residual stresses. Since the effect of residual stresses on the resulting fatigue strength generally depends on their stability during the load history it is often argued that under service conditions, e.g under loading with variable amplitudes (VAL) in the lifetime of a component several overloads must occur which may reduce the initial residual stresses due to local or entire plastifications almost completely. Then it must be expected that the benefit of such treatments, which is well proved by comprehensive investigations under constant amplitude loading (CAL), dissolves thoroughly due to the residual stress relaxation.

2. Influence of the HFMI-treatment on the fatigue performance

For the fatigue strength of welded steel structures, the local loading of the fatigue critical locations is of decisive importance. The fatigue strength and thus the durability can be increased by different postweld treatment methods. Numerous studies [1,3-10] using high-frequency mechanical impact (HFMI) called hammer peening methods have proven to be an effective method for increasing the fatigue strength of welded details. Results of these studies have been commonly accepted and incorporated into recommendations and guidelines [11, 12]. However, the effectiveness of HFMI treatments under real service loading is currently still questioned because initial residual stresses induced by the HFMI can relax due to cyclic loading and high peak stresses as part of the service load [13–15]. Then it is expected that the benefit of the initial residual stresses dissolves in practice [16.20]. However, the beneficial effect of the strain-hardened surface layer, as well as the reduction of notch stresses due to the HFMI treatment is not necessarily affected. These studies focus mostly on the additional influence of high mean stresses, high peak stresses as single incidents or the influence of randomly distributed VAL on the fatigue strength of HFMI-treated notch details. For untreated welded joints, first studies on the sequence effect were carried out in [21].

Recent works [21] reveal that the fatigue life obviously depends on the type of VAL-test and that means if the tests are performed once with blocked and once with random loading sequence. These results are in accordance with the literature [22], showing that due to random loading, lower fatigue life is to be expected compared to blocked loading sequence. However, there are no studies available for HFMI-treated welds that focus on sequence effects for discrete load spectra. Previous studies have focused on the sequence effect of individual overloads within a constant amplitude load or the influence of random service loading on the fatigue strength of HFMI-treated welds. For this purpose, tests and HFMI-treated welded T-joints of the steels S355 N and S700M with transverse stiffeners were carried out under different loading conditions. The welds were investigated in the as-welded condition and after a HFMI-treatment performed at the critical weld toes (Fig 1). The applied load spectra are schematically shown in Fig.2. The stress ratio was set uniformly to R=-1. Each spectrum included $2 \cdot 10^6$ load cycles and was repeated continuously during the lifetime. Details of the test programme and the evaluation of the damage calculation using the palmgren-miner rule are summarized in [22].

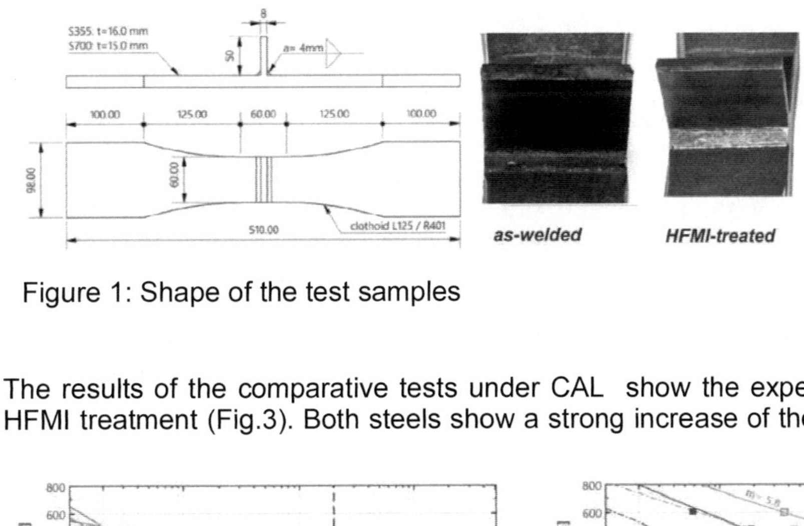

Figure 1: Shape of the test samples

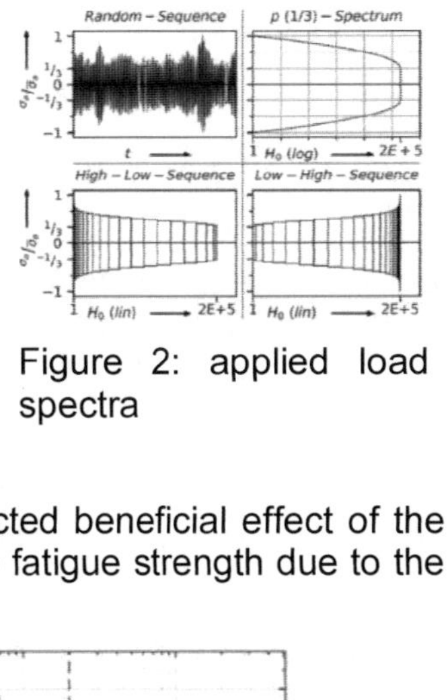

Figure 2: applied load spectra

The results of the comparative tests under CAL show the expected beneficial effect of the HFMI treatment (Fig.3). Both steels show a strong increase of the fatigue strength due to the

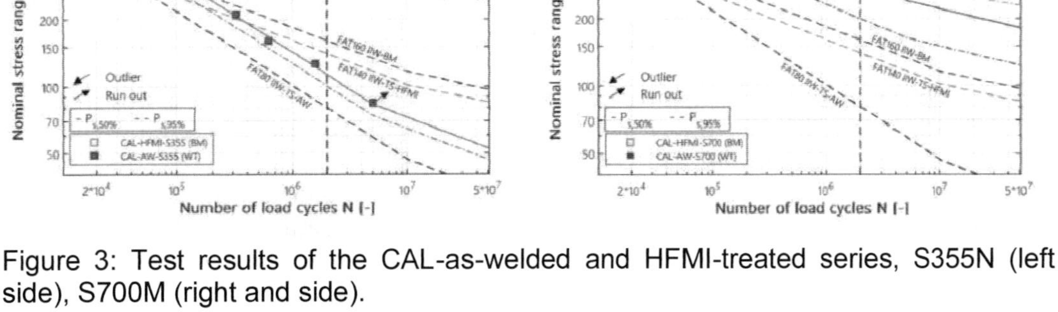

Figure 3: Test results of the CAL-as-welded and HFMI-treated series, S355N (left hand side), S700M (right and side).

treatment while the improvement is obviously higher in the high strength S700M. Noticeable is the different slope of the SN-curves. This effect is usually explained with the lower residual stress effectiveness in lower strength steels due to their stronger relaxation at higher stress levels. Therefore the fatigue strength improvement is expected to be higher at higher lifteimes. The same behavior is present under random-VAL (Fig.4) and under blocked VAL (Fig.5).

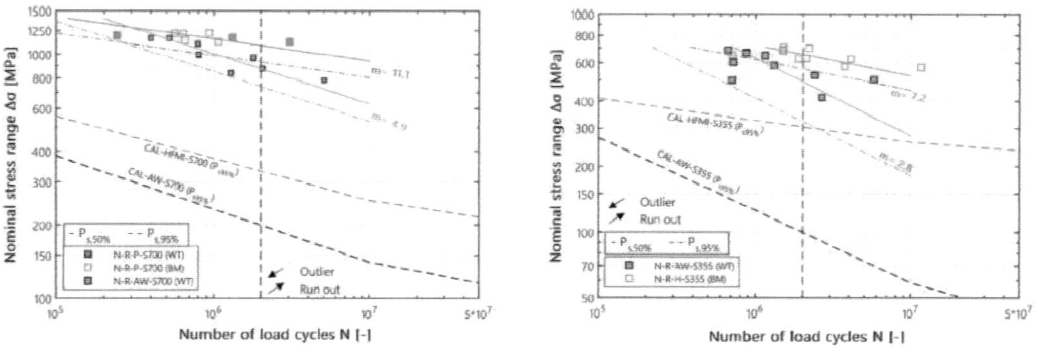

Figure 4: Results of the random VAL-tests. S355N (left hand side), S700M (right hand side)

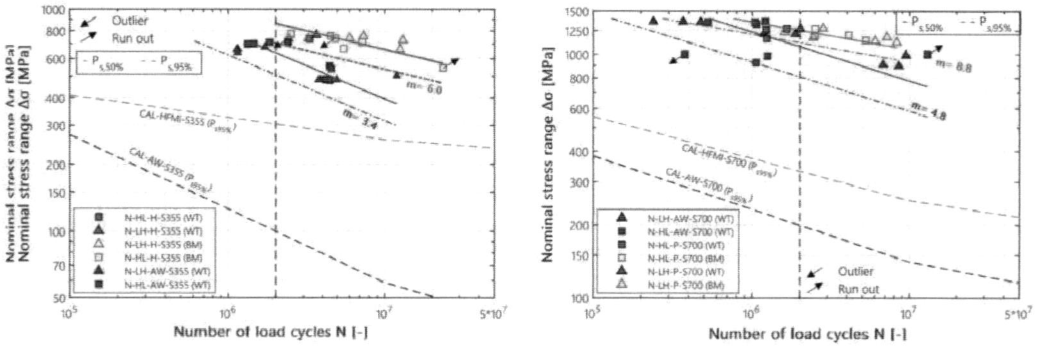

Figure 5: Results of the blocked VAL-tests. S355N (left hand side), S700M (right hand side)

The comparison of the VAL-test results obviously show that under VAL the fatigue life is obviously almost independent from the shape of the load spectrum. Stress peaks and overloads occurring during the first load cycles or at the end of the block as well as randomly do not lead to a significantly different lifetime or fatigue strength (e.g. the gassner-line). Assuming that the fatigue behavior is influenced by the initial residual stresses two conclusions are possible. Either the residual stresses are stable during the lifetime and not relaxed due to to single overloads or, if a relaxation is present the fatigue performance is not affected mainly by the residual stress condition.

3. Residual stress stability under different load conditions

3.1. Initial residual stresses

The initial residual stress condition after the local HFMI-treatment of both steels is presented in Fig.6. The residual stress profiles at the surface show the typical shape with maximum compressive residual stresses in the bottom of the peening track. In load direction (σ^{RS}_{trans}) the level of the compressive residual stresses is at approximately -300 MPa (S355N) and -450 MPa (S700M) while in longitudinal direction the residual stresses are uniformly at -450 MPa which is also typical for this kind of treatment. The integral widths of the related diffraction lines which are influenced by the local microstructure indicate that in the treated zone a strain hardening which is higher in the high strength S700M must be effective.

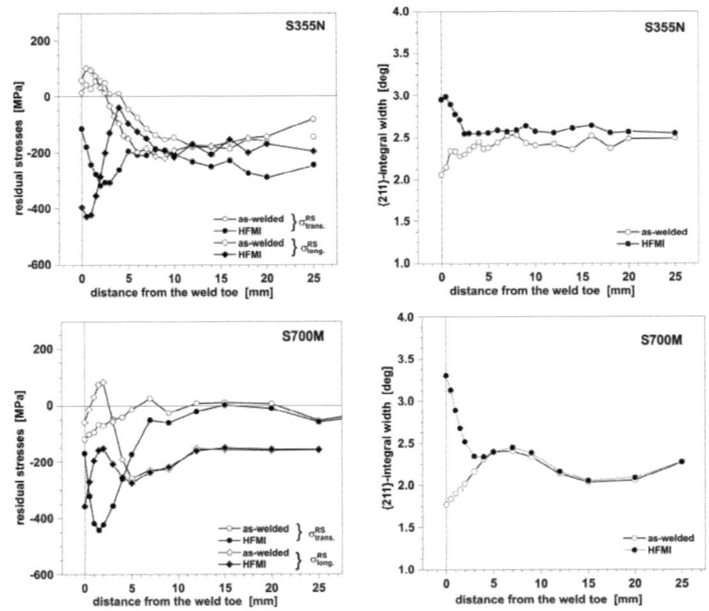

Figure 6: Initial residual stresses in the weld toe area at the surface before and after the HFMI-treatment

In depth direction both steels show a slight decrease of the compressive residual stresses without a steep gradient up to a depth of 0.5mm (Fig.7).

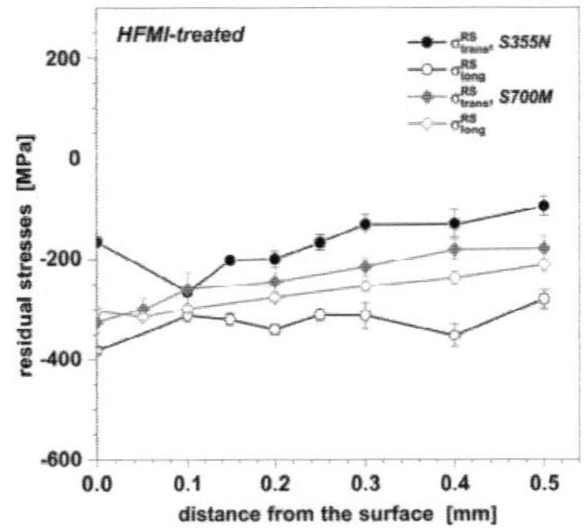

Figure 7: Initial residual stress depth profiles in the centre of the HFMI-zone

3.2. Residual stress relaxation due to static overloads

Figures 8 and 9 illustrate the effect of a static overload ocurring due to a single sress peak in the load history. In the S355N steel the loads were adjusted close to the nominal yield strength of the base material while in the S700M steel the load was varied from 55% R_e to 100% R_e. As Fig 8 shows the initial compressive residual stresses at the weld toe are released by a single load close to the yield strength almost completely. A slight increase of the load up to 377 MPa, which represents the yield strength of the related base material leads obviously to an overall plastic deformation in the near weld zone and the adjacent material. The shape of the residual stress profile changes completely and low tensile residual stresses are generated in the HFMI-treated zone. Nevertheless the constant amount of the integral widths indicates a stable hardening condition.

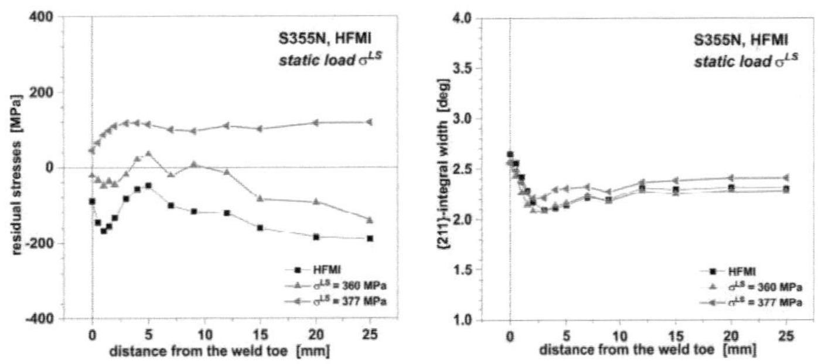

Figure 8: Residual stress surface distributions after single overloads, S355N

The high strength S700M shows a more sophisticated behaviour. The residual stress relaxation increases continuously with the maximum load but the plastic deformations are limited on the HFMI-treated zone. Even after the highest load representing the yield strength of the base material compressive residual stresses of 300 Pa remain in the treated zone as well as the shape of the profile is not changed remarkable. A slight decrease of the integral widths can be observed but the general shape of the strengthening condition seems to be not affected significantly (Fig.9).

It can be concluded in accordance with recent investigations [22] that the stability of the residual stresses is generally connected with the amount of single overloads. As expected the relaxation is the more pronounced the lower the yield strength of the material is. In low strength material probably an overall plastification is responsible for a general change of the entire residual stresses while in high strength materials the plastifications are more focused on the local residual stress peak at the notch due to the notch stress condition. However the strain hardening condition rarely affected in cases where the load stress does not exceed the yield strength significantly which seems also reasonable considering the particular stress-strain behaviour of the material.

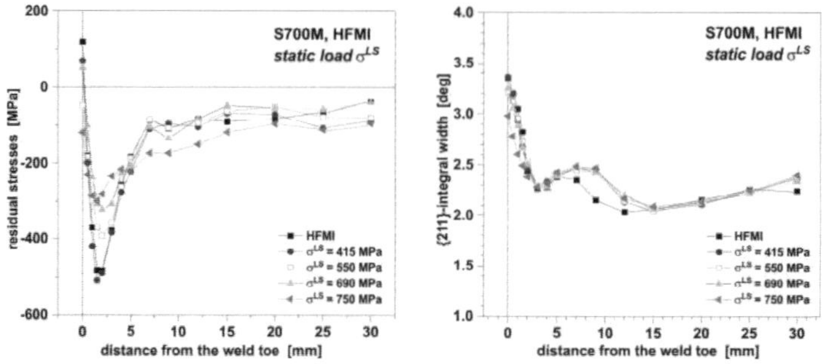

Figure 9: Residual stress surface distributions after single overloads, S700M.

3.2. Residual stress relaxation during variable amplitude loading

Figures 10 and 11 show the near weld residual stress profiles in different stages of a block loading programme with a North-Sea- representative load spectrum. The upper limit of the maximum load was set to 370 MPa for the S355N-steel and to 618 MPa for the S700M-steel. Blocks from 60% to 95% of the maximum load (200000 cycles each) were performed until the failure of the samples. Thus the number of block repetitions was 20 for the S355N and 6 for the S700M. The correlation between the amount of the observed relaxation of the residual stresses and the maximum load is less pronounced than determined under static overloads. However the relaxation takes place in the early stadium of the load history. After the first block almost stable residual stresses are present with neglectable remaining residual stresses in the low strength S355N and lowered compressive residual stresses in the S700M. Contrary to the static tests the S355N shows a slight cyclic softening with increasing number of load cycles while the strengthening condition of the S700M seems to be less affected. Similar results for the investigated weldments were found under a block sequence LowHigh with the same loading conditions and under random loading.

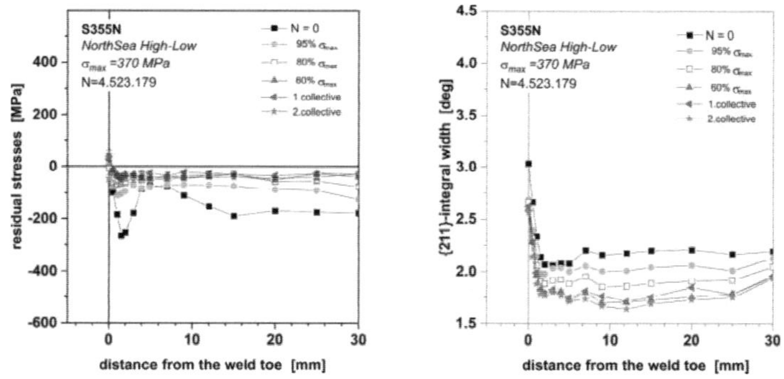

Figure 10: Residual stress surface distributions after spectrum loading (HighLow-Block programme), S355N.

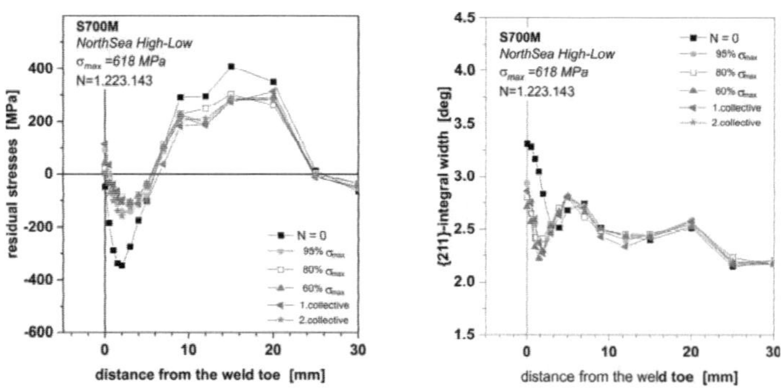

Figure 11: Residual stress surface distributions after spectrum loading (HighLow-Block programme), S355N.

4. Conclusions

The initial idea of the presented investigations was to clarify the use of a localized mechanical surface treatment on the fatigue performance under typical service conditions.

Since the benefit of such treatments is in practice primarily related to the influence of the generated compressive residual stresses the relaxation behaviour e.g. their stability in presence of single overloads was observed in order to evaluate the interaction of fatigue strength improvement and the changed local material condition at the crack initiation sites. As expected the residual stresses are affected significantly during fatigue loading whereas the most important factor is obviously the amount of the greatest load occuring in the load history. The strain hardening condition, which was evaluated by means of the integral width the measured diffraction lines, induced by the local plastic deformation is the more affected, the lower the yield strength of the material is. In high strength steels the strain hardening condition is more or less stable. Additionally it must be taken into account that the local HFMI-tretament affects also the shape of the local notch geometry because the sharp notch at the weld toe is smoothend by the strong plastic deformation resulting in a weld toe radius related to the tool diameter.

Both investigated steels showed a strong increase of the lifetime respectively the fatigue strength under all investigated load conditions. Concerning the observed complete residual stress relaxation and the slight softening the increased fatigue strength of the S355N-welds obviously is mainly related to the local notch shape smoothening. The higher stability of the generated residual stresses and the hardening condition in combination with the improved notch geometry is responsible for the improved fatigue strength of the higher strength S700M-steel.

Further investigation on other weld details and different steel grades should always take into account that the evaluation of the use of a mechanical post weld treatment requires necessarily the consideration of the change of the near surface conditions at the weld toe in combination with the notch geometry. Focusing only on the generated residual stresses may lead to misinterpretation of the use of such treatments and an underestimation of the potential benefit.

5. References

[1] Weich I, Ummenhofer T, Nitschke-Pagel Th: Welding in the World 53, 2009,11–12, S. 322–332.

[2] Hobbacher,A: Recommendations for Fatigue Design of Welded Joints and Components, New York: Welding Research Council, 2009.

[3] Ummenhofer T, Weich I.: Stahlbau 75, 2006, Heft 7, S. 605–607

[4] Yildirim H.C., Marquis G.B.: Welding in the World 56, 2012, 7–8, S. 82–96.

[5] Weich I.: Dissertation, 2009, Braunschweig: TU Braunschweig

[5] Berg J.: Dissertation (2017) RWTH Aachen, Shaker Verlag, Dissertation

[7] Durr A.: Dissertation (2006) Universitat Stuttgart

[8] Shams-Hakimi P, Yıldırım HC, Al-Emrani M.:Int J Fatigue 99, 2017, pp111–124

[9] Leitner M, Stoschka M, Eichlseder W.: Welding in the World 58, 2014, Heft1, S. 29–39

[10] Shams-Hakimi P, Zamiri F, Al-Emrani M et al: Eng Struct 155:251–266. 2017

[11] Stahlbau Verlags- und Service GmbH: Ermudungsbemessung bei Anwendung höherfrequenter Hämmerverfahren. Stahlbau Verlags-und Service GmbH, DASt-Richtlinie Heft 026, Stahlbau Verlags- und Service GmbH, Dusseldorf, 2019.

[12] Marquis G.B., Barsoum Z.: IIW recommendations for the HFMI treatment – for improving the fatigue strength of welded joints, IIW Collection, Springer Singapore, Singapore, 2017.

[13] Leitner M, Khurshid M, Barsoum Z.: Eng Struct 143:589–602, 2017

[14] Schubnell J, Carl E, Farajian M et al: Welding in the World 64, Heft 6, S. 1107–1117. 2020

[15] Ono Y, Yıldırım HC, Kinoshita K et al: Metals 12, Heft 1, S. 145, 2022.

[16] Leitner M, Gerstbrein S, Ottersbock MJ et al: Procedia Eng 101:251–258, 2015

[17] Leitner M, Stoschka M, Barsoum Z et al: Welding in the World 64, Heft 10, S. 1681–1689, 2020.

[18] Yıldırım HC, Marquis G, Sonsino C.M.: Int J Fatigue 91:466–474, 2015

[19] Yildirim HC, Marquis G.B.: Welding in the World 32, Heft 11, S. 1617, 2013.

[20] Nazzal SS, Mikkola E, Yıldırım H.C.: Engineering Structures 237, S. 111928, 2021.

[21] IGF-Vorhaben Nr. 18.848 N, 2017–2021.

[22] Schiller R, Loschner D, Diekhoff P et al: Welding in the World. 10. 1007/ s40194- 021, 2021.

A SED-based approach to predict the fatigue strength of notched and plain shot-peened parts

M. Benedetti[1], M. Bandini[2], S. Raghavendra[1], V. Fontanari[1], and F. Berto[3]

1 Department of Industrial Engineering, University of Trento, Italy
2 PeenService, Bologna, Italy
3 Dep. of Mechanical and Industrial Engineering, Norwegian University of Science and Technology, Norway

Abstract
A Strain Energy Density (SED) based fatigue criterion is developed to incorporate residual stresses (RS). Mechanical and residual stresses are treated separately. The formers are encapsulated in the form of SED associated to stress range and maximum SED. The work done when tensile mechanical strains fluctuate in the presence of compressive RS is added to the equivalent SED of the fatigue criterion. This is calibrated and validated using the results of an extensive experimental campaign aimed at exploring the effect of shot-peened induced RS on plain and notch fatigue strength of Al-7075-T651.

Keywords Strain energy density; Notch fatigue assessment; Residual stresses; Shot peening.

Introduction

Shot peening is a widely used surface treatment known to greatly improve the fatigue strength of machine elements, especially in the presence of stress raisers [1]. Its beneficial effect is mainly related to the introduction of compressive residual stresses (RS) into the surface layer. A reliable fatigue calculation method of shot-peened parts necessitates the incorporation of residual stresses into a suitable fatigue criterion. In stress-based approaches, RS are treated as mean stresses superimposed to the external stress field. Sometimes, the notch effect is accounted for according to the Theory of Critical Distances. In fracture mechanics approaches, RS are encapsulated into the definition of a closure stress intensity factor that reduces the effective stress intensity range available for the fatigue crack propagation. In recent years, the concept of strain energy density (SED) emerged as a very promising theoretical framework for accurate fatigue calculations [2]. However, to the authors' best knowledge, none attempted so far to incorporate shot-peening RS into the definition of SED.

Definition of the SED fatigue criterion

An exhaustive explanation of the theoretical framework of the SED approach to notch fatigue assessment can be found in [2]. In essence, the underpinning concept is that the fatigue damage is controlled by the SED averaged over a control volume Ω of material characteristic size. For plane or axisymmetric problems, the control volume simplifies into a circular domain, as schematically illustrated in Fig. 1a. In the classical formulation of the method, the fatigue damage is linked to the variation between maximum and minimum SED occurring during the load cycle. To take into account the mean stress effect, the authors proposed in [3] to incorporate the range $\Delta\overline{W}$ and the maximum value $\overline{W}_{max}$ of the average SED into a Walker-like equation to express an equivalent fatigue damage parameter:

$$\Delta\overline{W}_{eq} = \Delta\overline{W}^{\alpha}\,\overline{W}_{max}^{1-\alpha} = \Delta\overline{W}_{1} \qquad (1)$$

The SED components are graphically illustrated in Fig. 1b by the dashed and grey area, respectively, in the simple case of uniform uniaxial stress. The overbar sign indicates that the two SED components are averaged over the same domain Ω in the case of a non-uniform stress distribution occurring in the vicinity of notches. The mean stress sensitivity factor 1-α, the control radius R_1 and the fatigue strength characteristic $\Delta\overline{W}_1$ are material-dependent constants, which are assumed to depend only on the number of cycles to failure N_f. They must be calibrated using at least three experimental fatigue data. In this paper, we propose to conceive RS as a stress field independent of the external one. As schematically shown in Fig. 1c in the simple case of uniaxial stress, when tensile mechanical strains fluctuate in the presence of compressive RS, a negative specific work is done and expressed as:

$$W^{RS} = \sigma^{RS}\Delta\varepsilon^{+}$$
$$\Delta\varepsilon^{+} = \begin{cases} \varepsilon_{\max} - \varepsilon_{\min} & \varepsilon_{\min} > 0 \\ \varepsilon_{\max} & \varepsilon_{\min} \le 0 \end{cases} \tag{2}$$

The RS specific work $\overline{W}^{RS}$ expressed by Eq. (2) and averaged over the domain Ω reduces the SED associated to the external stresses, thus accounting for the beneficial effect of compressive RS. Therefore, we propose to incorporate RS into the fatigue criterion simply by including $\overline{W}^{RS}$ in $\Delta\overline{W}_{eq}$:

$$\Delta\overline{W}_{eq}^{RS} = \Delta\overline{W}^{\alpha}\,\overline{W}_{\max}^{1-\alpha} + \overline{W}^{RS} = \Delta\overline{W}_1 \tag{3}$$

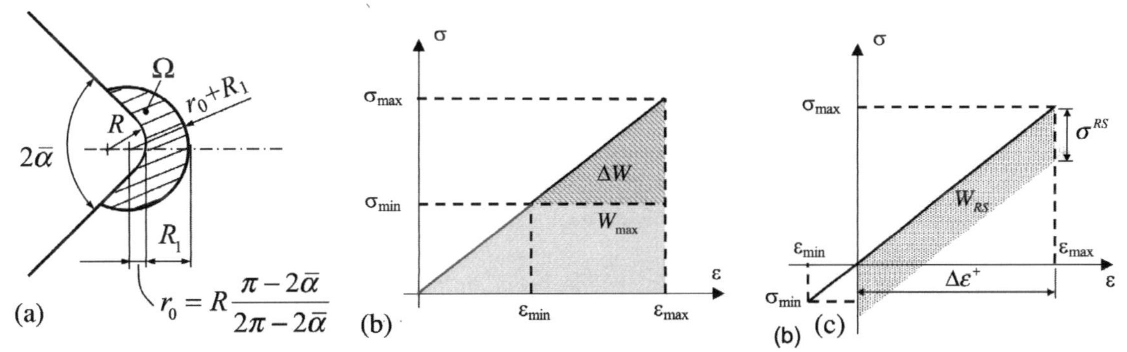

Fig. 1. (a) Strain energy density averaging domain Ω (dashed area) ahead of the notch root in the case of plane or axisymmetric problems. (b) SED components associated to the external mechanical loading, viz. SED associated to the stress range ΔW (dashed area) and maximum SED $W_{\max}$ (grey area). (c) Specific work W^{RS} done by fluctuating positive mechanical strains in the presence of residual stresses.

Material and experimental procedures

An extensive experimental testing campaign was carried out in the past with the aim of investigating the effect of shot peening on plain and fatigue strength of Al alloys. Details regarding the material and experimental procedures used to generate the fatigue data analyzed in this article can be found in [4,5,6,7]. In brief, the experimentation was performed on the aeronautical grade Al-7075-T651 aluminum alloy supplied in the form of 4 mm thick rolled plate. Characteristic dimensions are listed in Table 1. Plain (smooth) and notched prismatic specimens with three increasing levels of notch severity were machined. A schematic view of specimens' geometry is provided in Fig. 2. Part of the specimens were subject to a gentle shot peening treatment termed B120 (ceramic beads of 120 μm diameter, Almen intensity 10N). The specimens were fatigued under bending loading (about *y*-axis, see Fig. 3) at two load ratios R, namely -1 and 0.05.

RS measurements were carried out in [8] in the vicinity of the notch tip in notched samples and of the lateral edge in smooth samples (where fatigue crack initiation occurs) through two complementary experimental techniques, namely micro-XRD and micro-hole drilling/slot cutting. RS measurements were undertaken in non-tested samples as well as in samples fatigued at two load levels for a number of cycles equal to 50% of the expected fatigue life, viz. 2×10^5 and 2×10^6 cycles, respectively. Such measurements were used to reconstruct the RS field in the vicinity of the notch tip according to the eigenstrain method illustrated in [9]. In brief, this consists in the superposition of the misfit strains produced by two fictitious temperature distributions, whose intensity and spatial distribution were deduced by fitting the experimental data after being corrected for the radiation penetration depth.

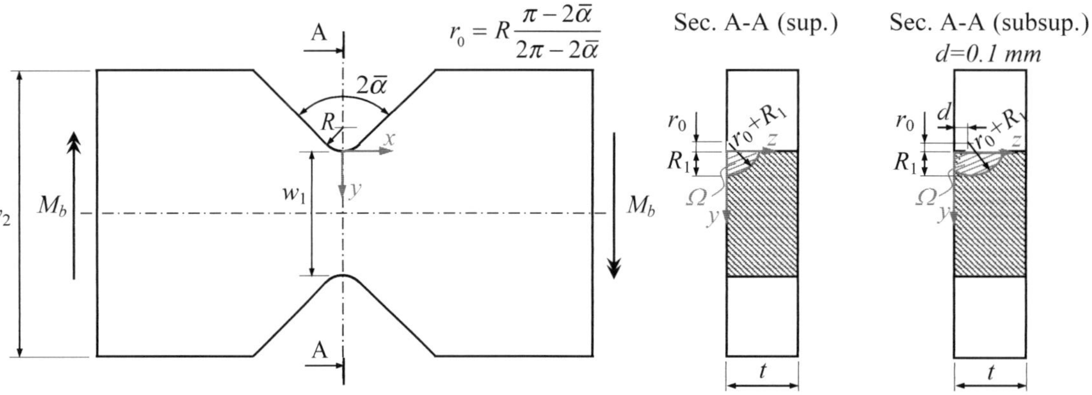

Fig.2. Geometry of the prismatic samples used for the fatigue tests. Characteristic dimensions are listed in Table 1. The SED-averaging control domain Ω is centered either on the frontal tensioned side of the specimen ("sup.") or at distance *d* from this plane ("subsup.").

Table 1. Fatigue sample designation and geometry. See Fig. 2 for the meaning of the geometrical parameters.

Designation	w_1 (mm)	w_2 (mm)	t (mm)	$2\bar{\alpha}$ (°)	R (mm)
Plain (R30)	6	12	4	90	30
Smooth (R2)					2
Sharp (R0.5)					0.5
Ultra-sharp (R0.15)					0.15

Experimental Results

The results of the fatigue tests are shown in Fig. 3a and b, for pulsating (R=0.05) and fully-reversed (R=-1) bending loading, respectively. Solid and open symbols refer to as-received and peened conditions, respectively. Run-out tests are marked by arrows. Solid lines represent best-fit curves, corresponding to 50% failure probability. The scatter of the fatigue data was estimated by computing the regression variance assumed to be uniform for the whole fatigue curve and used to draw the fatigue curves (dashed lines) corresponding to 10% and 90% failure probability. It can be noted that the shot peening treatment is beneficial to the fatigue strength, especially in the presence of stress raisers. Moreover, the fatigue improvement increases with notch severity and is higher under fully-reversed rather than pulsating bending fatigue. Importantly, in the very high-cycle fatigue (VHCF) regime explored in pulsating bending fatigue tests, the improvement imparted by shot peening at shorter fatigue lives is retained in the sharp and ultra-sharp notched conditions, whereas it is almost lost or reduced to 20% in plain and blunt-notched specimens, respectively.

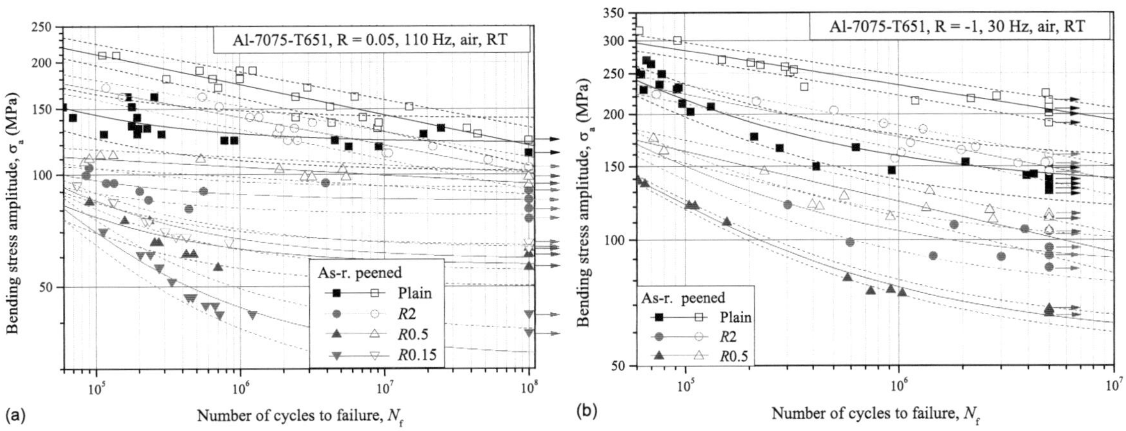

Fig. 3. Bending fatigue curves of the plain and notched specimens: (a) pulsating, (b) fully-reversed fatigue tests. Run-out tests are marked by arrows. Data of as-received samples are indicated as "As-r." in figures' legend. Solid lines represent 50% failure probability, while dashed lines refer to 10% and 90% failure probability.

Discussion and Conclusions

The obtained fatigue results are here reinterpreted on the base of the proposed SED fatigue calculation approach. For this purpose, a 2D domain consisting in a circular sector centered in the origin of the notch curvilinear reference frame, a circular domain Ω lying on the symmetry plane *yz* is considered. The location of this domain with respect to the specimen thickness (*z*-axis) is set according to the expected position of the fatigue crack initiation site. Fractographic inspections [6,7] revealed that the majority of cracks nucleated very close to the tensioned frontal surface of the samples. In this scenario, denoted as "sup." in Fig. 2, the center of Ω is coplanar with the tensioned frontal surface. Since plain and blunt-notched samples tested in the VHCF regime exhibited a tendency towards sub-superficial crack initiation (about 0.1 mm below the outer surface), the center of Ω is moved 0.1 mm beneath the tensioned surface to account for this scenario, indicated as "subsup." in Fig. 2. The range (maximum value) of the SED associated to mechanical loading is computed via FEM for each element lying in the control volume in terms of range (maximum value) of stress and elastic strain components. These element SED components are finally used to compute the corresponding SED components averaged over the control volume $\Delta\overline{W}$ and $\overline{W}_{\max}$. The RS specific work $\overline{W}^{RS}$ is calculated by means of FE models, wherein RS are introduced by means of two independent temperature fields. After introducing the bending load in the same manner described above, the positive range of elastic strains and the corresponding RS components are evaluated, whereby $\overline{W}^{RS}$ is averaged on the domain Ω. Fatigue calculations of shot peened variants will be done incorporating either initial (IRS) or stabilized (SRS, partially relaxed) residual stresses. The tri-parametric fatigue criterion is calibrated by least-square fitting the unpeened fatigue data to get a robust determination of material dependent parameters R1, α and $\Delta\overline{W}_1$. Figure 4a and b compare experimental data and SED predictions in the whole explored fatigue regime for load ratios R=0.05 and R=-1, respectively, considering either SRS (solid lines) and IRS (dash-dotted lines). Looking at Fig. 4b, the very good agreement between calculations (made under the assumption of superficial crack initiation) and experiments is evident in the whole explored fatigue range ($5\times10^4<N_f<5\times10^6$) for all the specimen geometries. SRS calculations are in better agreement with experimental data with respect to IRS, especially in the medium-to-high cycle fatigue (M-HCF) regime, where more intense RS relaxation takes place. Looking at Fig. 4a (R=0.05), the discrepancy of IRS from the experimental data of the notched variants is much more pronounced, suggesting that cyclic RS relaxation plays a crucial role in dictating the pulsating fatigue strength in the presence of stress raisers. If we look at SRS predictions reported in Fig. 4a, where the fatigue tests were prolonged until 1×10^8 cycles, the scenario is different from that shown in Fig. 4b.

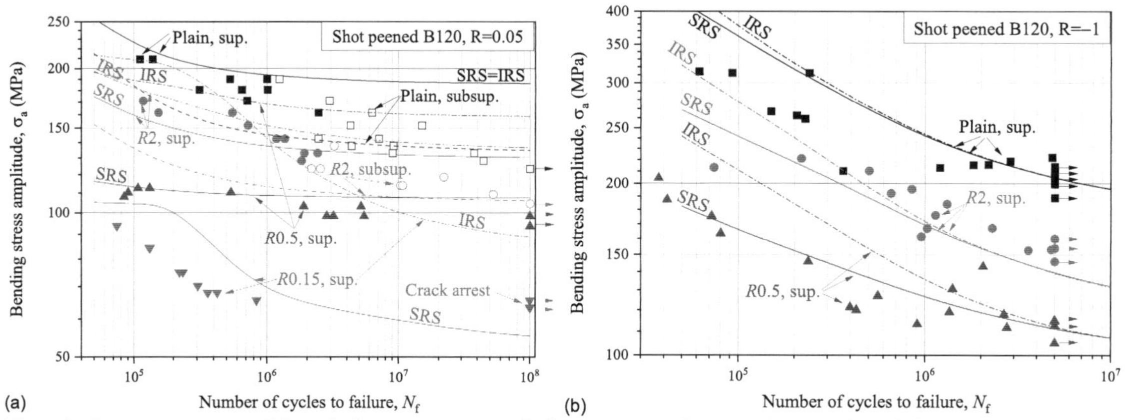

Fig. 4. Assessment of the fatigue curves of shot-peened samples under (a) pulsating and (b) fully-reversed bending fatigue. Solid and dash-dotted lines indicate fatigue predictions incorporating stabilized and initial RS. "Sup." and "subsup." refer to fatigue calculations done on or 0.1 mm below the tensioned frontal surface, respectively. "Crack arrest" denotes experimental evidences of non-propagating crack at the tip of the ultra-sharp-notched specimens (Fig. 5a). The points indicate to the experimental data.

Fig. 5. SEM micrographs of peened samples. (a) non-propagating crack nucleated at the tip of the ultra-sharp-notched specimen. (b)-(d) Fracture surface in the vicinity of the crack initiation site marked by a red arrow. (b) Superficial and sub-superficial (c) crack initiation in plain samples, (d) *R*2 blunt-notched. All samples were tested at R=0.05.

The sharp-notched (R0.5) variant is accurately assessed in the entire fatigue range, the ultra-sharp notched (R0.15) one is very well assessed in the central part of the SN diagram

($1\times10^6<N_f<1\times10^7$), while the medium and the very-high-cycle fatigue regimes are overestimated and underestimated, respectively. This underestimation can be ascribed to the presence of non-propagating cracks nucleated at the notch root in run-out specimens, as shown in the SEM micrograph reported in Fig. 5a. In other words, the devised fatigue criterion can accurately estimate the surface crack initiation in ultra-sharp notched specimens, but the intense RS stress field along with the pronounced external stress gradient ahead of the notch tip is able to arrest the crack after nucleation. In brief, notches of high severity favor crack initiation in the vicinity of the notch tip lying on the frontal tensioned surface of the specimen. Conversely, the plain and blunt-notched (R2) variants tested at R=0.05 presents a much more complicated situation. In the medium cycle fatigue regime ($Nf<1\times10^6$), fatigue cracks tend to nucleate in the vicinity of the tensioned surface, as shown in fractographs reported in Fig. 5b and 5d, respectively. Therefore, it is not surprising that assessments made according to the "sup." calculation method are in good agreement with the experimental data in this fatigue regime. In the HCF and mostly in the VHCF regime, this calculation method yields less and less accurate estimations of the fatigue strength. The experimental data progressively approaches the assessments done with the "subsup." SED-approach, thus suggesting that in this fatigue regime crack initiation is more prone to happen in the sub-superficial layers below that affected by compressive RS, as confirmed by fractographs depicted in Fig. 5c and d, for plane and blunt-notched specimens. This evidence highlights the importance of correctly identifying the crack initiation site for an accurate fatigue prognosis, given the tendency of shot peened components towards sub-superficial crack nucleation.

References

[1] S. Bagherifard, *Enhancing the Structural Performance of Lightweight Metals by Shot Peening*, Advanced Engineering Materials, Vol. 21 (2019), 1801140.

[2] F. Berto and P. Lazzarin, *Recent developments in brittle and quasi-brittle failure assessment of engineering materials by means of local approaches*, Mater. Sci. Eng R, Vol. 75 (2014), pp. 1–48.

[3] M. Benedetti, F. Berto, L. Le Bone and C. Santus, *A novel Strain-Energy-Density based fatigue criterion accounting for mean stress and plasticity effects on the medium-to-high-cycle uniaxial fatigue strength of plain and notched components*, Int J Fatigue, Vol. 133 (2020), 105397.

[4] M. Benedetti, V. Fontanari, P. Scardi, C.L.A. Ricardo and M. Bandini, *Reverse bending fatigue of shot peened 7075–T651 aluminium alloy: the role of residual stress relaxation*, Int J Fatigue, Vol. 31 (2009), pp. 1225–1236.

[5] M. Benedetti, V. Fontanari, C. Santus and M. Bandini, *Notch fatigue behaviour of shot peened high-strength aluminium alloys: experiments and predictions using a critical distance method*, Int J Fatigue, Vol. 32 (2010), pp. 1600–1611.

[6] M. Benedetti, V. Fontanari, M. Allahkarami, J.C. Hanan and M. Bandini, *On the combination of the critical distance theory with a multiaxial fatigue criterion for predicting the fatigue strength of notched and plain shot-peened parts*, Int J Fatigue, Vol. 93(2016), pp. 133–47.

[7] M. Benedetti, V. Fontanari, M. Bandini and E. Savio, *High- and very high-cycle plain fatigue resistance of shot peened high-strength aluminum alloys: the role of surface morphology*, Int J Fatigue, Vol. 70 (2015), pp. 451–462.

[8] B. Winiarski, M. Benedetti, V. Fontanari, M. Allahkarami, J.C. Hanan and P. J. Withers, *High spatial resolution evaluation of residual stresses in shot peened specimens containing sharp and blunt notches by micro-hole drilling, micro- slot cutting and micro-X-ray diffraction methods*, Exp Mech, Vol. 56 (2016), pp. 1449-1463.

[9] M. Benedetti, V. Fontanari, B. Winiarski, M. Allahkarami and J.C. Hanan, *Residual stresses reconstruction in shot peened specimens containing sharp and blunt notches by experimental measurements and finite element analysis*, Int J Fatigue, Vol. 87 (2016), pp. 102–111.

Cavitation Peening Using a Submerged Water Jet and a Pulsed Laser Comparing with Shot Peening

H. Soyama[1]

1 Department of Finemechanics, Tohoku University, Japan

Abstract

Improvement of fatigue strength treated by cavitation peening, in which cavitation impact at bubble collapse is utilized for peening, was demonstrated by using a plane bending fatigue test comparing with shot peening and non-peened. In the present paper, a submerged water jet and a pulsed laser were used for cavitation peening. Stainless steel SUS316L, aluminum alloy A2024-T3 and magnesium alloy AZ31B were treated by cavitation peening and shot peening. It was revealed that the best improvement of the fatigue strength comparing with non-peened was 25% for SUS316L by cavitation peening using the jet, 42% for A2024-T3 by cavitation peening using the laser and 55% for AZ31B by cavitation peening using the laser.

Keywords cavitation peening, submerged jet, pulsed laser, fatigue strength.

Introduction

Cavitation normally causes severe damage in hydraulic machineries such as pumps and valves because of intense impacts at bubble collapse. However, the impact at bubble collapse can be utilized for mechanical surface treatment in the same way of shot peening. A peening method using cavitation impact is named as "cavitation peening" [1-3].

At conventional cavitation peening, cavitation is produced by injecting a high-speed water jet into a water filled chamber through a nozzle, and cavitation is generated in a shear layer around the jet. The submerged high-speed jet with cavitation is called as "cavitating jet" [4]. Note that peening mechanism of cavitation peening is different from that of water jet peening, in which water column impacts in jet center are used. It was reported that efficiency of optimized cavitation peening was 3.2 times better than that of water jet peening [5]. In the case of cavitation peening using the submerged water jet, the optimum injection pressure is 40 MPa [5], then an expensive plunger pump is not required.

In the case of submerged laser peening, a bubble, which behaves like a cavitation bubble, is generated after laser ablation [6-8]. In the present paper, the bubble after the laser ablation was named as "laser cavitation". When amplitude of pressure wave in water was measured by a submerged shock wave sensor, the amplitude at laser ablation was larger than that of laser cavitation collapse [6-8]. On the other hand, when impact which propagated in the target metal was measured by a PVDF sensor, the impact of laser cavitation collapse was larger than that of laser ablation [7,8]. Namely, the submerged laser peening is a kind of cavitation peening using the pulsed laser.

When the fatigue poperies of stainless steel treated by cavitation peening were compared with that of shot peening, the fatigue strength of cavitation peening was larger than that of shot peening, and the relief of compressive residual stress introduced by cavitation peening during the fatigue test was smaller than that of shot peening [9]. At the equivalent peening intensity condition for stainless steel, the summation of compressive residual stress was nearly equivalent for cavitation peening and shot peening, however, the dislocation density of the sub-surface treated by cavitation peening was smaller than that of shot peening [10]. These differences might be caused by the impact characteristics such as strain speed and applied stress distribution of the impact [3]. Then, these effects would cause the differences on the fatigue properties.

In the present paper, in order to make clear the difference on improvement of fatigue properties by cavitation peening comparing with shot peening, stainless steel Japanese Industrial Standards JIS SUS316L, aluminum alloy JIS A2024-T3 and magnesium alloy JIS AZ31B were treated by cavitation peening using the submerged water jet and the pulsed laser and shot peening, then tested by a displacement-controlled plane bending fatigue test.

Experimental Methods

Figure 1 illustrates geometry of specimen. The thickness of specimen was 2 mm for SUS316L, 3 mm for A2024-T3 and 4 mm for AZ31B, respectively.

Figure 2 shows a schematic diagram of cavitation peening system using the submerged water jet. The water pressurized by a plunger pump was injected to the specimen, which was set in the recess, through a nozzle. The nozzle throat diameter d was 2 mm and it had an optimized outlet bore whose diameter D was 16 mm and the length L was 16 mm. It also had a cavitator whose diameter d_c was 3 mm and a guide pipe to enhance aggressive intensity of the cavitation impacts. The injection pressure was 30 MPa and the standoff distance was 222 mm.

Figure 3 reveals a schematic diagram of cavitation peening system using a pulsed laser. The used pulse laser source was a Q-switched Nd:YAG laser. The wave length was 1,064 nm, as laser cavitation impact was used for the peening at the present test to utilize heat effect caused by the pulse laser, as the laser cavitation is a kind of sub-cool boiling. Note that conventional submerged laser peening uses 2nd harmonics of Nd:YAG laser, i.e., 532 nm, to mitigate attenuation of the pulse energy due to water, and 40 % of source energy was lost at the wavelength conversion from 1,064 nm to 532 nm. The used pulsed laser energy was 0.35 J, the pulse width was 6ns, the beam diameter was 6 mm, the repetition frequency was 10 Hz. The pulsed laser was reflected by the mirrors, and expanded by a concave lens, then focused on the specimen, which was placed in a water filled glass chamber, by a convex lens to avoid the damage of the glass chamber. The standoff distances in air s_a and water s_w were optimized by measuring the peening intensity. The specimen was placed on the stage which was controlled by stepping motors.

Figure 4 shows the schematic diagram of shot peening system. In the preset experiment, a recirculating shot peening system accelerated by a water jet was used. Stainless steel shots, whose diameter were 3.2 mm were installed in the chamber. The number of shots was 500.

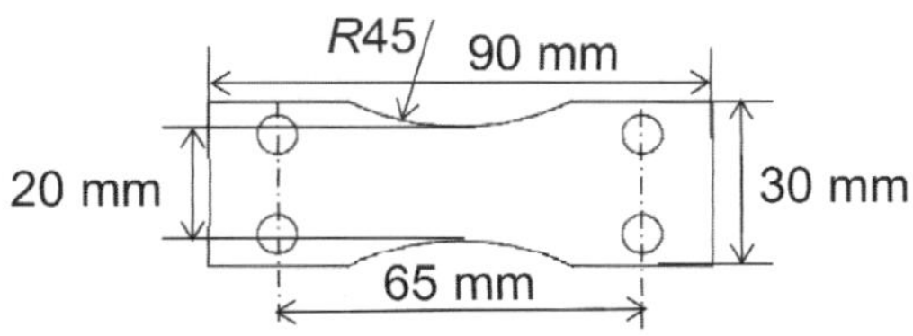

Fig. 1 Geometry of specimen

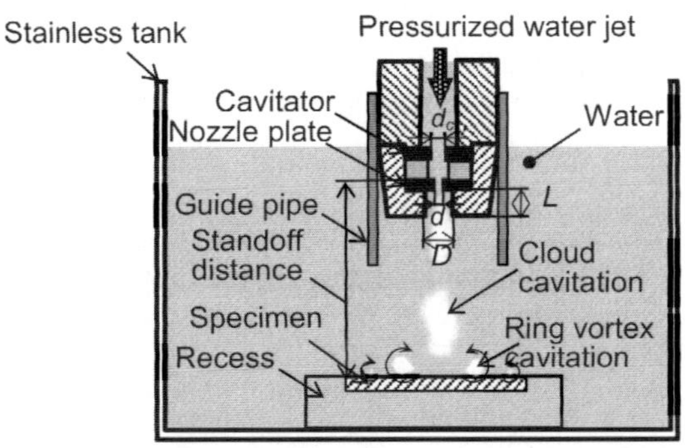

Fig. 2 Schematic diagram of cavitation peening system using submerged water jet

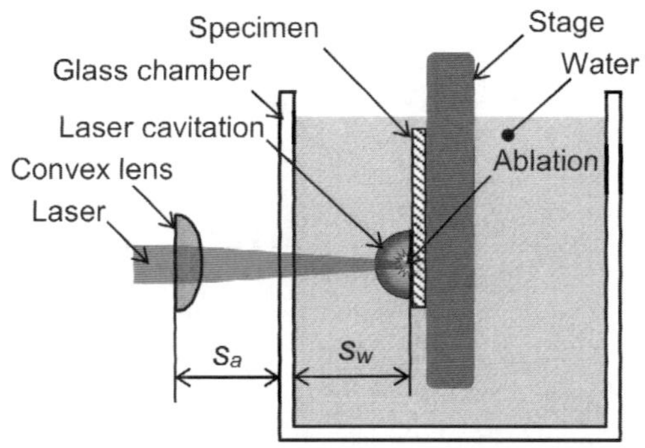

Fig. 3 Schematic diagram of cavitation peening system using pulsed laser

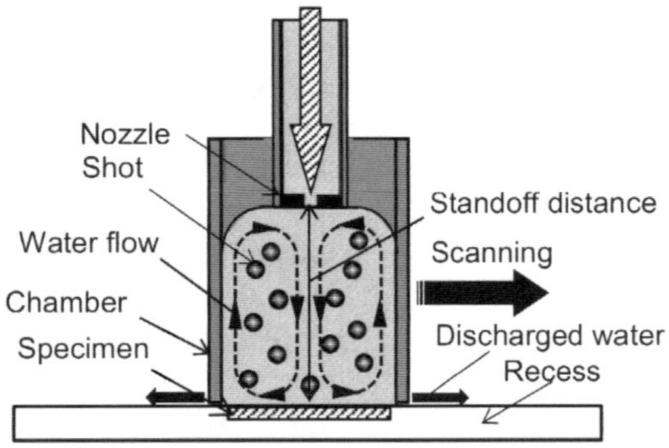

Fig. 4 Schematic diagram of shot peening system accelerated by water jet

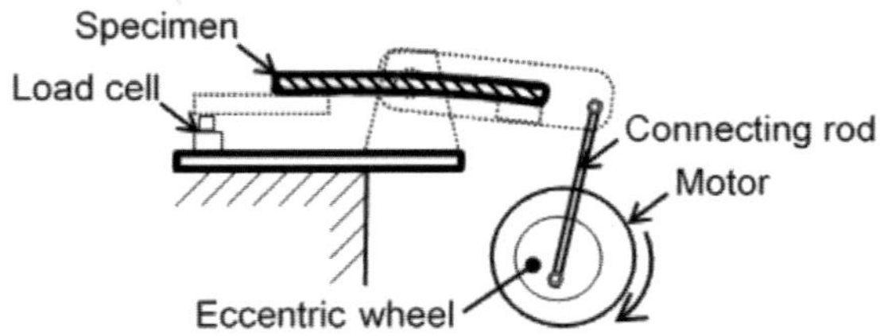

Fig. 5 Schematic diagram of a Schenk-type displacement-controlled plate bending fatigue tester

The standoff distance from the nozzle to the specimen surface was 50 mm. The water jet was injected into the chamber through three holes with a diameter of 0.8 mm. The injection pressure was 12-15 MPa. At the present condition, the water jet without shots did not introduce compressive residual stress into metallic surface.

Figure 5 illustrates a schematic diagram of a Schenk-type displacement-controlled plate bending fatigue tester. The fatigue properties of tested materials were evaluated at stress ratio R =-1. The test frequency was 12 Hz. In order to investigate mechanical properties of peened surface, the residual stress on the surface was evaluated by 2D method using X-ray diffraction [11]. The surface roughness and the surface hardness were also measured.

Experimental Results

In order to reveal effect of impact at laser cavitation collapse comparing with laser ablation, Fig. 6 shows the aspect of laser ablation and laser cavitation observed by a high-speed video, and Fig. 7 reveals signal from the PVDF sensor and the submerged shock wave sensor [7]. After the laser ablation, laser cavitation was developed and it was collapsed at t = 1 ms. The amplitude of laser ablation was larger than that of laser cavitation, when the amplitude of pressure was measured by the submerged shock wave sensor. On the other hand, when the impact passing through the target metal was measured by the PVDF sensor, the impact at

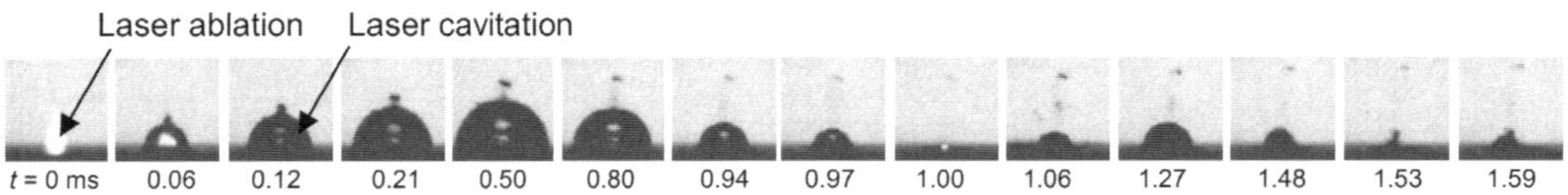

Fig. 6 Aspect of laser ablation and laser cavitation

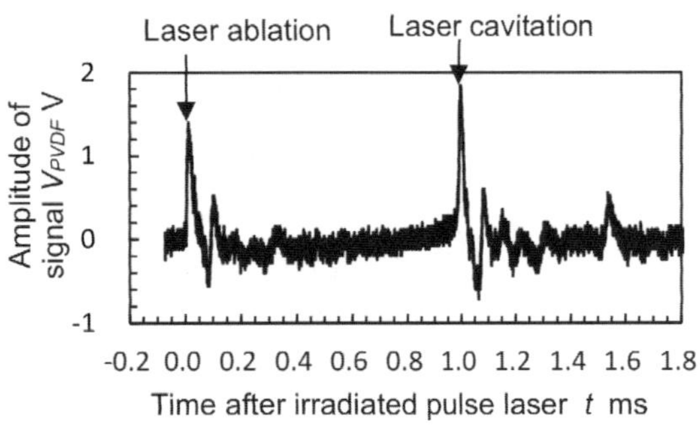

(a) Signal from PVDF sensor

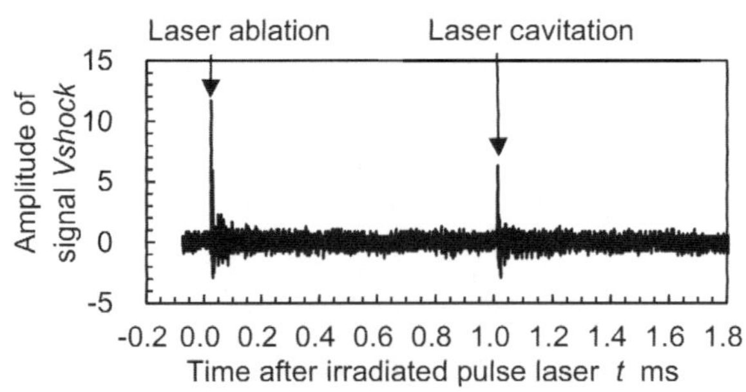

(b) Signal from submerged shockwave sensor

Fig. 7 Signals of laser ablation and laser cavitation

the laser cavitation collapse was larger than that of the laser ablation. Namely, the laser cavitation impact can be utilized for the peening.

In order to find optimum coverage for cavitation peening and shot peening, Fig. 8 shows the number of cycles to failure at constant amplitude of bending stress as a function of coverage *Cov* for (a) SUS 316L, (b) A2024-T3 and (c) AZ31B. The used pulse density or processing time per unit length to calculate coverage for each materials and peening method are shown in Table 1. As shown in Fig. 8, the number of failures at constant applied stress N_f was increased with *Cov*, and then saturated, except AZ31B treated by CP by laser. Although AZ31B was relatively soft material, the fatigue life was still increasing at 14 pulse/mm^2.

In order to demonstrate the improvement of fatigue properties of metallic materials by cavitation peening using the laser and the jet comparing with shot peening, Figs. 9 -11 reveal the *S-N* curves for SUS316L, A2024-T3 and AZ31B obtained by the plane bending fatigue test. Table 2 shows hardness of non-peed one and fatigue strength obtained by Little's method [12]. In Figs.9-11, the amplitude of bending stress was normalized by the fatigue strength of non-peened which was shown in Table 2.

As shown in Fig. 9, at bending stress σ_a = 400 MPa, the fatigue life of shot peening was better than that of cavitation peening using the jet. However, the fatigue strength of cavitation peening using the jet was larger than that of shot peening. In the case of AZ31B, $\sigma_a \approx$ 140 MPa, the fatigue life of shot peening was better than that of cavitation peening using the jet, and the fatigue strength of cavitation peening using the jet was slightly larger than that of shot peening, as shown in Fig. 11. In the case of A2024-T3, the fatigue life of shot peening at $\sigma_a \approx$ 280 MPa was better than cavitation peening using the laser, but the fatigue strength of cavitation peening using the laser was larger than that of shot peening. These might be caused by the difference on the increase of surface roughness by the peening methods.

In the case of the fatigue strength of SUS316L, cavitation peening using the jet was best, and it improved 25% comparing with non-peened. In the case of A2024-T3 and AZ31B, the improvement by cavitation peening using the laser was best, and it improved 42% for A2024-T3 and 55% for AZ31B.

Table 1. Used Pulse Density or Processing Time per Unit Length to Calculate Coverage.

Material	Peening method			
	Cavitation peening by laser	Cavitation peening by jet	Water jet peening	Shot peening
SUS316L	4 pulse/mm^2	8 s/mm	8 s/mm	0.88 s/mm
A2024-T3	10 pulse/mm^2	8 s/mm	—	0.9 s/mm
AZ31B	14 pulse/mm^2	16 s/mm	—	1 s/mm

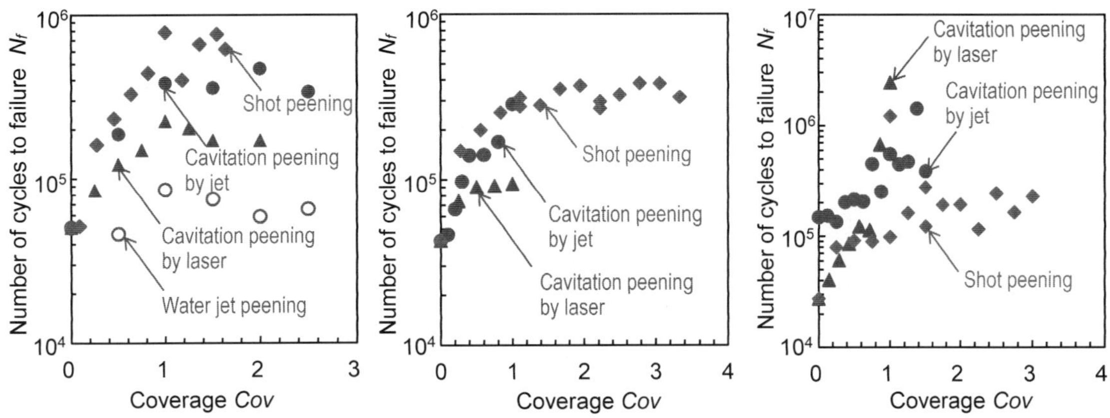

(a) SUS316L (b) A2024-T3 (c) AZ31B

Fig. 8 Fatigue life as a function of coverage

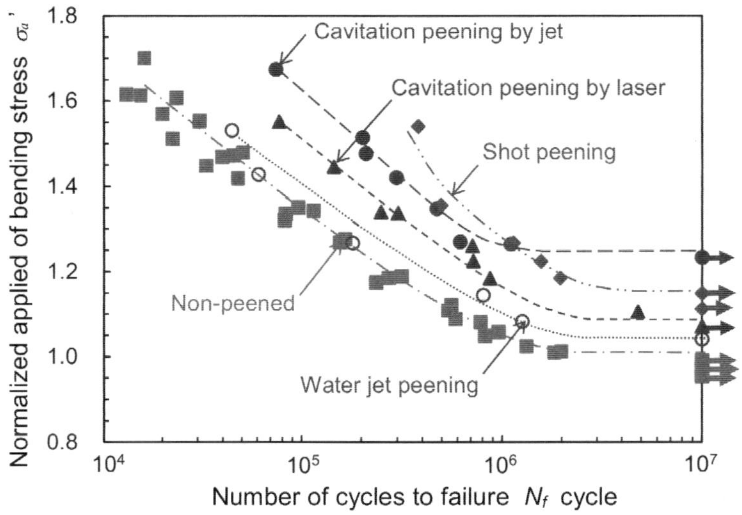

Fig. 9 *S-N* curves (SUS316L)

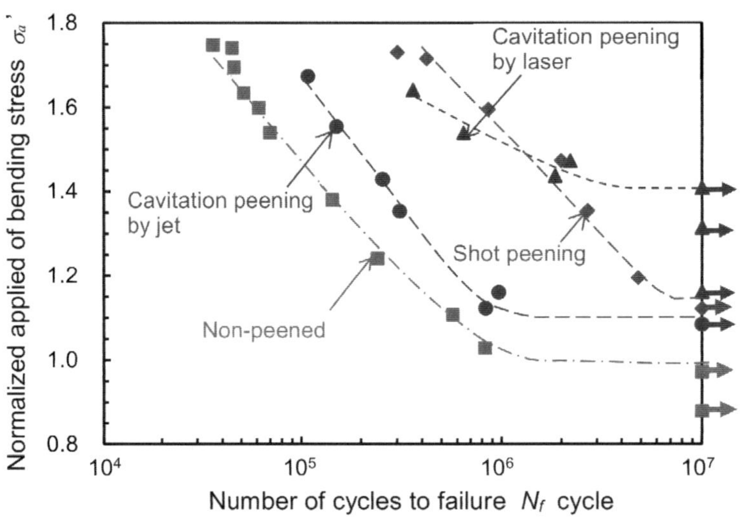

Fig. 10 *S-N* curves (A2024-T3)

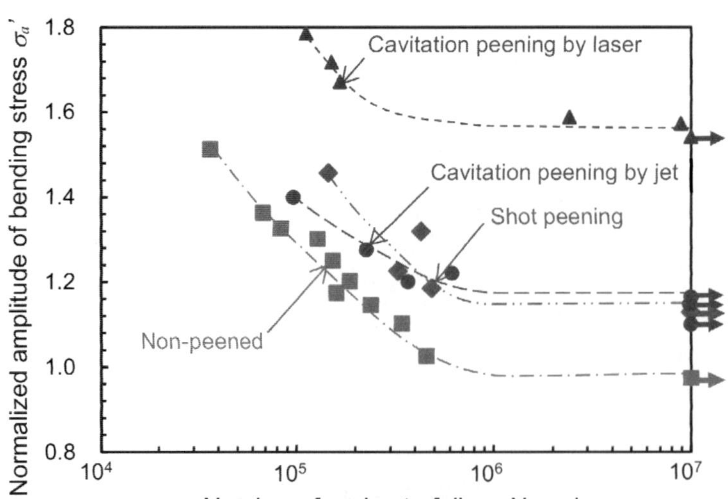

Fig. 11 *S-N* curves (AZ31B)

Table 2. Improvement of Fatigue Strength of Tested Materials.

Material	Yield strength [MPa]	Hardness		Fatigue strength [MPa]			
		HR15N	HR15T	NP	CP by laser	CP by jet	SP
AZ31B	-	38.6±0.6	74.3±0.4	97±3	151±2	115±2	112±3
A2024-T3	311	63.3±0.3	87.0±0.2	175±5	248±3	193±3	202±6
SUS316L	216	63.2±0.7	86.6±0.2	279±3	303±5	348±5	325±5

Conclusions

In order to investigate the fatigue properties of metallic materials treated by cavitation peening using the pulsed laser and the submerged water jet comparing with shot peening, stainless steel SUS316L, aluminum alloy A2024-T3 and magnesium alloy AZ31B were treated, and tested by the plane bending fatigue test. It was revealed that the fatigue strength of AZ31B and A2024-T3 treated by cavitation peening by the laser was better than shot peened one, and that of SUS316L treated by the jet was better than shot peened one.

Acknowledgments

This work was partly supported by JSPS KAKENHI, grant numbers 20H02021 and The Amada Foundation AF-2021219-B3.

References

[1] H. Soyama, *Key factors and applications of cavitation peening*, International Journal of Peening Science and Technology, Vol. 1 (2017), pp 3-60.
[2] H. Soyama, *Cavitation Peening: A Review*, Metals, Vol. 10 (2020), article No. 270.
[3] H. Soyama and A.M. Korsunsky, *A critical comparative review of cavitation peening and other surface peening methods*, Journal of Materials Processing Technology, Vol. 305 (2022), article No. 117586.
[4] H. Soyama, *Cavitating Jet: A Review*, Applied Sciences, Vol. 10 (2020), article No. 7280.
[5] H. Kamisaka and H. Soyama, *Effect of injection pressure on mechanical surface treatment using a submerged water jet*, Journal of Jet Flow Engineering, Vol. 33, No. 2 (2018), pp 4-10.
[6] A. Sasoh, K. Watanabe, Y. Sano and N. Mukai, *Behavior of bubbles induced by the interaction of a laser pulse with a metal plate in water*, Applied Physics, Vol. A 80 (2005), pp 1497-1500.
[7] H. Soyama, *Comparison between the improvements made to the fatigue strength of stainless steel by cavitation peening, water jet peening, shot peening and laser peening,* Journal of Materials Processing Technology, Vol. 269 (2019), pp 65-78.
[8] H. Soyama, *Laser cavitation peening and its application for improving the fatigue strength of welded parts*, Metals, Vol. 11 (2021), article No. 531.
[9] H. Soyama, Christopher R. Chighizola and Michael R. Hill, *Effect of compressive residual stress introduced by cavitation peening and shot peening on the improvement of fatigue strength of stainless steel*, Journal of Materials Processing Technology, Vol. 288 (2021), article No. 116877.
[10] M. Kumagai, M.E. Curd, H. Soyama, T. Ungar, G. Ribarik and P.J. Withers, *Depth-profiling of residual stress and microstructure for austenitic stainless steel surface treated by cavitation, shot and laser peening*, Materials Science and Engineering A, Vol. 813 (2021), article No. 141037.
[11] H. Soyama, C. Kuji, T. Kuriyagawa, C.R. Chighizola and M.R. Hill, *Optimization of residual stress measurement conditions for a 2D method using X-ray diffraction and its application for stainless steel treated by laser cavitation peening*, Materials, Vol. 14 (2021), article No. 2772.
[12] R.E. Little, *Estimating the median fatigue limit for very small up-and-down quantal response tests and for S-N data with runouts*, ASTM STP, Vol. 511, pp 29-42.

New Electronic Peening Intensity Sensor: Theory and Experimental Validation

Sylvain A. Forgues[1], Brigitte Labelle[1], Ramzi Ben Moussa[1], Adel Alouani[1]

[1] Shockform Aeronautique Inc., Canada, engineering@shockform.com

Keywords: Shot *Peening, Electronic Intensity Sensor, Theory, Experimental Validation*

Abstract

The method to measure peening intensity has not evolved significantly in decades. The current method is time intensive as it requires peening of several metal strips to different times in order to generate a saturation curve. Each strip must be pre-measured, installed on a holder, peened for a specific amount of time, removed from the holder, measured once more and the value recorded manually. These recorded values must then be made into a graph and the intensity calculated using what is called the "10% rule". All these steps to find intensity make peening a complicated process that is confusing for many operators.

A recently developed wireless sensor offers the possibility to simplify intensity calculations with real-time process measurement and continuous transfer and analysis of data. This sensor is part of a new trend known as the Internet of Things (IoT) and promises to allow companies to progress into the latest industrial revolution called Industry 4.0.

Objectives

The objective of the paper is to introduce the theory and experimental validation behind this new electronic peening intensity sensor. The concept of impact force will be presented and discussed as an alternative to Almen intensity for peening

Sensor Description

The sensor's characteristics and properties are first presented to provide a good understanding of its functions and capabilities. The sensors' exterior is illustrated in Figure 1. It features a load cell pin, a protective cover, a metal base with an on/off button and five screw locations for permanent installation.

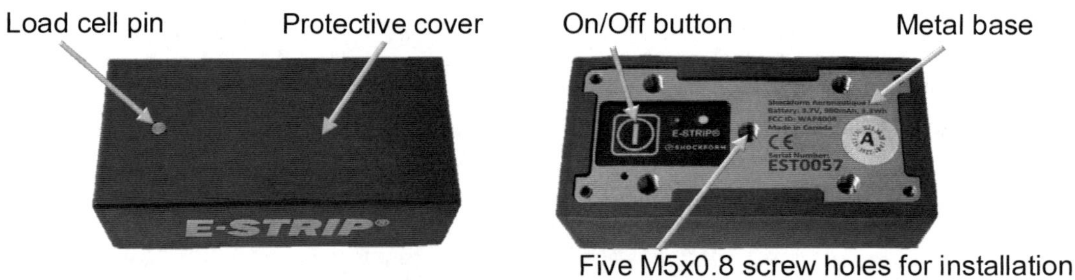

Figure 1 Sensor Features

The sensor has the size and screw pattern of an SAE J442 [1] Almen strip holder. This allows for easy installation on most jigs and fixtures currently used for peening intensity determination and verification. The load cell pin is made of tungsten carbide for durability and receives the impacts from the peening media. The sensor has a protective cover that absorbs the impact energy of the media and protects the internal components of the device. This protective cover can be replaced if necessary but has shown little wear from all the testing performed so far.

Each impact on the load cell is recorded and sorted and only a few hundred impacts are necessary to determine the intensity. Once the peening is completed, an equivalent intensity is calculated by the sensor using a proprietary algorithm. The equivalent intensity and the number of impacts are sent wirelessly to a computer where it is displayed in the form of a histogram (Figure 2). The sensor offers a huge time saving since no human interaction is required to load, unload and measure Almen strips to determine or verify intensity.

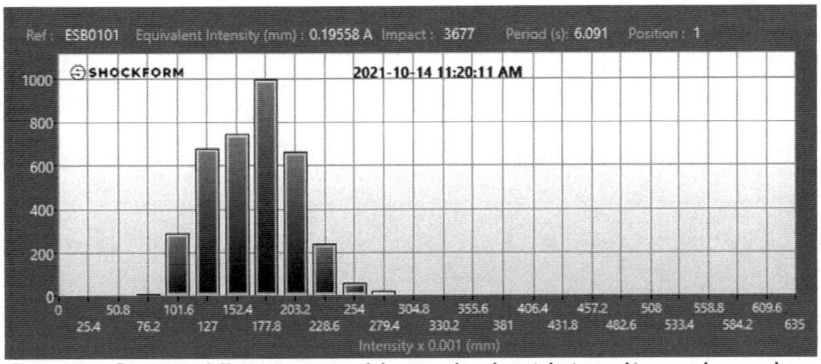

Figure 2 Sensor Output Histogram with equivalent intensity and number of impacts

Theory

The theoretical equations governing the utilisation of the sensor is found in most basic physics handbooks [2]. It is repeated here for completeness.

Consider a peening particle of mass m moving in a xy plane at speed $\vec{v}$ and acted on by a force $\vec{F}$ during a period of time t. From Newton's second law, the force acting on the particle can be defined as follows:

$$\vec{F} = m\frac{d\vec{v}}{dt} \quad (1)$$

By rearranging, we get

$$\int_{t1}^{t2} \vec{F}\,dt = \int_{v1}^{v2} m\,d\vec{v} \quad (2)$$

The Impulse–Momentum principle is derived from equation (2) with impulse on the left and change of momentum on the right of the equation. It states that the Impulse defined as the force acting on a particle over a period of time is equal to the change in momentum of the particle defined as the mass multiplied by the change in velocity of the particle.

In our application, the peening particle is hitting the load cell pin. The particle has a velocity v_{in} toward the sensor just before the impact and a velocity v_{out} away from the sensor just after the impact. We select v_{in} with a negative sign and v_{out} with a positive sign since they are in opposite directions. Since the mass of the particle is assumed constant during the impact, the above equation then becomes:

$$\int_{t1}^{t2} \vec{F}\,dt = m\,(v_{out} - (-v_{in})) \quad (3)$$

or

$$\int_{t1}^{t2} \vec{F}\,dt = m\,(v_{out} + v_{in}) \quad (4)$$

A relation exists between v_{in} and v_{out} known as the coefficient of restitution.

$$C_R = \frac{v_{out}}{v_{in}} \quad (5)$$

The coefficient of restitution is defined as the ratio of the velocity after impact to the velocity before impact. Coefficients of restitution depend on factors including material properties, body geometry, and impact velocity. During peening some kinetic energy is dissipated during the impact and 0 < C_R < 1.

Equation (4) can be re-written as

$$\int_{t1}^{t2} \vec{F} dt = m\,(C_R + 1)\, v_{in} \quad (6)$$

From testing, and from the literature [3-4], we also know that there is a direct relationship between shot velocity before impact and peening intensity.

$$v_{in} = K_B\; I_A \quad (7)$$

where I_A is the Almen intensity and K_B is a factor associated with particle properties.

Replacing equation (7) in equation (6) gives us a direct relation between the impulse and the peening intensity:

$$\int_{t1}^{t2} \vec{F} dt = m\,(C_R + 1)\, K_B\; I_A \quad (8)$$

where the mass m, the coefficient of restitution C_R and the factor K_B are properties of the peening particle involved in the impact of the peening process.

The integral on the left side of equation (8) can be evaluated when the force applied on the particle is known as a function of time. During impact, this interaction can be simplified as illustrated in Figure 3. As the particle hits the sensor, the force gradually increases as the velocity decreases until the force reaches a maximum value F_{max}. The force then gradually decreases as the particle departs in the other direction until the force reaches zero.

The area under the curve in Figure 3 is equal to the impulse of the impact. In this case, it can be calculated as the area of a triangle with:

$$F\Delta t = A = \frac{1}{2}(base)(height) \quad (9)$$

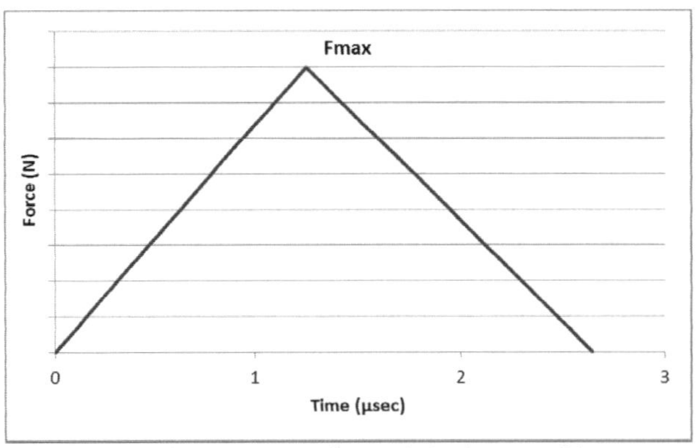

Figure 3 Graphical illustration of particle force of impact over time

Where the base of triangle is the duration of the impact and the height of the triangle is the maximum force of impact F_{max}. Equation (9) then becomes:

$$F\Delta t = A = \frac{1}{2}(t_{impact})(F_{max}) \tag{10}$$

Merging equations (10) and (6) we get a relationship between the maximum impact force and the velocity before impact.

$$\frac{1}{2}(t_{impact})(F_{max}) = m\,(C_R + 1)\,v_{in} \tag{11}$$

Rearranging, the maximum impact force can be calculated as follows:

$$F_{max} = \frac{2\,m\,(C_R + 1)\,v_{in}}{t_{impact}} \tag{12}$$

The max force F_{max} can be calculated since the mass m, the coefficient of restitution C_R, the initial velocity v_{in} and the time of impact t_{impact} can be measured. Merging equation (12) and (7), we see that F_{max} has a direct relationship with Almen Intensity I_A.

$$F_{max} = \frac{2\,m\,(C_R + 1)\,K_B\,I_A}{t_{impact}} \tag{13}$$

For our sensor, we believe the impulse is measured directly by the load cell. This will be confirmed if a linear relationship exists between the sensor output and Almen intensity. We also believe the Almen intensity can be represented very simply by the maximum impact force. Both of these concepts will be illustrated in the experimental validation.

Experimental Validation and Discussion

The experimental validation is performed with ACWR28 media meeting specification AMS2431/3 [5]. The average mass of one particle of media was determined by weighing 2000 particles. The average mass of one particle was found to be 1.85E-06 kg. This is within the 1.80E-06 to 2.50E-06 kg range specified in the specification.

A special apparatus was developed to calibrate the sensor. This apparatus shoots one or multiple particles at the sensor while measuring the velocity v_{in} of the particles just before impact. Since the sensor allows the measurement and recording of each impact, it is possible to see the change in output of the sensor over a period of time as illustrated in Figure 4. The area under the curve represents the impulse of the impact where the maximum value of the sensor is F_{max} and the impulse time is t_{impact} as per equation (10).

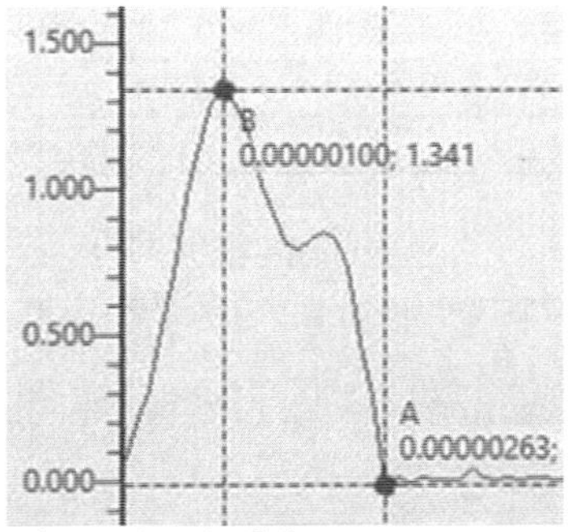

Figure 4 Change in output of the sensor over a period of time

The apparatus also allows the peening of Almen strips for the same set of initial velocities. It is then possible to plot sensor output in function of Almen intensity as illustrated by Figure 5. The graph confirms the linear relationship between the sensor output representing the impulse of the impact and the peening intensity as expressed in equation (8).

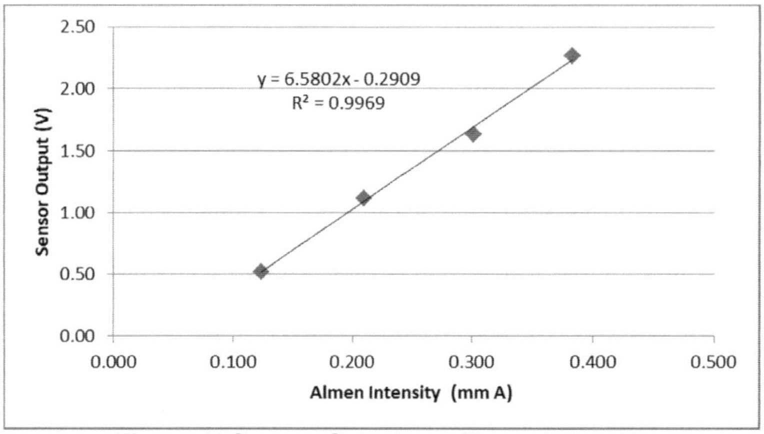

Figure 5 Sensor Output vs. Almen Intensity

Particle velocity before and after impact was determined using another apparatus developed internally and inspired from the work in Reference [6]. This apparatus uses a camera, LED lighting and an image analysis software to measure the velocity of the particle before and after impact with the sensor pin. A typical image is shown in Figure 6 where the velocity before impact is expressed in meters per second. The image for the velocity after impact is analysed separately.

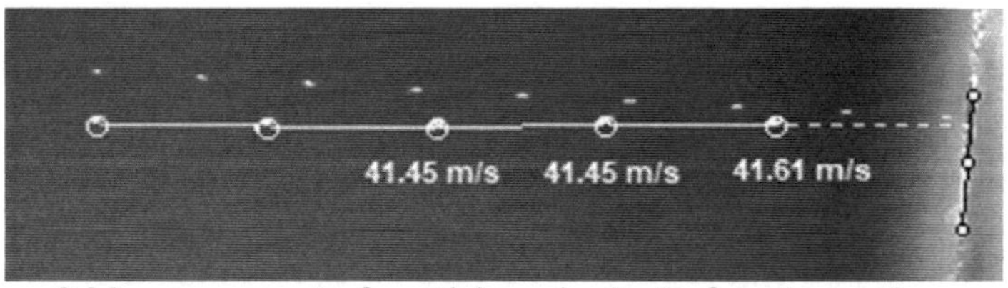

Figure 6 Measurement of particle velocity before impact on sensor

Knowing the particle mass, the velocity before impact, the coefficient of restitution and the impulse time taken from the sensor, it is possible to calculate the maximum force of impact F_{max} using equation (12). Calculations are presented in Table 1.

Table 1 Average data for ten measurements taken at four different initial velocities

Initial Speed Vin (m/s)	Final Speed Vout (m/s)	Coefficient Restitution	Sensor Max Output (V)	Calculated Fmax (N)	Intensity (mm A)
13.4	7.7	0.57	0.522	30.2	0.124
25.4	12.6	0.50	1.115	54.4	0.209
39.4	16.9	0.43	1.632	80.7	0.301
52.8	20.7	0.39	2.264	105.2	0.384

The coefficient of restitution changes with initial velocity since more energy is lost for impacts at higher velocities than at lower velocities [7]. However, the time of impact does not change with velocity. It is believed to be a function of particle material and size. For AWCR28 media, the average time of impact on the sensor was found to be 2.59E-06 seconds.

The data in the table shows that a particle of AWCR28 produces an impact force of 30 N for a peening intensity of 0.124 mmA. That force increases to 105 N for a peening intensity of 0.384 mmA. The relation between F_{max} and Almen intensity is shown in Figure 7.

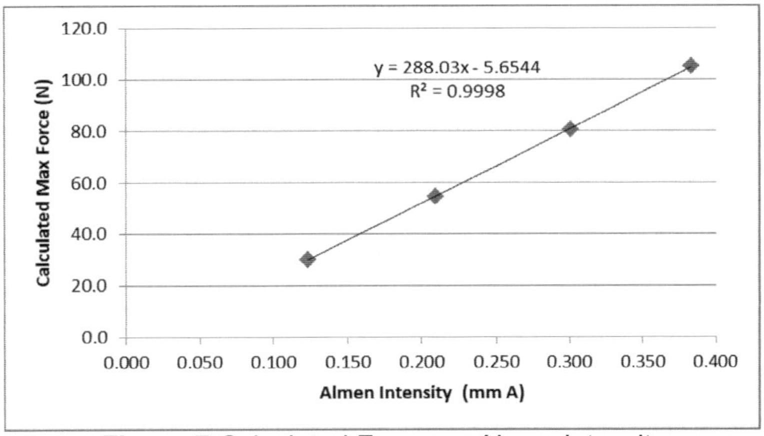

Figure 7 Calculated Fmax vs. Almen Intensity

The graph shows a nearly perfect linear relationship. Using the maximum impact force to define peening intensity would greatly facilitate the understanding of the process. The maximum impact force can be illustrated by a hammer hitting a surface. A higher impact force will generate a deeper level of compressive stress. This is much simpler to understand than the relation between arc height of a metal strip and Almen intensity. Especially since the peening time on the metal strip is important when generating the saturation curve. This leads to much confusion for many people in the industry.

Conclusion

A new electronic peening intensity sensor is presented in this paper. The sensor development and usage are backed by fundamental physics equations and illustrated by experimental data. The data shows that the output of the sensor correlates very well with Almen intensity. The sensors could therefore be a good alternative to Almen strips. The sensor output can provide a measure of maximum impact force which was also shown to have a nearly perfect linear relationship with Almen intensity. The use of maximum impact force to quantify the peening intensity offers the possibility to greatly facilitate and improve the understanding of the peening process. Testing of the force-intensity relationship for other types, size and hardness of media is under way and will be presented in future technical papers.

References

[1] SAE J442 Test Strip, Holder, And Gage For Shot Peening

[2] Sears, Zemansky, University Physics, 1970, Addison-Wesley Publishing Company

[3] Barker, Young, Pouliot, Particle Velocity Sensor for Improving Shot Peening Process Control, 2005, Conf Proc: ICSP-9 (pg 385-391)

[4] Könitzer, Polanetzki, Implementation of velocity measurement as intensity verification in Production, 2011, Conf Proc ICSP-11(pg 19-24)

[5] SAE AMS 2431/3 Peening Media - Conditioned Carbon Steel Cut Wire Shot, Regular Hardness (45 to 52 HRC) (AWCR)

[6] Badreddine, J., Modélisation du grenaillage ultrason pour des pièces à géométrie complexe, 2014, Thèse de doctorat de l'UTT

[7] Yardeny, Portnikov and Kalman, Experimental investigation of the coefficient of restitution of particles colliding with surfaces in air and waterVelocity vs Almen intensity relationship. Advanced Powder Technology Volume 31, Issue 9, September 2020, Pages 3747-3759

New Electronic Peening Intensity Sensor: Comparison with Almen Arc Height Verification

Sylvain A. Forgues [1], Brigitte Labelle[1], Adel Alouani[1], Ramzi Ben Moussa[1]
[1]Shockform Aeronautique Inc., Canada, engineering@shockform.com

Keywords: *Shot Peening, Electronic Intensity Sensor, Intensity Verification,* Intensity Histograms

Introduction

Almen strips are often used for peening intensity determination and verification. For intensity verification, one must measure the pre-bow of the Almen strip, install the strip on the holder, peen the strip for a period of time, remove the strip from the holder and measure the arc height of the strip with a dedicated gage. For intensity determination, these steps must be repeated at least four different times to generate a saturation curve. Intensity determination and verification can account for 10% to 30% of machine usage.

A new electronic peening intensity sensor was recently developed to help provide an easier, faster and better method for tracking the peening process. This wireless sensor offers the possibility for near real-time peening intensity measurement and verification. It also provides new insight into the process by counting the number of impacts on the sensor and showing the force of impacts with an intensity histogram. All data generated is digital which facilitates feedback for machine control as well as data analysis and transfer for process tracking and storage.

Objectives

The objective of the paper is to show the use of an electronic peening intensity sensor for daily machine parameter verification in comparison with the Almen strip arc height measurements. The measurements from both methods are illustrated and the variability over a 31 day period is discussed.

Sensor Description

The electronic peening intensity sensor was custom designed for peening applications [1]. The sensor is illustrated in Figure 1. It features a load cell pin, a protective cover and a metal base with five screw locations for installation. It has the size and screw pattern of a SAE J442 [2] Almen strip holder to easily be installed on most jigs and fixtures.

Figure 1 Sensor Exterior and Back

The pin of the load cell is made of tungsten carbide for durability. It is very small and is the only part of the sensor that needs to be subjected to the peening media stream. It is thus possible to design fixtures for the sensor that ensures realistic impact scenarios and representative peening intensity measurements even in difficult to reach locations such as gear roots or turbine disc fir tree slots (Figure 2). The rest of the sensor has a protective cover that absorbs the impact energy of the media and protects the internal components of the device. This protective cover can be replaced if necessary.

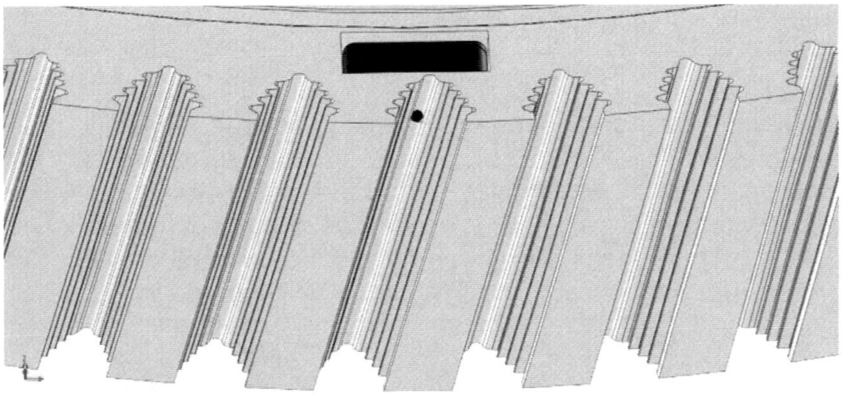

Figure 2 Example of a sensor usage in a difficult to reach location

The sensor records individual impacts during the peening process when media particles hit the load cell. Only a few hundred impacts are necessary to determine the intensity. Each impact has a different level of energy because of the size, hardness and velocity of the media particle. The impacts are recorded and sorted by a processor and a proprietary algorithm is used to calculate an equivalent Almen intensity. The sensor output is transferred via wireless communication to custom software on a computer. The software can accept data from an unlimited number of sensors. The software shows data from each sensor such as the calculated equivalent intensity, the battery level and the strength of the wireless communication (Figure 3).

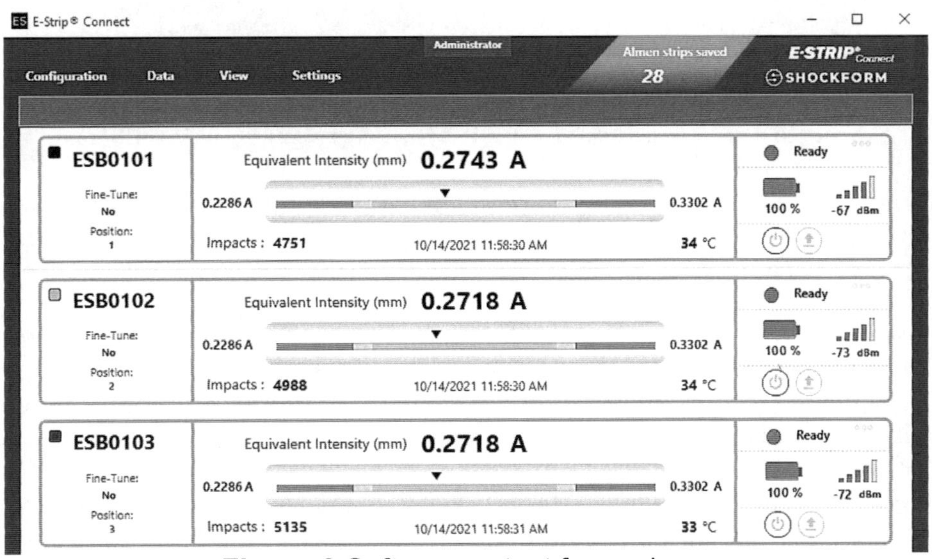

Figure 3 Software output for each sensor

Because the intensity is easy to measure and requires no human manipulation, it becomes feasible to perform measurements each time the machine operates. This data can be recorded in a database for quality tracking and could eventually be used in a feedback loop to adjust peening parameters in real-time.

The sensor also provides new insight into the peening process by counting the number of impacts on the sensor and showing the distribution of impacts with an intensity histogram. The number of impacts can be associated to the media flow rate, while the distribution of impacts illustrates the quality of the peening stream. A change in number of impacts or a change in the shape of distribution and equivalent intensity are signs the process is no longer stable. More insight into these functions will be discussed in future papers.

Experimental Method

To show the use of an electronic peening intensity sensor for daily machine parameter verification, five sensors and five Almen strips were placed alternatively on the turn table of a SmartPeen® peening machine. Peening was performed at five pressures 0.55, .83, 1.10, 1.38, 1.65 bar at a media flow rate of 9.1 kg/min using AWCR28 cut wire media meeting specification SAE AMS 2431/3D [3]. This machine features full control and recording of the pressure and media flow rate for accurate process control. The setup is illustrated in Figure 4.

Figure 4 Sensors and Almen Strips in Peening Machine

A 9.52mm internal diameter nozzle was used for peening. It was fixed perpendicular to the turn table at a distance of 152 mm. The turn table was rotated at 4 rpm which translates to a tangential velocity of 4775 mm/min.

Almen strips were type A-1 sourced from Reference [4]. Pre-bow compensation was performed on all strips before measurements. Almen intensities were determined using the Shockform Saturation Curve Solver (SSCS) [5] with five data points, the three parameter curve fit equation and the 10% rule as per SAE J443 [6].

On day one, the Almen intensity at five different pressures was determined along with the output for the five sensors. The sensor output was fine tuned to match the recorded Almen intensities. These values served as the baseline reference for the measurements taken the following days. Arc height values were then recorded every day for 30 additional days along with the sensors equivalent intensity. The variability of the arc height readings is compared with the variability of sensors equivalent intensity.

Results and Discussion

The output of the sensors is called “Equivalent Intensity” to differentiate from “Almen Intensity”. The relationship between the Almen intensity and the sensor equivalent intensity after calibration is illustrated in Figure 5. The sensors provide a good representation of Almen intensities. If the sensor output perfectly matched the Almen intensity, all the point would lie directly on the line. In this case, the goodness-of-fit measure R-squared has a value of 0.9952.

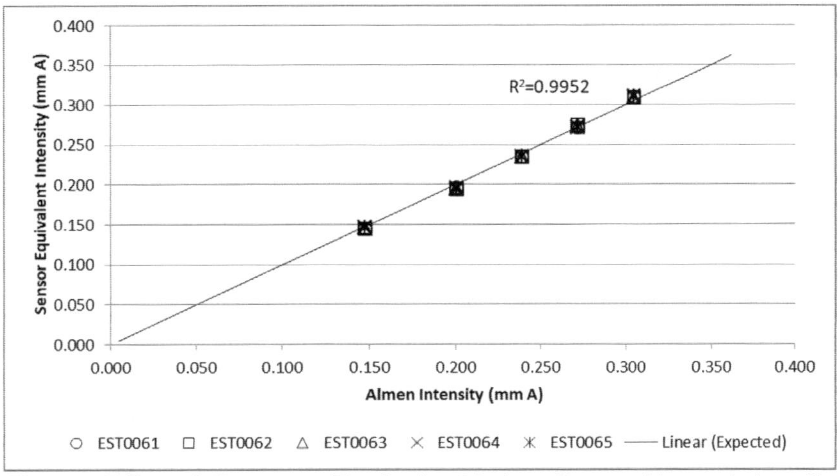

Figure 5 Day 1 Sensors Equivalent Intensity vs. Almen Intensity

The saturation curves obtained at the different pressures were used to find the target arc heights. These target arc height are the reference for the process verification performed the following 30 days. The target arc heights were determined at 4 passes of the turn table which is at or slightly after the number of passes where intensity was calculated.

This illustrates a major complication with Almen strips. Almen intensity cannot be measured with a single strip. Peening at different times on at least four Almen strips is required to generate a saturation curve. Since generating saturation curves is very time consuming, process verification is usually performed by measuring the arc height of a single strip at a pre-determined time. Many people are confused by the relationship between Almen intensity and target arc height for process verification.

For the sensors, the output equivalent intensity from day 1 was used as the reference for the verification performed the following days. The sensors measure intensity directly each time they are used. Target arc height values for the Almen intensity and target equivalent intensities for the sensors are shown in Table 1.

Table 1 Target Almen Arc Heights and Target Equivalent Intensities on Day 1

Pressure (bar)	Almen Intensity (mm A)	Target Arc Height (mm)	Target Equivalent Intensity (mm A)				
			EST0061	EST0062	EST0063	EST0064	EST0065
0.55	0.147	0.145	0.145	0.145	0.145	0.146	0.147
0.83	0.201	0.201	0.196	0.193	0.195	0.196	0.194
1.10	0.239	0.251	0.235	0.234	0.234	0.234	0.234
1.38	0.272	0.295	0.271	0.274	0.273	0.272	0.271
1.65	0.305	0.333	0.310	0.309	0.309	0.309	0.311

The variation of Almen arc height and sensor equivalent intensity was studied over a period of 31 days. Almen arc heights and sensor equivalent intensity were measured each day at the five different pressures and compared with the baseline values shown in Table 1. A total of 155 Almen strips were required for the study. The same five sensors were reused every day. Figure 6 illustrates a histogram of the 31 measurements at 0.55 bar pressure for the Almen strips and sensor EST0062. The histogram shows that the Almen strips have more variability than the sensor with several very low measurements compared with the target value. The sensor shows a tighter distribution closer to the target equivalent intensity.

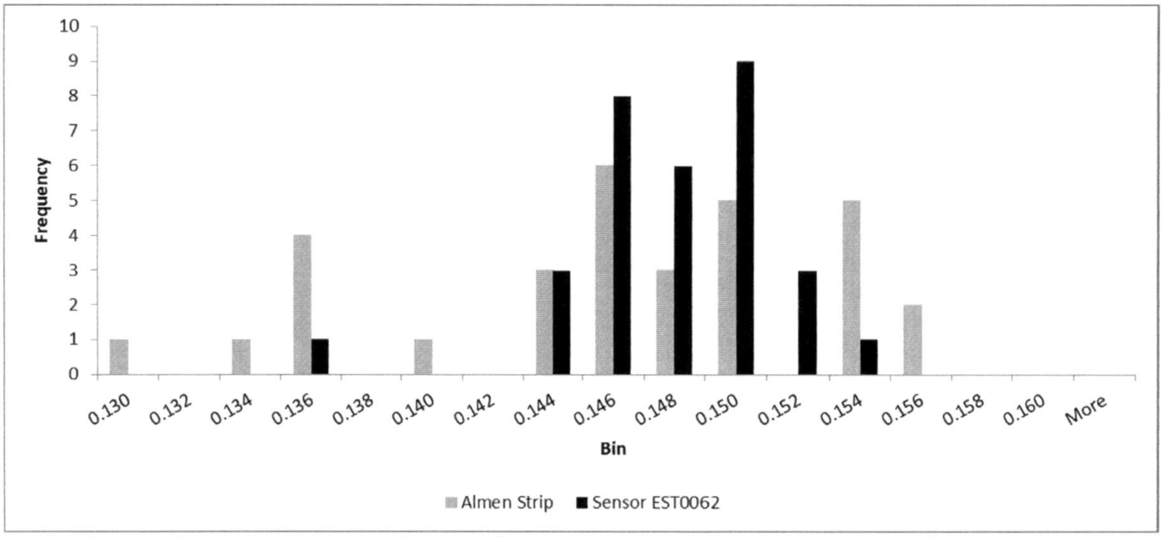

Figure 6 Almen Arc Height and Sensor Equivalent Intensity Variation over 31 days

The average of the measured values along with the standard deviation for the 31 days is presented in Table 2. The data shows that the standard deviation for both the Almen strips and the sensors is not affected by the different pressures. The variability seems constant at the different intensities. Overall, each sensor seems to have slightly less variability than the Almen strips at each of the five pressures. This is reflected in the average standard deviation for the five pressures found at the bottom of the table.

Table 2 Data Average and Standard Deviation over 31 days

Pressure (bar)	Almen Strips (mmA)		EST0061 (eq. mmA)		EST0062 (eq. mmA)		EST0063 (eq. mmA)		EST0064 (eq. mmA)		EST0065 (eq. mmA)	
	Avg.	Std. dev.	Avg.	Std. dev.	Avg.	Std. dev.	Avg.	Std. dev.	Avg.	Std. dev.	Avg.	Std. dev.
0.55	0.145	0.0072	0.146	0.0037	0.147	0.0036	0.147	0.0032	0.147	0.0043	0.149	0.0035
0.83	0.206	0.0070	0.195	0.0035	0.195	0.0042	0.195	0.0025	0.197	0.0048	0.196	0.0034
1.10	0.254	0.0069	0.237	0.0042	0.237	0.0036	0.235	0.0026	0.238	0.0050	0.237	0.0043
1.38	0.297	0.0067	0.278	0.0040	0.274	0.0059	0.275	0.0026	0.278	0.0060	0.277	0.0052
1.65	0.338	0.0073	0.318	0.0049	0.312	0.0057	0.314	0.0032	0.315	0.0068	0.314	0.0053
Avg.		0.0070		0.0041		0.0046		0.0028		0.0054		0.0043

Almen strip variability has been studied by a large OEM in Reference [7]. A sample of 153 Almen strips were tested on different machines, at different peening intensities and using different media type and size. For A type Almen strips, the overall variability for all tests was found to be 0.0056mm. This is similar to the 31 day variability presented here for the five different pressures.

Reference [8] compared three sources of Almen strips in a common machine setup using various shot types and sizes at different peening intensities. The standard deviation for the three different manufacturers for a total of 105 strips is 0.0142. This is twice the variability that was measured on the Almen strips in this study. Data for a single manufacturer was found to be 0.0054mm. Again this is similar to the variability found here.

Overall, the data shows that over the 31 days period, the sensors are as stable as the Almen strips to verify machine consistency. Sensors also offer the advantage of a digital output for easy data analysis and transfer for process tracking and storage. Furthermore, the output is directly in equivalent intensity instead of arc height thus avoiding much confusion in the process.

Conclusion

A new electronic peening intensity measurement sensor was presented. It was designed specifically for peening applications and allows for intensity measurements even on two and three dimensional surface curvatures difficult to represent accurately using Almen strips. The output of the sensor was shown to correlate very well with Almen intensity. Once correlated, the sensor provides a stable measure of peening intensity over 31 days with standard deviation smaller or equivalent to Almen strips. This new sensor offers the possibility to save considerable manipulation time when determining or verifying intensity. It avoids much confusion with intensity verification and is also a much better choice for the environment.

References

1. Forgues, S.A. et Al., New Electronic Peening Intensity Sensor Theory and Experimental Validation, Conf Proc: ICSP14, 2022, Milan, Italy
2. SAE J442, Test Strip, Holder, and Gage for Shot Peening, https://www.sae.org/standards/content/j442_202002/
3. SAE AMS 2431/3, Peening Media (AWCR), Conditioned Carbon Steel Cut Wire Shot, Regular Hardness, https://www.sae.org/standards/content/ams2431/3/
4. Electronics Inc., www.electronics-inc.com
5. Shockform Saturation Curve Solver (SSCS), version 3.2, http://shockform.com/
6. SAE J443, Procedures for Using Standard Shot Peening Test Strip, https://www.sae.org/standards/content/j443_198401/
7. Happ, M.B., Rumpf, D.L., Almen Strip Variability – Statistical Treatment, Conf Proc: ICSP-6, 1996, (p.302-311), San Francisco, CA, USA
8. Champaigne, J.M., A Review of ALmen Strip Performance by Electronics Inc. prepared for Surface Enhancement Division Fatigue Design and Evaluation Committee. SAE, April 20m, 1999. Auburn Hills, Mi, USA

New Electronic Peening Intensity Sensor for Pneumatic Needle Peening Repair Tools

Adel Alouani[1], Sylvain A. Forgues [1], Brigitte Labelle[1], Ramzi Ben Moussa[1]
[1]Shockform Aeronautique Inc., Canada, engineering@shockform.com

Keywords: *Rotary Flapper Peening, Pneumatic Needle Peening, Portable Tools, Repair Peening, Electronic Intensity Sensor*

Introduction

Controlled pneumatic needle peening is often used after blending surface damage to rebuild the residual compressive layer that is removed by the blend. Controlled pneumatic needle peening can be used to peen surfaces on assembled aircraft with minimal dismantling and minimal risk of Foreign Object Debris (FOD). This technique has the advantage of a small work-head to peen difficult to reach areas while maintaining optimal stand-off distance at all times.

Generating a saturation curve to find the Almen intensity with repair tools can be time consuming. Typically, generating a single saturation curve takes 15 minutes in the A scale and up to 1 hour or more in the lower N scale. The introduction of a new electronic sensor specifically developed for pneumatic needle peening promises to considerably simplify and shorten the intensity measurement and verification process. This sensor also facilitates user training on the equipment while being re-usable and save thousands of Almen strips which usually end up in the landfill.

Objectives

Evaluate the capability of an electronic peening intensity sensor for controlled pneumatic needle peening. Compare intensity values obtained by the sensor with methods currently used. Demonstrate potential time savings when using the sensor. Provide an example of the use of such a sensor for training purposes.

Pneumatic Needle Peening Tool Description

Manual peening repairs have been traditionally done using flapper peening [1], where tungsten carbide balls glued on a flexible flap, rotate and impact the surface. Due to the nature of the assembly of those flaps, the tungsten carbide balls tend to detach and fly off when they wear, causing potential FODs and greatly limiting peening applications. Therefore, a tool which eliminates this kind of risk was required. One such tool recently developed uses pneumatically actuated needles to perform peening without this risk. Figure 1 shows all the elements of this pneumatic needle peening tool, which features an embedded touch screen computer, a hand piece which hold the needles, an embedded electronic peening intensity sensor, as well as various cables and accessories, all stored in a portable case. [2]

The heart of operation of the pneumatic needle peening tool is located within the small work-head of the hand piece shown in Figure 1. Pneumatically actuated needles move back and forth to allow peening of a given surface. With flapper peening, the height of the flap/mandrel assembly depends on the operator's skill, thus introducing significant variability in the process. Additionally, flapper peening introduces semi-random impact angles and rubbing against the part, resulting in the typical surface linear roughness pattern seen after flapper peening. In the case of needle peening, the work head is rested against the surface and the needles impact perpendicularly to the surface, resulting in a much better surface finish.

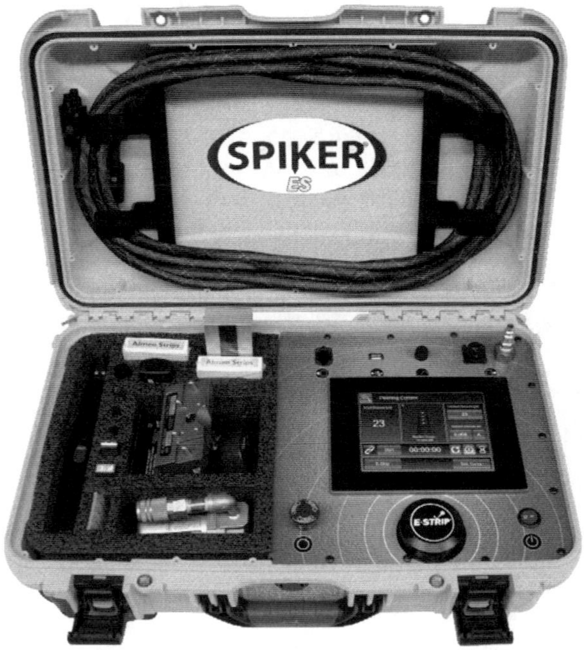

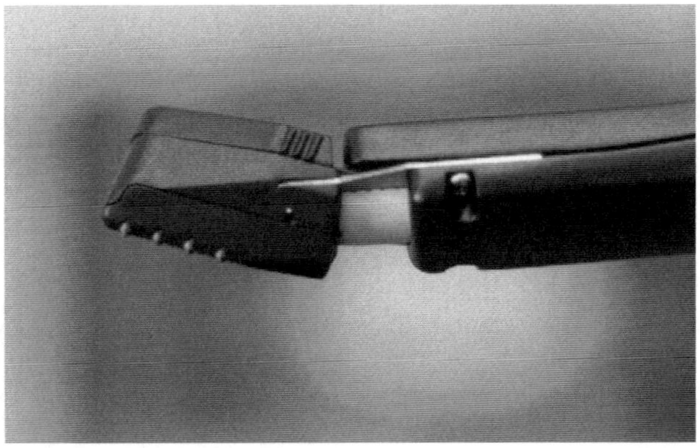

Figure 1 Spiker-ES® Pneumatic Needle Peening tool kit (Left) and Work-head (Right)

The needles are made of tungsten carbide and the tips are rounded to a diameter similar to shot sizes seen in the industry. The tips can be rounded to specification, thus needles can be supplied with different tip diameters depending on what type of surface finish or intensity is required. To match the surface finish, needle tip diameter must match the shot used as much as possible. The tip of the needles used here is similar to cast shot S230 with a diameter of 0.600mm. The needles are encased inside the work-head, hereby eliminating the risk of FOD. The needle's up-and-down movement is measured continuously during operation with small position sensors inside the work-head. An alarm is triggered in the event of needle malfunction. The needles and caps wear down with time and higher intensity peening, which is why it is possible to replace them after many hours of use.

The controller features a touchscreen computer that monitors and controls various parameters as well as store operational data. The controller also features a saturation curve solver to calculate Almen intensity and provides pressure requirements for a given Almen intensity. In typical repairs, the operators use these features to quickly calculate and determine intensities required.

Applications scenarios range from aero engine blades, and wheels repairs, as well as structural components. A typical application on a difficult to reach area is illustrated in Figure 2.

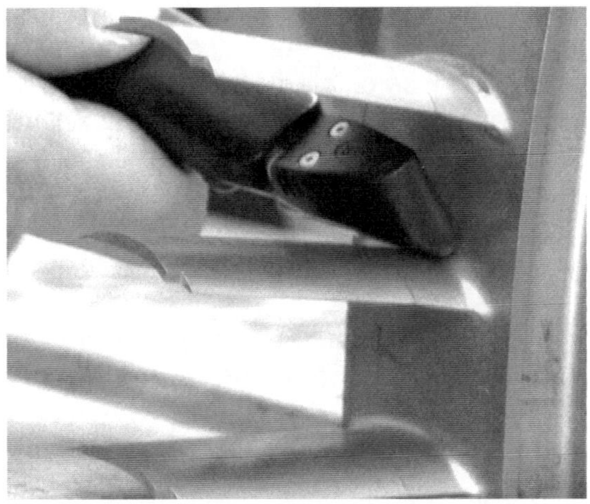

Figure 2 Pneumatic Needle Peening Application on an engine blade

Peening Sensor Description

In the controller box, right under the touch screen, a newly developed electronic peening sensor allows measurement of needle peening intensity in near real-time. (Figure 3) This integrated device features a circular protective cover placed over a load cell. It is 42mm in diameter and is big enough to simulate the motion required for needle peening. The cover is reusable and can be replaced when necessary.

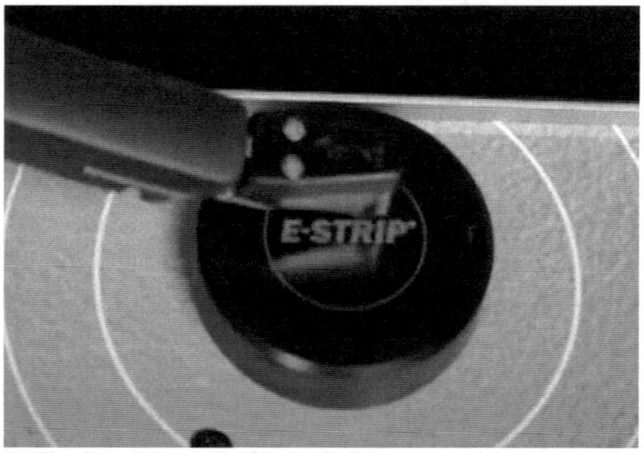

Figure 3 Embedded E-Strip® Electronic Intensity sensor

The sensor measures individual impacts of the needles and a custom algorithm calculates the average equivalent intensity after 10 seconds of continuous needle peening. The output is then displayed on the touch screen (figure 4), which allows for a quick read-out of the equivalent intensity.

Figure 4 Touch screen interface – Peening in progress (Left), Equivalent Intensity result (Right)

Experimental Method

To test the sensor, needle peening is performed successively on Almen strips installed on a standard magnetic block and then on the electronic peening sensor. An average of 3 output values of the sensor was used to compare output results to the Almen intensities. These intensities are calculated using the onboard saturation curve solver which respects the SAE J2597 specification [3]. Different intensities were established by setting the pressures at 0.69bar, 1.38bar, 2.07bar, 2.76bar, 3.45bar, and 4.14bar for the A range intensities. For the N range, pressures were set to 0.69bar, 1.03bar, 1.38bar, and 2.07bar.

Peening was performed using the 4-needle work-head. Peening on the Almen strips is performed doing circular motions while covering the whole Almen strips and making sure the work-head is well rested against it. A set of 5 arc-height points is used to calculate the intensity using the integrated Shockform Saturation Curve Solver [4].

For the electronic peening sensor, the same circular movements are performed, but there is no need to move left or right since the sensor doesn't need to be covered due to its reusable nature. Almen intensity was determined according to the saturation curve, whereas the electronic intensity sensor displays the result after 10 seconds of continued peening on its surface, no matter the intensity level. Sensor output and Almen intensities are then compared for the intensity ranges and peening times are recorded and compared.

Results and Discussion

Pneumatic needle peening has been performed on a wide range of Almen intensities varying from the N scale to the A scale. It was then possible to correlate the outputs between the electronic peening sensor and Almen intensities, and plot the graphs seen in figure 5. The response of the sensor provides a good representation of Almen intensities. A perfect match would show all the points lie directly on the line. In these results, the goodness-of-fit measure R-square has a value of 0.984 in the N range and 0.997 in the A range.

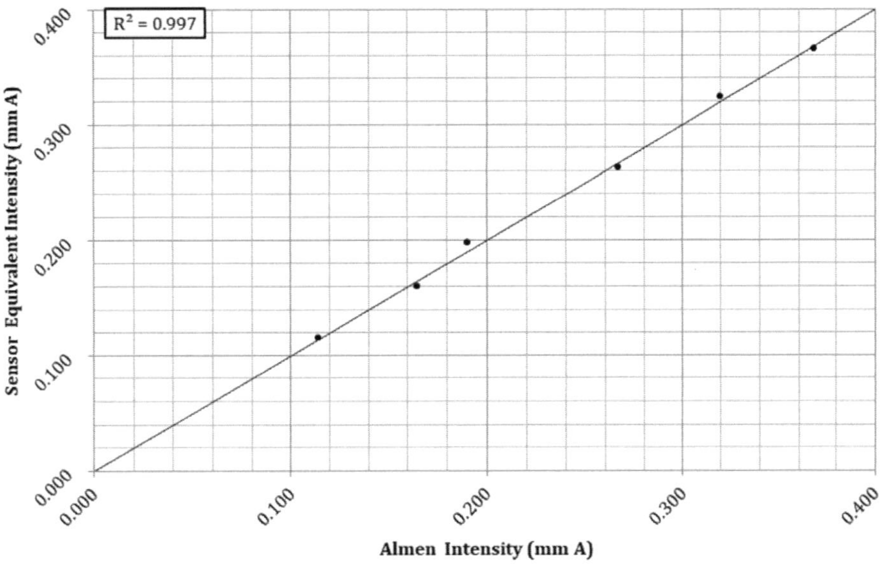

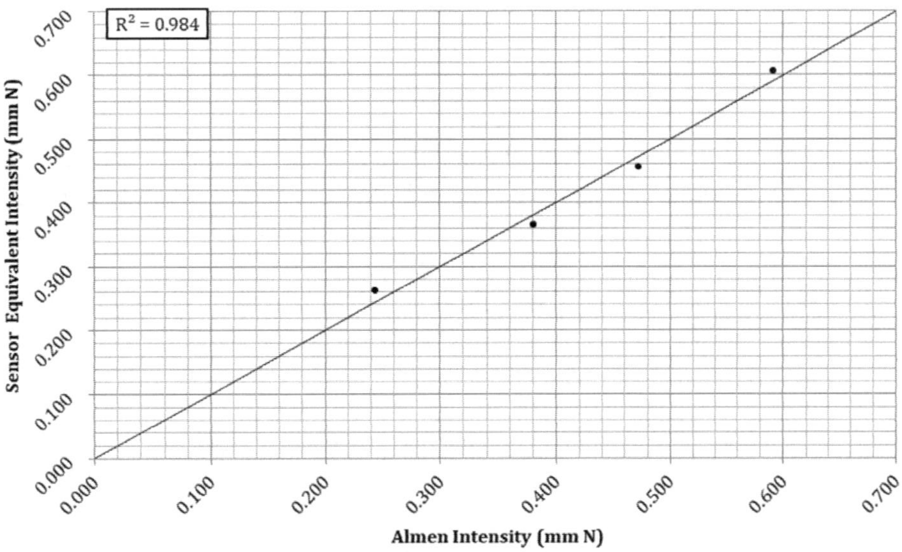

Figure 5 Sensor response plotted against Almen Intensity: N (Top) and A scale (Bottom)

A time study was also performed to see how much time could be saved using the sensor when determining intensity. At lower intensities in the N range, the longest total run time to complete a saturation curve was 96 minutes, or 1h36min., for a pressure of 0.69bar giving an Almen intensity of 0.244mm N (Figure 6). On the other hand, it took 10 seconds per test to verify the intensity with the electronic peening sensor.

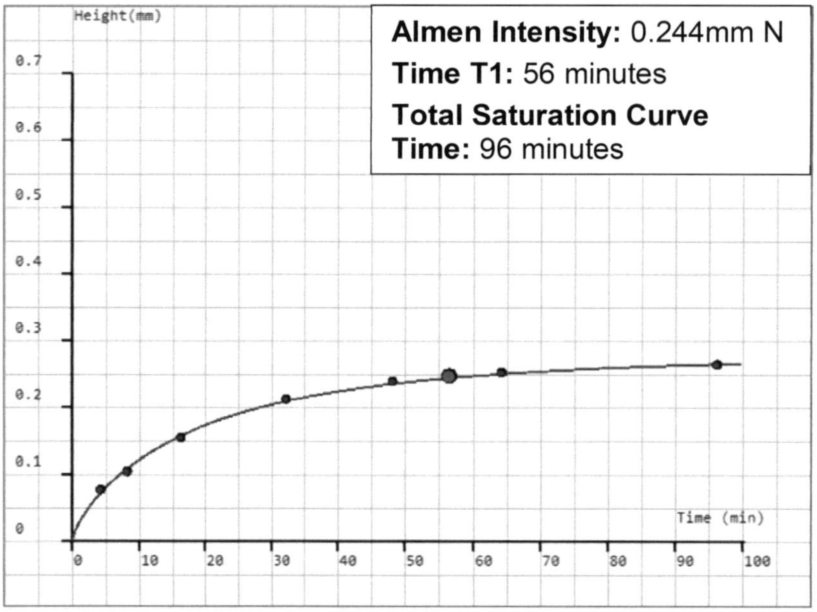

Figure 6 Saturation curve at 0.69bar.

Table 1 compares the total times for different intensity determination. Time savings range from 7.5 minutes at the high intensity in the A scale up to 95.5 minutes in the low N scale intensities. It becomes quite clear that the use of an electronic peening intensity sensor offers a important advantage when verifying intensities. In an organization, this could mean significant annual cost savings and increased productivity.

Table 1 Intensities and times results

Intensity range	Pressure (bar)	Almen intensity (mm)	Saturation time T1 (MM:SS)	Total Time taken to perform a full curve (MM:SS)	Sensor Equivalent intensity (mm)				Time taken on Sensor (MM:SS)	Time Saved (MM:SS)
					Test 1	Test 2	Test 3	Average		
N	0.689	**0.244**	56:09	96:00	0.265	0.254	0.266	**0.262**	00:30	95:30
	1.034	**0.381**	22:07	32:00	0.363	0.370	0.364	**0.366**	00:30	31:30
	1.379	**0.472**	07:42	16:00	0.459	0.452	0.457	**0.456**	00:30	15:30
	2.068	**0.592**	05:58	16:00	0.603	0.611	0.604	**0.606**	00:30	15:30
A	0.689	**0.114**	16:23	24:00	0.116	0.112	0.117	**0.115**	00:30	23:30
	1.379	**0.165**	06:32	12:00	0.158	0.161	0.159	**0.159**	00:30	11:30
	2.068	**0.191**	03:14	08:00	0.199	0.197	0.198	**0.198**	00:30	07:30
	2.758	**0.267**	03:28	08:00	0.261	0.265	0.261	**0.262**	00:30	07:30
	3.447	**0.320**	02:55	08:00	0.326	0.322	0.324	**0.324**	00:30	07:30
	4.137	**0.368**	02:21	08:00	0.374	0.360	0.364	**0.366**	00:30	07:30

In addition, using the sensor offers the possibility of reducing the complexity and risk of determining intensities since it is possible to very quickly determine if something is wrong in the process, such as the wrong setting or the wrong technique when needle peening. This is especially useful in the cases where users are not accustomed to the process. In a training environment for example, it becomes easy to adjust one's technique based on the output of the electronic peening intensity sensor and achieve optimal performance. For example, the sensor will show a low value if the tool is not held properly against the surface.

Conclusion

The paper introduced a new electronic peening intensity sensor for pneumatic needle peening repair tools. We showed the relationship between the output of the electronic peening intensity sensor and Almen intensity as well as the potential time saving for using the sensor in needle peening repair tools. It was determined that this new sensor responded closely to Almen strips while offering significant time savings. Furthermore, the sensor can help in the training a training environment to adjust one's technique and achieve optimal performance. A similar sensor will soon be available for flapper peening portable tools which would allow for the same advantages.

References

[1] Controlled Rotary Flap Peening for Repair Applications, 2011, S. Forgues, B. Labelle, M. King, N. Manor, ICSP-11, South Bend, In, USA, Pgs. 1-6.

[2] Shockform Spiker-ES controlled Pneumatic Needle Peening, 2022, https://shockform.com/product/spiker-controlled-pneumatic-needle-peening

[3] SAE J2597, Computer Generated Shot Peening Saturation Curves, 2010, https://www.sae.org/standards/content/j2597_201001/

[4] Shockform Saturation Curve Solver (SSCS), version 3.2, http://shockform.com/

Size and Shape Characterization for Shot-Peening Impingement Models

Paul Mort, Langdon Feltner, Mark Gruninger and David Bahr
Purdue University Department of Materials Engineering, Center for Surface Engineering Enhancement (CSEE)

Abstract

Size and shape characterization of shot is used to develop an impingement model that accounts for the effect of impact momentum and radius of curvature at the point of peening impact. The model parameters include the shot particle's size-dependent mass, shape-dependent radius of curvature (ROC), and a probability function for ROC selection based on possible orientations of the shot particle. The ROC model is based on shape archetypes of cast and cut-wire shot which are calculated based on relatively simple features obtained from 2D dynamic image analysis.

Introduction

Recent advances in modeling compressive stress fields achieved by shot peening have used idealized shot particles impinging on a substrate, the latter described as a mesh of finite elements (Xiao, Tong, Li, Yang, & Gao, 2017) (Xiao, et al., 2018). Finite Element Analysis (FEA) enables the prediction of surface coverage, work flux, and stress-depth profiles in the substrate; all are of interest in the industrial shot-peening community. Models using multiple discrete impacts can be used to provide statistical information on the distributions of coverage and stress field (Miao, Larose, Perron, & Lévesque, 2009) (Ghelichi, Crispiatico, Guagliano, & Bagherifard, 2018). Critical to the accuracy of the FEA model is the descriptive features of the shot particle size and shape distributions, as shot particle momentum and radius of contact have been shown to have a direct and significant impact on the induced stress state below the impact (Meguid, Shagal, Stranart, & Daly, 1999) (Larour, Bäumer, Dahmen, & Bleck, 2013).

The motivation for the current work is to expand the capability of FEM modeling to include distributions in shot characteristics: size and shape. The result is a distribution of shot particle mass and local radius of curvature (ROC) at the point of peening impact; when implemented in FEM, this provides additional modes of coverage and stress field distribution on the target substrate.

Experimental

Characterization was done using Dynamic Image Analysis (DIA);* the method has developed into a mature technology for particle characterization, capable of measuring size and shape distributions of particles based on their 2D projections in flows that can be manipulated to promote either aligned or random orientation. Random orientation is chosen to provide a statistical analogy of peening impacts. Both cast and cut-wire samples were characterized, typically imaging more than about 10k particles per sample to achieve robust statistical datasets. Data were collected for four grades of cast shot (S70, S110, S230, S330),† and four grades of twice-conditioned cut-wire shot (CW12, CW14, CW20, CW32).‡

Raw data collection included grayscale images (0-255) on a back-lit white (255) background; the grayscale images are binarized using a threshold (173 was used in this work) and then analyzed to provide size and shape information for each particle in the dataset. This procedure is summarized in Figure 1. Thresholding and analysis can be done in real-time during a measurement, or can be done by post processing the raw image data. This provides an output list having the size and shape characteristics of each particle; the detailed list can be further reduced to provide statistical distributions of size and shape features.

---

* SolidSizer, JM Canty, Lockport, NY, USA and Dublin, Ireland
† Ervin Industries, Adrian, MI, USA
‡ Toyo Seiko NA, South Bend, IN, USA

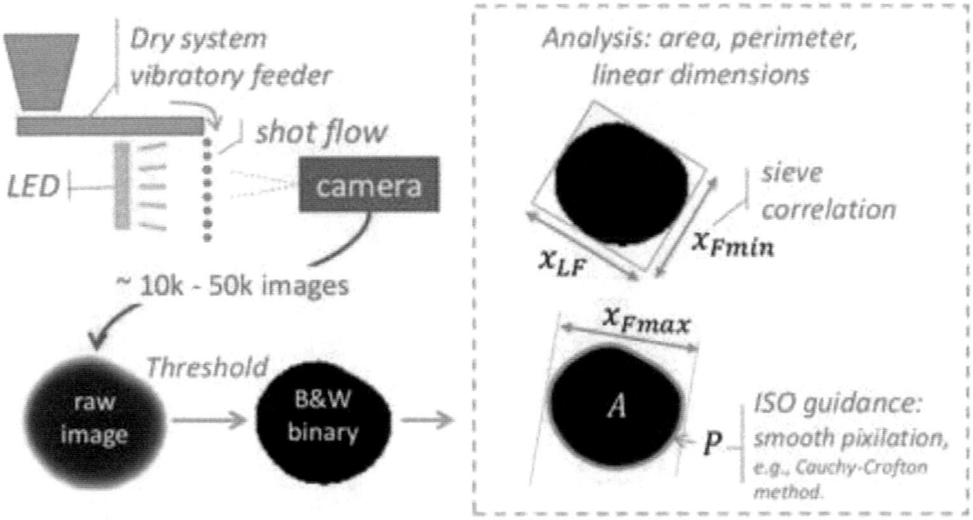

Figure 1. Dynamic image analysis experimental setup, image thresholding, and feature analysis.

DIA features include 2D area (A), perimeter (P), min and max Feret lengths (x_{Fmin}, x_{Fmax}) and the orthogonal to the min Feret, x_{LF}. Feret lengths are linear dimensions between two parallel tangents. Shape factors derived based on these features are summarized in Table 1. The elliptical form factor is defined as an elliptical analogy of the ISO-defined form factor, where the factor β is used to approximate the perimeter of an ideal ellipse having an aspect ratio AR. The reduced curvature diameter is defined using the archetype for cast shot, which is based on overlapping or "glued spheres" (Figure 2).

Table 1. Summary of size features and shape factors obtained using dynamic image analysis

Size Features	Impact curvatures & probabilities	Shape Factors (dimensionless)
• x_{Fmin} = min Feret length • x_{LF} = Feret length orthogonal to x_{Fmin} • x_{Fmax} = max Feret length • A = area • P = perimeter • x_A = equivalent area diameter, $x_A = \sqrt{4A/\pi}$ • V = volume, $\frac{4}{3}A^{3/2}/\sqrt{\pi}$	• x_{RC} = reduced curvature diameter, $\frac{-b+\sqrt{b^2-4ac}}{2a}$; $a = \frac{\pi-2}{8}$; $b = \frac{x_{Fmax}-x_{Fmin}}{2}$; $c = \frac{4 \cdot x_{Fmax} \cdot x_{Fmin}+(\pi-2)x_{Fmin}^2}{8} - A$ • $ROC1$ = major curvature $\sim x_{Fmin}/2$ • $ROC2$ = minor curvature $\sim x_{RC}/2$ • $d_{cc} = x_{Fmax} - (x_{Fmin} + x_{RC})/2$ • $cos(\alpha) = (ROC1 - ROC2)/d_{cc}$ • $\Phi1$ = major probability, $(\pi - \alpha)/\pi$ • $\Phi2$ = minor probability, α/π	• AR_{ISO} = aspect ratio, ISO definition, x_{Fmin}/x_{Fmax} • AR_{box} = aspect ratio, bounding box, x_{Fmin}/x_{LF} • FF = form factor, $\frac{4\pi A}{P^2}$ • EFF = elliptical FF, $\frac{\beta\pi A}{P^2}$; $\beta = \left(\frac{1.5\cdot(AR+1)}{\sqrt{AR}} - 1\right)^2$

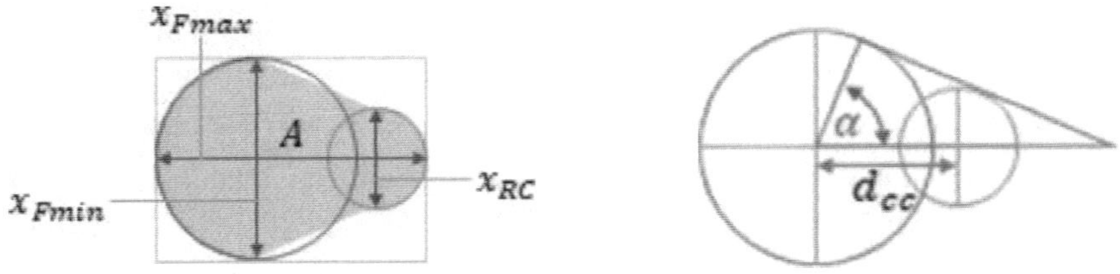

Figure 2. Archetype for reduced curvature: LHS) derivation of xRC; RHS) derivation of impact probability.

Results and Discussion, Characterization

Idealized archetypal particle shapes representing cut-wire and melt-atomization cast shot processes are shown on a shape map diagram (Fig. 3). The contour lines show parameters used to define the shapes. The shape map plots the Aspect Ratio and Elliptical Form Factors as orthogonal axes; the ordinate (AR) showing elongation, and the abscissa (EFF) showing angularity or other perimeter irregularity. In both cases, a value of (1,1) represents a circular projection of an ideal sphere.

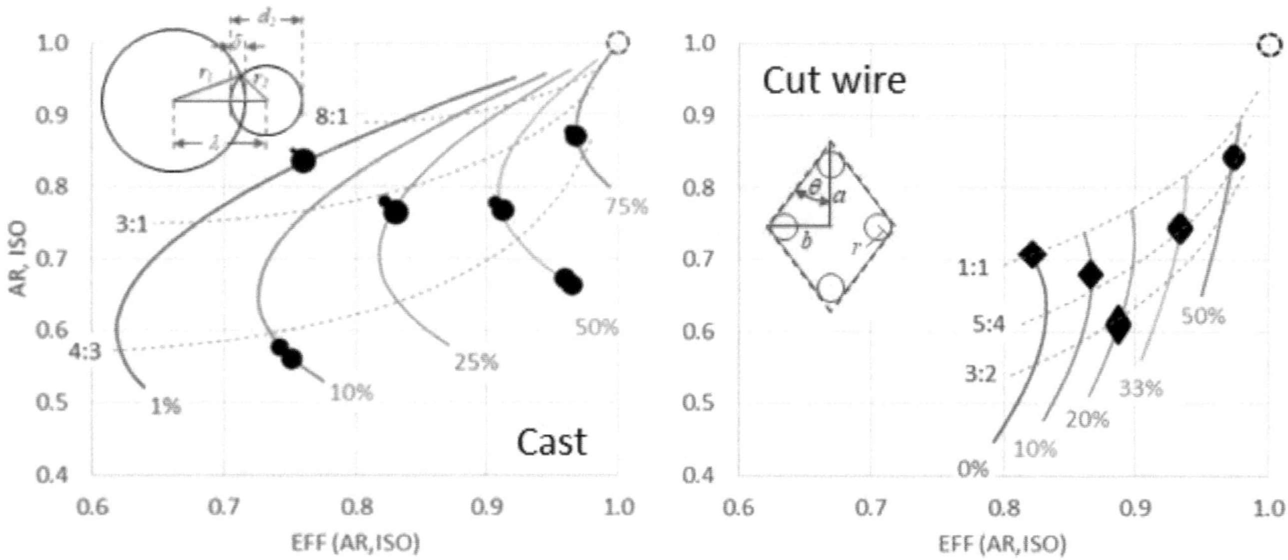

Figure 3. Shape mapping using conditioned cut-wire and atomized cast w/ satellites. Cast: colored contours represent the overlap of satellites; dashed contours represent their size ratio. Cut wire: colored contours refer to relative rounding of cut corners, r/b; dashed contours refer to cut orientation, a/b.

An example of shape mapping with experimental data is shown in Figure 4. The mapped particles are illustrated by their DIA images. Most of the particles (~90%) have a Form Factor that is > 0.95, i.e., highly spherical. The tail of the distribution has the more irregular shapes shown on the map.

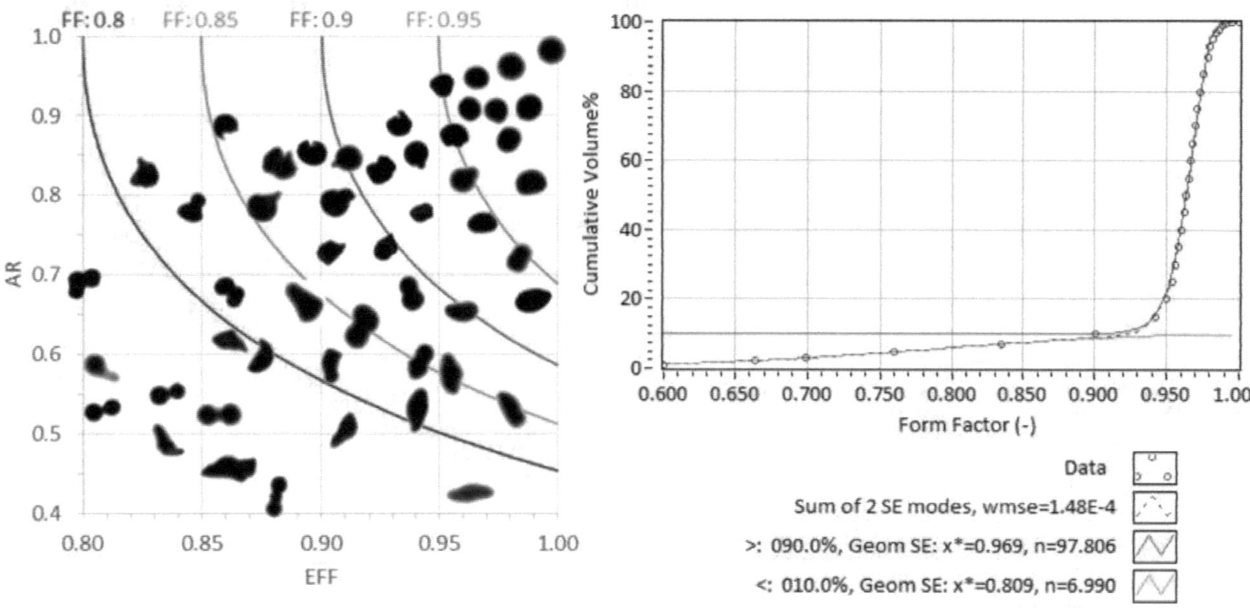

Figure 4. Shape mapping and distribution statistics for S70 shot.

Results and Discussion, Impact Modeling

The effect of the shape curvature analysis broadens the distributions of impact curvature versus conventional spherical shape assumptions (Figure 5). This is further illustrated in Figure 6, comparing a spherical assumption (black) with a glued-sphere representation.

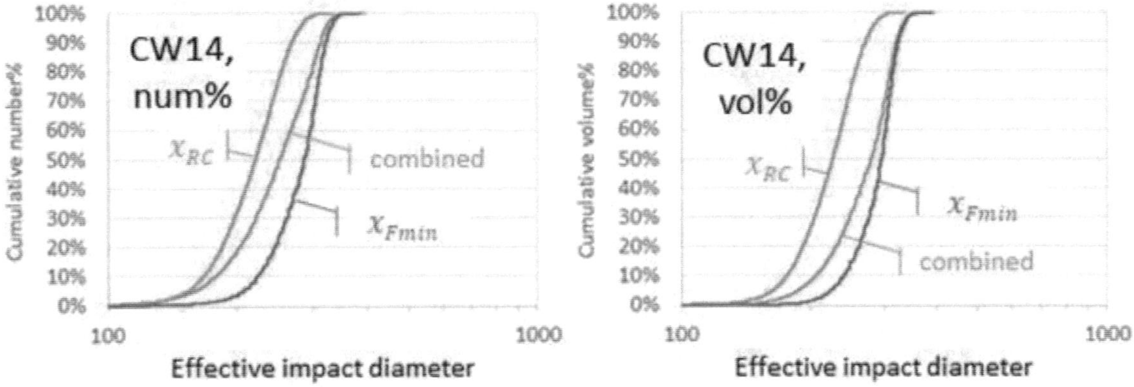

Figure 5. Effective impact curvature distribution for a cut-wire sample.

Combined (x_{Fmin} , x_{RC}) size distributions were created as inputs to a set of discrete-impact FEM models. These were compared to spherical equivalent models without shape effects, using the stress depth profile as a means of comparison. An Abaqus input file generation script was created that reads image data from the Canty SolidSizer imaging system. From this, area equivalent diameter and reduced radius of curvature are calculated for each particle in the distribution. From a user specified coverage percentage and shot size, an approximate number of impacts needed to reach the desired coverage is calculated.

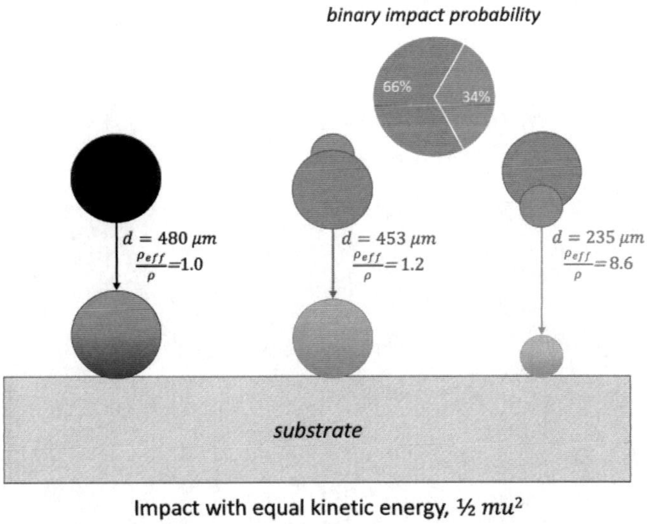

Figure 6. Area equivalent diameter (black) and glued sphere (orange and blue) representation of shot particles. Glued spheres have a smaller radius of contact with the same mass, translating to a higher effective density upon impact.

With each run of the input file generator, the number of necessary particles is sampled from the distribution and placed in random locations above the substrate. Each particle is assigned a radius of contact, based either on the area equivalent diameter (Fig 6, black), or a weighted probability of impacting with either $ROC1$ (Fig 6, blue) or $ROC2$ (Fig 6, orange). Mass was assigned to all particles based on the area equivalent diameter and the material density, ρ.

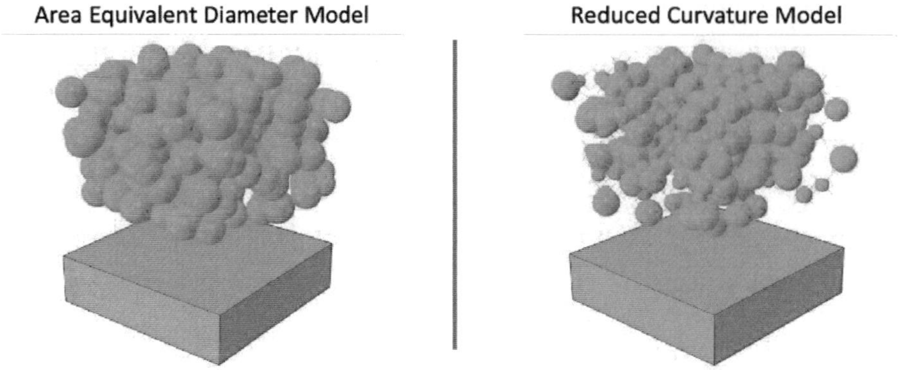

Figure 7. Area equivalent diameter representations of shot particles have a larger radius of contact for particles with similar mass.

A 5mm x 5mm representation of an Almen strip was used as the substrate, utilizing Johnson-Cook plasticity. Particles were not allowed to interact with each other, and all impacts were normal to the substrate surface at 65 m/s. As is shown in Figure 7, particles defined by area equivalent diameter have a larger radius of contact for particles of a similar mass. This means that there is a higher effective density multiplier for each of the smaller particle representations in the reduced curvature model.

This difference in effective density of particles translates to differences in both coverage/surface roughness and induced stress state in the substrate. As shown in Figure 8, the reduced curvature radius of contact definition tends to have deeper dimples with smaller radii, whereas the area equivalent dimples seem to be more evenly distributed in size and depth. By virtue of this, the area equivalent diameter seems to have a more uniform coverage, with less bare area (dark blue, Figure 8) where no particles have contacted.

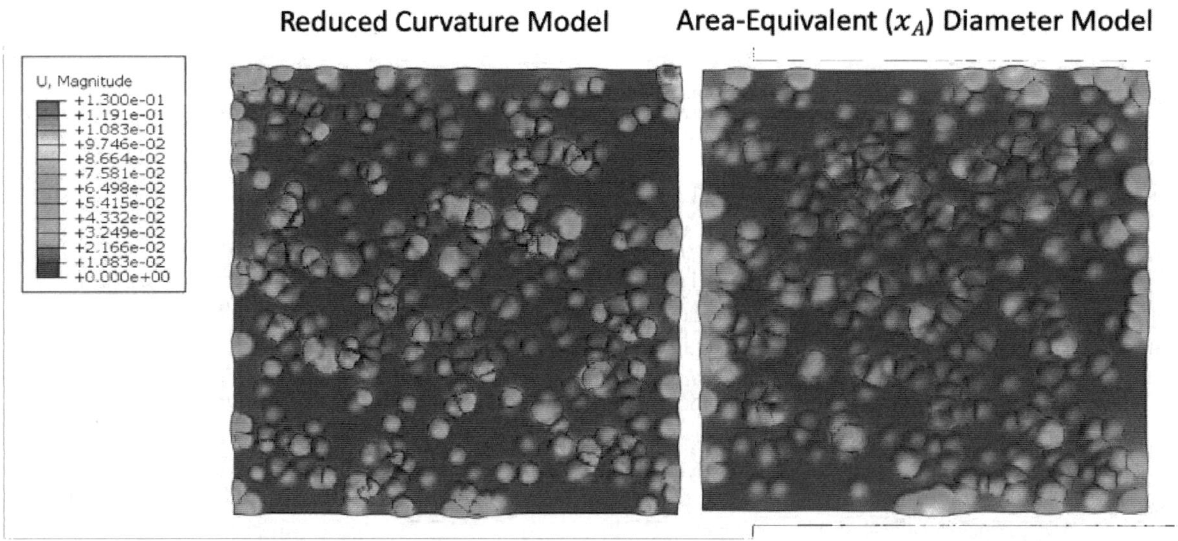

Figure 8. Deformed substrates with magnitude of displacement of surface nodes shown as the color gradient.

The differences between the particle radius of contact definitions are also shown in the induced stress-depth profiles through the thickness of the substrates. In both the 312 (~100 percent theoretical coverage) and the 937 (~300 percent coverage) trials, the area equivalent radius of contact definition produces a deeper cross over depth. Coupled with a similar peak stress and surface stress for both trials, this means that the area equivalent diameter particle definition translates to more cumulative work done on the substrate. Both trials have similar kinetic energies, only varying based on random sampling of the particle size distribution, and it can be inferred that there is more energy lost in the reduced curvature trials. Additionally, the curves show that the reduced curvature radius of contact definition is less sensitive to number of impacts since there is a relatively small difference in cross over depth between the 312 and 937 impact trials in comparison to the area equivalent radius of contact definitions.

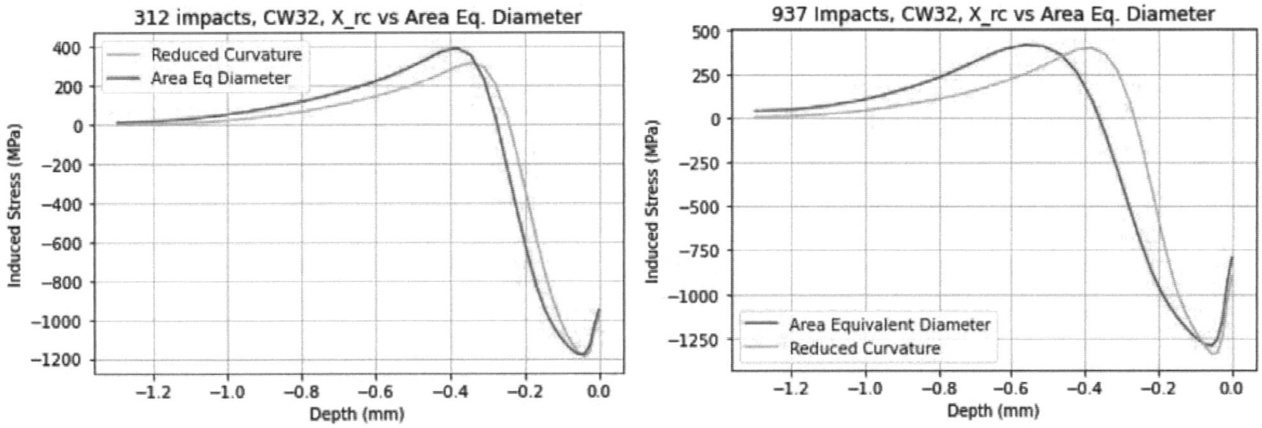

Figure 9. Induced stress states by way of area equivalent diameter and reduced curvature contact definitions in identical substrates, controlled for the number of impacts and impact velocity.

Conclusion

Measurement and analysis of shot media size and shape characteristics are of critical importance for model development. Media size and shape distributions affect surface coverage, work flux, and stress-depth profiles, all of which are of interest in the industrial shot-peening community. Both momentum of a particle and it's local radius of contact play a significant role in the formation of the induced stress state beneath it. Dynamic image analysis of shot media was used to develop an impingement model that accounts for the effect of impact energy and radius of curvature at the point of peening impact. Model parameters include the shot particle's size-dependent mass, shape-dependent radius of curvature (ROC), and a probability function for ROC selection based on possible orientations of the shot particle. Comparing finite element simulations having combined (x_{Fmin} , x_{RC}) radius of contact distribution versus an area-equivalent radius of contact, it is shown that the shape-dependent radius of contact has higher variability in local coverage and dimple size, and a reduced stress-depth profile. This means that non-spherical shapes reduce the work efficiency of the peening process. Future work will statistically quantify variability and work efficiency, and assess the effect of "working-mix" dynamics (i.e., media classification and recycle) in the peening process.

References

Ghelichi, R., Crispiatico, G., Guagliano, M., & Bagherifard, S. (2018). An energetic approach to predict the effect of shot peening-based surface treatments. *Metals*.

Larour, P., Bäumer, A., Dahmen, K., & Bleck, W. (2013). Influence of strain rate, temperature, plastic strain, and microstructure on the strain rate sensitivity of automotive sheet steels. *Steel Research International*, 426-442.

Meguid, S. A., Shagal, G., Stranart, J. C., & Daly, J. (1999). Three-dimensional dynamic finite element analysis of shot-peening induced residual stresses. *Finite Elements in Analysis and Design*, 179-191.

Miao, H. Y., Larose, S., Perron, C., & Lévesque, M. (2009). On the potential applications of a 3D random finite element model for the simulation of shot peening. *Advances in Engineering Software*, 1023-1038.

Xiao, X., Tong, X., Li, Y., Yang, M., & Gao, G. (2017). Numerical Research on the Effect of Boundary Constraint in Shot Peen Forming. *Journal of Materials Engineering and Performance*, 3839-3853.

Xiao, X., Tong, X., Liu, Y., Zhao, R., Gao, G., & Li, Y. (2018). Prediction of shot peen forming effects with single and repeated impacts. *International Journal of Mechanical Sciences*, 182-194.

CONTROLING SURFACE ROUGHNESS THROUGH SHOT MEDIA SIZE DISTRIBUTION

David Bahr[1] and Siavash Ghanbari[1]

1 Center for Surface Engineering and Enhancement, School of Materials Engineering, Purdue University, West Lafayette IN USA

Abstract
A simulation has been developed to predict the surface roughness profile and residual stresses as a function of media size distribution and shot velocity on steel samples. The simulation was validated by direct comparison to Almen strip tests. The simulation probed shot peening tests with different shot sizes and peening conditions such as individual impacts, sequential impacts, and mixed (i.e. concurrent) impacts. The simulation results are compared with experimental outcome measured by optical profilometer and x-ray diffraction in published experimental tests. Using a mixed size distribution of shot, rather than sequentially peening with larger and then smaller shot, appears to provide a smaller surface roughness than peening with larger shot alone, while providing a surface roughness similar to that with a finer peening size.

Keywords Roughness, hardness, shot size.

Introduction

Dual shot peening [1], where larger shot is first used to impart larger or deeper residual stresses and finer shot is used to minimize surface roughness or variations, is known to have an impact on the uniformity of residual stress [2] as well as the microstructural refinement on the surface layer [3]. Compressive residual stresses are developed by plastically deforming the metallic substrate and elastic deformation will be recovered by unloading of the impacting shot. Wohlfahrt showed two types of residual stresses that developed during the shot peening process, first a direct plastic elongation of the primary layer of the peened surface due to multiple shot dents and the second one is the elastic-plastic deformation that creates compressive residual stresses with a maximum intensity within the substrate [4]. This cold working process and residual stresses help to prolong the fatigue life, but too much peening may lead to surface flaws that are potential sites for fatigue crack initiation.

The purpose of this study was to develop a computational process to predict roughness and residual stress profiles for nominal coverages of 100% peening with the ability to vary the shot size. Previously published work by Bagherifard and co-workers [5] was used as the benchmark for model validity.

Computational Methods

The complete description of the processes used to simulate the peening process is provided in a recent publication by the authors [6]. In summary, we used conventional 1070 steel Almen strips for calibrating the model to ensure our stress and coverage matched experimentally measurable bulk peening conditions. The finite element simulation of the peening process was carried out in Abaqus/ Explicit 6-14. Shot particles were modeled as rigid balls with diameters between 0.3 and 0.60 mm, impacting the test piece at random locations until 100% of the surface had experienced plastic deformation. To obtain reasonable computational time the sample mesh was chosen as approximately 1/12th of the expected dimple diameter [7]. To determine coverage conditions, we simulated peening

Almen strips and calculated the Almen height as a function of number of impacts, and then converting those to coverage (shown in figure 1 for 0.5 mm shot at different velocities).

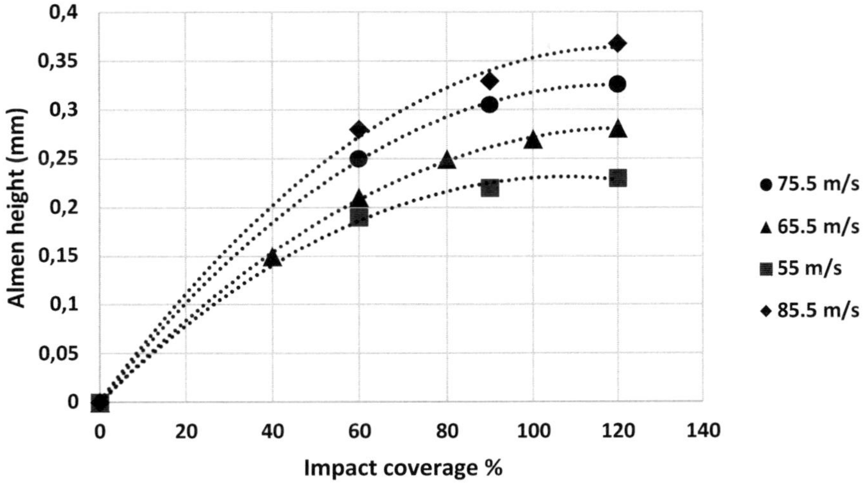

Figure 1. Almen height as a function of coverage for varying shot velocity using 0.5 mm shot to calibrate model conditions.

After calibration with experimental Almen strips, we then simulated a hardened steel system [5] using random shot impact with steel shot of 0.6, 0.43, and 0.35 mm. Shot velocity was fixed at 80m/s, and we used the simulation conditions in Table 1 were used to determine surface roughness and the subsequent residual stress profile.

Table 1. Simulation conditions for shot size and order processing.

FEM modeling	D (mm)	Distribution of Shots % (0.6 mm, 0.43 mm, 0.35 mm)	Sequence	Rz, μm
Shot peening model-1	0.6	100%, 0%, 0%	n/a	14
Shot peening model-2	0.43	100%, 0%, 0%	n/a	25
Shot peening model-3	0.35	100%, 0%, 0%	n/a	43
Shot peening model-4	0.6, 0.43, 0.35	33%, 33%, 33%	Mixed simultaneous	26
Shot peening model-5	0.6, 0.43, 0.35	33%, 33%, 33%	Sequential impact (0.6 mm→0.43 mm→ 0.35 mm)	25
Shot peening model-6	0.6, 0.43, 0.35	20%, 40%, 40%	Mixed simultaneous	24

Results

First, the effect of shot size distribution on the Almen height as a function of the number of impacts was determined. Two shot distributions were used, the wide distribution had 60% 0.6 mm, 20% 0.5 mm, 10% 0.4 mm., and 10% 0.3 mm shot, while the narrow distribution was 85% 0.6 mm, 10% 0.5 mm, and 5% 0.4 mm. The resulting mass of shot is shown in Table 1, and the resulting arc height of the Almen strip is shown in Figure 2.

Table 2. Mass of shot used in range simulation as a function of impacts

Number of shots	Mass of shots, wide range (g)	Mass of shots, narrow range (g)
12,000	3.9	4.76
28,000	9.11	11.09
40,000	13.02	15.83
52,000	16.93	20.69
68,000	22.13	26.95

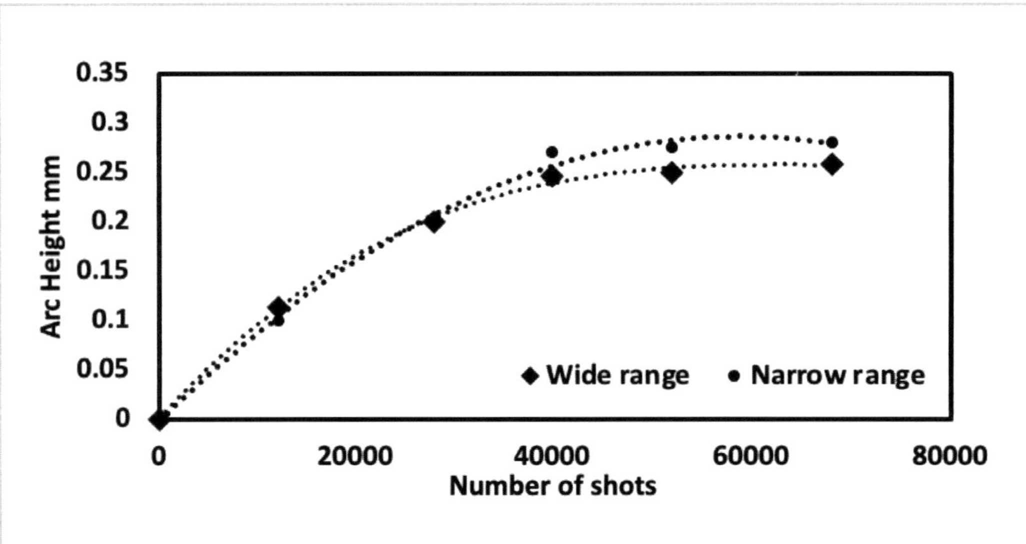

Figure 2. Almen arc height as a function of shot size distribution, showing the narrower (and larger average size) distribution leads to slightly larger arc heights for a similar number of impacts.

Next, the simulations described in Table 1 were carried out, a typical surface topology and stress distribution is shown in Figure 3. While different roughness parameters were determined (RMS, arithmetical of five highest conditions, and the maximum high to low peak) were determined, all scaled similarly. For this paper, we will consider only the Rz, calculated by measuring the vertical distance from the highest peak to the lowest valley within five sampling lengths, which is the largest (worst case) roughness measurement extracted in this study. Similarly, from the simulation we can extract a typical stress profile, shown for several of the conditions in Figure 4. Table 1 shows the Rz surface roughness as a function of the process condition.

Figure 3. Typical surface profile and von Mises stress of the hardenable steel substrate after peening (here for case 4, mixed shot size distribution).

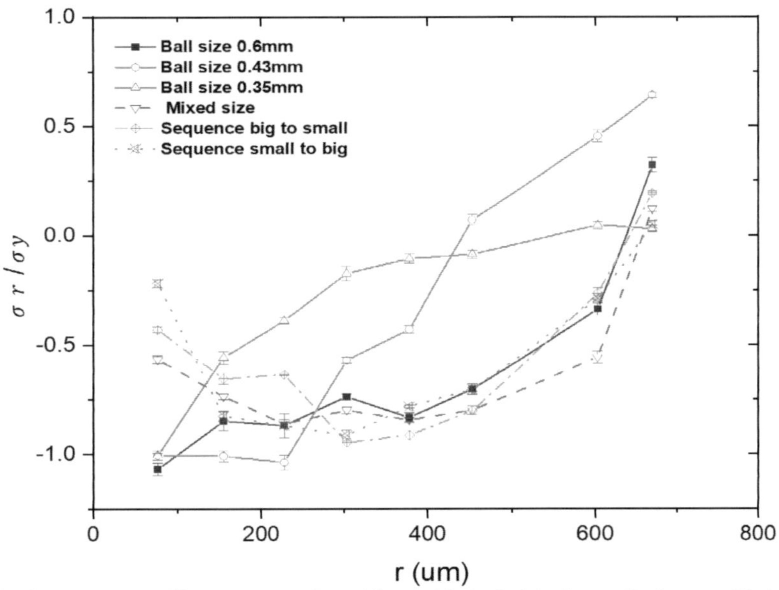

Figure 4. Typical stress profiles, normalized by initial yield strength for multiple size peening conditions. No "0" height is shown because of the variation in surface roughness, but an average height was determined for "0" by selecting the plane at which the integral of the mesh above and below that plane were equal.

Discussion and Conclusions

Sequential peening (three sizes, but not peening to 100% coverage in any given pass) decreased surface roughness over peening with the maximum size shot, as might be expected. The more interesting observation was that using a controlled distribution of shot sizes it appears to be possible to create a surface as smooth as the sequential peening in a single pass. Furthermore, mixed shot peening showed increased the depth of the total compressive residual stresses over sequential impacts (though the maximum value of that stress is slightly lower than the sequential peening or single large diameter peening condition). This suggests that controlling shot size distribution (beyond standard graded shot) can improve surface roughness and residual stresses simultaneously while reducing the risk of the over-peening during sequential impacts.

References

[1] D. Breuer, *Dual* Peening, The Shot Peener, Summer (2013), pp. 18-19.

[2] K. Zhan, C.H. Jiang, V. Ji, *Uniformity of residual stress distribution on the surface of S30432 austenitic stainless steel by different shot peening processes,* Materials Letters, Vol. 99 (2013), pp. 61-64.

[3] O. Unal, R. Varol, *Surface severe plastic deformation of AISI 304 via conventional shot peening, severe shot peening, and repeening,* Applied Surface Science, Vol 351 (2015), pp. 289-295.

[4] H. Wohlfahrt, "*The influence of peening conditions on the resulting distribution of residual stress,*" Proceedings of the 2nd International Conference on Shot Peening,*(1984)* p. 316-331.

[5] S. Bagherifard, R. Ghelichi and M. Guagliano, *A numerical model of severe shot peening (SSP) to predict the generation of a nanostructured surface layer of material*, Surface & Coatings Technology, vol. 204 (2010) pp. 4081-4090.

[6] S. Ghanbari and D.F. Bahr, *Predictoins of decreased surface roughness after shot peening using controlled media dimensions,* Journal of Materials Science and Technology, Vol. 58 (2020) pp. 120-129.

[7] J. Schwarzer, V. Schulze and O. Vohringer, *Evaluation of the Influence of Shot Peening Parameters on Residual Stress Profiles using Finite Element Simulation*, Materials Science Forum, Vol. 462 (2003) pp. 3951-3956.

Peening Intensity and Coverage Estimated by Simulation for Automotive Coil Spring

J. Shin[1], Y. Yoon[2]

1 R&D Center, Youngwire, Chungcheongnam-do, Republic of Korea
2 Solution Lab, Daejon-si, Republic of Korea

Abstract
The peening intensity and coverage is simulated by DEFORM® for the conventional shot peening facilities. Peening facilities are selected by two types; the one is the horizontal-type and the spring is free to move through the shot tunnel. The other is the turret-type and the spring is under the compression. Those facilities are conventionally used for the coil spring industry. Amen® strip is attached in the coil spring to measure the arc height and coverage under the current peening parameters. Passing the several springs with the strips into the peening tunnel or turret, the arc height and intensity will be measured and evaluated whether the peening parameters are adequate for the coil spring quality assurance. Typically, those facilities are very huge and therefore hard to freely change or optimize the peeing parameters. Instead, computer analysis model can be a useful and powerful tool. In this paper, the peening process are modelled and the results are compared with the measured data to show the validity of the simulation.

Keywords Automotive Coil Spring, Peening Intensity, Peening Coverage, DEFORM®

Introduction

Shot peening is a typical and important process for the improvement of product operating life in automotive industry. Especially, among those products, coil spring must have two or more shot peening processes for extremely high fatigue life expectation. As a tendency of lightweight design, coil spring, as is relatively light component in a car, also becomes lighter and therefore, the higher design stress would threaten its fatigue life reduction or failure.
With the spring design optimization to reduce the fatigue failure, shot peening process is also optimized by shot parameters such as a shot ball size, shape, material, blasting speed, angle and even peening time. Those parameter changes can be verified when the newly-peened coil spring will run the fatigue test directly. Practically, fatigue test takes several weeks to see its effect. The reason is that the shot peening facility is considered as the black-box and the change such as shot parameters can be indirectly checked by the fatigue test only. If the computer simulation model is purposely set for the peening facility, the improvement effect can be directly expected and therefore, reduce the cost and the time of fatigue tests. Also, the coil spring, which is newly designed, can be surely evaluated its fatigue performance before the real manufacturing.
The computer simulation model is set up for the conventional peening facilities made by Sinto® and Wheelabrator®. The peening performances are verified by the measurement and shows the validity of the analysis model. This result will be used for the improvement of fatigue life expectation of a newly-developed coil spring.

Peening Facilities to be Simulated

One of the peening facilities is installed in YOUNGWIRE by SINTO, Japan company. It still continued operating since 2016. The operation of the peening facility is briefly explained as follows; coil springs are continuously fed on one pair of the long liners, which are rotated at the same direction. It plays a role in rotating the coil springs on them. Shot balls are blasted by the impellers, which are rotated by the opposite direction. For coil spring, this facility is named as "First Peening" or just simply "Peening Process".

The other peening facility is called "Second Peening" or "Stress Peening", installed by Wheelabrator, Swiss company. It is already installed and continue operating. The operation is as follows; two coil springs simultaneously get into the cabinet by robot arms. In the cabinet, there are two sets of upper and lower holding fixtures. Between them, one coil spring is located by robot arm and subsequently, compressed by the upper holding fixture moving down. The door cabinet is closed and the two fixtures are rotated. Finally, the shot balls are blasted.

Several Assumptions for Computer Simulation

First and second peening facilities has an "impeller" type blasting mechanism. Typically, shot peening can be blasted by impeller and air-nozzle type. For coil spring in automotive suspension, impeller type is generally applied. But for the simulation, the impeller type is modelled as nozzle type. Two blasts are actually overlapped and shot balls in the region are collided each other. But in a simulation, each blast is applied sequentially, for example, upper blast is followed by lower one. These assumptions are quite sufficient for the current research level and also computational cost can be reduced.

Figure 1. Concept of the First Peening (Left) and the Second (Right) Facilities

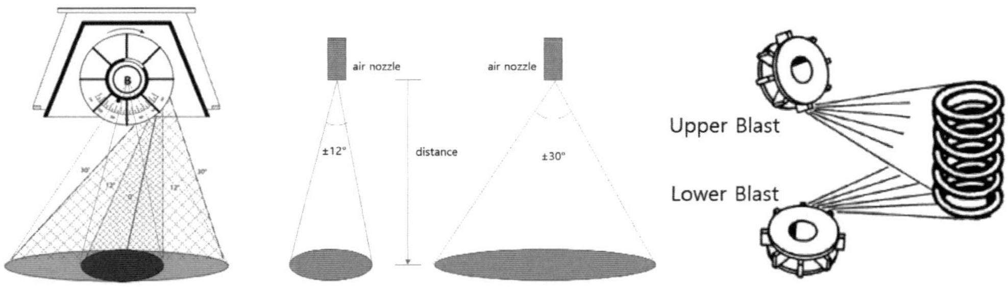

Figure 2. Assumption for Peening Simulation

Shot Peening Intensity and Coverage : Simulation and Measurement in Coil Springs

Shot peening is very important process for fatigue life of the coil spring. And therefore, it is necessary to measure the performance of the process for the spring to have a uniform quality of the peening. Practically, Almen Strip is fixed in coil spring and checked the arc height and the surface of the peened Almen strip. The arc height will become higher as a peening time is increased and eventually be saturated. The part of the peened surface of the Almen strip will be also increased and saturated. In a manufacturing point of view, the peening time must be reduced and also the peened coil spring must be satisfied by the fatigue test. That is, the peening time will be set by the coil spring design requirement and the manufacturing cost or cycle time in a factory.

For the coverage of the peening, the several samples of the peened Almen strips are prepared in advance. Those samples are specified by 100%, 90%, 70% etc. The newly-peened Almen strip is compared by the samples through the image analysis. The practical measurement of the peening intensity and coverage is based on the fatigue test of the coil spring and the ready-made samples.

In simulation, the coverage[1] is calculated by Eq. 1.

$$C = 100\{1-\exp(-\pi r2Rt)\} \quad (1)$$

where r is the radius of each impression, R is the uniform rate of creation of impressions and t is the time during which the impressions were being created.
From Eq. 1, the coverage can be changed from 0 to 100%. The simulation results in a theoretical coverage based on the size of the impression by 2D micro model.
The arc height can be estimated by "spring-back" analysis from the simulation. More specifically, the residual stress obtained from the simulation is the input for the spring-back analysis and the final elastic shape of the Almen strip or arc height can be calculated to satisfy the force equilibrium. The arc height from the simulation is directly compared with the measured value to show the simulation validity.

Shot Peening Simulation Procedure

Shot peening is simulated by three steps; axis-symmetric finite element analysis of 2D micro model, data analysis from the 2D micro model analysis, and impact simulation of the shot peening process.
First, 2D micro model simulation is performed to get the fundamental residual stress analysis. The impeller will accelerate shot balls to certain shot speed and hit on the surface. They will hit on the surface of the spring as an arbitrary angle, δ. And a specific region will be hit by any number of shot balls. To consider the angle and the hit number cases, 2D micro analyses are performed and the results are shown in Figure 4. As the increase of the number of hits, the minimum residual stresses are located in deeper, and the cross points are also deeper from the surface. As the increase of the hit angle, the residual stress is also increased as magnitude manner.

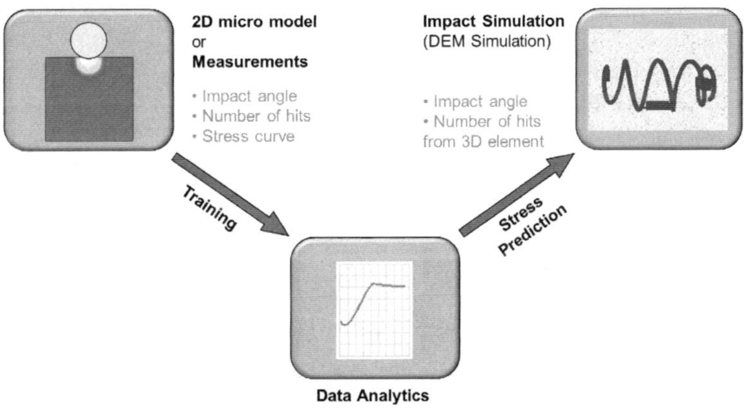

Figure 3. Shot Peening Simulation Procedure and Coil Spring Mesh Generation

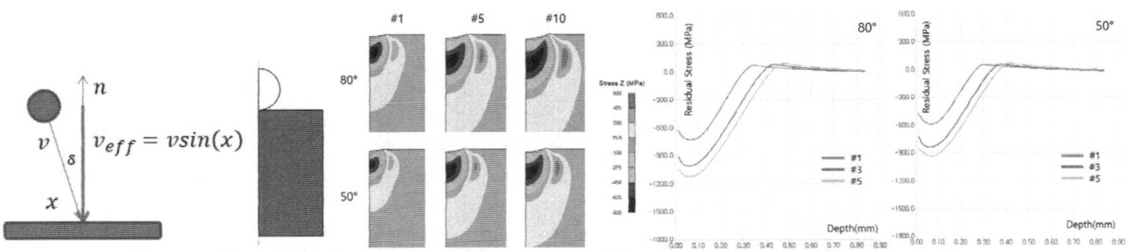

Figure 4. Analysis of the Data obtained by 2D micro model simulation

Finite Element Modelling of the Coil Spring and 2D Micro Model Simulation

The material of coil spring has flow stress as Eq. (2)

$$\bar{\sigma} = 607\bar{\varepsilon}^{0.23} + 1800 \quad (2)$$

The elastic material constants are that Young's modulus is 210 GPa and Possion's ratio is considered as 0.3. Shot ball diameter is 1mm and its velocity is 62 m/s, which is calculated by the impeller rotational speed and the radius. Coil spring is modelled by tetra elements as much as 800,000. The surface is fine-meshed to apply for accurate residual stress prediction in the depth since the residual stress varies greatly from the surface to less than 0.6mm as shown in Figure 6(a).

First Peening Simulation Results and Case Studies

Coil spring actually moves through the shot stream generated by the impeller. In simulation, the nozzle moves reversely with shot velocity 127mm/s, stream angle 50 degree and mass flow rate 1200 kg/min.

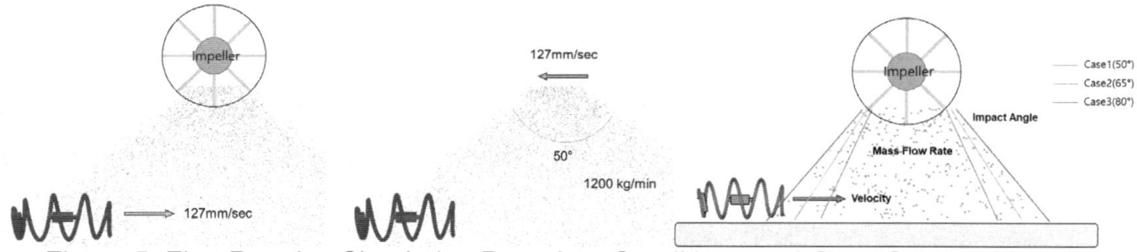

Figure 5. First Peening Simulation Boundary Conditions and Case Studies (Right)

Impact simulation results are shown in Figure 6(a). The average hit counts are 4.26 per unit area. Although this data may be hardly validated by the actual test, hit counts in the central area (A) is higher than in the upper or the lower area (B) in the coil spring. This phenomenon is expected since the shot ball stream from the impeller is perpendicular to the coil spring axis. It is a good information that there may exist some weak area in which the hit counts lower than 2. The region where hit counts are over 6 can be dangerous due to over-peening. The expected coverage is 100% in the entire surface as shown in Figure 6(b). The coverage is calculated by Eq. 1. The arc height of the Almen strip is compared with the ones from actual test data. For central and upper part of the coil spring, the actual arc heights are 0.6 and 0.3 mm. The spring-back analysis after the impact simulation will expect as 0.56 and 0.35 mm. It shows the current peening analysis can be used to optimize the peeing process.

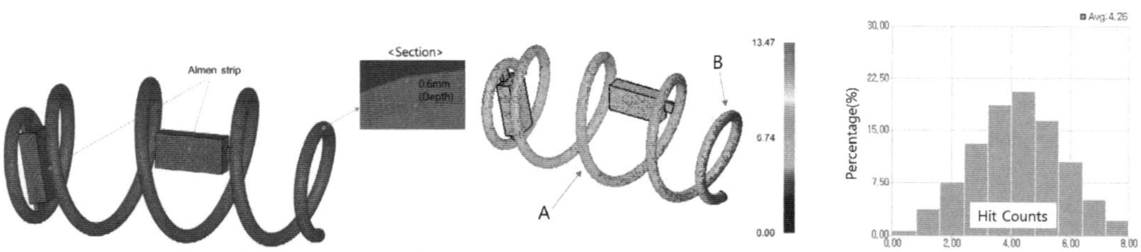

(a) Coil Spring Mesh Generation (Left) and Hit Counts Distribution

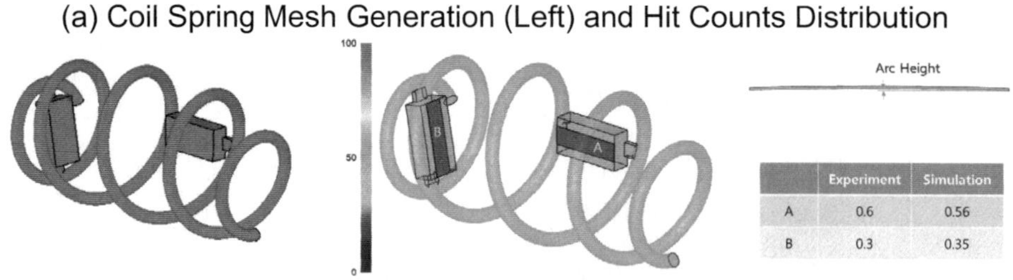

(b) Coverage and arc height

Figure 6. First Peening Simulation Results

Several peening parameters are considered as shown in Figure 5. The parameters are spray angle , mass flow rate and movement speed. Physically, the higher mass flow rate and movement speed, the lower hit counts. As the spring angle becomes obtuse, usable hit counts may be lower.

When the movement speed changes as 106, 127 and 152 mm/s with spring angle 65 degree and mass flow 1200 kg/min fixed, the results are shown in Figure 7. The average hit count is decreased as 2.31, 1.93 and 1.57. The results for other cases are shown in Figure 7.

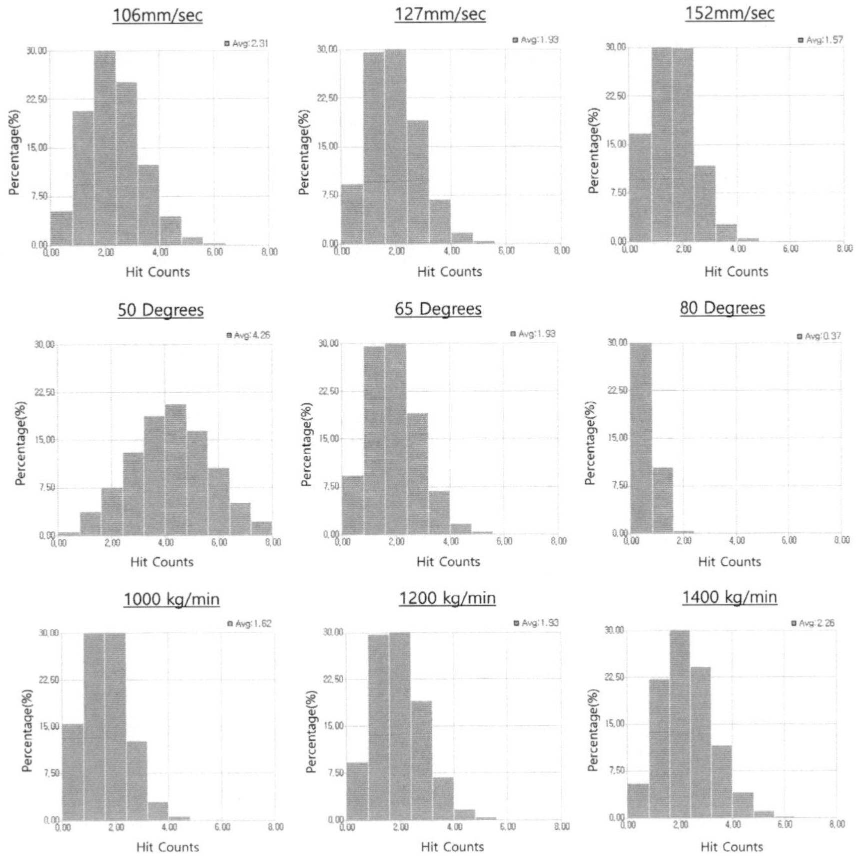

Figure 7. Case Studies Results of the First Peening

The coverage is 100% for all cases except 80 degrees. 25% area on the coil spring surface shows the coverage 0%. This is extreme case and does not happen in a coil spring manufacturing. But the simulation says if the spray angle cannot be controlled, the coverage is degraded and the coil spring must have a problem.

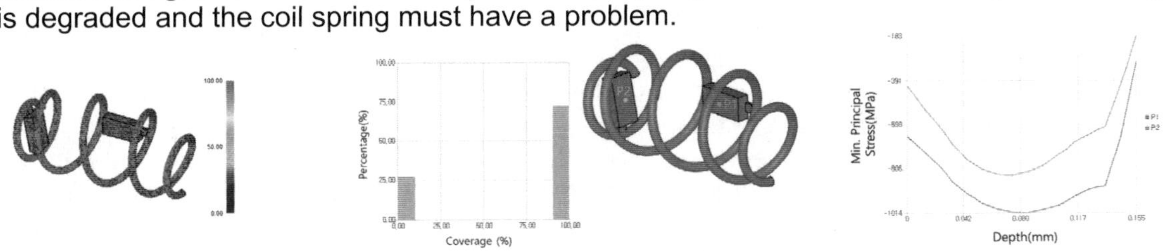

Figure 8. Extreme Case with Spray Angle 80 degree and Residual Stress at 65 Degree

The impact simulation can predict the residual stress based on 2D micro model and data analysis. The minimum principal stress is -1,014 MPa for central part in the coil spring. The residual stresses also show that the upper part of the coil spring tends to become weaker than the central part and this phenomenon is the same as the hit count density.

Second Peening Simulation

For stress peening simulation, the compressed coil spring model shape is applied with mass flow rate 350 kg/min, rotational speed 3 rad/s and process time 7 seconds. Although two impellers is actually located in the upper and the lower of the coil spring as shown in Figure 2, only one impeller is first applied due to analysis error, not converged issue.

Average hit counts in the stress peening is 70.04, which is very higher than 4.26 in first peening. In stress peening process, the distance between the coil spring and the impellers is shorter and the spray angle is more acute than first peening. The coverage is 100% except the lower part of the coil spring.

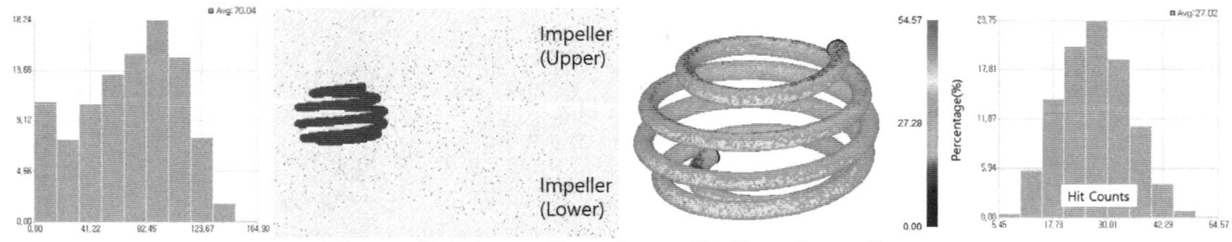

Figure 9. Stress Peening with Two impellers

To consider two impellers to shot the ball to the coil spring, mass flow rate is reduced to 85 kg/min for converged solution. The rotational speed is 3 rad/s and peening time is 6 seconds. The average hit count is 27.02, which is lower than one impeller case since the mass flow rate is decreased.

Conclusions and Discussions

Shot peening simulation was implemented to predict the peening intensity and coverage of the actual peening processes; first and stress peening. The commercial software, DEFORM, is used to model the process by three steps; 2D micro model, data analysis and impact simulation.

The arc height of Almen strip from simulation is correlated very accurately by comparing it with the experimental data. The coverage is also good correlated by actual visual inspection. From the simulation results, it is noted as follows;

1) The upper or the lower part of the coil spring tends to become easily weaker than the remaining part of the spring.

2) As the peening velocity and impact angle is increased, the hit counts will be decreased. Conversely, as the mass flow rate is increased, the hit counts are increased

3) The distribution of the hit counts tends to become "Normal Distribution".

Although the above information is amazingly interesting us, it should be improved as follows;

1) Residual stress from impact analysis for both peenings simulation shows a little unreasonable value locally.

2) Stress peening simulation does not consider the compression of the coil spring during the peening, but the compressed shape.

3) Current simulation deals with first and second peening separately but in fact, the second peeing is followed by the first one and therefore, in simulation, this must be considered.

Nevertheless, we think that the peening simulation by DEFORM® is reasonable and can be used for the optimization of the peening parameters for the conventional peening facilities. Some problems in the current version of the simulation can be soon fixed by the software development company and overcome by YOUNGWIRE engineer.

Acknowledgement

This research was supported by the Ministry of Trade, Industry and Energy and Korea Evaluation Institute of Industrial Technology(KEIT) in 2022. (20017248)

References

[1] Kirk, D. and Abyneh, M.Y., *Theoretical Basis of Shot Peening Coverage Control*, The Shot Peener Volume 9, Issue

[2] Ki-Hoon Shin, *Simulation of Surface Coverage Made by Impeller Type Shot-peening Machines*, Transactions of the Society of CAD/CAM Engineers, 19(1), pp.12-18.

[3] Majzoobi, G.H., Azizi, R. and Nia, A.A., 2006, *A Three-dimensional Simulation of Shot Peening Process Using Multiple Shot Impacts*, Journal of Materials Processing Technology, 164-165, pp.1226-1234.

Impact of Vibratory Peening on the Surface Integrity of Cemented Steel E16NCD13

A. Das[a], K. Chouchane[a], H.Y. Miao[a], B. Changeux[b], J. Badreddine[b], S. Turenne[a], M. Lévesque[a], E. Martin[a]

[a] Département de Génie Mécanique, École Polytechnique de Montréal, C.P. 6079, Succ. Centre-Ville, Montréal, Québec, H3C 3A7 Canada

[b] Safran Tech, Materials & Processes Department, Rue des Jeunes Bois, 78772 Magny-Les-Hameaux, France

Abstract

Vibratory peening is an alternative surface modification procedure that combines the shot peening and the vibratory finishing processes together. The process produces compressive residual stresses comparable to those obtained by shot peening, but with better surface finish. The process could enhance the fatigue life of the treated part and improve the industrial productivity by combining the two steps into one single operation. This study aims at determining the surface integrity properties of E16NCD13 steel after vibratory peening. Surface roughness, compressive residual stresses (CRS) and microhardness were evaluated after vibratory peening at different Almen intensities and compared with shot peening. It was found that vibratory peening decreased the surface roughness and was better than shot peening for specific Almen intensities. Moreover, vibratory peening induced deeper CRS and increased the micro-hardness of the cemented steel.

Keywords: Vibratory peening, Almen intensity, Surface roughness, Cemented steel, Residual Stress

Introduction

Vibratory peening is a recent development that combines the processes of shot peening and vibratory finishing into one step [1]. The surface roughness increases after shot peening, which raises the stress concentrations at the surface causing the fatigue life to decrease [2]. Vibratory finishing as an additional process is applied after shot peening to reduce the stress concentrations and improve the surface conditions [3]. Vibratory peening induces compressive residual stresses similar to those of shot peening but with a better surface finish. Consequently, the fatigue life could improve after vibratory peening, and the process could increase the industrial productivity since it combines two individual processes into one single operation. The key parameters of vibratory peening are processing time, vibration amplitude, media type, media mass, part position and orientation of the tub.

E16NCD13 is a case hardened cemented low alloy steel which is used widely in aerospace components like gears and gearboxes [1]. These components experience fatigue under contact loading and in the form of load reversals, for which the surfaces require treatments to improve their service lives. Case hardening is usually applied on E16NCD13, which increases the hardness of the surface around 700 HV [4]. Studies revealed that vibratory peening has the potential to alter the surface properties of E16NCD13 steel. Surface roughness was found to decrease with vibratory peening time at constant Almen intensity [1]. The maximum residual stress increased by 345% and the depth of CRS by 135%, when the Almen intensity was increased from 0.12 mmA to 0.25 mmA [1]. It was found that for higher Almen intensities, higher processing times and higher frequencies led to deeper and larger CRS in E16NCD13 steel after vibratory peening [1]. Hardness was also found to increase after vibratory peening, which is mainly affected by the reduction of media mass [1]. In other alloys, when compared with shot peening, the surface roughness after vibratory peening was lower, for similar Almen intensities [5,6]. Shot peening produces higher CRS for similar Almen intensities but higher depth of penetration is achieved by vibratory peening [5,6]. However, the current literature lacks any report on systematic investigation of the surface properties of E16NCD13 steel after shot and vibratory peening. Moreover, the influence of the vibratory peening parameters on the Almen intensity and surface integrity also needs emphasis.

The current work tries to fill up this gap. The objectives are primarily (a) calibrating the process parameters of a vibratory peening system using a DOE method and understand the influence of these parameters on the Almen intensity, and (b) compare the surface properties of E16NCD13 steel after vibratory and shot peening for different Almen intensities, where the vibratory peening is performed with process parameters found suitable from objective (a). This paper mainly discusses the comparison of surface properties after shot and vibratory peening of E16NCD13 steel for different Almen intensities, i.e., the objective (b).

Experimental Methods

The E16NCD13 steel selected in this study was obtained after case hardening with a case depth of 1.1 mm. The material is manufactured according to the ASTM 4911N standard with a chemical composition according to Table 1. Specimens of 60 x 19 x 8 mm^3 were cut from 100 mm round bars. The hardness of the steel at the surface ranged between 750 and 780 HV whereas the bulk hardness varied between 340 and 430 HV. The average surface roughness R_a of the E16NCD13 steel in the as-received condition was 0.2 μm in the longitudinal direction and 0.25 μm in the transverse direction.

Table 1: Chemical composition of the selected E16NCD13 cemented steel [7]

Element		Fe	C	Mn	Si	Cr	Ni	Cu	Mo	P	S
Wt%	Min.	Bal	0.13	0.30	0.15	0.80	3.00	-	0.20	-	-
	Max.	Bal	0.17	0.60	0.40	1.10	3.50	0.35	0.30	0.015	0.010

The vibratory peening machine was designed and manufactured by Vibra Finish in collaboration with Polytechnique Montreal and Safran Tech, France. The treated specimens were attached with a holder inserted into a tank resting on air bags and filled with carbon steel media (3 mm diameter). Two shafts with eccentric weights rotating in opposite directions on each side of the tub were used to produce oscillations in the vertical direction. The controlling parameters of vibratory peening were found out to be the eccentric weights on the shafts (X_Ecc), the media mass inside the tank (X_Mass), the rotating shaft frequency (X_Freq), the airbags pressure (X_Press), the media height above the part (X_Height), the horizontal part position (X_Pos) and the lubricant flow rate (X_Lub). The vibratory peening process was calibrated using a DOE method in which the effect of parameters on the Almen intensity was analyzed in detail. From these analyses, the process parameters selected for the vibratory peening of E16NCD13 steel at Almen intensities of 0.1, 0.15 and 0.2 mmA are: X_Freq = [20, 22.5, 25, 27.5 and 30 Hz], X_Height = 10 and 25 cm, X_Mass = 500 kg, X_Ecc = 24 kg/shaft, X_Press = 2.8 bars, X_Pos = centre and X_Lub = 20 rpm.

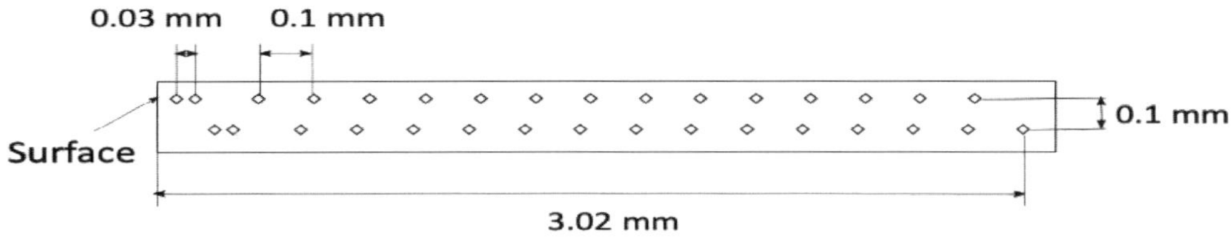

Figure 1: Schematic of the microhardness profile measurement points undertaken on the peened specimens for evaluation of the microhardness profiles.

2D surface roughness measurements were performed after peening using a Mitutoyo SV-C4000 series profilometer (resolution of 0.05 μm). The roughness parameters were calculated in accordance to ISO 4287 standard. The measurements were conducted at multiple locations on a peened surface in both longitudinal and transverse directions.

The residual stress profiles measurements were carried out using X-ray diffraction method with a goniometer Stresstech Xstress 3000 G2R, 7257/7258 according to ASTM E915-16.

Microhardness profiles were measured after peening with a MVK-H0 hardness tester using a 300 gf according to ASTM E384 standard [8]. In each case the profiles were evaluated using 32 indentations across the surface in a pattern shown in Figure 1. Two measurement lines spaced 0.1 mm apart with 16 indentations in each line were formed to measure the micro-hardness profiles.

Experimental Results

Surface roughness measured after vibratory peening showed a decrease in the surface roughness after peening at all intensities, where the decrease was more pronounced in the longitudinal direction, when compared to the transverse direction. This is because the grinding marks were along the longitudinal direction for which the surface roughness values are higher in transverse direction. Table 2 lists the average surface roughness R_a (μm) obtained for the E16NCD13 steel after vibratory peening at different Almen intensities and direction of measurement, when peened for two time periods, T_{sat} and $2T_{sat}$. T_{sat} is the processing time which corresponds to the point between arc height vs processing time where doubling the peening time produces no more than a 10% rise in arc height. Further details can be found in SAE J443 [9]. In each case three measurements were used to calculate the average values along with the standard deviation, as reported in Table 2.

Table 2: Average surface roughness, R_a (μm), of E116NCD13 steel measured for two different specimen directions and saturation time after vibratory peening at various Almen intensities

Almen Intensity (mmA)	Average Surface Roughness, R_a, (μm)					
	Longitudinal			Transverse		
	As-received	T_{sat}	$2T_{sat}$	As-received	T_{sat}	$2T_{sat}$
0.1	0.19 ± 0.03	0.09±0.02	0.08±0.009	0.25 ± 0.05	0.27±0.03	0.28±0.04
0.15		0.105±0.009	0.109±0.02		0.25±0.07	0.22±0.04
0.2		0.13±0.014	0.13±0.011		0.30±0.011	0.19±0.007

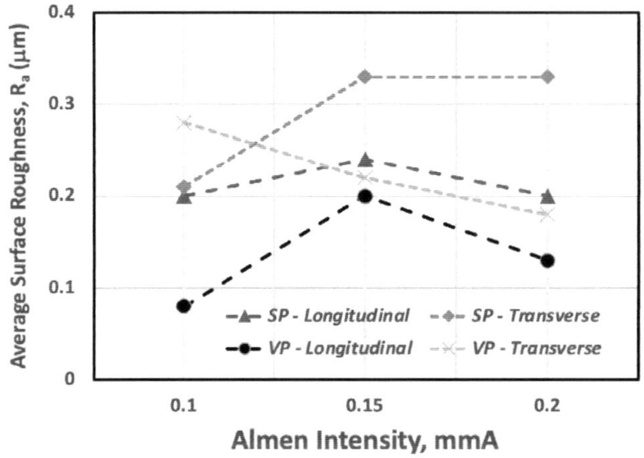

Figure 2: Variation of average surface roughness, Ra (μm), along longitudinal and transverse directions after both shot and vibratory peening at different Almen intensities.

It can be seen that the surface roughness has the tendency to increase with increasing Almen intensity in both directions. In the longitudinal direction, the surface roughness values become stable after T_{sat}, irrespective of the Almen intensity. In the transverse direction, the values decrease after T_{sat}, and the saturation of surface roughness is found to be dependent on the process parameters. When comparing with shot peening, it is observed that the surface roughness was smaller in almost all cases after vibratory peening. Figure 2 shows the average surface roughness R_a (μm) along the longitudinal and transverse directions at various Almen intensities after shot and vibratory peening. For almost all the Almen intensities, the surface roughness is lower for vibratory peening in both the longitudinal and transverse directions. The results reflect the efficacy of vibratory peening in lowering the surface roughness in comparison to shot peening.

Compressive residual stresses were evaluated at the surface up to a certain depth for E16NCD13 steel after vibratory peening at different Almen intensities. Figure 3 shows the variation of CRS with depth from surface at three different Almen intensities. The variation of CRS between Almen intensity 0.1 and 0.15 mmA is marginal, with the CRS slightly higher in the 0.15 mmA Almen intensity. However, the CRS increases substantially at 0.2 mmA Almen intensity, where the maximum CRS is -845 MPa. Interestingly, the depth at which the maximum CRS occurs is also higher in case of 0.2 mmA Almen intensity, when compared to others.

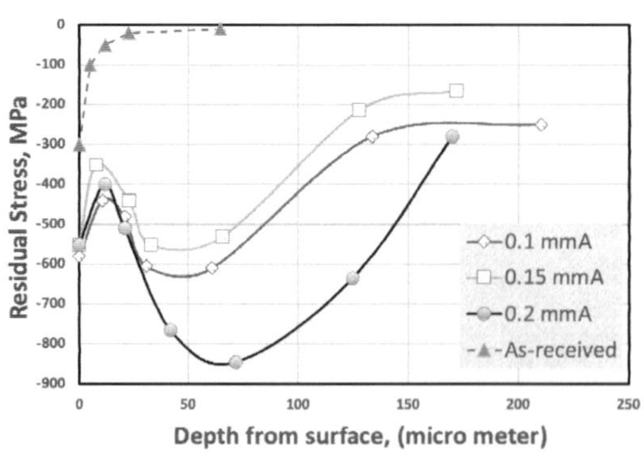

Figure 3: Variation of CRS with depth from surface after vibratory peening at different Almen intensities where the processing time was $2T_{sat}$ and X_Height is 25 cm.

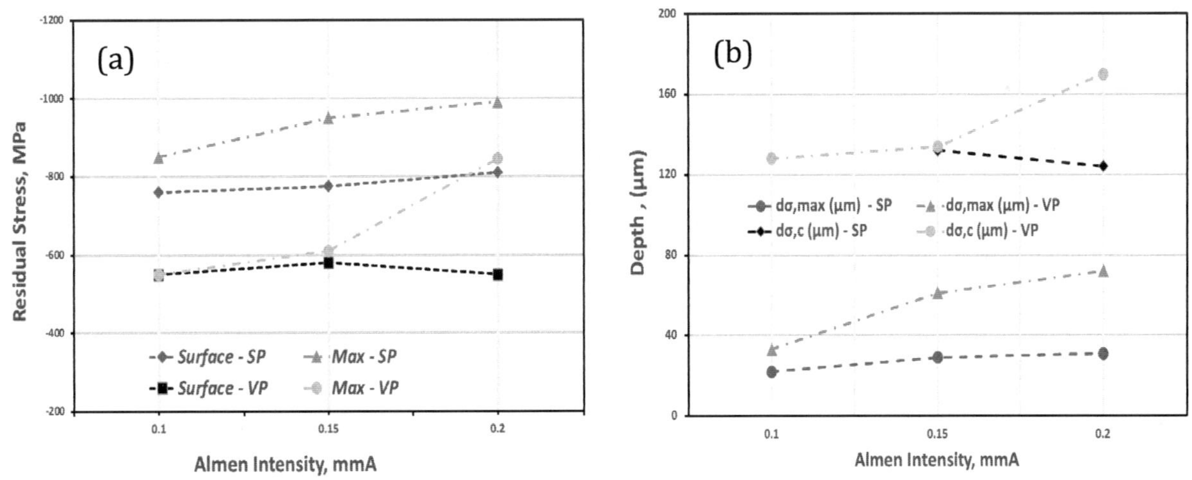

Figure 4: (a) Residual stress values measured at the surface of E16NCD13 steel and the maximum compressive residual stress after shot and vibratory peening at different Almen intensities, and (b) depth of maximum CRS values obtained from CRS distribution after shot and vibratory peening at different Almen intensities.

When comparing with shot peening, it was found that the residual stress at the surface and maximum CRS are always higher after shot peening at all the Almen intensities, as shown in Figure 4(a). However, the depth of the compressive residual stress is always higher for vibratory peening at all intensities, as shown in Figure 4(b). This comparative assessment of the distribution of CRS shows that although shot peening produces higher CRS, which will be effective in retarding the fatigue crack growth, this effect might be more pronounced up to higher depth in vibratory peening.

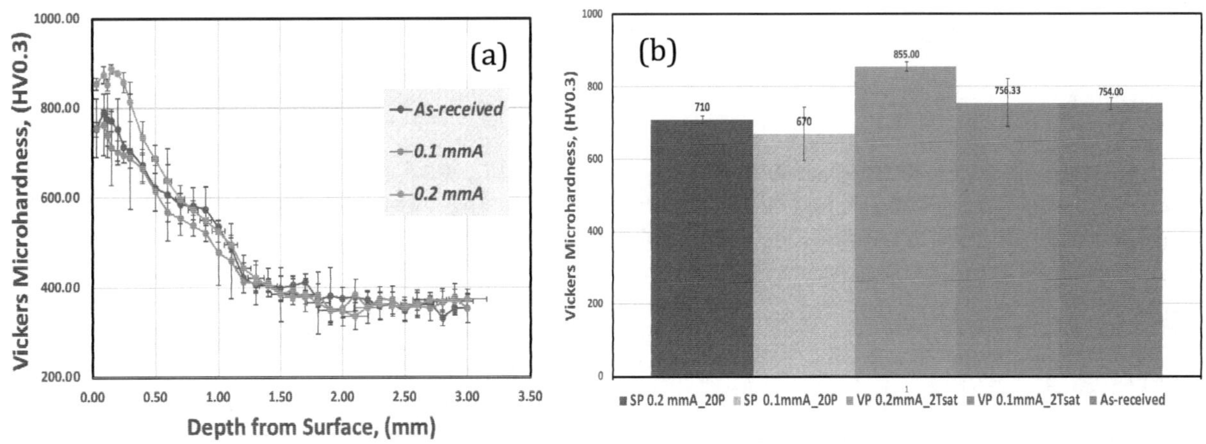

Figure 5: (a) Vickers microhardness profiles from the surface of E16NCD13 steel after vibratory peening at various Almen intensities and (b) average microhardness values at the surface after shot and vibratory peening for two different saturation time for 0.1 and 0.2 mmA Almen intensity.

Microhardness profiles were measured on the cross-section, from the surface of peening, according to Figure 1. The results show that the microhardness at the surface changes marginally from the as-received state when vibratory peened at 0.1 mmA Almen intensity. In fact, the change in hardness was not so pronounced after peening at 0.15 mmA Almen intensity. Figure 5(a) shows the hardness profiles from surface in as-received and after vibratory peened at 0.1 and 0.2 mmA Almen intensity. The increase in hardness is only observed at the surface and a few micrometers (0.6 μm) below the surface for 0.2 mmA Almen intensity. Even after shot peening, the microhardness did not show any significant variation, as shown in Figure 5(b). The maximum increase in microhardness at the surface was only found after vibratory peened at 0.2mmA Almen intensity.

Concluding Remarks

The surface of a EN16NCD13 cemented steel specimen was treated using the vibratory peening method at three different Almen intensities, which are 0.1, 0.15 and 0.2 mmA. The surface properties were evaluated after the peening and the results showed that vibratory peening can produce a better surface finish. The surface roughness was reduced after vibratory peening, although the degree of reduction depended on the direction of measurement and Almen intensity. Moreover, the surface roughness after vibratory peening was always lower than that of shot peening. In terms of residual stress, vibratory peening did produce compressive residual stresses beneath the surface, but the maximum CRS was lower than what is obtained after shot peening. However, through vibratory peening, deeper depth of CRS was achieved. Microhardness, on the other hand, showed negligible change from the as-received state and the maximum increase in microhardness was only observed at the highest Almen intensity of 0.2 mmA. Together with better surface finish and deeper CRS, vibratory peening has the potential to limit crack initiation at the surface and restrict crack growth at the sub-surface layers.

The current work only discussed the results obtained from experiments related to surface integrity. These findings need further experimental validation through observation of microstructure of the peened layers, which is being currently undertaken at Polytechnique Montreal with the technical and financial support of Safran Tech, France.

References

[1] L. Canals, J. Badreddine, B. McGillivray, H. Y. Miao, and M. Levesque, *Effect of vibratory peening on the sub-surface layer of aerospace materials Ti-6Al-4V and E- 16NiCrMo13* , Journal of Materials Processing Technology, vol. 264, pp. 91–106, 2019.
[2] S. Gangaraj and G. Farrahi, "Side effects of shot peening on fatigue crack initiation life," International Journal of Engineering, vol. 24, no. 3, pp. 275–280, 2011.
[3] D. A. Davidson, "Mass finishing processes," Metal finishing, vol. 100, pp. 104–117, 2002.
[4] Aubert and Duval, *Acier FADH 14NiCrMo13-4*, unknown year.
[5] D. Kumar, S. Idapalapati, W. Wang, D. J. Child, T. Haubold, and C. C. Wong, *Microstructure-mechanical property correlation in shot peened and vibro-peened Ni- based superalloy*, Journal of Materials Processing Technology, vol. 267, pp. 215–229, 2019.
[6] H. Miao, L. Canals, B. McGillivray, and M. Lévesque, *Comparison between vibratory peening and shot peening processes*, in International Conf. Shot Peen, 2017, pp. 521–526.
[7] SafranTech, Document matériau -Ma-0405-12-C, 2019.
[8] ASTM E384, Standard Test Method for Microindentation Hardness of Materials. ASTM International, 2017.
[9] ASTM J443, Procedures for Using Standard Shot Peening Test Strip, J443. ASTM International, 2017.

Experimental investigation of controlled single and repeated impact testing of shot peening on Ti-6Al-4V, 300M and AA7050-T7451

H. Y. Miao, M. Singhal, M. Lévesque and E. Martin

Département de Génie Mécanique, École Polytechnique de Montréal, C.P. 6079, Succ. Centre-Ville, Montréal, Québec, H3C 3A7 Canada

Abstract

During the shot peening process, multiple spherical shots impact the component at high velocity, causing plastic deformation in the component surface. This plastic deformation introduces beneficial subsurface compressive residual stresses. The shot peening process is governed by several important parameters that determine the shot peening results. To better understand the shot peening process and peening results on different materials, it is critical to study the effect of individual process parameters for different target materials. This paper investigates the effects of individual shot peening parameters such as shot velocity, shot diameter, angle of impact, and material properties on the diameter of the indentation and the change of the coefficient of restitution (CoR) value after a single and three repeated impacts on aerospace alloys: Titanium Ti-6Al-4V, 300M Steel and Aluminum Alloy 7050-T7451, using a single shot peening cannon. For the three studied materials, the resulting indentation diameters increase with the increase of shot diameter and shot velocity. The CoR decreases with the increase of the impact velocity and the shot diameter has very limited effect on the CoR. For multiple impacts at the same location, the CoR value increases with the increase of the number of impacts due to the previous plastic deformation.

Keywords: shot peening, controlled single impact, indentation, coefficient of restitution.

Introduction

Shot peening is a cold working surface treatment to produce compressive residual stresses and to increase fatigue properties of the treated components by inducing plastic deformation on the surface of metallic components [1]. Many shot peening process parameters influence the shot peening results, which include the size, the velocity and the impacting angle of the shot, the mass flow and the air pressure of the media flow, the exposure time, the distance between the nozzle and the target surface, and the material properties of the target component, etc. Almen intensity is the most widely used method to control the impact energy [2]. Miao et al. [3] developed an analytical model to relate the shot peening process parameters to Almen intensity. More recently, sequentially coupled discrete element modeling (DEM) and finite element modeling (FEM) approach has been applied to simulate the complicated shot peening process [4]. The DEM model usually requires the knowledge of the shot/target coefficient of restitution, which is defined as the ratio of the final and initial relative speed between the two impacting objects [4].

Breumier et al. [6] performed an experimental single shot peening using an in-house developed shot peening canon. The authors developed a 3D trajectory reconstruction to analyze the impacting and rebound trajectories of the shot and obtain the impacting and rebounding velocities to compute the CoR. The impacting indentations and the CoR values on a copper specimen were measured and compared with the FEM simulation. Seifried et al. [7] computed the evolution of the CoR on aluminum 6060 specimens over multiple impacts using a pendulum dropped by a magnet. The simulation of repeated impacts on aluminum rods shows an increase of the CoR with the increase of the number of impacts, until a stationary value is reached.

In this paper, the same shot peening cannon used by Breumier et al. [6] is applied to perform single and repeated impact tests of shot peening on Ti-6Al-4V, 300M and AA7050-T7451. The diameters of the indentation on each material after different shot diameters, shot velocities were obtained and compared to the analytical predictions. The CoR after single impact and repeated impacts were analyzed.

Experimental Methods

A shot peening cannon set up as shown in Figure 1 was used to perform single and repeated impacts on Ti-6Al-4V, 300M and AA7050-T7451 specimens using bearing balls having diameters of 1.19 mm and 2.50 mm. Three impacting angles of 0º, 15º and 30º with respect to the normal direction were considered. Two highspeed Phantom V310 cameras positioned at the top and left of the specimens were used, together with specially built 3D-eye software, to compute the impact and rebound velocity of the shot [6]. A Keyence microscope was used to examine the indentation diameters due to its efficiency for large number of measurements. A Bruker's Contour GT-X 3D profiler was used to observe the 3D surface after repeated impacts. The GT Contour method can measure both the diameter and the depth of the indentations, which could be very useful for future validation of FEM simulation.

Experimental Results

Relationship between impact velocity and impact pressure.

Figure 2 shows the relationship between the cannon pressures and the computed impact velocities on the three tested materials for both 1.19 mm and 2.50 mm single shot peening with a total number of 183 tests. Shot having a diameter of 2.50 mm produce larger impact velocity compared to the shot having a diameter of 1.19 mm. The relationship between the impact velocity V and cannon pressure P can be fitted with equation:

$$V = \alpha \times P^{0.4}, \quad (1)$$

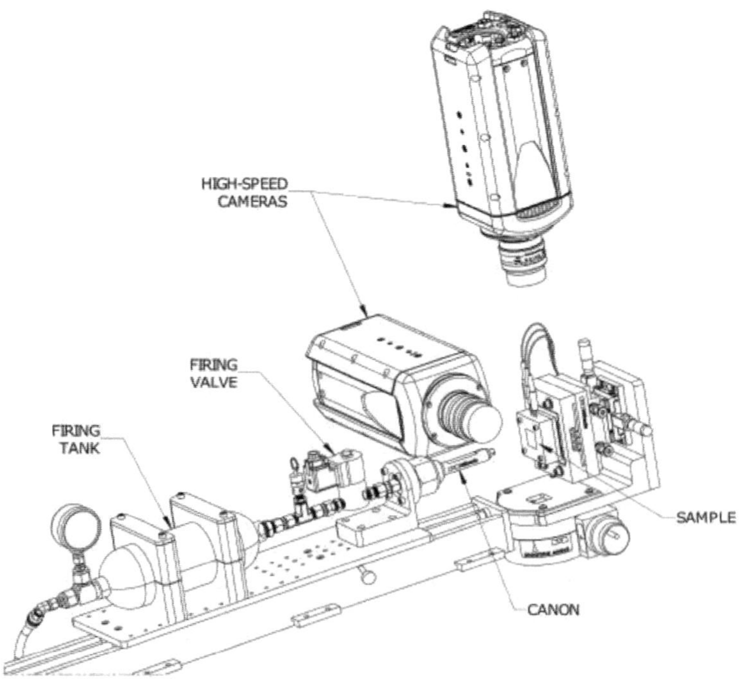

Figure 1. Shot peening cannon setup consisting of two high speed Phantom v310 cameras at top and left position to capture the 3D trajectory of the shot, nozzle, cannon and sample holding device [5].

where α is the fitting parameter, which equals to 76.28 for shot diameter of 1.19 mm and equals to 108.8 for shot diameter of 2.5 mm. This fitting equation shows that $v \propto P^{0.4}$ can be used to describe the relationship between the impact velocity and the pressure in this cannon setup.

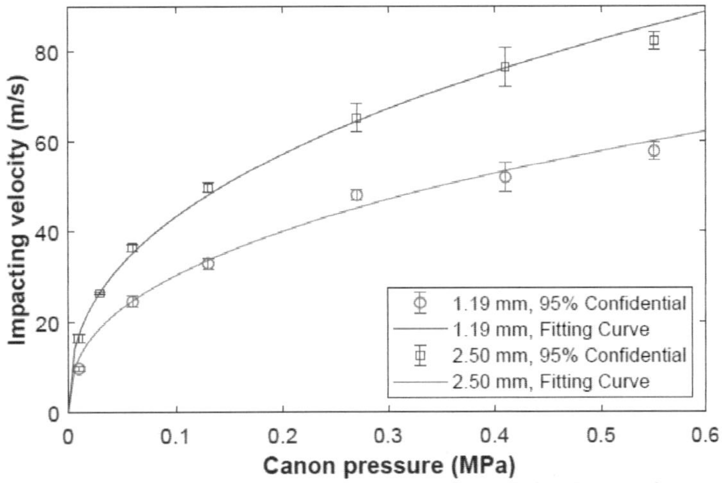

Figure 2. Relationship between the cannon pressures and the impacting velocities for two impact media having diameters of 1.19 mm and 2.50 mm.

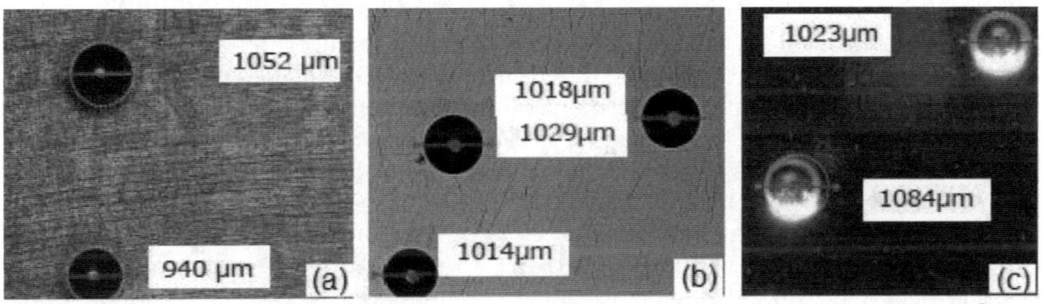

Figure 3 Surface indentation diameter measurement using Keyence microscope after one single impact for (a) Ti-6Al-4V, (b) 300M and (c) AA7050-T7451 with 2.50 mm media.

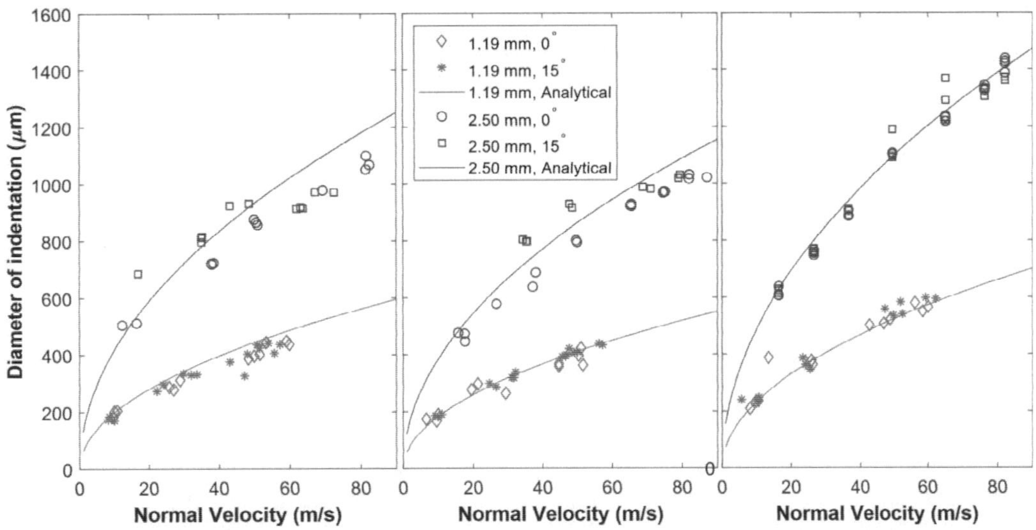

Figure 4. Relationship between the diameter of the indentations and the impact velocities for different impact shot size and angles. (a) Ti-6Al-4V, (b) 300M and (c) AA7050-T7451.

Diameter of the impact indentation.

Figure 3 shows an example of the indentation measurements using Keyence microscope for Ti-6Al-4V, 300M and AA7050-T7451 after impacts with shot diameter of 2.50 mm. The diameters of 183 indentations were measured for the two shot sizes with different impact velocities.

Error! Reference source not found. shows the relationship between the measured diameters of the indentation for the three tested materials with respect to the normal impact velocities, respectively.
An analytical model developed by Miao et al. [3] is used to compute the indentation diameter as:

$$D = 2R\left(8\rho V^2 / 9\sigma_s\right)^{1/4} \tag{2}$$

where D is the diameter of the indentation, R is the radius of the shot, ρ is the density of the shot, which equals to 7650 Kg/m^3, V is the normal impact velocity of the shot, σ_s is the yield stress of the target materials, which equals to 900 MPa for Ti-6Al-4V, equals to 1240 MPa for 300M, and equals to 470 MPa for AA7050-T7451 [8], respectively. The analytical prediction of the diameters of the indentation using equation (2) shows that the analytical model predicts well the relationship between the indentation diameters and the impact velocities for the single impact. Both experimental measurements and the analytical predictions show that the indentation diameters increase with the increase of the impact velocities and the shot diameters.

Influence of shot diameter and shot velocity on the Coefficient of restitution.

Figure 5 presents the evolution of the CoR with respect to the impacting velocity for two shot diameters, three impacting angles and different impacting velocities for Ti-6Al-4V. Figure 5(a), (b) and (c) show the total CoR, normal CoR and tangential CoR, respectively. The total CoR includes the effect of both the normal and tangential components. Both the total CoR and the normal CoR decrease with the increase of the impact velocity. The diameter has very limited influence on the normal CoR. For the same initial impacting velocity, the total CoR increases with the increase of the impact angles. This is because during the angle impact, a smaller energy is absorbed by plastic deformation, when compared to normal impact. The impacting angle has no influence on the normal CoR. The tangential CoR shows a larger dispersion than the normal CoR in the lower range of the tangential velocity smaller than 10 m/s. For larger impact angle, the tangential CoR increases with the increase of the initial tangential velocity.

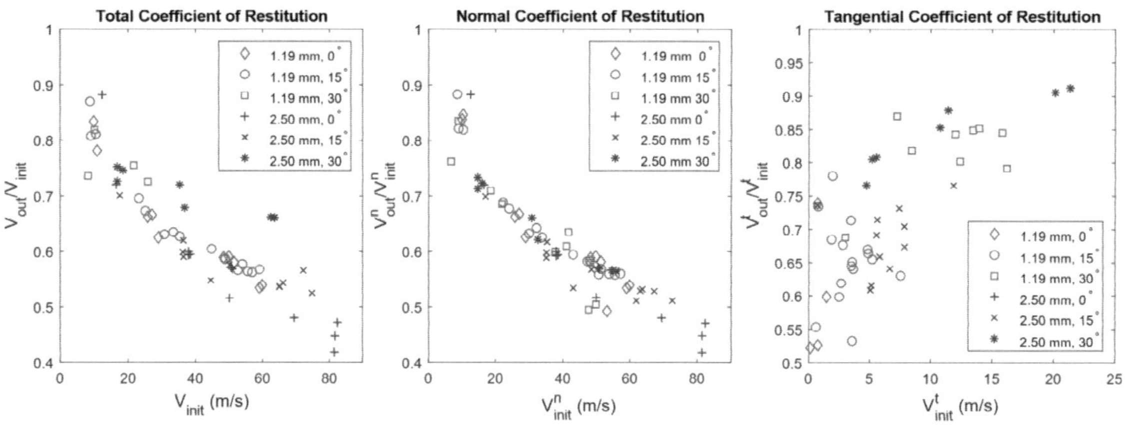

Figure 5 Relationship of CoR with the velocity before the impact for different shot diameters and impact angles for Ti-6Al-4V after single impact for: (a) Total CoR, (b) Normal CoR and (c) Tangential CoR.

It should be noted that only the values of Ti-6Al-4V are presented as examples, similar figures for 300M and AA7050-T7451 have been obtained in the test with different CoR ranges. In general, the CoR values on Ti-6Al-4V and 300M steel are higher than the values on AA7050-T7451 due to the lower yield stress value of the aluminum alloy component.

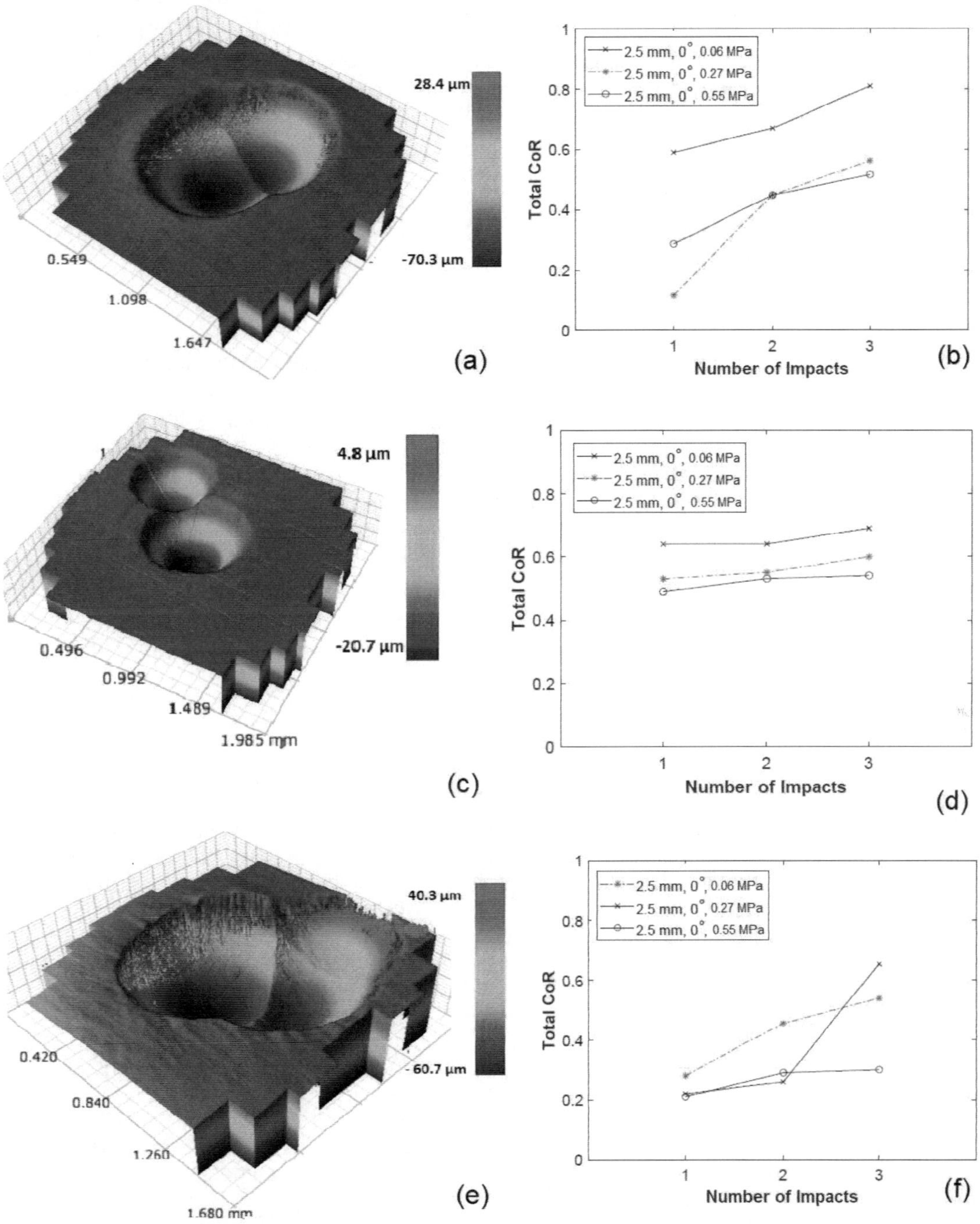

Figure 6 Contour analysis and CoR for shot diameter of 2.5 mm and impact angle of 0°, pressure of 0.27 MPa after multiple impacts, (a), (c) and (e) are for Ti-6Al-4V, 300M and AA7050-T7451 after the third impact, respectively. (b), (d) and (f) are for the relationships between the Total CoR value with respect to the number of impacts for Ti-6Al-4V, 300M and AA7050-T7451, respectively. The red curves in (b), (d) and (f) are corresponding to the indentation images in the left side for each material.

CoR after repeated impact.

Figure 6(a), (c) and (e) show the 3D surface contour measurements after three repeated impacts with 2.5 mm shot, impact angle of 0° and pressure of 0.27 MPa on Ti-6Al-4V, 300M and AA7050-T7451 specimens, respectively. Figure 6(b), (d) and (f) present the normal CoR values after one, two and three repeated impacts at the same location for shot diameter of 2.5 mm, impact angle of 0°, and pressure of 0.06, 0.27 and 0.55 MPa on Ti-6Al-4V, 300M and AA7050-T7451 specimens, respectively. In these figures, the dashed red lines are consistent with the impacting images by the contour 3D measurements in their left side, the other two cases have not been measured by the 3D contour profiler. In Figure 6(a) and (e), the three are almost coincident. Therefore, the measured CoR increases significantly with the increase of the number of impacts, as shown in Figure 6(b) and (f). This increase of the CoR is due to the plastic deformation of the previous impact which is hardening the material. However, in Figure 6(c), the repeated impacts did not completely coincide, resulting in a slightly increase of the CoR as shown in Figure 6(d).

Conclusions

For the current shot peening cannon set up, a quantitative relationship between the shot impacting velocity and the pressure were obtained for shot media having diameters of 1.19 mm and 2.50 mm. The diameter of indentations of a single impact after different shot diameters, shot velocities, and impact angles have been measured by Keyence microscope and compared with analytical predictions. Both experimental measurements and analytical predictions show that with the increase of the shot diameters and shot velocities, the indentation diameters increase. For the three studied materials, the total and normal CoR increase with the increase of the impact velocities. Shot diameter and impact angle have no influence on the normal CoR. The increase of the impact angles increases the total CoR and the tangential CoR. For repeated impact at the same location, with the increase of the number of impacts, the CoR increases due to the hardening effect of the previous impact.

References

[1]. Lieurade, H. P., and A. Bignonnet. 1987. Fundamental aspects of the effect of shot peening on the fatigue strength of metallic parts and structures. In *3rd International Conference on Shot Peening, Garmisch-Partenkirchen, Germany*, pp. 343-360.

[2]. SAEJ443, Procedures for Using Standard Shot Peening Test Strip, 2017, SAE International.

[3]. Miao, H. Y., S. Larose, C. Perron, and Martin Lévesque. 2010. An analytical approach to relate shot peening parameters to Almen intensity. *Surface and Coatings Technology* 205, no.7, 2055-2066.

[4]. Tu, F., Delbergue, D., Miao, H., Klotz, T., Brochu, M., Bocher, P. and Levesque, M., 2017. A sequential DEM-FEM coupling method for shot peening simulation. Surface and Coatings Technology, 319, pp.200-212.

[5]. Ahmad, M., Ismail, K. A., and Mat, F. 2016. Impact models and coefficient of restitution: a review. J. Eng. Appl. Sci., 11(10), 6549-6555.

[6]. Breumier, S., Trudeau-Lalonde, F., Lafrance, T., Robert, E., Villani, A., Kermouche, G., and Lévesque, M. 2021. Controlled single and repeated impact testing for material plastic behavior characterization under high strain rates. Strain, 57(6), e12399.

[7]. Seifried, R., Schiehlen, W., & Eberhard, P. 2005. Numerical and experimental evaluation of the coefficient of restitution for repeated impacts. International Journal of Impact Engineering, 32(1-4), 508-524.

[8]. M. Singhal, 2022.Experimental Investigation of Controlled Single and Multiple Impacts of Shot Peening on Titanium Ti-6Al-4V Alloy, 300M Steel and Aluminum Alloy 7050-T7451, Msc Thesis. Ecole Polytechnique de Montréal.

Acknowledgement

The authors would like to appreciate Safran SA for funding part of the work and providing Ti-6Al-4V and 300M test samples.

The effect of high temperature shot peening on the surface integrity of 2024-T351 aluminum alloy.

A. Sadallah[1], J. Badreddine[2], B. Changeux[2], S. Turenne[1], H. Y. Miao[1], M. Lévesque[1], and E. Martin[1]

1 Département de Génie Mécanique, École Polytechnique de Montréal, C.P. 6079, Succ. Centre-ville, Montréal, Québec, H3C 3A7, Canada.
2 Safran Tech, Materials & Processes department, Rue des Jeunes Bois, 78772 Magny-Les-Hameaux, France.

ABSTRACT

Shot peening is considered to be an important surface modification procedure for structural materials to improve their fatigue properties. The surface properties can be further improved by changing the temperature of the surface during the process of peening. In this study, shot peening at room temperature and at high temperatures ranging from 100°C to 250°C was conducted using an Almen intensity of 0.12 mmA to study the influence of the process on the arc height, the surface roughness, and the work hardening of AA2024-T351 specimens. Arc height measurements, 2D roughness measurements, and Rockwell B hardness tests were performed on each shot peening condition. The results show that the high temperature shot peening process at 200°C had an effect on increasing the arc heights of specimens when compared to specimens peened at the other high temperatures, the surface roughness increased with the increase of the peening temperature, and the highest value of surface hardness was seen at a temperature of 250°C.

Keywords High temperature shot peening, Almen intensity, 2024-T351 aluminum alloy, Surface roughness, Surface hardness, Arc height.

Introduction

Shot peening (SP) is a surface treatment where a metallic surface is impacted with high velocity shot to introduce plastic deformation. The plastic deformation induces compressive residual stresses (CRS) and grain refinement that can delay crack initiation at the surface and crack propagation beneath the surface, and therefore increase the fatigue life of the component [1]. However, the induced CRS in shot peened specimens relax under cyclic loading [2], which reduces the effectiveness of the SP process at increasing fatigue life. When compared to SP at room temperature (RT), SP performed at higher temperatures leads to better fatigue strength for steels [3-6] , titanium and aluminum composites [7-8], as well as for magnesium alloys [9]. The fact that the CRS span over a larger depth, are more compressive and relax less than those induced by room temperature (RT) SP under cyclic loadings [3-6]–[6] could explain this increased performance. The publicly available literature related to the mechanical performance after high temperature surface treatments of aluminum alloys focusses on high temperature laser shock peening (HTLSP) [10] and high temperature deep rolling (HTDR) [11]. High temperature surface treatments such as HTLSP and HTDR increase the surface hardness of aluminum alloys by 17% to 73%, respectively. This surface hardening phenomenon is due the dynamic strain aging (DSA) that occurs during the high temperature deformation of aluminum alloy. The DSA combines both strain hardening effect through the surface plastic deformation and the precipitation hardening effect through the generation of nano-precipitates in the surface of treated materials. No study has been reported in the publicly available literature on the effect of high temperature shot peening HTSP on the surface integrity of aluminium alloys. In this research, we investigate the effects of the HTSP on AA2024-T351 specimens having dimensions of Almen strips. The AA2024-T351 aluminum alloy is widely used in the aerospace industry for manufacturing airplane components due to its high strength to weight ratio and excellent fatigue resistance [12]. The HTSP temperature investigated range lies between RT and 250°C. HTSP specimens are characterized in terms of arc height, surface roughness, and hardness. The relationships between the treated temperatures and the resulting arc height of the strips, the surface roughness and hardness were obtained.

Experimental methods

The material used for this study is AA2024-T351, supplied from 12.9 mm thick rolled plate. Its composition range is listed in Table 1. The T351 treatment for aluminum alloy indicates that it is solution heat treated followed by a stress relieving operation where it is stretched by 1-1.5% and then naturally aged for 96 hours [4]. AA2024-T351 specimens having dimensions of 76 mm × 19 mm × 2 mm were prepared from rolled plates and the as-rolled surface was mechanically polished to a mirror finish with a 0.05 µm colloidal silica solution prior to the shot peening process.

AA2024-T531 specimens were peened for a single Almen intensity of 0.12 mmA with full coverage, at RT and at high temperatures ranging from 100 to 250 °C with an interval of 50°C. Fifteen tests (5 temperatures x 3 repetitions) in total were performed. The AA2024-T351 strips were fixed to a steel holder. A high temperature setup for performing HTSP tests was installed inside a conventional SP machine, which includes an induction coil and a pyrometer. The induction coil during HTSP tests heats the steel holder by induction, and the holder heats the sample by conduction. The pyrometer is used to record the surface temperature during the tests. The determination of the SP process conditions consisted in performing several tests based on a previous experience on the SP machine. Ceramic media Z425 (0.425 mm diameter), a stand-off distance of 152.4 mm, a peening angle of 90° between the nozzle and the specimen holder, an air pressure of 48 kPa, and a mass flow of 4.5 kg/min were applied to obtain an Almen intensity of 0.12 mmA. The specimens were heated simultaneously during the SP process and the temperature profiles were recorded using the pyrometer. Figure 1 shows a typical surface temperature profile for a target temperature of 200 °C. To obtain a stable temperature during the shot peening process, the induction heating was programmed at a temperature (induction temperature) of 10 °C lower than the target SP temperature to avoid the overheating of the steel holder. The HTSP process included 3 main steps, as shown in Figure 1. Step 1 is a simple heating phase where the induction is activated to heat the holder and specimen at the target peening temperature. Only air is blown on the surface without media. At the end of this step the nozzle moves above the specimen to activate the media. Step 2 is the HTSP process. In this step, the induction heating is always activated, and the media are impacted onto the surface of the specimen. At the end of this step, the nozzle is moved above the specimen for one second to deactivate the media, which explains the increase in the temperature at that stage. Step 3 is a cooling phase. The induction heating and the mass flow are deactivated. However, the air still blows to cool down the specimen.

Table 1: Chemical composition (wt%) of the 2024-T351 aluminum alloy [3].

Element		Cu	Mg	Mn	Fe	Si	Zn	Ti	Cr
% weight	Min	3.8	1.2	0.3	-	-	-	-	-
	Max	4.9	1.8	0.9	0.5	0.5	0.25	0.15	0.1

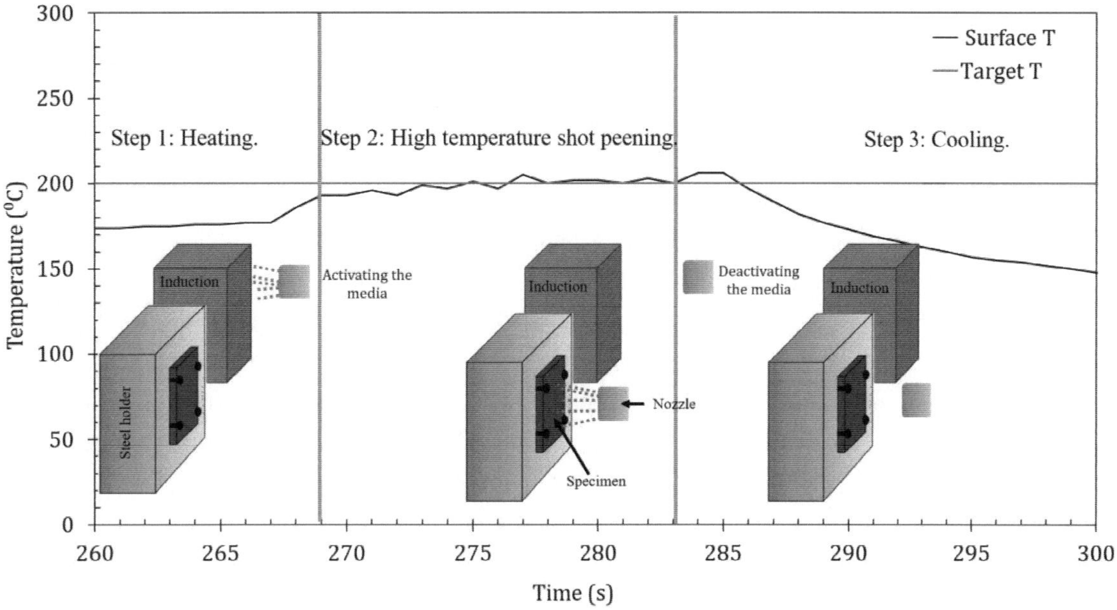

Figure 1: Typical temperature profile on the surface of the specimen while shot peening at 200°C. Step 1: the induction heating and the air pressure are activated to heat the sample at the desirable temperature. Step 2: start of the high temperature shot peening phase where the induction, the air pressure, and the media flow are activated. Step 3: an air-cooling phase where the induction and the media flow are deactivated, and the air flow is kept activated.

Three repetitions were made for each condition to evaluate the discrepancy by considering the initial deflection of the specimens. The deflections of the specimens were measured using an Almen gauge. A simple computation was made to obtain the arc height of the specimens by subtracting the initial measured arc height from its value after peening to consider the initial deflection of the specimens. A Mitutoyo profilometer was used to measure the surface roughness profiles along the longitudinal direction of the specimens. Rockwell B hardness tests using a 1/16 inch diameter ball indenter with a 100 kg load were carried out on the 15 AA2024-T531 peened specimens.

Results and discussion

Figure 2 (a) presents the resulting arc height values from shot peening of the AA2024-T351 specimens at different temperatures. The SP condition resulting in the highest arc height is at RT with an average arc height of 0.222 mm, followed by SP at 200°C with an average arc height of 0.195 mm. Whereas, SP at 100°C had the smallest average arc height of 0.105 mm, SP at 150°C and SP at 250°C, result in average arc heights of 0.113 mm and 0.120 mm, respectively. The resulting arc height is caused by the plastic deformation stored in surface layer of the specimen during SP [13]. It is thus not surprising to observe the maximum value at RT where the strain hardening is maximum. At higher temperature, the arc height decreases in most of the cases. This is due to strain softening mechanism that occurs at low temperature in aluminum alloys [14]. The outliers at 200°C, is due to the larger compressive residual stresses that have been produced at this peening temperature. This could be due to a change in the deformation mechanism operating in this temperature range. More investigation is required to understand the behavior of the alloy under HTSP conditions.

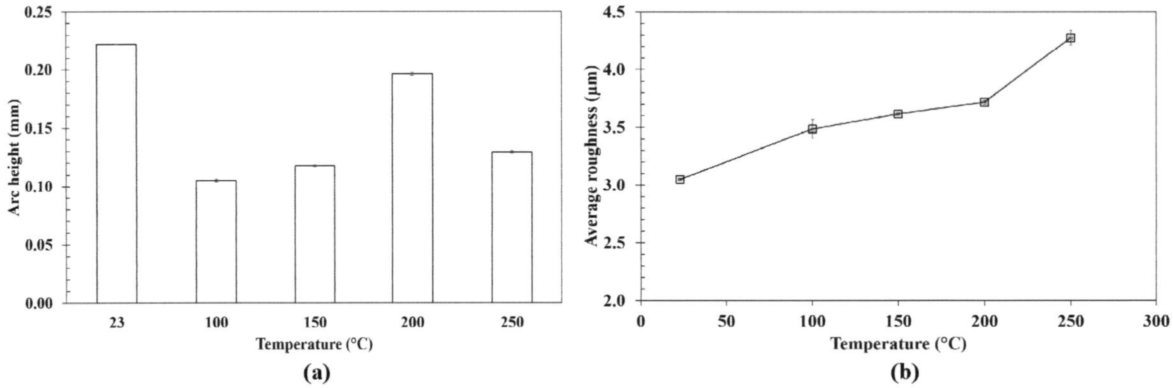

Figure 2: (a) Arc heights resulting from peening of AA2024-T351 strips at an Almen intensity of 0.12 mmA and at RT (23°C), 100, 150, 200, and 250°C. (b) The surface roughness profile under the same peening conditions.

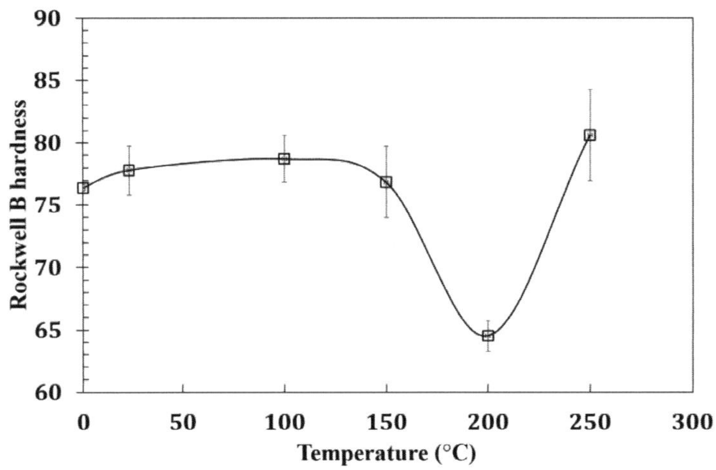

Figure 3: The Rockwell B hardness of the AA2024-T531 specimens treated using an Almen intensity of 0.12 mmA and at temperatures of 23, 100, 150, 200, and 250°C.

The change of surface roughness after shot peening is critical for the fatigue properties. Arithmetic average (Ra) was extracted from the roughness profiles to characterize the mean depth of the peening indentations. Figure 2 (b) shows the Ra of the AA2024-T351 specimens peened at different temperatures. The average surface roughness increases with the increase of temperature. When the temperature increases, the yield point and the hardening mechanisms of the material decrease. Therefore, the media indentation increases, which relates to larger surface roughness.

Figure 3 shows the relationship between the surface hardness and the SP temperature. The as-received specimens' hardness is shown by 0 °C in the temperature axis. From RT to 100°C, no significant change in hardness is observed. After 100°C a decrease in the surface hardness is shown. The hardness has a minimum value at 200°C and a maximum value at 250°C. We assume that the hardness evolution in the case of HTSP of AA2024-T351 comes from the two competing mechanisms: hardening and softening as shown in the case of HTLSP of AA6160 [10], HTLSP of AA7075 [15], and HTDR of AA6110 [11]. The decrease in the surface hardness is due to the softening mechanism activated in the temperature range of 100 to 200°C. We also assume that the increase in hardness at 250°C is maybe due to the hardening mechanism activated at this temperature as seen in the case of AA7075 after HTLSP at 250°C [15].

Conclusion

SP and HTSP treatments on AA2024-T351 specimens were carried out to reveal the effects of temperatures ranging from RT to 250°C on the resulting arc height of the specimen arc height, surface roughness, and the hardness of the aluminum alloy. At RT and at 200°C, the AA2024-T351 specimens showed high deflections, which indicate the large CRS. The average surface roughness increased with the increase of the peening temperature that can be related to the large surface plastic deformation induced during the HTSP process, which led to larger dents. When compared to the as-received specimen and the SP at RT, HTSP at 250°C led to highest material hardness. CRS measurements and microstructural observations will be conducted on the AA2024-251 specimens peened at RT and high temperatures to understand the material behavior after SP at different temperatures.

References

[1] C. Bianchetti, D. Delbergue, P. Bocher, M. Lévesque, and M. Brochu, "Analytical fatigue life prediction of shot peened AA 7050-T7451," *Int. J. Fatigue*, vol. 118, no. July 2018, pp. 271–281, 2019.

[2] M. Benedetti, V. Fontanari, and B. D. Monelli, "Numerical simulation of residual stress relaxation in shot peened high-strength aluminum alloys under reverse bending fatigue," *J. Eng. Mater. Technol. Trans. ASME*, vol. 132, no. 1, pp. 0110121–0110129, 2010.

[3] A. Wick, V. Schulze, and O. Vohringer, "Influence of the shot peening temperature on the relaxation behaviour of residual stresses during cyclic bending," *7th Int. Conf. SHOT Peen.*, vol. 4140, pp. 102–109, 1999.

[4] R. Menig, V. Schulze, and O. Vöhringer, "Optimized warm peening of the quenched and tempered steel AISI 4140," *Mater. Sci. Eng. A*, vol. 335, no. 1–2, pp. 198–206, 2002.

[5] and G. G. M Schilling-Praetzel, F Hegemann, "Influence of temperature of shot peening on fatigue life," in *Influence of temperature of shot peening on fatigue life*, 1993, vol. 5, pp. 227–238.

[6] and J. S. Akira Tange, Hiroshi Koyama, Hiroto Tsuji, "Study on warm shot peening for suspension coil spring.," *SAE Trans.*, pp. 463–467, 1999.

[7] W. Luan, C. Jiang, V. Ji, Y. Chen, and H. Wang, "Investigation for warm peening of TiB2/Al composite using X-ray diffraction," *Mater. Sci. Eng. A*, vol. 497, no. 1–2, pp. 374–377, 2008.

[8] W. Luan, C. Jiang, and V. Ji, "The texture effect of warm peening on TiB2/Al composite," *Mater. Sci. Eng. A*, vol. 504, no. 1–2, pp. 124–128, 2009.

[9] Y. Huang, W. C. Liu, and J. Dong, "Surface characteristics and fatigue performance of warm shot peened wrought magnesium alloy Mg-9Gd-2Y," *Mater. Sci. Technol. (United Kingdom)*, vol. 30, no. 12, pp. 1481–1487, 2014.

[10] C. Ye, Y. Liao, and G. J. Cheng, "Warm laser shock peening driven nanostructures and their effects on fatigue performance in Aluminum Alloy 6160," *Adv. Eng. Mater.*, vol. 12, no. 4, pp. 291–297, 2010.

[11] P. Juijerm and I. Altenberger, "Effect of high-temperature deep rolling on cyclic deformation behavior of solution-heat-treated Al-Mg-Si-Cu alloy," *Scr. Mater.*, vol. 56, no. 4, pp. 285–288, 2007.

[12] G. Chen, L. Lu, C. Ren, and X. Ge, "Temperature dependent negative to positive strain rate sensitivity and compression behavior for 2024-T351 aluminum alloy," *J. Alloys Compd.*, vol. 765, pp. 569–585, 2018.

[13] W. Cao, R. Fathallah, and L. Castex, "Correlation of Almen arc height with residual stresses in shot peening process," *Mater. Sci. Technol. (United Kingdom)*, vol. 11, no. 9, pp. 967–973, 1995.

[14] H. Z. Li, H. J. Wang, X. P. Liang, H. T. Liu, Y. Liu, and X. M. Zhang, "Hot deformation and processing map of 2519A aluminum alloy," *Mater. Sci. Eng. A*, vol. 528, no. 3, pp. 1548–1552, 2011.

[15] C. Ye, Y. Liao, S. Suslov, D. Lin, and G. J. Cheng, "Ultrahigh dense and gradient nano-precipitates generated by warm laser shock peening for combination of high strength and ductility," *Mater. Sci. Eng. A*, vol. 609, pp. 195–203, 2014.

Discrete elements method simulation of the movement of media and the velocity between the media and treated parts during the vibratory peening process

L. De La Torre [a], H. Y. Miao [a], B. Changeux [b], J. Badreddine [b], S. Turenne [a], M. Lévesque [a]

[a.] Département de Génie Mécanique, École Polytechnique de Montréal, C.P. 6079, Succ. Centre-Ville, Montréal, Québec, H3C 3A7 Canada

[b.] Safran Tech, Materials & Processes department, Rue des Jeunes Bois, 78772 Magny-Les-Hameaux, France

Abstract

Vibratory peening combines the shot peening and vibratory finishing processes into one process. It has the same effect as shot peening that increases fatigue life by inducing compressive residual stresses while polishing the treated part, which is usually done by vibratory finishing. In this work, the vibratory peening process was investigated to assess if a correlation can be made with the shot peening process using Almen strips. A vibratory peening machine was equipped with a data acquisition system that can measure the displacement of the whole machine in a three-dimension space. Then, Almen strips were treated using 3 mm steel bearing balls while the data acquisition system recorded the vibrations modes of the tub from 10 Hz to 30 Hz. The vibratory peening process was simulated using a model based on the discrete element method (DEM) and its main output was the impact velocity of the media on the Almen strip. Simulations were compared to the vibratory peening treatments on Almen strips and to the literature on conventional shot peening.

Keywords: Vibratory peening, DEM simulation, impact velocity

Introduction

Shot peening (SP) is a largely studied surface enhancement process on metallic parts. Indeed, it can increase fatigue life up to 20 times, when compared to the as-machined parts [1]. SP consists in blasting spherical media particles at high velocity on a metal part, which generates plastic deformation by inducing a layer of compressive residual stresses and cold working. The shot peening process increases fatigue life because its compressive residual surface stresses delay crack initiation. However, SP increases surface roughness that is unfavorable to fatigue life since it creates local stress concentrations. A post finishing process is usually applied onto the shot peened part to reduce surface roughness [2]. The vibratory finishing (VF) process is a mass finishing process developed for polishing, deburring and cleaning small workpieces. This process consists in vibrating a tub filled with polishing media. The workpieces are inserted inside the polishing media, and the vibrations induce a relative motion between the workpiece and the media particles that polishes the workpieces [3].

The vibratory peening (VP) process is a manufacturing finishing process that aims to combine both SP and VF by fixing the workpiece in the tub filled with media. The kinetic energy transmitted to the treated surface by the impacts is higher, when compared to the vibratory finishing process, because the fixed workpiece is submitted to that of higher velocity impacts. VP induces similar effects to those of SP on the workpiece, such as compressive residual stresses, while decreasing the surface roughness simultaneously. Therefore, it couples the same effects of both SP and VF.

The purpose of this study was to assess the motion of the media inside a vibratory peening machine, which is essential to understand the behavior of both the vibratory peening machine and the impacts between the media and the treated part. It is a preliminary study that aims to further guide the modelling of the vibratory peening process.

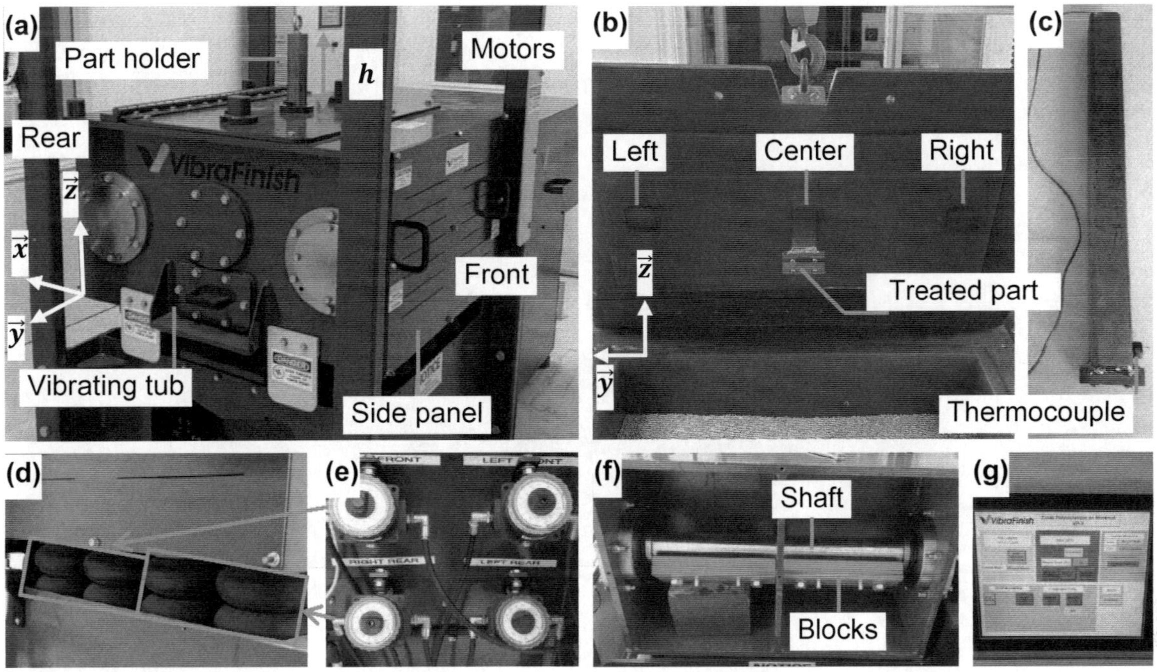

Figure 1 Images of the vibratory peening machine illustrating its operating parameters. (a) The vibrating tub is filled with media and the depth of the part inside the media is controlled by **h**, which is the height of part holder above the tub. (b) The part holder is tightened at the center of the tub lid and two other positions (left and right) are available. (c) The part holder is equipped with a thermocouple probe to measure the temperature during the operation. (d) Two-pair airbags are located on each side of the tub. (e) The pressure in each pair is adjusted by a manometer. (f) The rotating shafts behind the side panels. The number of blocks installed on the shafts determines the eccentricity. (g) The motor drive panel controls the motor speed.

Experimental Methods

The vibratory peening machine was equipped with a setup of six laser sensors that provided the 3D displacement of the vibratory peening machine. The VP machine was filled with 3 mm diameter steel particles. The workpieces in the VP machine were Almen A strips having dimensions of $19 \times 76 \times 1.27$ mm^3. The main modes of vibrations determined experimentally were used as input of the discrete element model.

The operation consists first in fixing an Almen strip on an aluminum holder that is inserted into media inside the tub and fixed on the cover as shown in Figure 1 (a)-(b). The tub lays on airbags for which the inflating pressure can be changed (Figure 1 (d)-(e)). The vibrating motion is induced by eccentric masses on rotating shafts (Figure 1 (f)) controlled by the motor drive panel (Figure 1 (g)). A thermocouple was installed to measure the temperature near the Almen strip during operation.

A DEM model was developed to simulate the media movement inside the vibratory peening tub to obtain the impact velocities of particles against the Almen strip, and to study the correlation between Almen intensity and the corresponding impact velocity against the strip. The simulations were based on the vibratory peening tests with shaft frequencies of 17.5, 20, 22.5, 25, 27.5 and 30 Hz, a media mass of 544 kg, an eccentricity on shafts of 24 kg/shaft, an airbag pressure of 2.8 bar, a height of media above the workpiece of = 150 mm, the workpiece in the middle of the tub and a lubrication rate of 20 rpm.

Figure 2 illustrates the locations of the six sensors on the three surfaces of the tub. Two sensors were installed on the top of the tub, one sensor in the front surface and three sensors on the right surface, as shown in Figure 2 (a). The vibration amplitudes of the tub were computed with the measurements recorded from these six sensors. Figure 2 (b)-(e) shows the location and installation of each sensor on the actual vibratory peening machine.

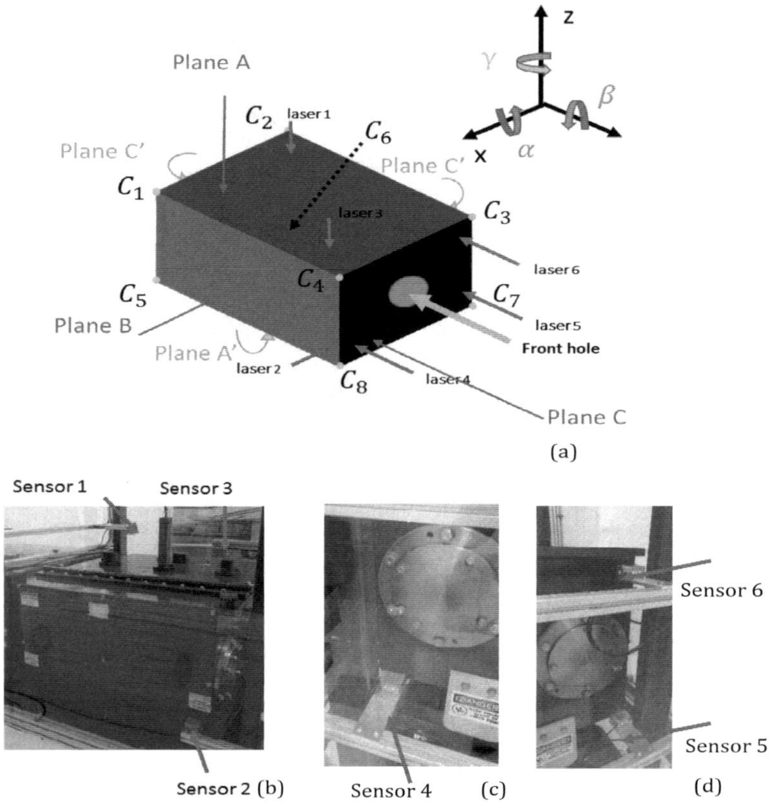

Figure 2 Installation of the six laser sensors at the three surfaces of the vibratory tub. (a) Location of the six sensors around the tub, (b) Installation of lasers 1, 2 and 3, (c) Installation of laser 4, (d) Installation of lasers 5 and 6.

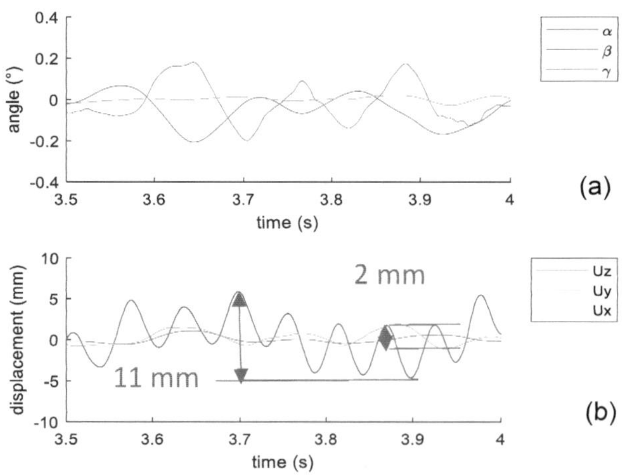

Figure 3 Computed rotations (a) and displacement (b) of the vibratory peening machine.

Experimental Results

Figure 3 presents the computed rotation and displacement of the center of the tub as a function of time for a media mass of 544 kg, an eccentricity on shafts of 24 kg/shaft, an airbag pressure of 2.8 bar, a height of media above the workpiece of 150 mm, the workpiece in the middle of the tub and a lubrication rate of 20 rpm. The figure shows that the maximum amplitude of *Uz* (the displacement in the *z* direction) is 11 mm, which is larger when compared to the maximum amplitudes in the *x* and *y* directions, which are of the order of 2 mm. The three rotation angles (α, β, γ) lie in the range of [−0.3°, 0.2°], which suggests that the tub remains relatively at the same orientation during operation. Therefore, only *Uz* is considered as the vertical amplitude of the movement of the tub for further simulations of the vibratory peening process.

Figure 4 shows the (*y, z*) cross-section of the DEM model with an Almen holder in the middle of the tub with same steel properties as those of the media. The dimensions of the holder are of $20 \times 100 \times 20$ mm^3 and the Almen strip dimensions are of $20 \times 80 \times 1$ mm^3. The depth between the bottom of the tub and the Almen strip is 250 mm. The thickness of the DEM model in the *x* direction in Figure 4 is 20 mm. The restitution coefficient between two spherical steel impacting media is defined as 0.95, according to Lecornu [4]. The restitution coefficients between the media and the Almen holder and between the media and the Almen strip were also assumed at 0.95. The restitution coefficient between particles and the tub was estimated manually by dropping one media onto the same rubber surface as the line of tub. The position of the initial drop and the position after one rebound were measured and used to compute the restitution coefficient. The restitution coefficient between the media and rubber material is defined as:

$$\boldsymbol{C_r} = \sqrt{\frac{\boldsymbol{H_{re}}}{\boldsymbol{H_{ini}}}} \tag{1}$$

where $\boldsymbol{H_{ini}}$ is the height measured at the initial position and $\boldsymbol{H_{re}}$ is the measured height after the first rebound. The drop tests were carried out 10 times and an average of restitution coefficient $\boldsymbol{C_r}$ = 0.71 was obtained. The maximum value was max($\boldsymbol{C_r}$) = 0.72 and the minimum value obtained was min($\boldsymbol{C_r}$) = 0.70 from Equation 1.

Figure 5 (a) presents the simulated average normal impact velocities to the Almen strip for six shaft frequencies of 17.5, 20, 22.5, 25, 27.5 and 30 Hz, with a media mass of 544 kg and vibrating for two seconds. Figure 5 (b) shows the Almen intensity experimentally measured after VP. In the post treatment of the simulation, every collision is tracked between all particles and the Almen strip. Every collision is associated to a normal impact velocity so all the impact velocities are saved. The normal impact velocity presented in Figure 5 (a) is the average velocity of all the impacts against the Almen strip. There is a linear correlation between velocity and shaft frequency.

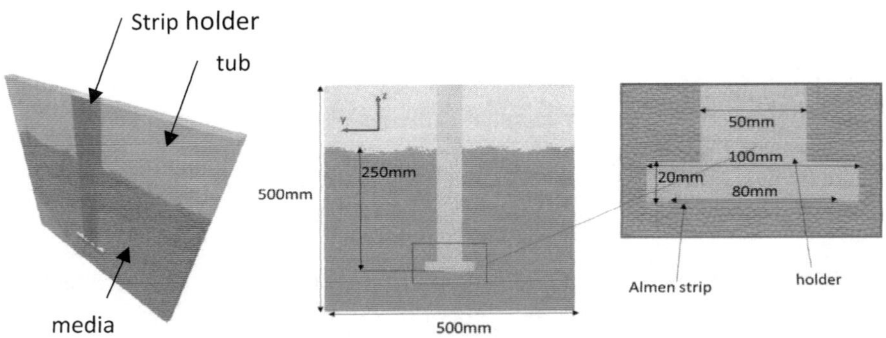

Figure 4: DEM model with 110,000 steel particles with diameter of 3 mm.

Figure 6 reports the experimentally measured Almen intensity as a function of the predicted impact velocity of the media. Figure 6 also plots the analytically predicted Almen intensity by Miao et al. [5]. The figure shows that the range of Almen intensities reached is almost the same for the six measurements (0.1 mmA to 0.2 mmA) and conventional shot peening (0.095 mmA to 0.200 mmA), which means that the vibratory peening process can achieve the same Almen intensity range as conventional shot peening. From both simulations and experimentations, an empirical linear correlation can be obtained between impact velocities and Almen intensity:

$$\boldsymbol{I} = \mathbf{0.2564} \times \boldsymbol{V} - \mathbf{0.0724} \tag{2}$$

where $\boldsymbol{I}$ is the Almen intensity (mmA) and V the normal impact velocity of the particles (m/s).

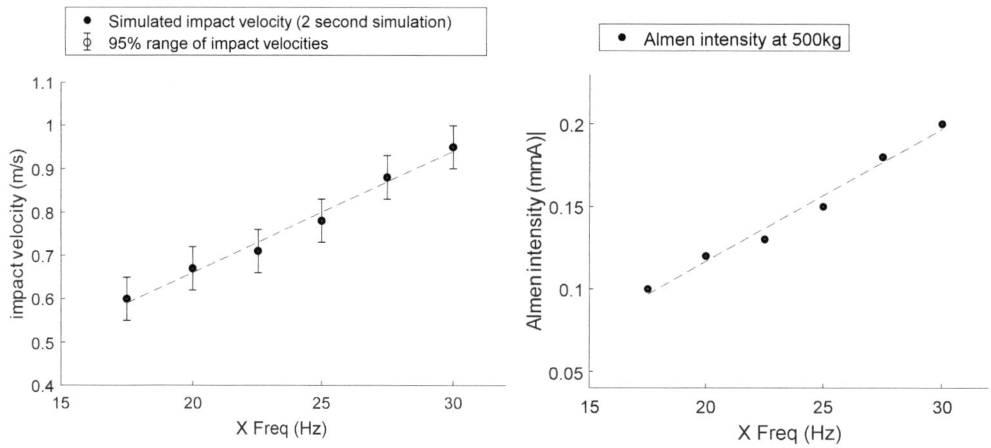

Figure 5: Impact velocity and Almen intensity according to shaft frequency of the VP machine. (a) Simulated impact velocity (b) experimentally measured Almen intensity after VP treatment.

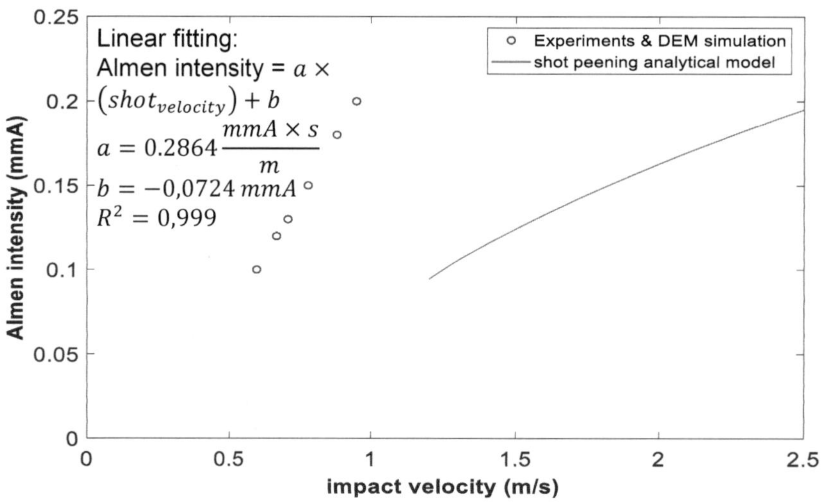

Figure 6: Experimental Almen intensity vs simulated impact velocity from DEM simulations.

A coefficient of determination R^2 = 0.999 was obtained for the experiments and DEM simulation points. In the case of the impact of a round steel particle on a steel surface, the relation between Almen intensity and impact velocity is also linear, as shown in Figure 6. It can be seen that larger impact velocities (1.2 m/s to 2.5 m/s) are needed to reach the same Almen intensities as those obtained for lower velocities (0.6 m/s to 0.95 m/s) during vibratory peening. This could partly result from the several layers of particles that increase the inertia of the shot impacting the strip, and therefore the energy transmitted to the Almen strips. This could explain the lower range of impact velocities for the vibratory peening process to reach the same Almen intensity range than the shot peening process. Further investigation is required for a deeper understanding of the difference of the two processes.

Discussion and Conclusions

A vibratory peening machine was equipped with a data acquisition system to measure its overall displacement. A discrete element method (DEM) model was developed to simulate the media movement inside the tub. The experimental results show that the vibratory peening machine has a dominant vibration mode in the vertical direction while the lateral and angular modes can be neglected. The range of achievable Almen intensities is the same as that of shot peening. However, the impact velocity range with the vibratory peening process is lower than the shot peening process, which means that energy is not only transferred by the first range of media impacting the part, but also by the other layers of particles under the first layer that transmit their inertia. Moreover, those simulations have permitted to create an empirical relation between shaft frequencies and Almen intensity. This allows to master the intensity that needs to be reached just by choosing the frequency of the vibratory peening machine. The limitation of this work is that it has been done with specific machine parameters. Therefore, it cannot directly be extrapolated to other machine parameters. Nevertheless, it provides a window of operation in which the Almen intensity can be predicted.

References

[1] T. Klotz, D. Delbergue, P. Bocher, M. Lévesque and M. Brochu, *Surface characteristics and fatigue behavior of shot peened inconel 718*, International Journal of Fatigue, vol. 110, pp. 10–21, 2018.

[2] D. E. Semones and W. H. Safranek, *Chemically accelerated metal finishing process*, Sep. 14 1976, US Patent 3,979,858.

[3] F. Hashimoto, S. P. Johnson, and R. G. Chaudhari, *Modeling of material removal mechanism in vibratory finishing process*, *CIRP Annals*, vol. 65, no. 1, pp. 325–328,2016.

[4] L. F. A. Lecornu, Dynamique appliquée. O. Doin, 1908, vol. 5.

[5] H. Y. Miao, S. Larose, C. Perron, M Lévesque, *An analytical approach to relate shot peening parameters to Almen intensity*, *Surface and Coatings Technology*, vol. 205, no. 7, pp. 2055–2066, 2010.

Control of the vibratory peening machine using the Almen intensity procedure

M. Paques[1], K. Chouchane[1], B. Changeux[2], L. De La Torre[1], J. Badreddine[2], H. Y. Miao[1], M. Levesque[1], S. Turenne[1] and E. Martin[1]

1 Département de Génie Mécanique, École Polytechnique de Montréal, C.P. 6079, Succ. Centre-Ville, Montréal, Québec, H3C 3A7, Canada
2 Safran Tech, département de Matériaux & Procédés, Rue des Jeunes Bois, 78772 Magny-Les-Hameaux, France

Abstract

The effects of the vibratory peening process parameters on Almen intensity were investigated using the Design of Experiments (DOE) method. A specific vibratory peening machine was built. The tub vibrates vertically and rests on airbags. A non-standard fractional screening design was performed to evaluate the effects of the seven adjustable controlling parameters of the machine on Almen intensity. The Analysis of variance (ANOVA) showed that the eccentricity and frequency of the rotating shafts, and the media height above the treated specimen, were the parameters to adjust for the control of Almen intensity. A first order empirical model was fitted. The model included both primary factors and two-factor interactions. Its predictions showed that Almen intensity increases with the increase of the eccentricity, frequency and media height above the part. On the other hand, the media mass, airbag pressure, part position and lubrication rate did not show a significant influence on Almen intensity.

Keywords Shot peening, Vibratory peening, Almen intensity, DOE method.

Introduction

Vibratory peening is a surface treatment process that combines the effect of vibratory finishing and shot peening. The process aims to increase both the service life, as well as the surface finish of the treated component. The specimen to be peened is fixed inside a vibrating tub [1] which is filled with media. Plastic deformation results from the impacts of the media on the surface. This produces compressive residual stresses like in the shot peening process [2], but with much lower average surface roughness, close to that obtained from vibratory finishing [1].

Different designs of vibratory peening machines were used to investigate the benefits of the process on industrial components or on Almen strips [1-8]. The parameters controlling the peening are dependent on the individual design of the vibratory peening system and cannot be compared with other designs. Eventually, it is necessary to find out the controlling parameters for this design. In this study, we used the DOE method to build an empirical model, which relates the vibratory peening parameters such as the rotating shafts eccentricity, frequency and media height above the part, to Almen intensity.

Experimental Methods

A new vibratory peening machine was designed and built for this study, similar to the vibratory peening machine used in the study of [3]. The tub rests on airbags and is filled with 3 mm diameter ball bearing media. Two shafts with eccentric weights rotating in opposite directions at each side of the tub are used to produce a vibration in the vertical direction. The treated specimen is fixed onto a holder, which is mounted on a rail through the cover to allow the adjustment of the media height above the specimen. The holder is clamped on the rail during the treatment. Seven parameters were identified which can be adjusted on this machine, namely: the eccentric weights on the shafts (X_Ecc), the media mass inside the tub (X_Mass), the rotating shaft frequency (X_Freq), the airbag pressure (X_Press), the media height above the part (X_Height), the horizontal part position (X_Pos) and the lubricant flow rate (X_Lub). The range of variation of these parameters is listed in Figure 1.

The DOE method was used to screen the effect of the seven primary factors on Almen intensity. A non-standard fractional screening design was selected to consider the constrains on the modalities between X_Ecc and X_Freq, and between X_Mass and X_Height, such as described in Figure 1 (b) and (c). A 16-run matrix design was generated by the JMP Pro software [9]. Two replications were performed, and additional runs were done when the Almen intensity values were different between the replications. A total of 41 runs were performed.

(a)

Factors	i, j or k	Min	Max
X_Ecc (kg/shaft)	1	15	24
X_Mass (kg)	2	300	500
X_Freq (Hz)	3	15	30
X_Press (bars)	4	2	2.8
X_Height (cm)	5	10	25
X_Pos (-)	6	Left	Center
X_Lub (rpm)	7	20	50

(b) (c)

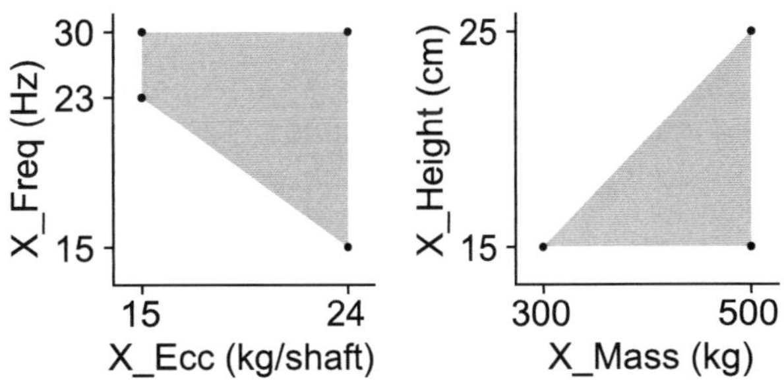

Figure 1. Factors ranges for the DOE. (a) The upper and lower bounds of the seven operating parameters with their associated variable name i, j or k for the identification in Equation (2). (b) Limited modality variation domain of X_Freq as a function of X_Ecc. (b) Limited modality variation domain of X_Height as a function of X_Mass.

The shot peening standard procedures of J442 and J443 were applied to construct the saturation curves with a minimum of 4 arc height measurements after different processing times. A standard Almen gauge was used for the arc height measurements. Almen A strips were peened for intensities that produced arc heights above 0.1 mmA or 0.3 mmN, and Almen N strips for intensities below 0.3 mmN. Almen A type of intensities were converted into Almen N type to have the same scale for the analysis of the DOE output. Three vibratory peening conditions from the design matrix were peened for both Almen A and N strips. The following equation was fitted on the results for the conversion:

$$N\ intensity = A\ intensity \times (7.05 \times A\ intensity + 0.76) \qquad (1)$$

The factors of major importance on Almen intensity were identified using an empirical relationship between the vibratory peening process parameters and Almen intensity, such as:

$$y = \beta_0 + \sum_{i=1}^{m} \beta_i x_i + \sum_{j=1}^{m} \sum_{k=1}^{m} \beta_{jk} x_j x_k + \varepsilon, \qquad for\ j < k \qquad (2)$$

where y is the Almen N intensity, β_0 is the intercept, β_i, β_{jk} are the partial regression coefficients associated to the factor effects, x_i, x_j and x_k are the variable modalities on a coded scale from -1 to +1, and ε is the error term. Table 1 details the i, j or k variable indices to identify the associated vibratory peening parameters. For $m = 7$, Equation (2) accounted for seven β_i and twenty-one β_{jk}, plus β_0, for a total of 29 regression coefficients. However, the 16-run design matrix was optimized to only fit the seven β_i of the primary factors with a minimal partial confusion between the factors. Thus, the least square method was not able to fit the 29 regression coefficients. An iterative regression approach was selected, in which empirical models with a limited number of partial regression coefficients were fitted.

It was considered that a maximum of 3 factors would be of major importance on Almen intensity, i.e. a maximum of $n_p = 6$ factor effects (3 primary factors + the 3 associated two-factors interactions). The empirical model of Equation (2) could be simplified with a set of maximum 6 non-negligible β_i, β_{jk} to neglect the remaining partial regression coefficients. An iterative regression routine was coded in Python to identify the set of non-negligible β_i, β_{jk} for $2 \leq n_p \leq 6$ which fit the data with the best regression quality.

At each iteration, an empirical model was built from Equation (2) with a selected set of β_i, β_{jk} and the remaining partial regression coefficients were set to zero. The least-squares method was used to fit the as-built model and the regression quality R^2 was computed. Eventually, all possible combinations of sets for $2 \leq n_p \leq 6$ β_i, β_{jk} , which are listed in Table 2, were tested. The list of sets was ranked by decreasing R^2 values and the factors of major importance were identified from the first set of the list.

Table 2. Partial regression coefficient sets selected in the iterative regression routine to identify the factors of major importance on Almen intensity. n_p stands for the number of factor effects in the selected sets, which includes both primary factors β_i and two-factors interactions β_{jk}.

	$n_p = 2$	$n_p = 3$	⋯	$n_p = 6$
	β_1, β_2	$\beta_1, \beta_2, \beta_3$	⋯	$\beta_1, \beta_2, \beta_3, \beta_4, \beta_5, \beta_6$
	β_1, β_3	$\beta_1, \beta_2, \beta_4$	⋯	$\beta_1, \beta_2, \beta_3, \beta_4, \beta_5, \beta_7$
	⋮	⋮	⋱	⋮
	β_7, β_{67}	$\beta_7, \beta_{57}, \beta_{67}$	⋯	$\beta_7, \beta_{46}, \beta_{47}, \beta_{56}, \beta_{57}, \beta_{67}$
Number of sets	168	1 946	⋯	322 476

Experimental results

The 41 Almen intensity values resulting from the screening DOE are shown in Figure 2. The range of Almen intensities were found to be from 0.084 mmN to 0.194 mmA (converted to 0.427 mmN). Figure 2 (a) and (b) shows that the upper modalities of X_Ecc or X_Freq did not limit the range of Almen intensities. On the contrary, low modalities of X_Ecc and X_Freq allowed a limited Almen intensity range of [0.084, 0.2] mmN and [0.095 – 0.192] mmN, respectively.

The iterative regression routine identified the set $\beta_1, \beta_3, \beta_{13}, \beta_{15}, \beta_{35}$ to fit the experimental results with the best regression quality. An empirical model was built to relate the parameters to Almen intensity. To respect a hierarchical model, β_5 was added into the model. The associated empirical model to predict Almen intensity was:

$$y = 0.758 + 2.67 \times 10^{-2} X_{Ecc} + 2.27 \times 10^{-2} X_{Freq} + 3.06 \times 10^{-3} X_{Height} + 1.07 \times 10^{-3} X_{Ecc} X_{Freq} + 8.03 \times 10^{-4} X_{Ecc} X_{Height} + 6.41 \times 10^{-4} X_{Freq} X_{Height} + \varepsilon, \quad (3)$$

where X_i are the vibratory peening parameters expressed in physical units. A p-value threshold of 0.1 was selected to state the importance of a variable [10]. The ANOVA analysis confirmed the model significance with a model *p*-value below 10^{-4}. Similarly, the non-significance of the error was showed with an error *p*-value of 0.57 [10]. The residuals analysis showed a normal, centered on zero and independent distribution. Besides, the residuals were found in the range of [-0.051, 0.041] mmN and no outliers were identified. Therefore, the assumption of the ANOVA was admitted and the model adequacy was confirmed.

The ANOVA analysis showed the importance on Almen intensity of the primary factors X_Ecc and X_Freq, as well as their associated two-factors interactions with X_Height with p-values below 10^{-4}. In addition, the importance of the two-factors interaction X_Ecc*X_Freq was confirmed with a p-value of 0.005. On the contrary, the primary factors X_Height had not a statistical influence on Almen intensity since its associated p-value was 0.524. However, the parameter had an influence on Almen intensity through its interaction with X_Ecc and X_Freq. Therefore, the three parameters X_Ecc, X_Freq and X_Height was considered to be the vibratory peening parameters of major importance on Almen intensity.

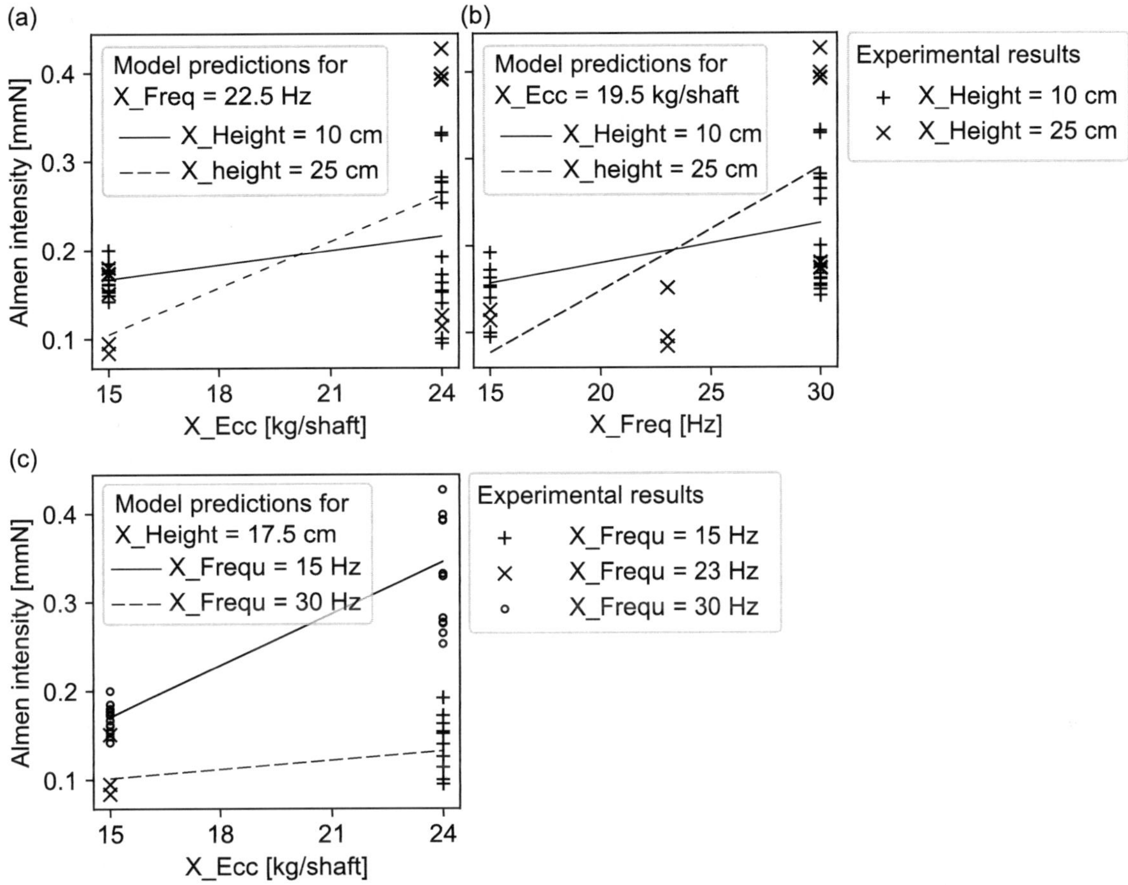

Figure 2. Comparison between the experimental results and the empirical model prediction. Interactions between X_Height and X_Ecc, X_Height and X_Freq and between X_Freq and X_Ecc can be observed by the difference of the gradients in the model predictions in (a), (b) and (c), respectively.

Almen intensities from the model predictions of Equation (2) are shown in Figure 2, as function of each important parameter. Almen intensity generally increases with all three parameters. At low levels of X_Height, the increase of Almen intensity as a function of X_Ecc and X_Freq is of lower importance, as shown in Figure 2 (a) and (b). This phenomenon is described by the two-factor interaction effects X_Height*X_Ecc and X_Height*X_Freq. However, higher Almen intensities are observed and expected for X_Height of lower importance when the level of X_Freq or X_Ecc is low. Similarly, the increase of Almen intensity is expected to be low for low frequencies, which is described by the X_Ecc*X_Freq two-factor interaction effect.

The model prediction showed that similar Almen intensities can be expected for different combinations of parameters. For example, predictions in Figure 2 (b) for X_Height =10 cm and X_Height =25 cm would both deliver an Almen intensity level of 0.195 mmN for X_Freq = 23.3 Hz. In addition, experimental results showed that Almen intensities in a range of [0.150, 0.175] mmN were observed for all levels of X_Freq.

Discussion and Conclusions

An iterative regression routine was used to identify that rotating shaft eccentricity (X_Ecc), rotational shaft frequency (X_Freq) and media height above the part (X_Height) are the factors of major importance on Almen intensity. It was also found that X_Mass, X_Press, X_Pos and X_Lub are non-influential on Almen intensity. This observation is limited to the range of factor modalities tested in this study. Besides, the media mass was adjusted without modification of the tub geometry, which had a direct consequence on the media height inside the tub. The adjustment of media mass from partitions of the tub, which would not influence the media movement near the treated part, was not tested.

X_Height did not have a direct statistical influence on Almen intensity. However, the adjustment of X_Height controlled the vibratory peening Almen intensity because of its important interaction with the factors X_Ecc and X_Freq. However, a sole effect of X_Height on the media-component interaction was shown in [4].

An empirical model was built to describe the relationship between the vibratory peening parameters and Almen intensity. Both experimental results and model predictions showed that higher Almen intensities were expected for higher eccentricities, frequencies, and media height above the part. Besides, similar Almen intensities were obtained for different vibratory peening conditions. However, the conclusions were based on a linear variation of Almen intensity. Yet, intermediate modalities of X_Freq led to a curvature in the relationship.

References

[1] D. H. Gane, Y. S. Rumyantsev, H. T. Diep, et L. Bakow, *Evaluation of Vibrostrengthening for Fatigue Enhancement of Titanium Structural Components on Commercial Aircraft*, Ti-2003 Science and Technology, Proceedings of the 10th World Conference on Titanium, Hamburg, Germany (pp. 1053-1058)

[2] H. Miao, L. Canals, B. McGillivray and M. Levesque, Comparison between vibratory peening and shot peening process, International Conference on Shot Peening (pp. 521-526)

[3] L. Canals, *Effect of vibratory peening on the sub-surface layer of aerospace materials Ti-6Al-4V and E-16NiCrMo13*, Journal of Materials Processing Technology, p. 16, 2019

[4] J. Y. Alcaraz, *Numerical modeling of residual stresses during vibratory peening of a 3-stage Blisk - A multi-scale discrete element and finite element approach*, Journal of Materials Processing Technology, p. 14, 2022.

[5] M. D. Sangid, J. A. Stori, et P. M. Ferriera, *Process characterization of vibrostrengthening and application to fatigue enhancement of aluminum aerospace components—part I. Experimental study of process parameters*, International Journal of Advanced Manufacturing Technology, p. 16, 2011.

[6] D. Kumar, *Microstructure-mechanical property correlation in shot peened and vibro-peened Ni-based superalloy*, Journal of Materials Processing Technology., p. 15, 2019.

[7] W. L. Chan, K. Ahluwalia, and A. Gopinath, *Parametric Study of Fixtured Vibropeening*, Metals, p. 10, 2019.

[8] G. Feldmann, *Application of Vibropeening on Aero – Engine Component*, Procedia CIRP, p. 6, 2014.

[9] JMP®, Version 16. SAS Institute Inc., Cary, NC, 1989–2021

[10] D. C. Montgomery, Design and analysis of experiments, Eighth edition. Hoboken, NJ: John Wiley & Sons, Inc, 2013.

Rolling Contact Fatigue Simulation Considering Surface Integrity

B. Zhang[1], G. Xiang[1], H. Liu[1*], P. Wei[1], J. Wu[1]

1 State Key Laboratory of Mechanical Transmissions, Chongqing University, China

Abstract

Rolling contact fatigue prediction is significant for the design of heavy-duty, high-reliability components, such as gears and bearings. For accurate prediction, it is vital to consider surface integrity parameters, such as surface roughness, hardness, and residual stress in the fatigue life model. Herein, an elastic-plastic finite element contact model of AISI 9310 rollers was developed, considering measured surface integrity parameters before and after shot peening. Subsequently, the Brown-Miller-Morrow multiaxial fatigue criterion was utilised to calculate fatigue life at each material point. By comparing with experimental results, the rolling contact fatigue model exhibits accurate simulation results and offers an effective tool for evaluating the effect of surface integrity on fatigue performance.

Keywords Rolling contact fatigue, shot peening, surface integrity, multiaxial fatigue

Introduction

As a primary failure mode of mechanical elements in aviation, wind turbines, ships, and automotive industries, rolling contact fatigue has become a crucial issue limiting operational reliability. Strengthening techniques enhance the anti-fatigue performance by altering various surface integrity parameters, including surface roughness, residual stress, hardness gradient, and material microstructure [1,2]. However, conducting fatigue tests uses large amounts of time and money. Furthermore, classical fatigue life model and parameters are difficult to meet strengthened components. Therefore, proposing an innovative rolling contact fatigue life model, which considers surface integrity parameters, is required.

The impact of one or two surface integrity parameters on contact fatigue has been well studied. W. Wang et al. [3] applied the Dang Van multiaxial fatigue criterion to discuss the evaluation of rolling contact fatigue of a carburised wind turbine gear, with consideration of residual stress and hardness gradient. L. Flour et al. [4] developed a numerical model to predict residual stress relaxation, and estimated the localisation of contact fatigue crack initiation, considering surface roughness. B. Zhang et al. [5] analysed the influence of surface roughness on fatigue and wear of an aviation gear, based on the Brown-Miller-Morrow multiaxial fatigue criterion and the continuous damage material constitutive behaviour. However, the contact fatigue behaviour under a complex combination of various surface integrity parameters has yet to be investigated. L. Cui et al. [6] established an elastic-plastic finite element fatigue damage accumulation model, which considered surface roughness, residual stress, and hardness of bearing rollers. Unfortunately, this model cannot be applied to surface-strengthened components, as gradient characteristics of hardness and residual stress induced by such surface engineering were not considered.

In this article, the surface topography, hardness gradient, and residual stress gradient of carburised rollers before and after shot peening were measured and applied in the finite element model. The kinematic hardening constitutive equation and the Brown-Miller-Morrow multiaxial fatigue criterion were employed to calculate the rolling contact fatigue life of each material point. Subsequently, rolling contact fatigue tests were conducted on a twin-disc rig. Experimental and simulation results were contrasted.

Experimental Method

Specimen: The specimen material is AISI 9310 steel with high surface hardness, which is widely utilised in the aerospace sector. Its chemical composition requirements are presented in Table 1. The specimens underwent rough turning, ultrasonic flaw detection, hardening, tempering, carburising, quenching, and grinding. The driving and driven specimens are illustrated in Figure 1.

Table 1. Chemical Compositions of AISI 9310.

C	0.07~0.13	Si	0.15~0.35
Mn	0.4~0.7	P	≤0.015
S	≤0.015	Cr	1.0~1.4
Mo	0.08~0.15	Ni	3~3.5
Cu	≤0.35	B	≤0.001

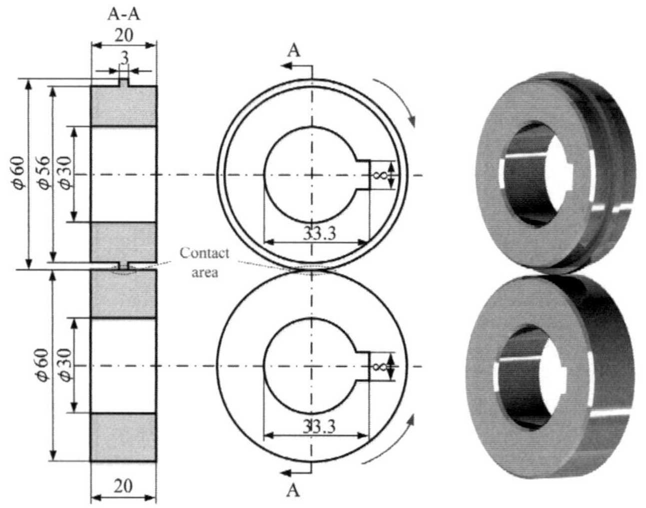

Figure 1. The Test Specimens.

Shot peening: The ground specimens were strengthened by the process of shot peening, using a pneumatic shot peening machine. Steel cut wire shot, with 0.6 mm diameter and 55~62 HRC average hardness, was adopted. The impact angle was 90°, and the diameter of the nozzle was 8 mm. The nozzle was maintained 150 mm away from the roller surface. The coverage rate reached 200% and shot peening reached 0.35 mmA.

Surface integrity test: The residual stress gradient, surface roughness, and microhardness gradient of specimens was measured, both before and after shot peening. The PULSTEC μ-360s portable residual stress detector was employed to measure residual stress at several depths. The surface micro-topography was measured using a white light interferometer. The average value of surface roughness was calculated following the removal of the curvature of rollers. The specimen microhardness was measured by a digital display automatic rotary microhardness tester (MHVS-1000AT). The loading force was 0.5 N and the load holding time was 10 seconds. The measurement results are displayed in Figure 2.

Rolling contact fatigue tests: Rolling contact fatigue tests were conducted for specimens before and after shot peening. A rolling contact fatigue testing machine (CQHH-RCF-A) was used for fatigue testing, which primarily consisted of the test system, the lubrication and cooling system, the control system, and the machine vision system. It can detect fatigue failure online through machine vision technology, enabling automatic shut-down. The rolling contact fatigue tester is illustrated in Figure 3. The contact stress was 2500 MPa (normal load of 2602 N), and the slip ratio was 10%. The driving speed was set at 1800 r/min, and the driven speed was 1980 r/min. The jetted lubricating oil was a commercial lubricant (Mobil 600 XP100).

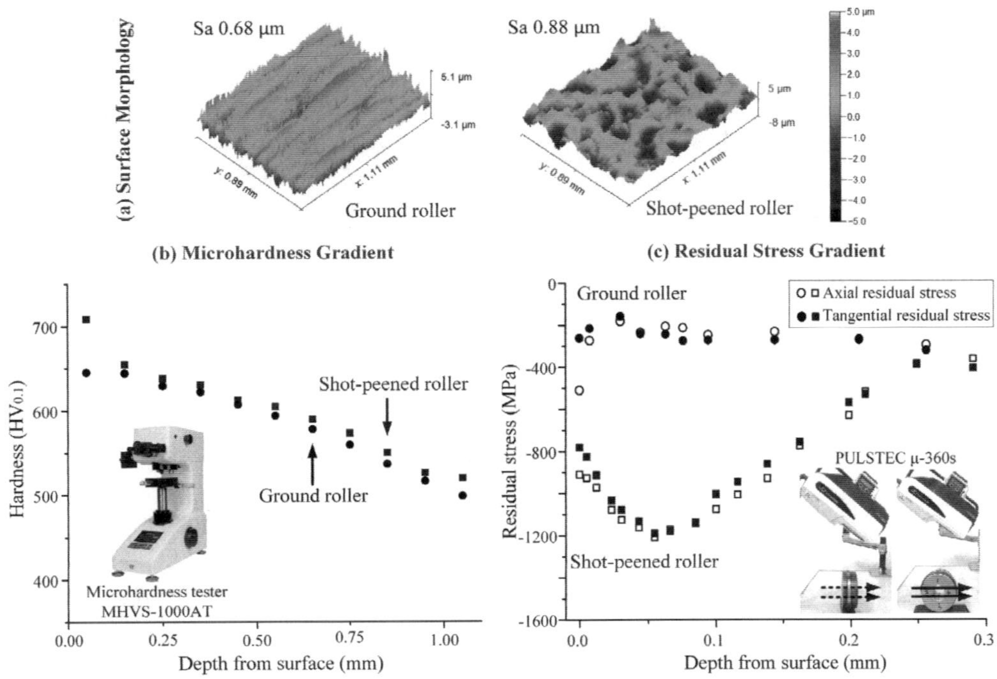

Figure 2. Measurements of Surface Integrity Parameters.

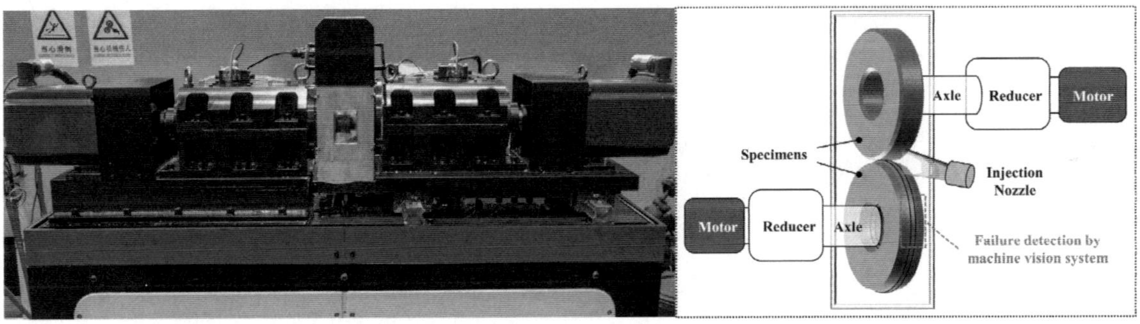

Figure 3. Rolling Contact Fatigue Tester.

Simulation Method

The elastic-plastic finite element model: The elastic-plastic finite element model of rollers was established, considering measured surface roughness, residual stress gradient, and hardness gradient. Following this, according to the stress-strain results, the fatigue life of the driving roller was calculated based on the Brown-Miller-Morrow fatigue life model. The modelling process of rolling contact fatigue is as follows:

1) Subsequent to removing the curvature of rollers, a rough curve was extracted along the tangential direction of the roller, which was further imported in the finite element model.

2) The finite element model was established in ABAQUS; the surface roughness was imported on the surface of the driving specimen by Python, and the surface of the driven specimen remained smooth.

3) Young's modulus E of specimens was 210 GPa, and the Poisson's ratio was 0.3. The kinematic hardening constitutive equation was applied, with the hardening modulus having been set to 10.5 GPa. The material yield strength was converted from measured hardness by the Pavlina-Tyne formula [7] as:

$$YS = -90.7 + 2.876HV \tag{1}$$

Where *YS* is the yield strength at a certain depth, and *HV* represents the measured hardness gradient.

4) The measured residual stress gradient was incorporated into the prestressed field.
5) The mesh size on the surface was 2 μm * 2 μm. The mesh size of the driven specimen was 0.5 mm * 0.5 mm.
6) A normal load of 2602 N was applied at the centre of the driven specimen. The coefficient of friction was assumed as 0.1, and the slip rate was 10%.
The elastic-plastic finite element contact model of AISI 9310 rollers is presented in Figure 4.

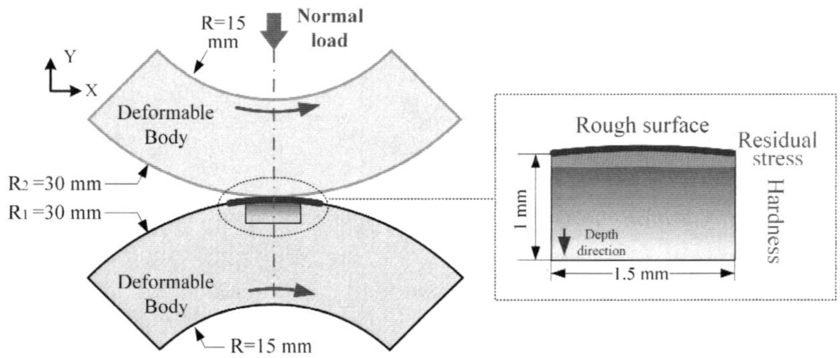

Figure 4. Elastic-plastic Finite Element Contact Model of AISI 9310 Rollers.

The Brown-Miller-Morrow fatigue life model: Considering the influence of surface roughness, residual stress, and hardness, plastic deformation may occur during the loading process, particularly at the near-surface area. Therefore, a strain-based multiaxial fatigue criterion was chosen. The Brown-Miller criterion assumes that the fatigue crack initiates along the plane with the maximum shear strain, before propagating along the normal strain direction on this plane. It is usual to define the plane with the maximum shear strain as the critical plane, on which the maximum shear strain and the normal strain are considered as damage variables. The Brown-Miller-Morrow fatigue life model [8] can be expressed as:

$$\frac{\Delta\gamma_{\max}}{2}+\frac{\Delta\varepsilon_n}{2}=1.65\frac{(\sigma'_f-\sigma_m)}{E}(2N_f)^b+1.75\varepsilon'_f(2N_f)^c \tag{2}$$

Where $\Delta\gamma_{\max}/2$ signifies the amplitude of the maximum shear strain, and $\Delta\varepsilon_n/2$ and σ_m represent the normal strain amplitude and the mean stress on the critical plane, respectively. σ'_f and ε'_f denote the fatigue strength coefficient and the fatigue ductile coefficient, while b and c represent the fatigue strength exponent and the fatigue ductile exponent, respectively. $2N_f$ portrays the life of the material. In this work, the fatigue parameters were assumed as *b*=0.057, *c*=0.27, $\sigma'_f=2894$, and $\varepsilon'_f=0.134$.

As the critical plane at each material point is unknown, the shear strain, the normal strain, and the normal stress at each candidate plane should be calculated; subsequently, the plane with the maximum shear strain should be defined as the critical plane. The angle, θ, represents the angle of the critical plane to normal direction. The normal stress σ_θ, the normal strain ε_θ, and the shear strain γ_θ on each plane can be expressed as [9]:

$$\begin{cases} \sigma_\theta=\sigma_x\cos^2\theta+\sigma_z\sin^2\theta+2\tau_{xz}\sin\theta\cos\theta \\ \varepsilon_\theta=\varepsilon_x\cos^2\theta+\varepsilon_z\sin^2\theta+\gamma_{xy}\sin\theta\cos\theta \\ \gamma_\theta=\gamma_{xz}(\cos^2\theta-\sin^2\theta)+2(\varepsilon_z-\varepsilon_x)\sin\theta\cos\theta \end{cases} \tag{3}$$

The range of the maximum shear strain $\Delta\gamma_{max}$, the range of normal strain on the critical plane $\Delta\varepsilon_n$, and the mean normal stress σ_m on the critical plane can be expressed as:

$$\begin{cases} \Delta\gamma_{max} = \gamma_{max} - \gamma_{min} \\ \Delta\varepsilon_n = \varepsilon_{max} - \varepsilon_{min} \\ \sigma_m = \dfrac{\sigma_{max} + \sigma_{min}}{2} \end{cases} \tag{4}$$

Where γ_{max} and γ_{min} denote the maximum and minimum shear strain on the critical plane, respectively, ε_{max} and ε_{min} signify the maximum and minimum normal strain on the critical plane, respectively, and σ_{max} and σ_{min} represent the maximum and minimum normal stress on the critical plane, respectively.

Following the determination of the critical plane at each material point, the fatigue life at the corresponding material point can be iteratively calculated according to the Brown-Miller-Morrow fatigue life model.

Results and Discussion

Experimental results: Experimental data of rolling contact fatigue of AISI 9310 rollers, before and after shot peening, were processed following the two-parameter Weibull distribution. For rollers subjected to contact stress of 2500 MPa and slip ratio of 10%, test points and fitting curves are presented in Figure 5. The rolling contact fatigue life with 50% failure probability of ground rollers is 4,601,039, and fatigue life with 10% failure probability is 3,316,870. While the fatigue life under 50% and 10% failure probability of shot-peened rollers reach 5,861,465 and 3,922,395, respectively. It is evident that shot peening greatly heightens rolling contact fatigue life.

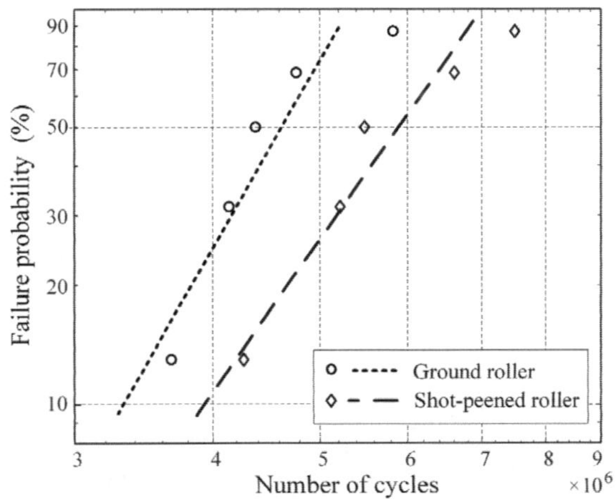

Figure 5. Experimental Results of AISI 9310 Rollers.

Simulation results: The simulated fatigue life of ground and shot-peened rollers are illustrated in Figure 6. Their minimum fatigue life occurs near the surface, and are 2,779,713 and 3,999,447, respectively. Compared to the 10% failure probability experimental results, the errors of simulated life before and after shot peening are 16.19% and 1.96%, respectively. This demonstrates that the proposed contact fatigue model enables an accurate prediction of rolling contact fatigue life. Due to the limited depth of residual stress introduced by shot peening, residual stress below the black dotted line is not considered. The fatigue life of material points above and below the dotted line is clearly different, which is evidence for the significance of residual compressive stress on fatigue life [3]. The reduction of fatigue life near the surface is shown in the red dotted circle, which is caused by surface roughness. It is generally accepted that, as surface roughness increases, the near-surface fatigue life decreases significantly [10]. Although surface roughness Sa of rollers is enhanced from 0.68 μm to 0.88 μm after shot peening, the hardness gradient and residual compressive stress layer are significantly heightened by shot peening. Thus, the near-surface material points are subject to complex combined impacts of surface integrity parameters.

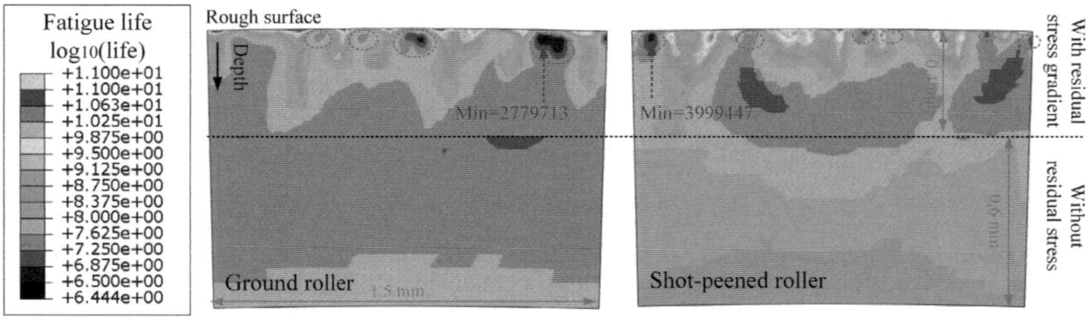

Figure 6. Simulation Results of AISI 9310 Rollers.

Conclusions

An elastic-plastic finite element rolling contact fatigue model of AISI 9310 rollers was established, considering measured surface integrity parameters before and after shot peening. By comparing findings with rolling contact fatigue tests, the proposed rolling contact fatigue model enables an accurate life prediction and provides an effective tool to evaluate the impact of surface integrity on contact fatigue life, with various manufacturing techniques. These techniques include shot peening, fine particle peening, superfinishing, and other surface treatments.

References

[1] I F Pariente, M Guagliano, *Influence of shot peening process on contact fatigue behavior of gears*. Materials and Manufacturing Processes, Vol. 24 (2009), pp 1436-1441.

[2] J Wu, H Liu, P Wei, et al., *Effect of shot peening coverage on residual stress and surface roughness of 18CrNiMo7-6 steel*. International Journal of Mechanical Sciences, Vol. 183 (2020), pp 105785.

[3] W Wang, H Liu, C Zhu, et al. *Evaluation of rolling contact fatigue of a carburized wind turbine gear considering the residual stress and hardness gradient*. Journal of Tribology, Vol. 140 (2018), pp 061401.

[4] L Flour. *Contact fatigue of automotive gears: evolution and effects of residual stresses introduced by surface treatments*. Fatigue & Fracture of Engineering Materials & Structures, Vol. 23 (2000), pp 217-228.

[5] B Zhang, H Liu, C Zhu, et al., *Simulation of the fatigue-wear coupling mechanism of an aviation gear*. Friction, Vol. 9 (2021), pp. 1616-1634.

[6] L Cui, Y Su. *Contact fatigue life prediction of rolling bearing considering machined surface integrity*. Industrial Lubrication and Tribology, Vol. 72 (2022), pp 73-80.

[7] E J Pavlina, C J V Tyne. *Correlation of Yield Strength and Tensile Strength with Hardness for Steels*. Journal of Materials Engineering & Performance, Vol. 17 (2008), pp 888-893.

[8] J D Morrow. *Cyclic plastic strain energy and fatigue of metals*. Internal Friction, Damping, and Cyclic Plasticity, (1965), pp 45-87.

[9] S. P. Timoshenko, J. N. Goodier, and H. N. Abramson, Theory of elasticity (3rd ed.). New York: NY: McGraw-Hill, 1970.

[10] B Zhang, H Liu, C Zhu, et al. *Numerical simulation of competing mechanism between pitting and micro-pitting of a wind turbine gear considering surface roughness*. Engineering Failure Analysis, Vol. 104 (2019), pp 1-12.

UPDATED THEORY OF THE X-RAY DIFFRACTION RESIDUAL STRESS MEASUREMENT COMPARING COS-ALPHA AND SIN²PSI TECHNIQUES – BIAXIAL VERSUS TRIAXIAL

Mohammed Belassel[1], James Pineault[1], Michael Brauss[2]

[1] PROTO MANUFACTURING LIMITED, Canada
[2] PROTO MANUFACTURING INC, USA

Abstract
When applying the $Sin^2\psi$ technique to x-ray diffraction (XRD) based residual stress measurements, both the normal and shear stress components can be easily and accurately evaluated in the direction(s) of interest. The $Sin^2\psi$ technique was developed and refined over several decades along with many other advances so as to achieve high accuracy residual stress measurement results in a wide variety of circumstances and material conditions. With the introduction of the $Cos\alpha$ technique, only the biaxial stress field was considered in the calculation of the stress tensor components and as such, is susceptible to significant errors often resulting in unreliable residual stress measurements. In the present study, the $Sin^2\psi$ and $Cos\alpha$ techniques were compared and evaluated. It is shown that the biaxial based $Cos\alpha$ technique leads to erroneous stress data when in-plane shear stress is present, and results are found to be equivalent to the two-tilt $Sin^2\psi$ technique. In other words, the $Cos\alpha$ technique used in only one orientation could not reproduce the correct results as obtained via the $Sin^2\psi$ technique with multiple tilts. Thus, the only viable solution is to assume a triaxial stress tensor and to collect data with tilts distributed evenly over the half-space when using the $Cos\alpha$ technique.

Keywords: Residual Stress, Shear stress, CosAlpha, Single Exposure, Sin^{2}Psi, Standard

Introduction

The development of residual stress measurements using x-ray diffraction techniques went through many steps over time. Initially, the residual stress field was assumed to be biaxial because of the stress boundary conditions and the shallow penetration of x-rays. Therefore, all stress components in the third direction were assumed to be zero, i.e., σ_{i3}=0. [1]. At the time, stress measurements were only executed using positive inclination angles (ψ^+) while a linear regression was sufficient to determine the normal stress in the direction of interest (ϕ=0°). As a result, the design of instruments was tailored to accommodate the biaxial stress tensor assumption. However, it was known that misalignment of the goniometer resulting in a beam shift would cause PSI-Splitting when measuring both the (ψ^+) and (ψ-) branches [2]. The common practice was to flip the component (ϕ=180°) exclusively to verify the misalignment of the goniometer.

The first time PSI-Splitting was recorded on a component, it was reported in a task force residual stress meeting [2]. After double verification of the alignment, it was established that the shear stress was indeed real, and that it originated from the processing of the component itself, not the misalignment of the goniometer. Since then, the scientific community has established a new reliable procedure to measure residual stresses in materials using the $Sin^2\psi$ technique assuming a triaxial stress tensor. Many advances have been introduced over the last several decades which have resulted in the development of thorough and well documented residual stress standards and best practice guidelines that enable high accuracy and reliable residual stress measurements using XRD [3,4].

With the emergence and development of the $Cos\alpha$ technique using a 2D detector, the biaxial stress tensor was the prime assumption included in all marketed instruments [5]. However, this technique was found to exhibit many limitations and weaknesses in addition to the ones already

known for all XRD techniques in general. Moreover, it has been shown that the Cosα technique suffers from the exact same limitations as the two-tilt Sin²ψ technique (SET- Sin²ψ) [6,7]. The Cosα technique cannot reliably measure stresses in materials when anisotropy and in-plane shear stresses (τ_{13} or τ_{23}) exist because their presence produces a significant effect that can easily mislead a measurement practitioner [7].
To demonstrate the need to use a triaxial stress tensor in the stress model, a simulation comparing the following techniques was developed: 1) the Sin²ψ technique with multiple tilt angles (MET- Sin²ψ), 2) the Sin²ψ technique with two tilt angles (SET- Sin²ψ), and 3) the Cosα technique with a single tilt (SET-Cosα). This work emphasizes the limitations of the SET- Sin²ψ and the Cosα techniques along with the effect of different measurement conditions on the stress results obtained.

Shear stress induced processes
The in-plane shear stress is a result of non-uniform processes applied to materials exhibiting heterogeneous microstructure including ferrous alloys. These processes are many, among them, machining, grinding, roll-peening, friction and directional peening. Other shear stresses are fictitious and can be generated from a curved geometry or tilted surface. All of these shear stresses will affect the calculation of the normal stress component [2].

Cos-Alpha techniques
The Cosα residual stress measurement technique is XRD based and uses the distance between crystallographic planes (i.e., the d-spacing) as a strain gauge. It can only be applied to crystalline, polycrystalline, and semi-crystalline materials [1,2]. When the material is in tension, the d-spacing increases and when the material is in compression, the d-spacing decreases. The d-spacing is calculated from the measured Bragg angle 2θ for a known x-ray wavelength λ using Bragg's law:

$$n\lambda = 2d\sin\theta \quad (1)$$

Subsequently, the strain ε is calculated from unstressed (d_0) and stressed ($d_{\psi\varphi}$) values for each orientation defined by the angles ϕ and ψ. Hence the strain is calculated using the following relationship:

$$\varepsilon_{\varphi\psi} = \frac{d_{\varphi\psi} - d_0}{d_0} \quad (2)$$

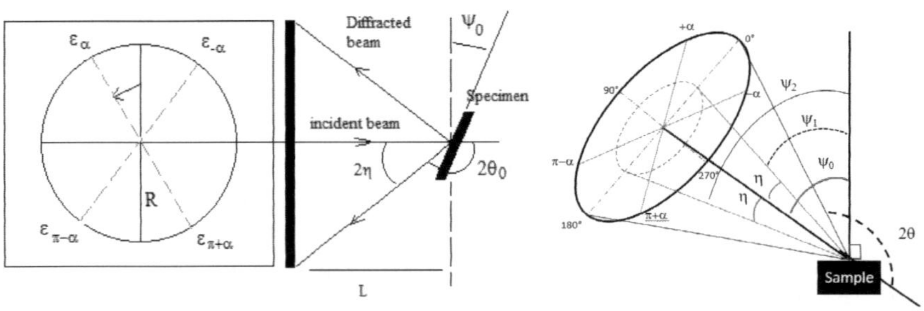

Figure 1: SET-Cosα technique geometry using full Debye ring.

Cosα technique uses the whole Debye ring collected with a 2D detector, and determines the peak position for each angle α around the ring [5]. ψ_0 is the inclination of the sample relative to the incident beam, ϕ_0 is the direction of the measurement, and 2η is the angle subtended by the incident and diffracted beams. The tilting angles are therefore defined as: $\psi_1 = \psi_0 + \eta$, and $\psi_2 = \psi_0 - \eta$, where $\eta = (\pi - 2\theta)/2$ (see Figure 1). Strain quantities can be defined as shown in the Eq. 3 and 4:

$$a_1 = \frac{1}{2}[(\varepsilon_\alpha - \varepsilon_{\pi+\alpha}) + (\varepsilon_{-\alpha} - \varepsilon_{\pi-\alpha})] \quad (3)$$

$$a_2 = \frac{1}{2}[(\varepsilon_\alpha - \varepsilon_{\pi+\alpha}) - (\varepsilon_{-\alpha} - \varepsilon_{\pi-\alpha})] \quad (4)$$

where ε_α, $\varepsilon_{\pi+\alpha}$, $\varepsilon_{-\alpha}$, and $\varepsilon_{\pi-\alpha}$ are the measured strains from the four sectors of the Debye ring. For ϕ_0 =0, the direction of the strain can be reduced to the following expression:

$$n_i = \begin{pmatrix} cos\eta \; sin\psi_0 + sin\eta \; cos\psi_0 \; cos\alpha \\ sin\eta \; sin\alpha \\ cos\eta \; cos\psi_0 - sin\eta \; sin\psi_0 cos\alpha \end{pmatrix} \quad (5)$$

Considering Triaxial stress tensor

Considering a triaxial stress field, new expressions are derived for a_1 and a_2:

$$a_1 = -\frac{(1+v)}{E}(\sigma_{11} sin2\psi_0 - 2\tau_{13} cos2\psi_0) sin2\eta . Cos\alpha \quad (6)$$

$$a_2 = \frac{2(1+v)}{E}(\tau_{12} sin\psi_0 + 2\tau_{23} cos\psi_0) sin2\eta . Sin\alpha \quad (7)$$

And the stress terms are calculated for the slopes of a_1=f(Cosα) and a_2=f(Sinα) as follows:

$$(\sigma_{11} sin2\psi_0 - 2\tau_{13} cos2\psi_0) = -\frac{E}{(1+v)sin2\eta}\frac{\partial a_1}{\partial Cos\alpha} \quad (8)$$

$$(\tau_{12} sin\psi_0 + 2\tau_{23} cos\psi_0) = \frac{E}{2(1+v)sin2\eta}\frac{\partial a_2}{\partial Sin\alpha} \quad (9)$$

Since most polycrystalline materials are considered anisotropic, Young's modulus, E, and Poisson's ratio ν, are replaced by ½S_2 and S_1 using the following expressions [1]:

$$\frac{1+v}{E} = ½S_2 \text{ and } \frac{v}{E} = S_1 \quad (10)$$

To solve Eq. 8 and 9 for all 4 components (σ_{11}, τ_{13}) and (τ_{12}, τ_{23}), a set of 4 equations are necessary. This will require at least two measurements, i.e., 2-Debye rings [10]. For better statistical sampling, more tilts are recommended.

Considering the Biaxial stress tensor

For the biaxial stress tensor, the second terms in Eq. 8 and 9 are dropped therefore the new equations are:

$$(\sigma_{11}) = -\frac{E}{(1+v).sin2\psi_0.sin2\eta}\frac{\partial a_1}{\partial Cos\alpha} \quad (11)$$

$$(\tau_{12}) = \frac{E}{2(1+v).sin\psi_0.sin2\eta}\frac{\partial a_2}{\partial Sin\alpha} \quad (12)$$

The components σ_{11} and τ_{12}, in this case, can be calculated from a single Debye ring data. Cosα instruments that use by default only one Debye ring assume a biaxial stress tensor and therefore, cannot calculate the shear components (τ_{13}, τ_{23}).

Sin²ψ Technique – Multiple Exposure Technique (MET-Sin²ψ)

The constitutive equations for the Sin$^2\psi$ technique are as follows [1]:

$$\varepsilon_{\varphi\psi} = \frac{1+v}{E}(\sigma_{11} cos^2\varphi + \sigma_{22} sin^2\varphi + \tau_{12} sin2\varphi) Sin^2\psi$$

$$-\frac{v}{E}(\sigma_{11} + \sigma_{22}) + \frac{1+v}{E}(\tau_{13} cos\varphi + \tau_{23} sin\varphi) Sin2\psi \quad (13)$$

The normal stress in the direction of interest, φ, can be defined as:

$$\sigma^\varphi = \sigma_{11} cos^2\varphi + \sigma_{22} sin^2\varphi + \tau_{12} sin2\varphi \quad (14)$$

$$\tau^\varphi = \tau_{13} cos\varphi + \tau_{23} sin\varphi \quad (15)$$

All stress components can be calculated from a triaxial measurement with multiple (φ,ψ) directions or (σ_{11}, τ_{13}) from a single measurement with multiple tilts. The normal and shear stresses are independent and can be calculated using either a linear or elliptical fit. It is recommended that when using this technique, both positive and negative ψ-angles ranging from +45° to -45° are covered, and that at least 9 tilts distributed approximately equally be sampled [4].

Two-tilt method Sin²ψ (SET-Sin²ψ)

The SET-$Sin^2\psi$ technique in this case uses a single goniometer tilt with two detectors, which is equivalent to using 2 goniometer tilts with one detector (i.e. both scenarios sample two different ψ angles). In general, a single tilt is used with two detectors where the angles are defined as: ψ_1 and ψ_2 defined in SET-Cosα paragraph [1]. These angles correspond to α =0° and α =180° on the Debye ring (see Figure 1). For the SET-$Sin^2\psi$, Eq. 13 can be reduced to the following equation with $\sigma_{22}=\sigma_{33}=\tau_{13}=0$ and $\phi=0°$:

$$\sigma^{\varphi} = \sigma_{11} = \frac{\partial\varepsilon_{\psi}}{\partial Sin^2\psi}\frac{E}{1+\upsilon} \tag{16}$$

Given these assumptions, when using the SET-$Sin^2\psi$, only one normal stress component (σ^{φ}) can be determined.

Residual stress simulation

In order to demonstrate the limitations of both the SET-Cosα technique and the SET-$Sin^2\psi$ technique, a theoretical approach was opted for using simulation since it is advantageous for studying a variety of parameters. The simulation method was described in [7] and it consists of calculating the stress values from a known initial input stress tensor σ_{ij} then the strains $\varepsilon_{\varphi\psi\alpha}$ in every direction for all 3 techniques: SET-Cosα, SET-$Sin^2\psi$ and MET-$Sin^2\psi$. This simulation illustrates the effect of the inclination angle ψ_0 on the normal stress, the stress error due to the presence of shear stress, stress evolution under load, and the residual stress value offset due to the presence of shear stress.

Parameters influencing the SET-Cosα technique

Tilting angle:

As illustrated in Eq. 8 and 9, the stress value is directly dependent on the values of the tilt angle ψ_0 and the magnitude of shear stress present. The normal stress (σ_{11}) is directly affected by the magnitude of the τ_{13}, and τ_{23} shear stress components as illustrated in Figure 2. These residual stresses were simulated for both the positive and the negative ψ-tilts and for two different magnitudes of shear stress as shown in the stress tensors below:

$$\sigma_{ij}(MPa)=\begin{pmatrix}500 & 100 & 100\\ 100 & 500 & 100\\ 100 & 100 & 0\end{pmatrix} \text{ and } \sigma_{ij}(MPa)=\begin{pmatrix}500 & 100 & 50\\ 100 & 500 & 50\\ 50 & 50 & 0\end{pmatrix}$$

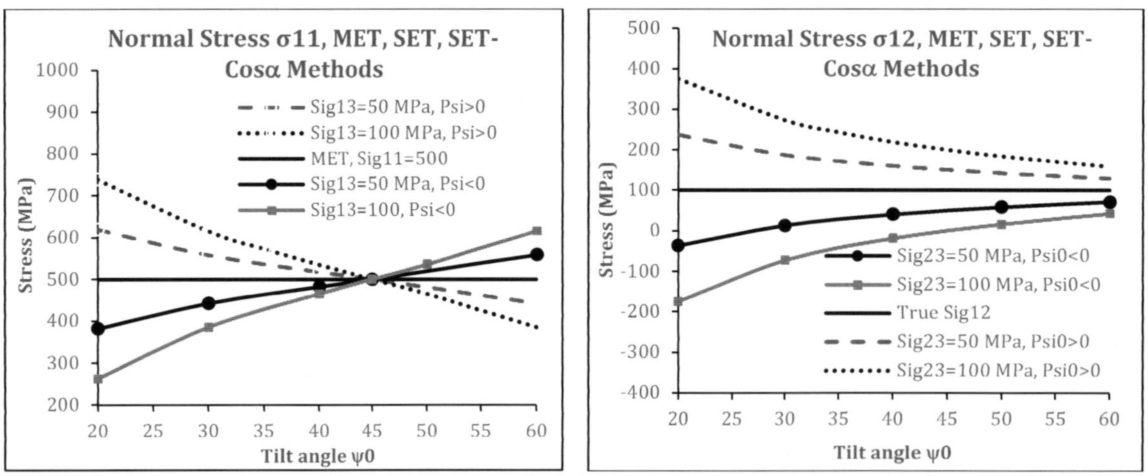

Figure 2: Simulated stress tensor terms σ_{11} and τ_{12} versus ψ_0, for both $\psi_0<0$ and $\psi_0>0$ values using: SET-Cosα,SET-$Sin^2\psi$, and MET-$Sin^2\psi$, for 2 different stress tensor inputs.

These simulation results indicate that the presence of the shear stresses τ_{13} and/or τ_{23} are expected to offset both the σ_{11} and τ_{12} stress components that the SET-Cosα technique produces. This effect has been confirmed experimentally [6,9]. The offset for σ_{11} varies with both the direction and the magnitude of ψ-tilt angle. It is only at ψ_0=45° where the shear stress effect is eliminated. The magnitude of residual stresses can be either overestimated or underestimated, depending on the orientation of the part and the sign of the shear stress component. Both components σ_{11} and τ_{12} converge to their true values of 500 MPa and 100 MPa when the ψ-tilt angles are ±45° and ±90° respectively. The influence of τ_{23} on τ_{12} is most significant when the stress levels are on the same order of magnitude. The effect of the tensor terms τ_{13} and τ_{23} on σ_{11} and τ_{12} respectively are independent, however, the presence of τ_{23} will affect σ_{22} when measured in the ϕ_0=90°direction.

Shear stress levels:

The absolute errors on the normal stress σ_{11} and the shear stress τ_{12} are proportional to the magnitude of the shear stress only (not the magnitude of the normal stress), and the relative error is inversely proportional to the normal residual stress magnitude. Therefore, the relative stress error is inversely proportional to the magnitude of the shear stress (see Figure 3).

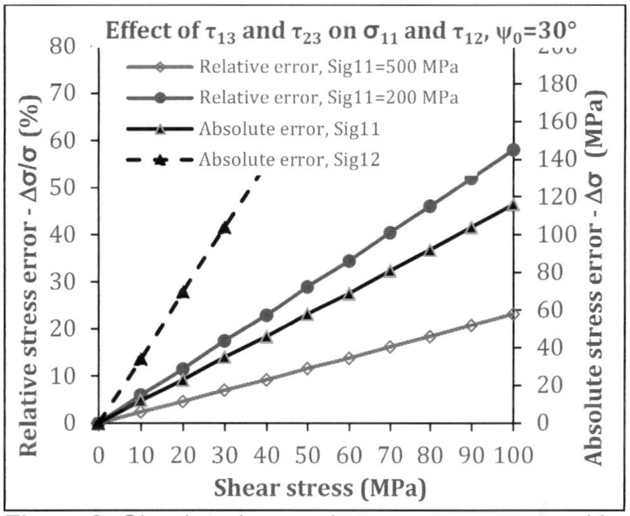

Figure 3: Simulated normal stress errors caused by the presence of the τ_{13} and τ_{23} shear stress components at different levels [8].

High shear stress magnitudes can easily be generated in ground and machined materials or components while still having relatively low magnitude normal stress values. The stress errors in such cases can be significant and cannot be discounted. Moreover, the shear stress τ_{12} is highly altered by the presence of the shear stress τ_{23} and as such, can have huge implications when considering principal stress calculations.

Presence of Anisotropy

All residual stress measurement techniques (including XRD) are sensitive to the microstructure present in the material including grain size and preferred orientation. As such, accurate residual stress measurements must be able to account for, and consider, the effect of the material microstructure. $\sin^2\psi$, state-of-the-art goniometers and software are generally equipped with tools to mitigate and/or eliminate these effects. However, the SET-Cosα technique does not enjoy these advantages and it is often either very difficult or impossible to equate SET-Cosα to MET-$\sin^2\psi$ results when grain size and preferred orientation effects are present. Depending on the severity of these factors, SET-Cosα stress results can be completely erroneous. These effects are illustrated in the example shown in Figure 4. When comparing these two techniques side by side, the MET-$\sin^2\psi$ technique has been shown to produce reasonably accurate results, whereas the SET-Cosα technique fails completely [7,11].

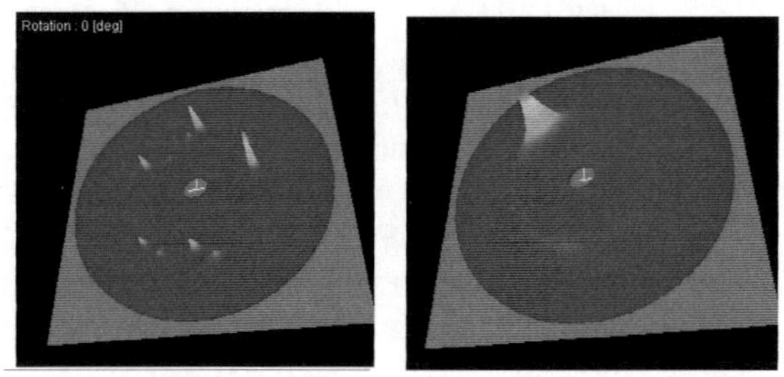

Figure 4: 3D Debye rings data for Stellite sample (coarse grain size) and stainless-steel sample (preferred orientation) [7].

Discussion and Conclusions

Many advances in residual stress measurements obtained using the $\mathrm{Sin}^2\psi$ technique have been made over recent decades. These have helped ensure that accurate stress results are obtained on a variety of materials processed under a wide spectrum of conditions. Many aspects of XRD based residual stress measurements have been fully developed and refined including: stress gradients, grain size effects, texture effects, shear stresses, triaxial stresses, two-phase materials, stress in single crystals, and many others. The amount of knowledge acquired over these decades is significant and has helped the scientific community vastly improve the accuracy of their residual stress measurements. All these advances were achieved using the $\mathrm{Sin}^2\psi$ technique assuming a triaxial stress tensor.

The SET-Cosα technique however, even today, suffers from the same difficulties the scientific community had experienced decades ago. The SET-Cosα technique theory reported in this paper clearly shows why it is essentially antiquated and relegated to the knowledge level of stress measurements performed in the 1970s, when the effects of shear stress were unknown. Results obtained using the SET-Cosα technique are extremely sensitive to the tilt angle selected when shear stress are present in the material as shown in Figure 2 [7]. Assuming the presence of a biaxial stress state is a huge and unnecessary risk that can lead to erroneous results, especially on materials with large grain size or texture as compared to $\mathrm{Sin}^2\psi$ technique [11,12]. The errors associated with SET-Cosα technique can be especially significant when the ratio between the shear stress and the normal stress is high. The solution for Cosα is to use the same assumptions used for the MET-$\mathrm{Sin}^2\psi$ technique, i.e., triaxial stress tensor with an adequate number of tilts to cover the full half-space and adequately solve for all of the unknown stress tensor components, and sufficient sampling statistics of d-spacing from the maximum number of grains possible. As this paper has clearly demonstrated, the practitioner cannot assume a priori that the stress field in the material is biaxial before the measurement is performed. Only by assuming a triaxial stress tensor will XRD based residual stress measurements be reliable and accurate using any technique.

References

[1] I. C. Noyan and J.B. Cohen, *Residual Stress, Measurement by Diffraction and Interpretation*, (Springer-Verlag, New York), (1987).

[2] V. Hauk, *Structural and Residual Stress Analysis by Nondestructive Methods*, (Elsevier, Amsterdam), (1997).

[3] ASTM E2860-12, *Standard Test Method for Residual Stress Measurement by X-Ray Diffraction for Bearing Steels*, (2012).

[4] EN 15305, NDT - *Test Method for Residual Stress Analysis by X-ray Diffraction*, (2008).

[5] T. Sasaki, Y. Hirose, K. Sasaki, S. Yasukawa, *Influence of image processing conditions of Debye Scherrer ring images in x-ray stress measurement using an imaging plate*, Adv. X-ray Anal. Vol. 40, (1997), pp.588–594.

[6] N. Peterson, Y. Kobayashi, B.Traeger, P. Sanders, *Assessment and Validation of Cosα Method for Residual Stress Measurement*, Proc. of the ICSP13, (2017), pp. 80-86.

[7] M. Belassel, J. Pineault, M. Bolla, M Brauss, *Evaluation of x-ray diffraction stress instruments using cos-alpha and sin²ψ techniques in relation to existing standards*, Adv. in X-ray Anal., Vol. 64, (2020), pp.1-10.

[8] M. Belassel, J. Pineault, M. Bolla, M Brauss, *Introducing a triaxial stress field for residual stress measurements using cos-alpha techniques,* Proc. of ICRS11, Nancy, France (2022), (in publication).

[9] J. Ramirez-Rico, S.-Y. Lee, J. J. Ling, and I. C. Noyan, *Stress Measurement Using Area Detectors: A Theoretical and Experimental Comparison of Different Methods in Ferritic Steel Using a Portable X-Ray Apparatus*, J Mater. Sci., Vol. 51, (2016), pp.5343-5355.

[10] K. Tanaka, *The cosα method for X-ray residual stress measurement using two-dimensional detector*, Bulletin of the JSME, Vol. 6, No.1, (2019), pp. 1-15.

[11] C. R. Chighizola, C. R. D'Elia, D. Weber, B. Kirsch, J. C. Aurich, B. S. Linke, M. R. Hill, *Intermethod Comparison and Evaluation of Measured Near Surface Residual Stress in Milled Aluminum*, Exp. Mech., (2021) Vol. 61, pp.1309–1322.
[12] H. Soyama, *Comparison of Various XRD Methods on Evaluation of Residual Stress Introduced by Submerged Laser Peening*, MFN Vol. 19, (2018), pp. 54-56.

X-Ray Diffraction and Hole-Drilling residual stress measurements of shot peening treatments validated on a calibration bench

M. Beghini[1], T. Grossi[1]*, C. Santus[1], A. Torboli[2], A. Benincasa[3] and M. Bandini[4]

[1] Dipartimento di Ingegneria Civile e Industriale, Università di Pisa, Pisa, Italy
[2] GNR Srl, Novara, Italy
[3] SINT Technology Srl, Firenze, Italy
[4] Peen Service Srl, Bologna, Italy

*Corresponding author: tommaso.grossi@phd.unipi.it

Abstract

The inverse problem of determining residual stresses from diffraction or relaxation methods is notoriously affected by a high sensitivity to errors in input data. A particular care must be devoted to ensuring that their input errors are minimized, and results shall come with a quantification of the corresponding uncertainties.
Residual stress measurements are often validated by comparing the results of different techniques. Although this approach can strengthen the measurement confidence, it does not highlight potential biases of the methods.
The authors presented a calibration bench [1, 2] which can impose a known bending distribution on a specimen while simultaneously performing an X-Ray Diffraction (XRD) or Hole-Drilling Method (HDM) residual stress measurement. Since the external load can freely be applied and removed, Bueckner's superposition principle [3] can be exploited to simultaneously identify both the reference bending distribution and the actual residual stress distribution with the same experimental setup. As the first is accurately known, the bench provides a direct estimation of the achieved accuracy. Moreover, it can reveal systematic errors in the chosen procedures.
Two shot peening treatments were analyzed on the calibration bench with both XRD and HDM. First, residual stresses on the surface were evaluated with XRD measurements, then electrochemical material removal was performed to investigate stresses at higher depths. After that, HDM measurements were carried out and compared with the results of XRD. Both methods were also used to identify the known bending stresses: that provided an additional validation of the residual stress results.

Keywords Residual stress, X-Ray Diffraction, Hole Drilling, Shot-peening, Calibration bench

Introduction

The hole drilling method (HDM) [4] is a well-established technique for measuring residual stresses, due to its low cost and its in-field applicability. It is standardized by ASTM E837 procedure [5], which can be carried out with commercially available devices, such as the MTS3000-Restan by SINT Technology. It is known to be affected by a high sensitivity to errors in input data, particularly near the surface, where stresses depend only on few measurement points and are affected by errors in the zero-depth datum, and at high depths, where the method loses sensitivity. Due to this, it is often complemented with X-Ray Diffraction (XRD), which exploits Bragg's law to identify residual stresses near the specimen surface [6], with a typical penetration depth of about 10 μm. Stresses at higher depths can be investigated by removing layers of material, generally through electropolishing.

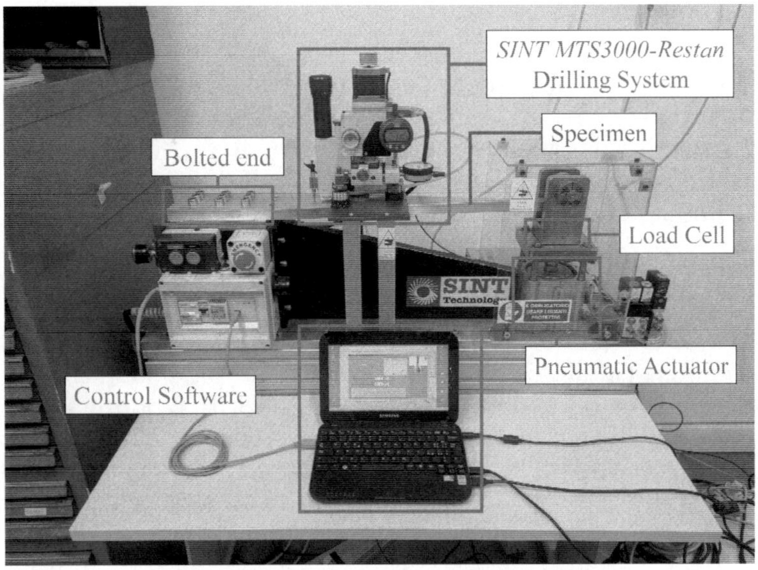

Fig. 1. Description of the calibration bench.

XRD is also sensitive to the crystallographic properties of the material under investigation and to the strategies used to identify the diffraction peak. Therefore, every application requires a careful evaluation of the instrumentation setup, in order to gain confidence on the obtained measurements.

To this aim, the authors presented a calibration bench [1, 2] which can impose a known bending distribution on a specimen while *simultaneously* performing an X-Ray Diffraction (XRD) or Hole-Drilling Method (HDM) residual stress measurement. Any issue regarding setup, instrument calibration or material crystallographic properties gives rise to notable differences between the reference bending distribution and its identified counterpart. Since the identification of the applied bending distribution is carried out with the same setup used for the residual stresses, the achieved accuracy in the identification of bending stresses is a direct validation of the residual stress measurements.

The bench is shown in Fig. 1. Details of the system are available in [1]. The identification of the bending stress distribution is based on two measurements, respectively in the unloaded and in the loaded configuration. Assuming linear elasticity, the difference between the two corresponds to the effect of the bending distribution alone, while measurements in the unloaded configuration involve residual stresses in the specimen.

In this work, the bench was used to validate XRD and HDM measurements on a shot peened 7075-aluminum specimen, reported in Fig. 2.

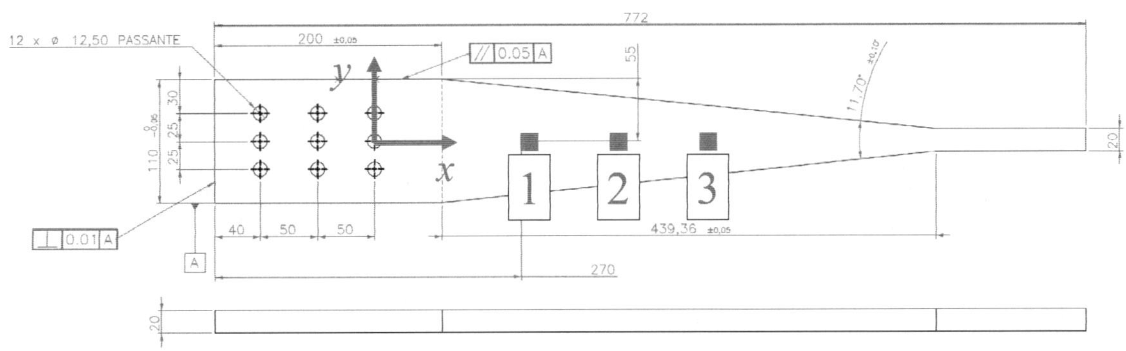

Fig. 2. Technical drawing of the aluminum specimen. The three measurement points on each specimen side are reported with red squares. The reference system is also shown.

Experimental Methods

The two sides of the specimen were shot peened with two different treatments, commercially known as AZB425 and CEB120, carried out at Peen Service srl. Details on their corresponding shot peening parameters are available in [1].
Three measurement points were marked on each side of the specimen (as in Fig. 2), then the following test campaign was conducted for each treatment:

- Surface XRD residual stress measurements were carried out at Peen Service srl on all three points, with a Stresstech Xstress3000 diffractometer. These measurements are labeled as Lab. 2 in Tables 1-2.
- Surface XRD residual stress measurements were repeated in a double-blind setting at University of Pisa on all three points, with a GNR SpiderX Edge diffractometer, shown in Fig. 3(a). These measurements are labeled as Lab. 1 in Tables 1-2. The experimental setup was validated on the calibration bench by identifying near-surface stress values of the bending distribution at two increasing load levels.
- HDM measurements were performed with a MTS3000-Restan equipment on points 1 and 2. For each drilling step, strains were sampled in both the unloaded and the loaded state, to simultaneously identify both the bending and the residual stress distribution.
- XRD depth profiling through electropolishing was carried out on point 3, up to a depth of about 0.1 mm. Diffraction measurements were performed with the GNR SpiderX Edge diffractometer.

XRD measurements were carried out with the $\sin^2\psi$ method, using a chrome anode and looking for the lattice spacing of {311} planes. In order to obtain the full stress tensor without any simplifying assumption, each XRD measurement was actually performed in three directions at angles of 0°, 45° and 90° with respect to the *x* axis. The etching depth obtained with electropolishing was measured with a micrometer dial gauge, shown in Fig. 3(b).
HBM RY61K strain rosettes (shown in *Fig. 4*) were used in HDM measurements. Holes were drilled with an inverted cone tool having a diameter of 1.8 mm, powered by an air turbine rotating at more than 300.000 rpm. A maximum depth of 1.2 mm was reached in 0.01 mm steps. Then, the hole diameter and eccentricity were measured through an optical microscope with crosshairs, included in the MTS3000-Restan system. Eventually, the stress distributions from 0 to 1 mm depth were obtained with the Influence Functions method [7], Tikhonov regularization and the Morozov discrepancy principle [8].

(a) (b)

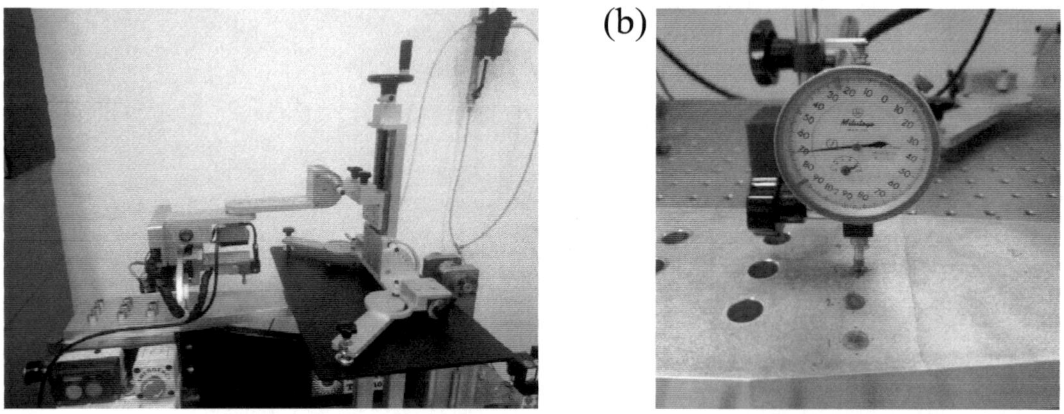

Fig. 3. (a) GNR SpiderX Edge Diffractometer, mounted on the calibration bench.
(b) Measurement of etching depth with a dial gauge.

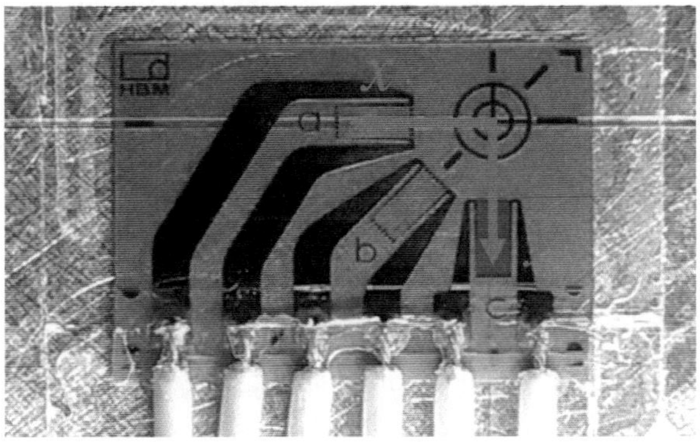

Fig. 4. Strain rosette (HBM RY61K) applied on the upper face of the specimen, aligned with the specimen reference system.

Experimental Results

Results of the XRD surface measurements are reported in Tables 1-2. For comparison, the near-surface stress values obtained through HDM measurements are also reported therein. For each treatment, point 3 was used to carry out the XRD depth profiling, so the corresponding HDM value is not available.
In *Fig. 5*, the validation data for XRD surface measurements are reported. Each graph refers to a component of the applied (plane) stress tensor, extracted from the three measurements at different orientation.
In *Fig. 6*, the bending stress distributions identified during HDM measurements are shown. The applied stress distributions were accurately characterized with the procedure described in [2].
In *Fig. 7*, the residual stress distributions obtained with HDM measurements on points 1-2 are compared with the XRD depth profiling performed on point 3.

Table 1. Summary of surface residual stress measurements – Normal stress σ_{xx} [MPa]

	HDM	**XRD (Lab. 1)**	**XRD (Lab. 2)**
AZB425 - 1	-259	-275	-270
AZB425 - 2	-255	-280	-274
AZB425 - 3	//	-246	-258
CEB120 - 1	-238	-246	-255
CEB120 - 2	-275	-242	-268
CEB120 - 3	//	-235	-249

Table 2. Summary of surface residual stress measurements – Normal stress σ_{yy} [MPa]

	HDM	**XRD (Lab. 1)**	**XRD (Lab. 2)**
AZB425 - 1	-293	-261	-262
AZB425 - 2	-276	-241	-265
AZB425 - 3	//	-251	-259
CEB120 - 1	-260	-249	-256
CEB120 - 2	-232	-244	-260
CEB120 - 3	//	-230	-256

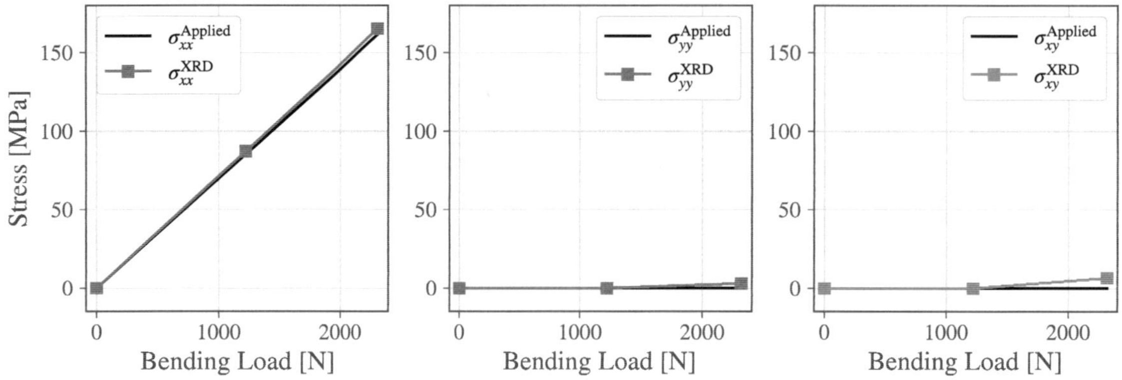

Fig. 5. Validation of XRD surface stress measurements, through identification of an accurately known bending stress distribution. The bending load is applied in two steps.

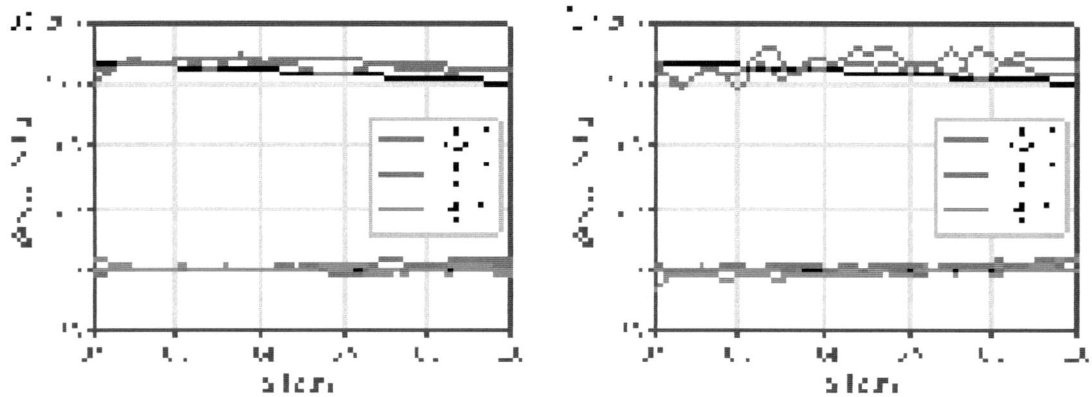

Fig. 6. Bending stress distributions identified during the four HDM measurements. The reference distributions are reported as black lines. (a) AZB425. (b) CEB120.

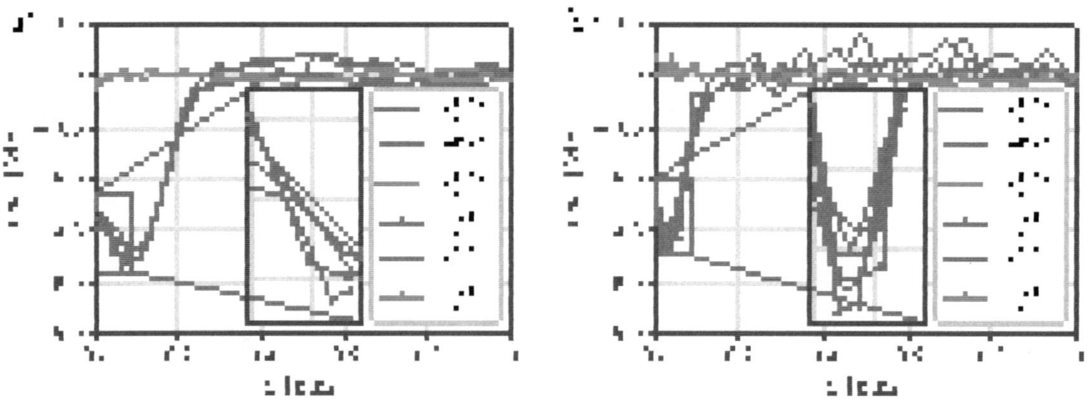

Fig. 7. Obtained residual stress distributions. The HDM measurements on points 1-2 are reported as dashed lines, while the XRD measurements on point 3 are reported as squares connected by solid lines. (a) AZB425. (b) CEB120.

Discussion and Conclusions

The validation phase confirmed that the instrumentation setup had been properly carried out. In particular, XRD surface measurements captured the stress fields produced by the bending load with notable accuracy (see Fig. 5). A slightly bigger error (in the order of 10 MPa) was obtained for σ_{xy} at the biggest load. That is still consistent with the fact that σ_{xy} is obtained as a linear combination of XRD measurements at the three orientations, so its precision is expected to be lower.
On the other hand, HDM measurements were able to capture the bending stress distribution along the entire hole depth (see Fig. 6). The achieved absolute error is about ±25 MPa, which is aligned with the state of the art [9]. Note that in this case the error is not particularly pronounced at the surface, where HDM is instead known to be prone to higher uncertainties.
Surface measurements reported in Tables 1-2 show comparable results across different measurement techniques and between the two diffractometers. Measurements show a variability which is compatible with the typical uncertainties of the methods [6].
The two residual stress profiles obtained with HDM for each shot peening treatment (shown in Fig. 7) are consistent with the ±25 MPa error achieved in the identification of the bending distribution. The XRD stress profiles obtained with electropolishing correctly mimic the trend of HDM curves. For a given depth value, the discrepancy can be up to 50 MPa. Uncertainties in the etching surface planarity and in the measurement of etching depth likely affect the achieved accuracy, which is still reasonable for engineering applications.
The presented calibration bench allows one to validate the chosen setup before and during residual stress measurements, both with XRD and HDM. For example, an unacceptably textured specimen material, insufficient grain statistics, or issues in the strain gauge bonding would immediately result in huge errors in the identification of the bending distribution.
Since the specimen can be designed with different materials and geometry, the calibration bench is particularly useful to investigate residual stresses produced by new surface treatments, with an increased measurement confidence provided by the validation phase.

References

[1] E. Valentini, M. Beghini, L. Bertini, C. Santus, and M. Benedetti, *Procedure to Perform a Validated Incremental Hole Drilling Measurement: Application to Shot Peening Residual Stresses*, Strain 47: e605–e618, Jun. 2011, doi: 10.1111/j.1475-1305.2009.00664.x.
[2] M. Beghini, T. Grossi, C. Santus, and E. Valentini, *A calibration bench to validate systematic error compensation strategies in hole drilling measurements*, ICRS 11 – 11th International Conference on Residual Stresses, Mar 2022, Nancy, France.
[3] H.F. Bueckner, *Novel principle for the computation of stress intensity factors*, Zeitschrift fuer Angewandte Mathematik & Mechanik 50(9), 1970.
[4] G.S. Schajer, *Measurement of Non-Uniform Residual Stresses Using the Hole-Drilling Method. Part I—Stress Calculation Procedures*, J. Eng. Mater. Technol., vol. 110, no. 4, pp. 338–343, Oct. 1988, doi: 10.1115/1.3226059.
[5] American Society for Testing and Materials, *Test Method for Determining Residual Stresses by the Hole-Drilling Strain-Gage Method*, West Conshohocken PA, 2020, doi: 10.1520/E0837-20.
[6] I.C. Noyan, and J.B. Cohen. *Residual stress: measurement by diffraction and interpretation*. Springer, 2013.
[7] M. Beghini, and L. Bertini. *Analytical expressions of the influence functions for accuracy and versatility improvement in the hole-drilling method.* The Journal of Strain Analysis for Engineering Design 35.2 (2000): 125-135.
[8] G.S. Schajer, *Hole-drilling residual stress profiling with automated smoothing.* J Eng Mater Technol 129(3):440–445, doi:10.1115/1.2744416
[9] G.S. Schajer, ed. *Practical residual stress measurement methods.* John Wiley & Sons, 2013.

RESIDUAL STRESS ANALYSIS BY CONTOUR METHOD ONTO COLD EXPANDED HOLE ASSEMBLY– COMPARISON WITH FEA

Vincent DELAGE [1], Jean-Michel DUCHAZEAUBENEIX [1], Laurence BESNAULT [1]

1 Meliad, www.Meliad-sas.com, 6, rue des Orfèvres 44840 Les Sorinières, France
v.delage@Meliad-sas.com, jm.duchazeaubeneix@Meliad-sas.com, l.besnault@Meliad-sas.com

1 Abstract

Introducing high residual stress in a hole is critical for many fatigue life components, especially in aerospace industry. The process introduces deep residual stress on hole surface providing fatigue life enhancement.
Cold expansion should be controlled to guaranty both the homogeneity and the repeatability of the compressive stresses generated by the process. However, it is difficult to control the stresses generated into the component with usual residual stresses measurement methods.
This is why Meliad and Capaero with Mat-In-Meca developed the contour method allowing to control the deep stress profile generated in the part.
This study shows how experimental measurement by the contour method and numerical model through finite element analysis gives effective characterization of the residual stresses introduced after hole expansion.

Material: 35NCD13 Steel

2 Objectives

This study shows how experimental measurement by the contour method and numerical model through finite element analysis gives effective characterization of the residual stresses introduced after hole expansion.

3 Processes and methods

3.1 Split sleeve cold expansion process

Split-sleeve coldworking was developed in the early 1970's as a way to improve joint fatigue performance by expanding or compressing the material around the fastener hole. In the split-sleeve process, this is accomplished by inserting a mandrel that has been prefitted with a disposable sleeve, through a fastener hole, and pulling the mandrel back through the sleeve.
The internal surface of the sleeve is pre-lubricated with a special dry lubricant, protecting the mandrel from excessive wear, and reducing the amount of force required to pull the mandrel.

3.2 Split sleeve expansion procedure

1. Drill start hole with start drill.
2. Ream hole to proper starting size with start hole reamer.
3. Verify start hole with hole gage.

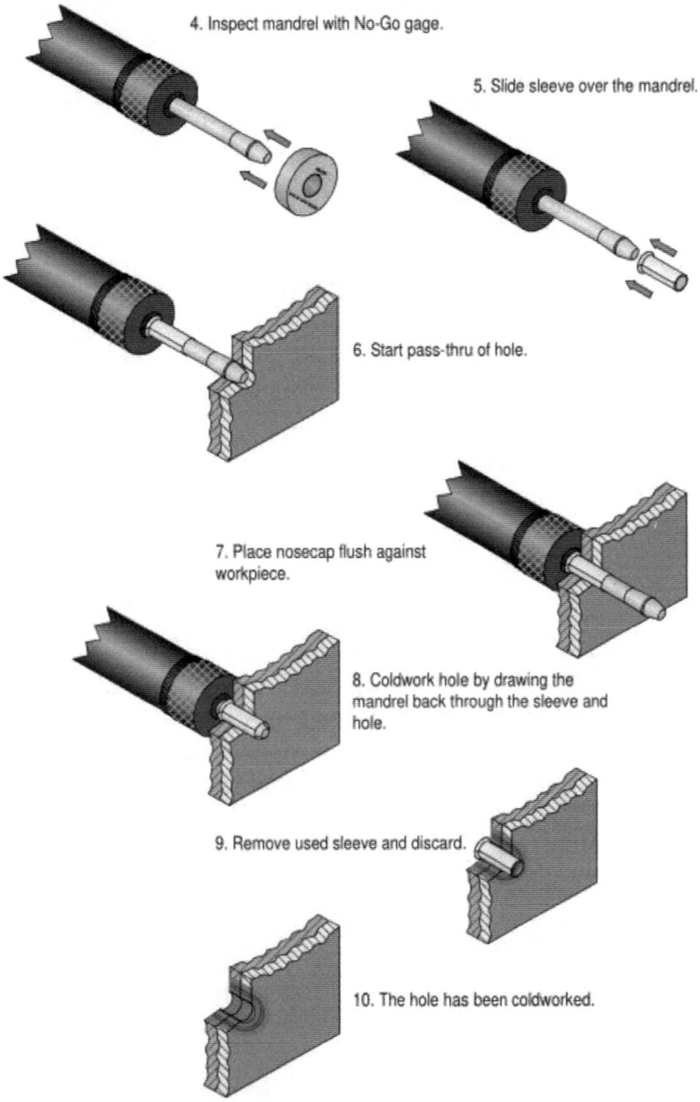

11. Inspect coldworked hole with hole gage.
12. Ream hole to final size with piloted reamer.
13. Inspect final reamed hole with hole gage. Countersink if necessary.

Figure 1 Split sleeve expansion procedure

3.3 Contour method

The contour method is a destructive method to obtain the residual stress map on the section of a part. This method allows a very different approach to stress measurements by giving a complete map and not only a spot value.

The method was firstly introduced in 2000[1] and has seen a strong development in the 2010 decade.

This method determines residual stress by carefully cutting the specimen into 2 pieces and measuring the resulting deformations dur to residual stress redistribution. The displacement measured is used to run a finite element analysis model allowing to obtain the original stress state inside the specimen along the cut plan.

Contour method is widely used on big parts at early manufacturing stages such forging, heat treatment and pre-machining.
The progress made during the last decades on the result uncertainty and the cutting technics allows to work on smaller and more finished part.

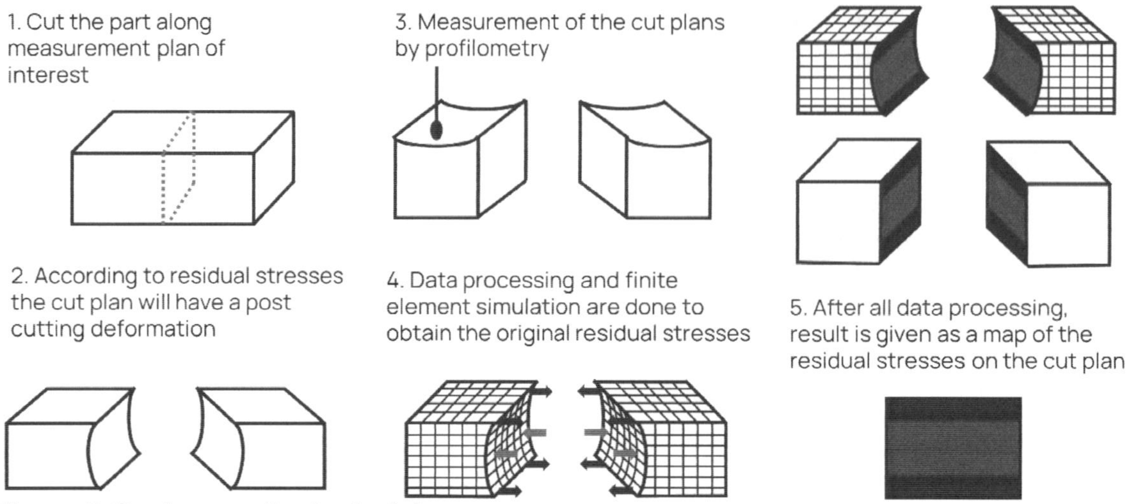

Figure 2 Contour method principle

4 Study

In this study we intended to assess the reliability of contour method on cold expanded hole.
Measurement of residual stresses is particularly difficult on cold expanded holes due to accessibility reasons:

- X-Ray measurement can measure close to the hole but on surface only
- Hole drilling method is usually quite far from hole edge because of the gauge size and can't capture enough residual stresses

Contour method offers a new opportunity to get residual stresses map around the hole and confirm the residual stresses introduced by the process.
After a finite element analysis to get an evaluation of the residual stress map, a representative sample has been cold expanded before being evaluated with the contour method.
This offers a unique opportunity to compare both methods result.

4.1 Finite element analysis

4.1.1 Method

Capaero simulation was made with Abaqus software using a 3D model including expansion tools (Nosecap, mandrel and split sleeve) with the original part geometry.
Contacts management and meshing parameters are parts of Capaero know-how to obtain reliable results in a limited computing time.
Experience shown that 3D model is more accurate than 2D ones and allows to fully simulate the sleeve deformation as well as the split impact on the residual stresses.
Simulation is done in 3 main steps:

1. Finite element modelling of the structure for the different expansion configurations. A convergence study has been made for each configuration.
2. Numerical simulation of the hole split sleeve and its stabilized state after expansion for all studied configurations.
3. Analysis of the numerical simulations and the impact of the split sleeve

1. 2. 3.

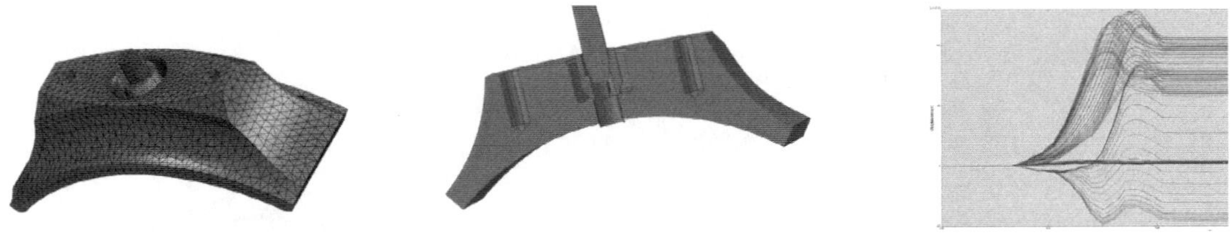

Simulated configurations were based on minimum and maximum expansion ratios created by tooling tolerances.
For this study, only one result is shown: the one corresponding to the sample measured in contour method.
The dynamic explicit simulation was done taking special care of contact properties. This kind of simulation tend to increase the stresses oscillations after spring back but **Capaero** experience allows to lower the uncertainness of the results.

4.1.2 Results

In this study, the measurement plan was taken at 90 degrees of the split location to avoid local behaviours due to the split.

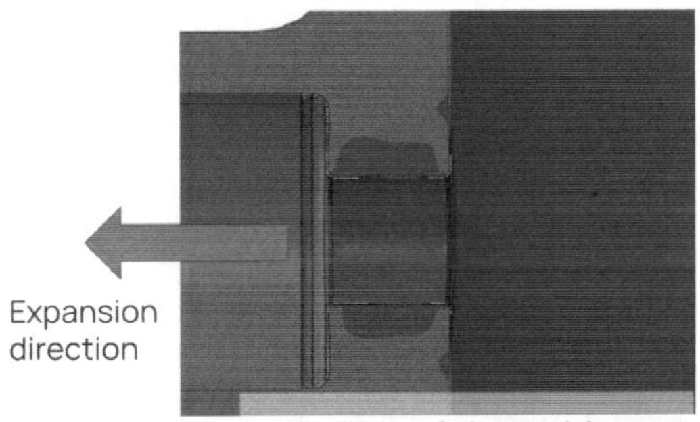

Figure 3 plastic strain map shown in a cut view of the cold expanded hole

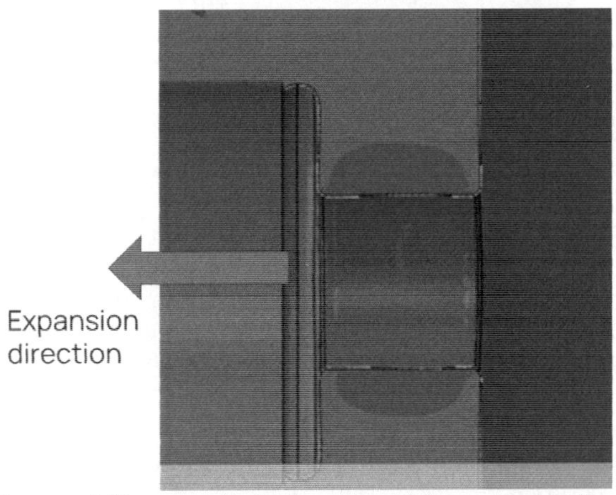

Figure 4 Tangential stresses after expansion

4.2 Residual stresses measurement using contour method

4.2.1 Method

The sample measured has a simplified geometry compared to the original part used for the simulation. However, the local stresses are expected to be similar.

First, the part was cut using and EDM machine and a thin cutting wire to prevent has much has possible effects of the cutting operation on the part.

Then the 2 cut surfaces are measured using a microscope.

Figure 5 one side of the sample being measured to obtain the surface geometry

Figure 6 3D measurement result

Both surfaces are the combined to obtain an average surface. Data are filtered using an interpolation parameter aiming to reduce roughness effect and keep only the variations dues to residual stresses relaxation.
At the final step, a final element analysis is performed with deformation data to obtain the original residual stresses state.

4.2.2 Results

Tangential residual stresses measured by the contour method are sown on the following figure:

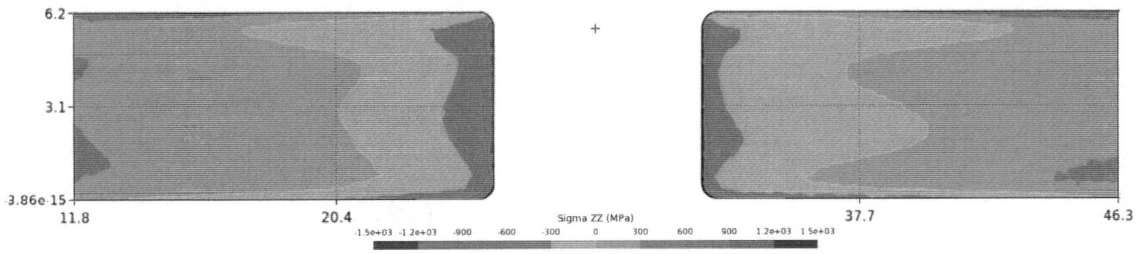

Figure 7 Tangential stresses measured with contour method (MPa)

4.3 Comparison between FEA and measurement

FEA and contour method agree on some points. They show that maximum compressive stress near the surface is higher than 1000 MPa and depth-to-zero is between 3 and 5 mm.
Significant differences are also clearly visible. FEA shows a stress plateau while contour method describes immediate decrease.
Uncertainty for the contour method is in the order of ±10%. It is also known that the contour method is hardly capable to measure very high gradient (1000 MPa/mm) close to surfaces.
FEA results is directly dependent of mechanical properties such as yield strength and work hardening considered for the elasto-plastic model. Now, standard properties have been considered for the present steel grade.
Observed differences will be discussed in further study in the light of reproducibility measurement for the contour method, sensitivity study for the FEA, and additional measurements with XRD at Meliad.

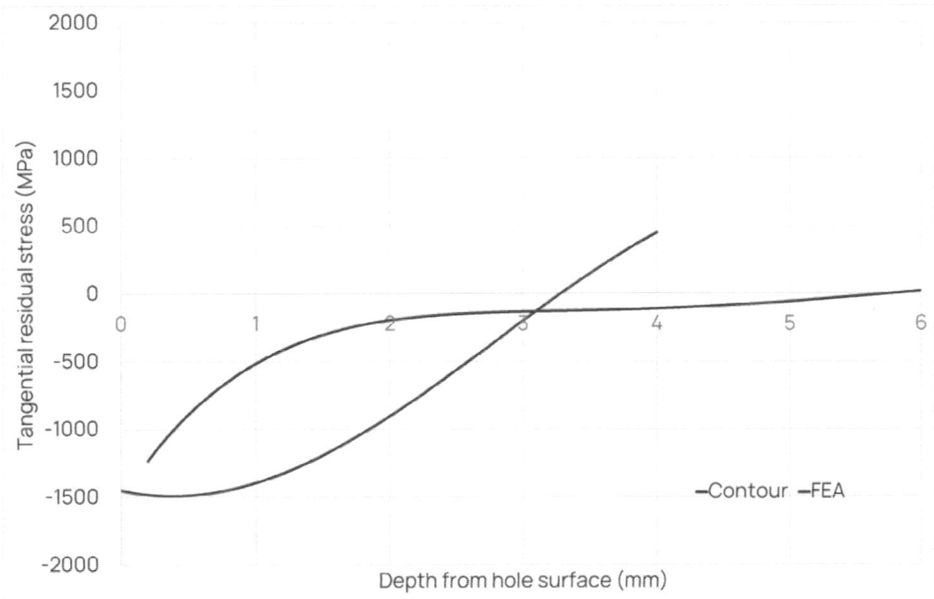

5 Conclusion

Contour method shown interesting results to allow residual stresses analysis on a cold expanded hole. This approach offered data we can't get with other conventional methods.
However, further work is required to improve the exactness of the values obtained and the compression depth obtained by cold expansion.
Working on the dimensional data acquisition close to hole edge could improve the final result.

6 List of references

[1] Prime, MB, Gonzales AR, 2000, *The Contour Method: Simple 2-D Mapping of Residual stresses*, The 6th International Conference on Residual Stresses, IOM Communications, London, UK, Oxford, UK 617-624.

THE EFFECT OF SHOT PEENING ON TRIBO-CORROSION CHARACTERISTICS OF ADDITIVELY MANUFACTURED Ti-6Al-4V USED IN THE MARINE INDUSTRY

K.A. Vella[a], J. Buhagiar[a], G. Cassar[a], L. Bonnici[a], M.M. Pizzuto[a], J. Chen[b], X.Y. Zhang[b], M.Y. Liu[b], Z.Q. Huang[b], A. Zammit[a]

a Department of Metallurgy and Materials Engineering, University of Malta, Msida, Malta – ann.zammit@um.edu.mt
b School of Materials Science and Engineering, Southeast University, Nanjing, Jiangsu Province, China – j.chen@seu.edu.cn

Abstract
Ti-6Al-4V is a long-established titanium alloy, however, its use is often limited by its poor behaviour under friction conditions. Such behaviour is more pronounced in environments where corrosion occurs in conjunction with wear. This limitation can be mitigated through surface engineering techniques. In this investigation, the titanium substrates, manufactured via selective laser melting, were subjected to a shot peening treatment. The effect of the surface treatment on the microstructure, microhardness and roughness of the samples was studied via a number of characterisation techniques. To provide an insight into the corrosion-wear response, the samples were tested under reciprocating sliding wear conditions at an anodic potential in simulated ocean water. The shot-peened samples exhibited no improvement in the corrosion-wear resistance when compared to the untreated printed samples. Tribocorrosion calculations suggest degradation occurred due to mechanical wear with a marginal contribution from corrosion.

Keywords additive manufacturing, shot peening, tribocorrosion, hardness, roughness

Introduction

Ti-6Al-4V (Ti64) is a widely used titanium alloy having various ideal properties, such as high specific strength, low elastic modulus and good fatigue strength [1]. Furthermore, manufacturing the alloy via additive manufacturing is beneficial due to the provided ability of rapidly producing complex parts with minimal post-processing. The alloy's employment in the marine industry is not uncommon, especially since its use provides the ability of producing lightweight components having excellent corrosion resistance. However, the use of Ti64 is limited by its poor behaviour under friction conditions observed as, high unstable coefficients of friction with a tendency to gall and fret. Surface engineering techniques such as shot peening (SP) can be used to combat this limitation.

SP is a cold working process which induces beneficial residual compressive stresses at the surface via high-velocity impingement of shots, intended to increase the fatigue strength and reduce stress corrosion cracking. It has been recognized that the tribological performance of the shot peened component can be enhanced by the surface hardening and lubrication effects of the dimples [2]. However, the effect of SP on the tribocorrosion behaviour of additively manufactured Ti64 is not yet extensively studied. So far some authors have reported that the SP treatment decreased the material loss due to both wear and corrosion, others reported that the induced roughness has a negative effect on the corrosion-wear resistance [3], [4].

In the current investigation, the effect of this SP treatment on the corrosion-wear characteristics of a 3D printed Ti-6Al-4V alloy, was studied. This was done by subjecting an untreated and a shot peened sample to linearly reciprocating ball-on-flat sliding wear testing conducted in simulated ocean water. The obtained wear tracks were characterised via profilometry and micrographic techniques, and material losses due to corrosion-wear were quantified. The samples were also analysed via microscopy, microhardness testing, profilometry and X-ray diffraction to investigate the effect of the surface treatment.

Experimental Methods

Sample preparation

The 20 mm diameter and 6 mm thick Ti64 samples were manufactured by selective laser melting (SLM) a Ti-6Al-4V powder, having a maximum particle size of 35 µm. The samples were printed layer by layer at a laser speed and power of 600 mms^{-1} and 100 W respectively, each layer having a thickness of 30 µm. The samples, having the composition listed in Table 1, were ground and polished up to a mirror finish. Prior to grinding and polishing, the SLM printed samples were heat treated, using a TAV Dualjet TPH-200 (Italy) furnace, in a nitrogen atmosphere till 800°C at a ramp rate of 5°$Cmin^{-1}$ for 2 hours followed by furnace cooling.

Table 1. Composition of 3D printed Ti64 obtained by EDS analysis

	Ti	Al	V	Fe
wt%	Bal.	5.37	3.94	1.15

Surface treatments

The SP treatment was carried out using an Almen intensity of 0.24 mmA, 6 bar pressure and a nozzle-to-specimen distance of 100 mm. The samples were peened using Zirshot Z300 ceramic shots for a duration of 60 seconds to obtain 500% coverage.

Surface and near-surface characterisation

Micrographic analysis of the samples was carried out using the Carl Zeiss Axioscope 5 (Germany) optical microscope and the Carl Zeiss Merlin Gemini (Germany) scanning electron microscope (SEM) in conjunction with an Ametek EDAX (USA) energy dispersive spectroscopy (EDS) analyser. Prior to visualisation, the substrates were etched for 10 seconds using Kroll's reagent composed of 5% HF, 13.5% HNO_3 and 81.5% H_2O.

For phase analysis, X-ray diffraction was performed via the Bruker AXS D8 Advance (USA) machine. Scans were performed over a 2θ range of 20° to 90° with a step size of 0.02° and a dwell time of 2 seconds using a CuKα radiation source. Stress measurements were obtained using the Rigaku Ultima VI (Japan) diffractometer. Measurements were taken at seven ψ tilts, from 0° to 60°, stresses were then obtained via the $Sin^2\psi$ method.

Surface roughness measurements were carried out via an AEP Technology NanoMap-500LS (USA) contact profilometer. Scans were performed over a distance of 2500 µm at 25 μms^{-1} and a resolution of 1 µm. Five scans were performed for each sample condition.

Using the Mitutoyo MVK-H2 (Japan) microhardness tester in conjunction with a pyramidal diamond indenter, Vickers microhardness measurements were obtained. A load of 100 gf was applied, the application lasting for 10 seconds. A series of five indentations were applied.

Tribocorrosion testing

For tribocorrosion testing, a Bruker UMT TriboLab (USA) set up with a reciprocating drive and a 3-electrode tribocorrosion cell was used. The cell was connected to a Gamry Interface 1000™ (USA) potentiostat to induce an anodic potential of 0.5 V with respect to the Ag/AgCl reference electrode. This potential ensures that all samples are in the passive regime. The cell was filled with approximately 150 ml of artificial seawater formulated in accordance with ASTM D1141-93 *Standard Practice for the Preparation of Substitute Ocean Water*. A 4.76 mm diameter Al_2O_3 counter-face was utilised. The voltage was induced for 600 seconds without sliding, for stabilisation, followed by sliding for 2000 seconds at a static Hertzian contact pressure of 678 MPa, frequency of 1Hz and a stroke length of 3.5 mm. During sliding, the dynamic anodic current and COF values were recorded. Once sliding stopped, the anodic potential was applied for a further 600 seconds. All tests were repeated three times at room temperature.

Wear track characterisation

Following testing, elemental analysis of any debris and artefacts present was carried out. All wear track depths were measured via profilometry, each wear track measured using an average of three linear scans. From data collected during testing and measurements taken, the material loss rates were quantified [5].

Results and Discussion

Characterisation of samples

Following shot peening, the effect of grain refinement was observed. The grains appear denser at the surface and assume a basket-weave network further away as observed in Figure 1.

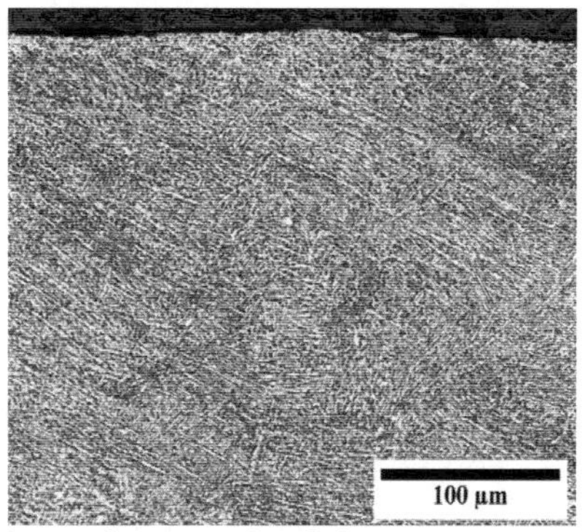

Figure 1. Low magnification micrograph of the cross-sectioned shot peened sample

As observed in Figure 2, a shift towards lower 2-theta values is observed for the printed and shot peened (SP) sample. This shift in the spectra is due to the induced compressive residual stresses, thus confirming that the peening treatment was effective. At the surface, a compressive residual stress (CRS) equivalent to 432 MPa was measured suggesting that a significant plastic deformation occurred. Similar stress magnitudes were measured by Zhang *et al.* [6] for SLM Ti64 following SP and furthermore, experienced no phase transformation. The stress magnitude is observed to be dependent on shot velocity and on the shot diameter. Impingement at higher speeds induces greater plastic deformation thus an increase in the CRS at the surface [6]. In the current study, no new peaks were detected thus, phase transformation due to SP also did not occur. It is likely that the peening intensity is too low to induce phase transformation. Moreover, broader and less intense peaks are observed for the SP sample as a consequence of the roughened surface as the surface asperities absorb the diffracted rays [7].

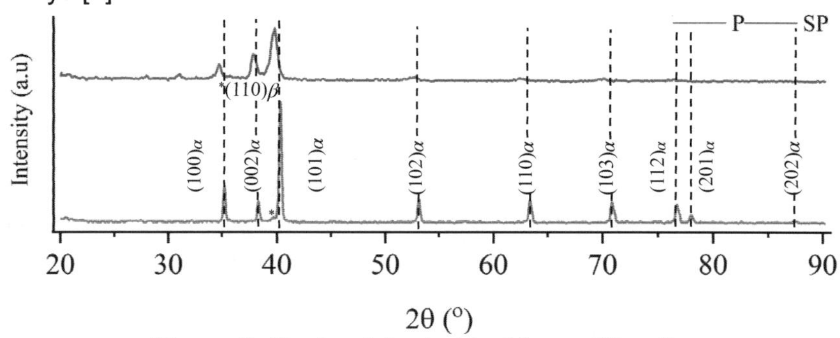

Figure 2. Peaks detected by X-ray diffraction

As observed in Table 2, the R_a and R_z roughness values were the highest for the samples which had been printed and subsequently shot peened (SP). This was expected due to the roughening effect imparted by the SP treatment. The impact of shots results in sharp protrusions due to the interaction of the dimples.

Table 2. Hardness and roughness measurements. For both measurements, uncertainty is represented by 95% confidence interval for a population size of 5.

Sample	Hardness ($HV_{0.1}$)	Roughness	
		R_a (µm)	R_z (µm)
P	331 ± 2.31	0.05 ± 0.01	1.33 ± 0.52
SP	375 ± 4.47	1.47 ± 0.13	11.60 ± 0.95

Following SP, the hardness of the printed sample increased by 13% thus confirming the occurrence of a degree of work hardening. A similar increase is observed by Zhang *et al.* [6] for SP SLM Ti64. Such hardening is due to grain refinement, which results in an increase in grain boundaries and thus a greater hindrance to dislocation movement, and induced compressive residual stresses. The hardness values were highest at the surface and decreased gradually to the bulk hardness resulting in an affected depth ranging from 180 to 200 µm; also in line with what was observed by other authors [6].

Corrosion-wear response of untreated and treated AM Ti-6Al-4V

Dynamic anodic current and coefficient of friction measurements

For both sample conditions, upon sliding, an increase in anodic current was observed which dropped back to initial values once sliding stopped. This increase is due to the removal of the passive film exposing the underlying substrate to oxidation [8]. Both samples underwent Type I corrosion-wear which is the removal or damage of the passive layer present at the coating surface and its subsequent re-passivation. This mechanism is reflected in the constant fluctuations of corrosion-current, corresponding to the continuous re-passivation and de-passivation, as observed in Figure 3(a,b).

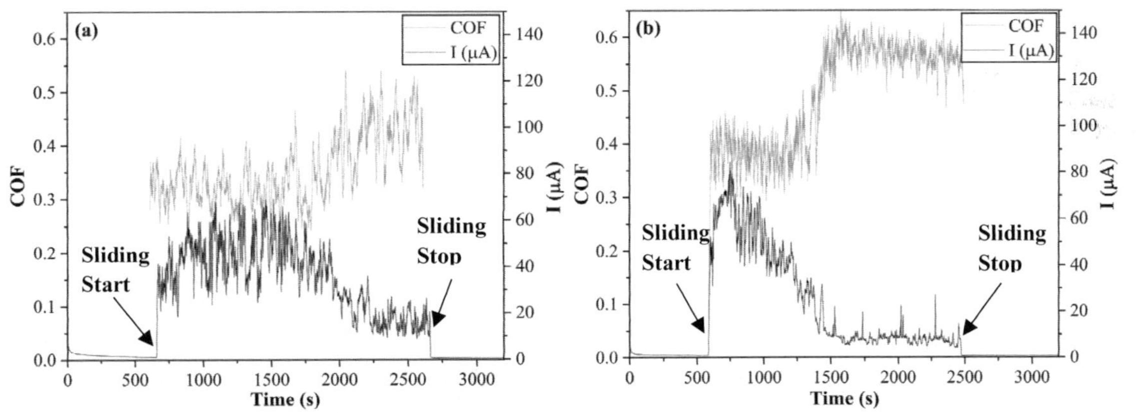

Figure 3. Dynamic anodic current and COF plots a) printed and b) printed + shot peened Ti-6Al-4V

Furthermore, as observed, both the P and SP sample exhibited a similar behaviour which was observed for all three repeats carried out for both sample conditions. This decrease in current, which implies a decrease in the amount of new substrate exposed to the electrolyte, is simultaneously accompanied with an increase in the coefficient of friction (COF). This behaviour is attributed to the oxidised areas formed on the surface, confirmed to be composed of TiO_2 by EDS analysis. These form due to the strong adhesion between the substrate and the ceramic ball at asperities which results in high frictional forces. With sliding, the asperities are removed resulting in generation of debris, which is either compacted into re-adhered material or free third body particles contributing to abrasion [8]. Thus, these patches protect the surface from being exposed to the corrosive solution and hence from Type I corrosion-wear. Therefore, the current values decreased greatly. However, these patches also result in an uneven wear track and increase the COF due to roughening the track [8], [9]. This behaviour could also be related to the formation of asperities which decrease the contact area. Thus, the contact pressure at such areas increases resulting in roughening. Meanwhile, due to a decrease in contact area, less area is being exposed to dissolution thus, the decrease in the current values [10].

For the SP sample, currents as high as 80 µA were recorded immediately upon sliding, suggesting that the substrate was exposed at once and that the SP treatment was ineffective. Moreover, the overall high COF values observed are typical for SP treated Ti64 since the treatment increases the surface roughness [11]. This increased roughness was most likely detrimental to the tribocorrosion behaviour. Several spikes, once current values stabilised at lower values, were also observed (Figure 3b). Such spikes are related to the development of wear particles where their formation results in de-passivation and subsequent re-passivation of the substrate [12].

Material Losses

Figure 4 depicts a plot of the material losses calculated from the anodic current and wear track depth measurements. From the plot no significant difference between the material losses the P and SP sample can be observed. Thus, the SP treatment appears to be ineffective.

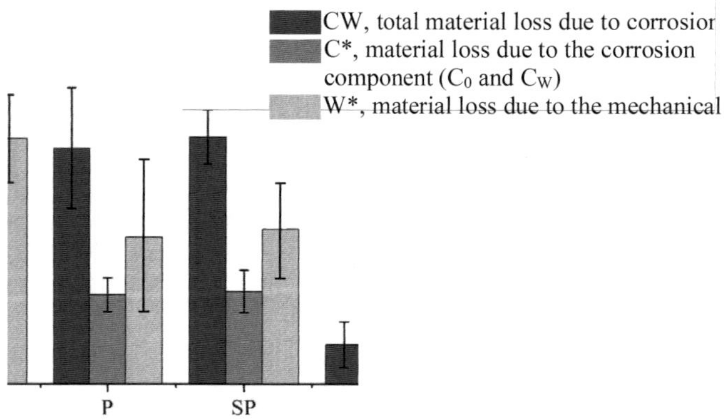

The ratio of the loss due to the corrosion component and due to the wear component (C*/W*) provides an insight in the dominant degradation mechanisms present [8],[13]. The ratios obtained were 0.61 ± 0.20 and 0.60 ± 0.12 for the P and SP sample respectively. Since both values are between 0.1 and 1, this suggests that synergism is present and the most dominant mechanism is wear-corrosion. Therefore, the greatest losses are caused by the total mechanical wear component, which is also reflected in the results obtained. The former contributed to 62 ± 7% and 63 ± 4.5% of the corrosion-wear for the P and SP samples respectively. However, this method does not clearly identify whether the loss was due to pure mechanical wear (W_0) or due to mechanical wear due to corrosion (W_C).

Mechanical wear occurred via abrasive and adhesive mechanisms. Abrasive wear was evidenced by thencing in the direction of the sliding direction. Whilst, adhesive wear was evidenced by micro-asperity shearing. The developed oxidation patches were also a consequence of adhesion. Its occurrence was also reflected by the unstable COF and high wear rates of the P and SP samples.

Conclusions

The following conclusions can be drawn from the results obtained when studying the effect of shot peening on 3D printed Ti-6Al-4V substrate:

1. Broader X-ray diffraction peaks shifted toward lower 2-theta values confirming the occurrence of grain refinement and presence of residual stresses due to the SP treatment. A compressive surface stress equivalent to 432 MPa was obtained.
2. Due to the shot impingement, the R_a and R_z values of the P sample increased by approximately 3 orders and 1 order of magnitude respectively.
3. Following SP, the hardness of the P sample increased by 13% and the affected depth reaches a maximum of 200 μm.
4. The corrosion-wear resistance did not increase following SP. The similar behaviour of the P and SP sample indicates that the peening treatment was ineffective in this case.
5. Both sample conditions obtained a C*/W* ratio in the range of 0.1 to 1 confirming that synergism was present and the wear-corrosion mechanism was the most dominant.
6. Type I corrosion-wear was observed. Mechanical wear was observed as abrasion in term of grooves. Adhesive wear was evidenced by micro-asperity shearing and reflected by the unstable COF values and significant wear rates obtained.

Acknowledgements

The research carried out is part of a Sino-Malta project regarding Surface Engineering of Additive Manufactured (SEAM) parts used in marine transportation and was possible due to funding provided by the Malta Council for Science and Technology (MCST).

References

[1] A. Loureiro *et al.*, "Properties and Applications of Titanium Alloys: A Brief Review," *Rev. Adv. Mater. Sci.*, no. 32, pp. 14–34, 2012.

[2] A. Zammit, "Shot Peening of Austempered Ductile Iron," in *Advanced Surface Engineering Research*, M. A. Chowdhury, Ed. Rijeka: IntechOpen, 2018.

[3] Y. Sun and R. Bailey, "Improvement in tribocorrosion behavior of 304 stainless steel by surface mechanical attrition treatment," *Surf. Coat. Technol.*, vol. 253, pp. 284–291, Aug. 2014.

[4] A. S. Hammood *et al.*, "Tribocorrosion Behaviour of Ti–6Al–4V Alloy in Biomedical Implants: Effects of Applied Load and Surface Roughness on Material Degradation," *J. Bio- Tribo-Corros.*, vol. 5, no. 4, Dec. 2019.

[5] B. Mallia and P. A. Dearnley, "The corrosion-wear response of Cr-Ti coatings," *Wear*, vol. 263, pp. 679–690, Sep. 2007.

[6] Q. Zhang *et al.*, "Microstructure change and corrosion resistance of selective laser melted Ti-6Al-4V alloy subjected to pneumatic shot peening and ultrasonic shot peening," *Surf. Topogr. Metrol. Prop.*, vol. 10, no. 1, Mar. 2022.

[7] A. Zammit *et al.*, "Enhancing surface integrity of titanium alloy through hybrid surface modification (HSM) treatments," *Mater. Chem. Phys.*, vol. 279, p. 125768, Mar. 2022.

[8] F. Toptan *et al.*, "Corrosion and tribocorrosion behaviour of Ti6Al4V produced by selective laser melting and hot pressing in comparison with the commercial alloy," *J. Mater. Process. Technol.*, vol. 266, pp. 239–245, Apr. 2019.

[9] I. Çaha *et al.*, "Corrosion and tribocorrosion behaviour of titanium nitride thin films grown on titanium under different deposition times," *Surf. Coat. Technol.*, vol. 374, pp. 878–888, 2019.

[10] R. Chetcuti *et al.*, "Tribocorrosion response of duplex layered CoCrMoC/CrN and CrN/CoCrMoC coatings on implant grade 316LVM stainless steel," *Surf. Coat. Technol.*, vol. 384, p. 125-313, Feb. 2020.

[11] D. G. Bansal *et al.*, "Surface engineering to improve the durability and lubricity of Ti–6Al–4V alloy," *Wear*, vol. 271, no. 9, pp. 2006–2015, 2011.

[12] D. Landolt, "Electrochemical and materials aspects of tribocorrosion systems," *J. Phys. Appl. Phys.*, vol. 39, no. 15, pp. 3121–3127, Aug. 2006.

[13] M. J. Runa *et al.*, "Tribocorrosion response of the Ti6Al4V alloys commonly used in femoral stems," *3rd Int. Conf. Tribocorrosion – Atlanta 2012*, vol. 68, pp. 85–93, Dec. 2013.

Improvement of the fatigue life of lattice structures covered by a thin envelope for biomedical applications

A. T. Sola[1], L. Weiss[1], P. Laheurte[1] and E. Fleury[1]

1 LEM3, Université Lorraine, 7 Rue Félix Savart, Metz, France

Abstract
Additive manufacturing allows the production of parts of a high geometric shape complexity hardly realisable by other manufacturing techniques. Additive manufacturing is particularly suitable for the fabrication of scaffold-type structures, also named lattice structures, that are of interest in various fields. By covering these lattice structures by a thin envelope, the realised parts can be applied in the biomedical sector and more particularly as prostheses.
The application of manufactured parts is still hindered by the presence of microstructural defects reducing considerably the mechanical properties. The objective of this study was thus two-fold: to investigate the fatigue properties of Ti-6Al-4V samples with lattice structures produced by additive manufacturing and to study the effect of SMAT (Surface Mechanical Attrition Treatment) on the improvement of the fatigue life.

Keywords *lattice, SMAT, additive manufacturing, fatigue life.*

Introduction

While conventional manufacturing processes enable the fabrication of engineering components of complex geometries by either moulding and/or material subtractive processes, additive manufacturing (AM) consists in building 3 dimensional components by adding specific thickness layers of material from a CAD and CAO data model. AM technology has been around for more than 20 years for the purpose of prototyping and today is used as a mass production technique with applications in different fields such as aeronautics, medicine, automotive, etc.
In AM, a heat source is applied to melt metal powder or wire that are used as base materials. The manufacturing is carried out layer by layer. The three most common types of technologies are powder bed fusion, material extrusion and material deposition from a concentrated energy source [1].

The most remarkable advantages offered by AM [2], are:

- the simplicity and flexibility to manufacture any complex part,
- the reduction in waste material,
- the reduced manufacturing times thanks to the optimization of the design,
- the reduction in the cost per part.

The interest in maintaining optimal mechanical properties while reducing weight is of great interest in the design of parts. This can be achieved by the introduction of lattice structures, Triply-Periodic-Minimal-Surfaces (TPMS) or topological optimization methods. The first two are based on spatial patterns that are repeated in space. The mechanical properties, such as Young's modulus, thus depend on the lattice and size and also on the strut radius [3]. The most widely investigated lattice structures are based on Bravais lattices or crystal structures, BCC or FCC and modifications of these, z-struts (BCCZ and FCCZ) [4]. These types of structure allow for better mechanical adaptation and osseointegration as well as good fixation of implants [4].

Several investigations have already studied possible post-treatments with the aim of improving the quality of the parts and have particularly addressed the problems of porosities as well as the surface condition that are inherent to the AM processing. Some examples are heat treatment, high isostatic pressure treatment, machining, chemical polishing, etc. The SMAT (Surface Mechanical Attrition Treatment) post-treatment consists in the application of a very severe plastic deformation at the surface of the component that refines the grain size and introduces compression-type residual stresses in the sub-surface. The microstructure modifications result in the enhancement of the fatigue limit by delaying the propagation of cracks owing to the crack-closing effect thus avoiding fast propagation rate. In addition, SMAT provides a good surface condition and increases wear resistance [5], [6].

Experimental Methods

Ti-6Al-4V spherical powder was used for the fabrication of cylindrical dogbone-shape samples (Figure 1a). XRD analyses (not shown here) confirmed the presence of α' martensite in the as-additive manufactured samples. Two geometries of tubular specimens were manufactured by PBF (Powder Bed Fusion), one in the form of a pipe with a variable thickness and the other with a homogeneous envelope of 400 µm thickness along the entire gauge length of the specimen (Figure 1a). The total length of the fatigue test samples was 64 mm, with a gauge length of 20 mm and a minimum diameter of 3,8 mm. The difference in fatigue life between specimens without lattices and specimens filled with cubic lattices (struts radius 0.26 mm and lattice size of 1 mm) has been investigated. In addition, the effect of severe deformation (SMAT) after fabrication was compared for each case. After manufacturing, each sample has been polished with a grid size of 1200, 2400 and 4000 #. For some of the specimens, this surface preparation was followed by a SMAT treatment.

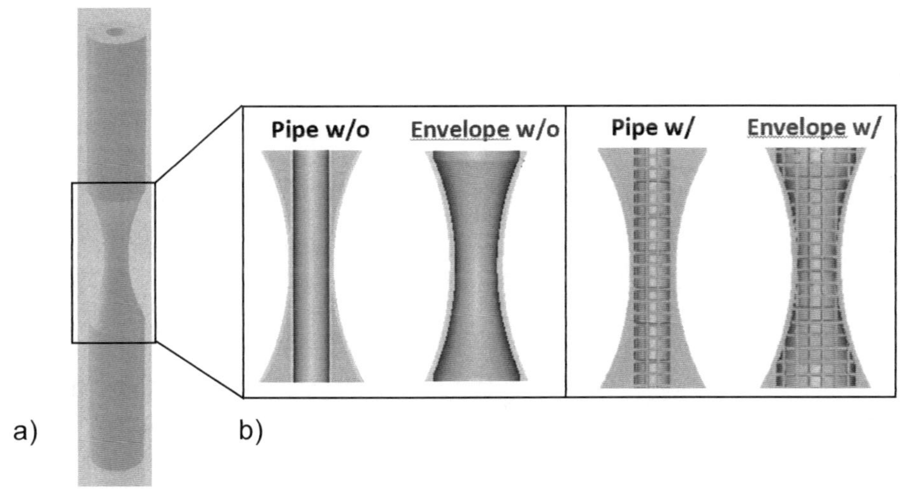

Figure 1. Schematics showing the geometry of the sample; a) an entire sample and b) details of the gauge length samples (w/o: without lattices, w/: with lattices).

The SMAT treatment was carried out by impacting 100C6 balls of 2 mm diameter out at room temperature for 10 min and at a distance of 20 mm from the vibrating part (frequency of 20 kHz for an amplitude of 60 µm). Rotational bending fatigue tests were performed at room temperature and at a frequency of 30 Hz until fracture. This made it possible to stress the material superficially and to observe the effect of the SMAT treatment.

Experimental Results

Figure 2 shows the fatigue resistance (number of cycles to failure) at different stress amplitude. Without SMAT treatment, the fatigue life of the envelope-type and pipe-type samples, with or without lattice structures, can be described by an unique Wohler-type curve. In general, SMAT post-treatment significantly improves the fatigue life. However, a thorough observation indicates that the gain in fatigue life is greater for the samples without lattice than for the sample having lattice structures.

The fatigue behaviour depends essentially on the mechanical properties of the material defined by the thickness of the envelope, the lattice shape and by the relative density, geometry and topology of the unit cell for the lattices on the other hand [4].

Before application of SMAT and, for any geometry, the samples with lattice structures have a considerably superior fatigue life than the samples without inner lattice structures. However, once the SMAT treatment is applied, the fatigue behavior for the samples with lattices tends to become similar to that of the samples of envelope geometry without lattices. One hypothesis that could explain these observations is that several small cracks are formed during the fatigue test within the lattice structure that eventually coalesce to lead to the catastrophic failure of the thin envelope.

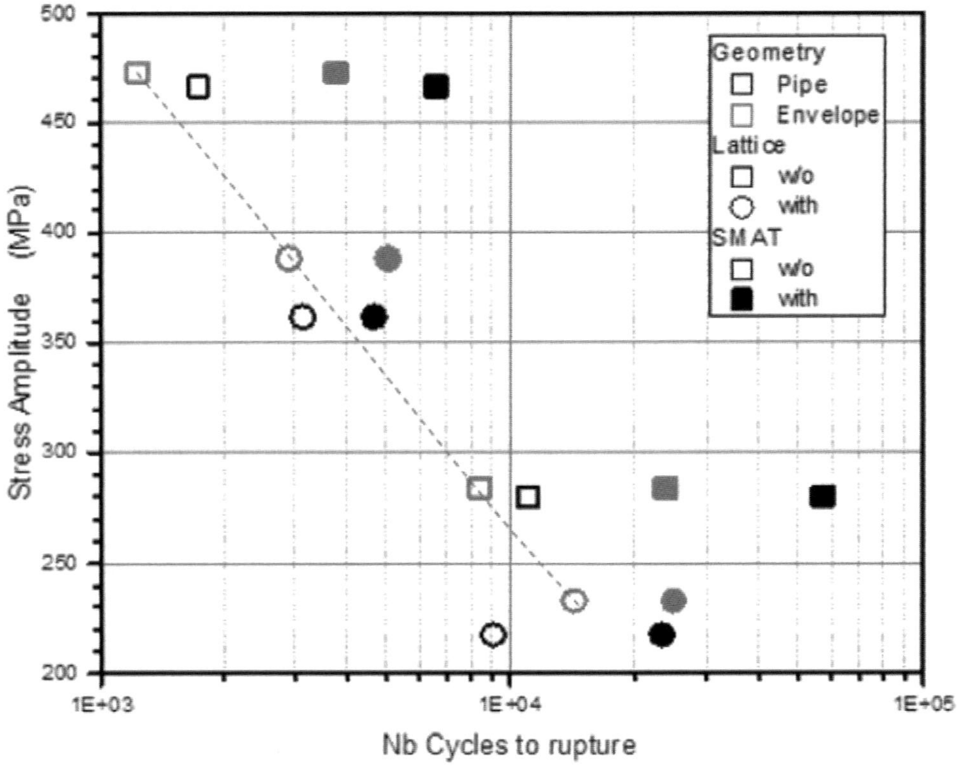

Figure 2. Fatigue performance for different geometries and processing conditions.

Additively manufactured samples contain defects and particularly porosities that are sometimes difficult to control. These defects can result from the lack of powders, lack of energy from the heat source, presence of gas bubbles, oxide, etc. and they affect premature crack formation in sample tested under cyclic loading. However, SMAT helps to close surface defects and to introduce compressive residual stresses slowing down the propagation of possible cracks formed.

The application of SMAT post-treatment improves the number of cycles to failure by 40 to 400% (Table 1), without affecting the integrity of the thin envelope (400 µm). This fact is more remarkable as it can also be observed under the most unfavourable mechanical conditions (part consisting of the envelope only, without the inner lattices). For samples with lattice structure, the theoretical stress amplitude values were obtained from the numerical simulation by Abaqus.

On 2 mm diameter specimens prepared by Electron Beam technique, Persenot et al. have compared the effect of SMAT with the effect of surface machining on the fatigue properties after HIP treatment [8]. Though the application of SMAT surface treatment considerably improved the numbers of cycles to failure of HIPed samples, these authors demonstrated that surface machining was more efficient to enhance the fatigue properties because of the reduction in the surface roughness that appears to be the most critical parameter.

Table 1. Fatigue gain (number of cycles) after SMAT.

Lattice	Stress Amplitude	Envelope geometry gain	Pipe geometry gain
With	Low > 350 MPa	75 %	157 %
Without	Low > 250 MPa	185 %	418 %
With	High > 200 MPa	74 %	49 %
Without	High > 450 MPa	205 %	280 %

SEM observations of the sample after fatigue tests revealed the origin of the fractures. Figure 3 shows the lack-of-fusion defects present on the surface and inside the envelope. Defect size also affects the fatigue resistance, and it is important to note that many of these defects are interconnected or separated by a very thin thickness of material which would lead to premature cracking.

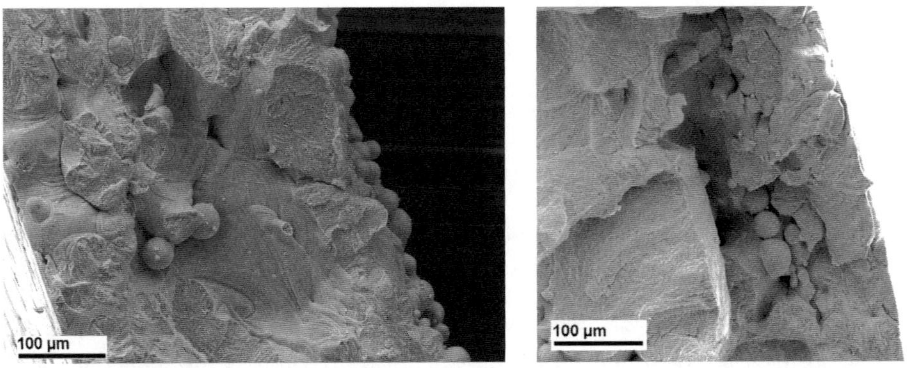

Figure 3. SEM images of the rupture of SMATed envelope sample (left) and SMATed pipe sample.

Figure 4 shows the cross-section of a SMATed sample. The modification in the grain size and morphology (initial needle shape) of α'Ti-martensite can be clearly observed in this figure. The thickness of the modified region is about 52 µm and it is composed of nanograins of α'Ti-martensite. Similar thickness size of the SMATed zone has been reported in Kumar et al.'s work [7].

The EBSD map shows black areas whose Kikuchi patterns are blurred or overlapped because of poor indexing. Poor indexing is due to highly strained regions or highly deformed microstructure that has formed nanograins. Indexing becomes more favourable deeper from the surface [9].

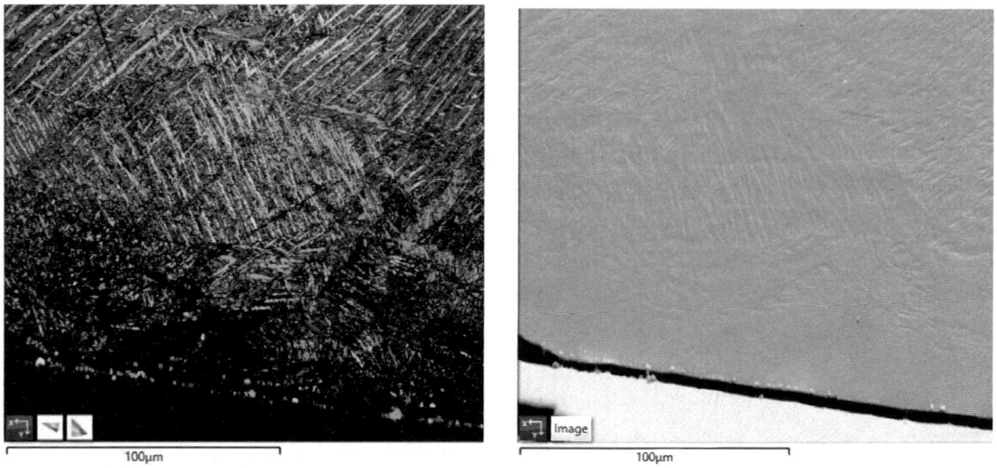

Figure 4. EBSD (left) and SEM (right) cross-section image of a SMATed sample.

The results presented in this article will be confirmed with additional tests in order to improve the data statistic.

Conclusions

The final goal of this study is to improve the quality of the new generation prostheses by lightening their weight while improving the mechanical resistance and the condition of the external surface in contact with the biological tissues (SMAT avoids machining the raw parts by homogenising the roughness).

In this work, two specimen geometries were tested and the effect of inner filling with lattice structures was examined under cyclic solicitation. The thin envelope surrounding these structures have been SMATed to compare the effect of post-treatment on the fatigue life via rotary bending tests performed at room temperature and high frequency.
By taking into account of the reinforcement resulting from the lattice on the stress calculation, the fatigue life of the specimens with and without inner lattice structures, if expressed in the Stress amplitude versus Number of cycles to failure, can be described by a unique Wohler curve. In general, the SMAT post-treatment enhances considerably the fatigue properties. The fatigue life enhancement was found to be more pronounced for pipe-life specimens in comparison to envelope-type samples.

A quantitative investigation is currently undertaken to optimize the SMAT conditions and to correlate the size and position of the defects with the fatigue life.

Acknowledgements

The authors would like to thank the Ministry of Higher Education, Research and Innovation for the grant obtained for this project. We would also like to thank Paul Lohmuller to produce the samples.

References

[1] F. Laverne, F. Segonds, and P. Dubois, '*Additive Manufacturing - General Principles*', Techniques de l'ingénieur, 2016. (in french)
[2] M. Srivastava, S. Rathee, S. Maheshwari, and T. K. Kundra, *Additive Manufacturing: Fundamentals and Advancements*. Boca Raton: CRC Press, 2019.
[3] P. Didier, G. L. Coz, G. Robin, P. Lohmuller, A. Moufki, and P. Laheurte, '*Consideration of additive manufacturing media for machining completion*', 24ème Congrès Français de Mécanique, 2019. (in french)
[4] T. Maconachie *et al.*, 'SLM lattice structures: Properties, performance, applications and challenges', *Mater. Des.*, vol. 183, p. 108137, Dec. 2019.

[5] P. Maurel, L. Weiss, P. Bocher, and T. Grosdidier, '*Effects of SMAT at cryogenic and room temperatures on the kink band and martensite formations with associated fatigue resistance in a β-metastable titanium alloy*', *Mater. Sci. Eng. A*, vol. 803, p. 140618, Jan. 2021.

[6] C. Dureau *et al.*, '*On the Influence of Ultrasonic Surface Mechanical Attrition Treatment (SMAT) on the Fatigue Behavior of the 304L Austenitic Stainless Steel*', *Metals*, vol. 10, no. 1, p. 100, Jan. 2020.

[7] S. Kumar, K. Chattopadhyay, and V. Singh, '*Effect of ultrasonic shot peening on LCF behavior of the Ti–6Al–4V alloy*', *J. Alloys Compd.*, vol. 724, pp. 187–197, Nov. 2017.

[8] T. Persenot, A. Burr, E. Plancher, J.-Y. Buffière, R. Dendievel, and G. Martin, '*Effect of ultrasonic shot peening on the surface defects of thin struts built by electron beam melting: Consequences on fatigue resistance*', *Addit. Manuf.*, vol. 28, pp. 821–830, Aug. 2019.

[9] Y. Samih, B. Beausir, B. Bolle and T. Grosdidier, *'In-depth quantitative analysis of the microstructures produced by Surface Mechanical Attrition Treatment (SMAT)', Materials Characterizatoin.*, vol. 83, pp. 129–138, Jun. 2013.

Evolution of the controlled shot peening process in the automotive field

E. Morgano SILCO - Rivalta TO

Keywords: *Shot peening – Fatigue results – Residual stress – Roughness – Stream finishing*

Introduction

The controlled shot peening process, on a mechanical component, represents the most robust solution to increase its resistance to fatigue; moreover, since it is usually the last operation in the production chain, it does not change the parameters of the previous operations, such as forging, mechanical processing and surface hardening treatments.

The choice of shot peening parameters and minimum values of residual stresses in compression are essential to keep the operation under control; the optimal residual stress, from surface up to deep, profile depends on the type of damage to be solved.

The advantages reached with controlled shot peening are achievable only if the process parameters are strictly predetermined and kept under control, such as:

- Shot speed and range
- Impact angle of the shot flow
- Distance of the piece to be treated
- Exposition time

To be reported on the drawing:

- **Shape, size and hardness of the shot**
- **Coverage**
- **Minimum intensity of shot peening**
- **Residual stresses MPa**

The controlled shot peening process gives high residual stresses in compression, both on the surface and in depth (Fig. 1); however, even using optimal parameters, the surface roughness undergoes a small deterioration (Fig. 2)

The choice of the correct profile depends on the damage mechanism that must be opposed in the images below, the curves for:

- **Bending**
- **Pitting**
- **QT – Micro alloyed**
- **Sheet products**

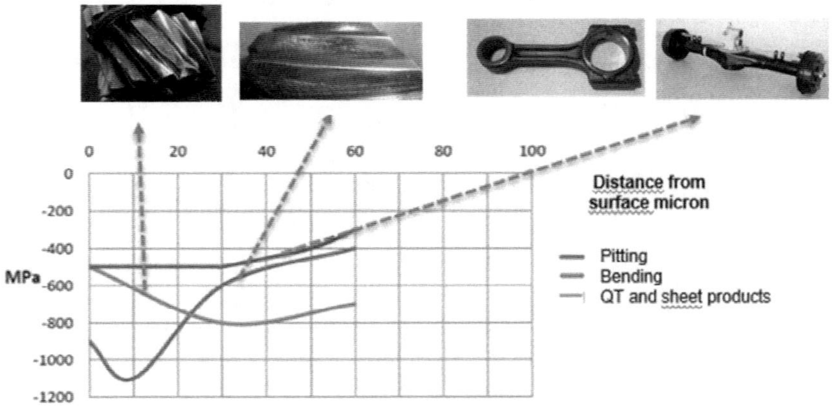

Figure 1 – Residual stress profile in function of the damage mechanism.

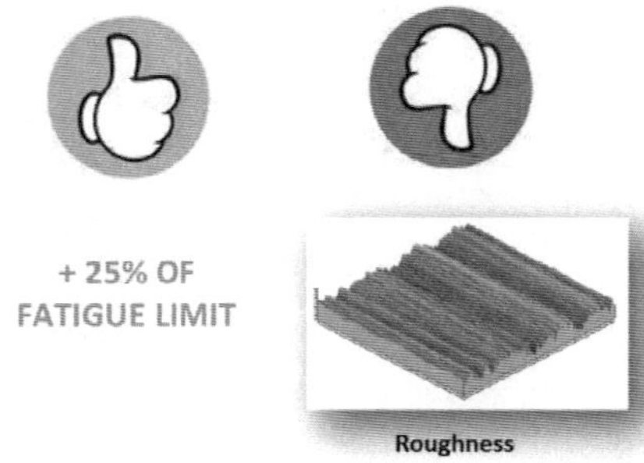

Figure 2 – Worsening of surface roughness after shot peening

The next steps of controlled shot peening, using compressed air equipment, concern the optimization to reduce the surface roughness and further improve the resistance to sub-surface fatigue (pitting).

In this regard, there are some solutions ranging from a final super-finishing to a double shot peening, with a second softer operation in terms of intensity, using both steel and ceramic shot.

The super-finishing process, after shot peening, guarantees a removal of material from the surface of max 15 microns (Fig. 3).

Different details of the gearbox with different geometries were tested, such as gear wheels, secondary and primary shafts and cylindrical crowns (Fig. 4)

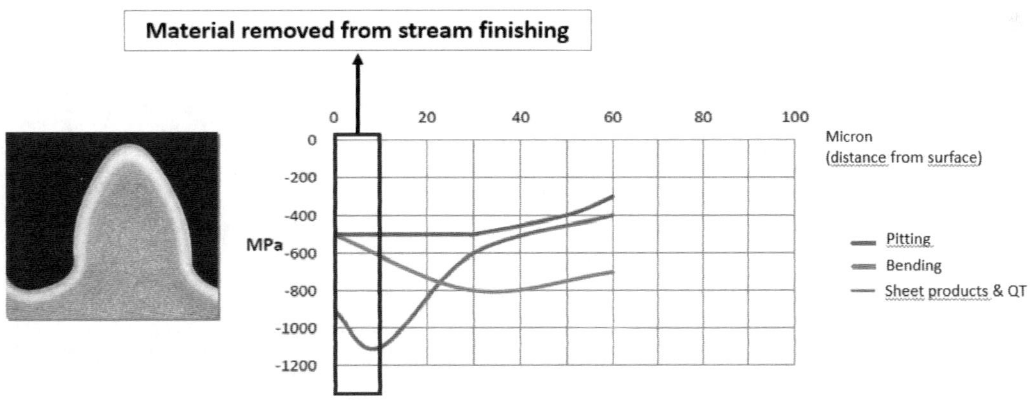

Fig. 3 – Appearance of the depth of material removed during the stream finishing operation.

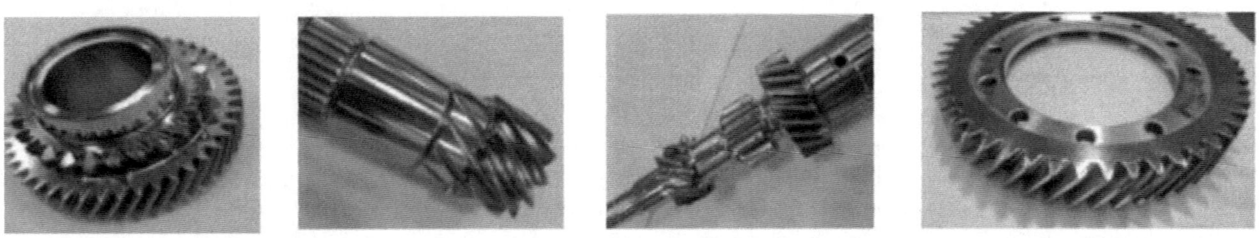

Pict. 4 – Parts tested with the stream finishing process

The metrological measurements of surface roughness have shown reductions in Ra of about 65% (Fig. 5); this guarantees a considerable increase in resistance to pitting, such as sub-surface fatigue.

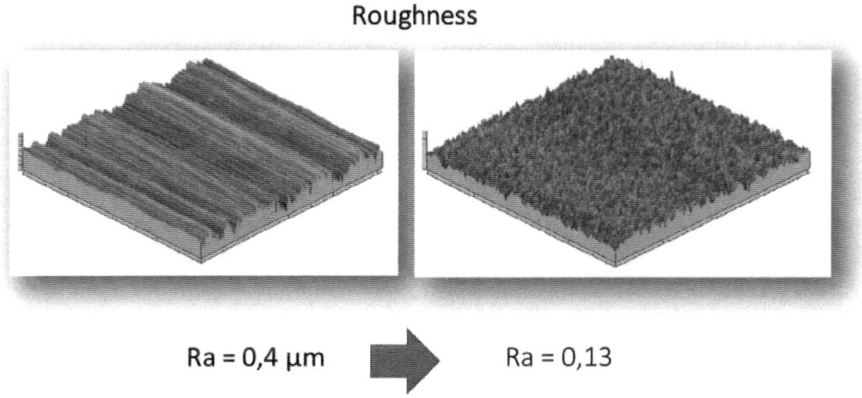

Fig. 5 – Magnification of surfaces, before and after the stream finishing operation.

As for the residual stresses, measured on non-shot peened parts and after the stream finisching operation, it emerged that high values are reached on the surface which are reduced and fade under the surface (Tab. 1). This is very important, since the roughness Ra is very low, as it is possible to evaluate, on some projects, even the possibility of using the superfinishing alone. This obviously depends on the application and the mechanical stresses, due to fatigue, during operation.

Table 1 – Residual stress values measured on all tested components, both on the surface and at a depth of 30 microns.

SAMPLE	RESIDUAL STRESS MPa	
	Surface	At 30 micron
Secondary shaft output	-1082	-412
Driven gear 1^ speed	-892	-461
Driven gear 2^ speed	-1012	-389
Reverse gear	-994	-401
Cilynder crown	-998	-398

Finally, fatigue tests, performed on test samples representative of the application with a shape that has a fitting that simulates the bottom tooth of a gear, have shown that controlled shot blasting guarantees an excellent increase in fatigue resistance (+ 20-25%); since this is an operation performed after the surface hardening process, it allows not to change the parameters that are upstream of the production chain. The subsequent stream finishing further improved the fatigue behavior by about 30% (Fig. 6). The equipment used for the test is a vibrofor with a run-out of 5 million cycles (Fig. 7).

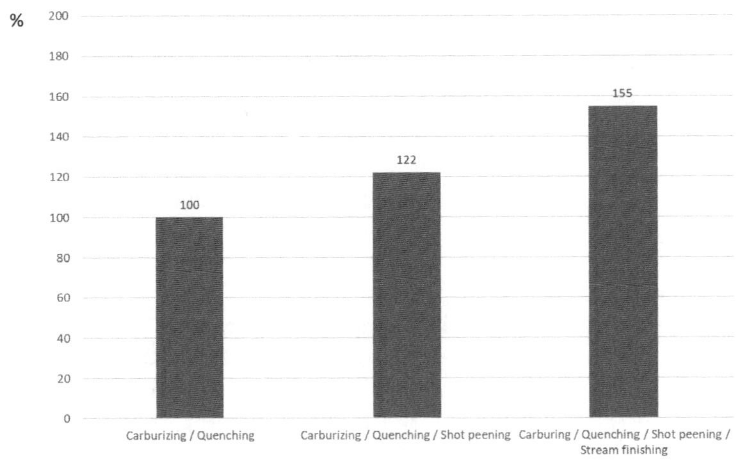

Fig. 6 – Results of fatigue test.

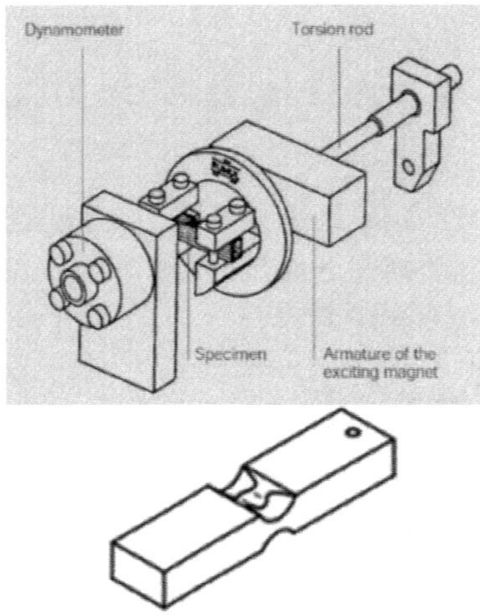

Fig. 7– Craktronic resonant machine and sample

Next steps

The activity will go ahead and includes:

- ***Double shot peening (*) - Steel + Steel***
- ***Double shot peening (*) - Steel + Ceramic***

- ✓ Preparation of 60 fatigue samples - 30 April 2022
- ✓ Vacuum carburizing – Quenching in gas - May 31, 2022
- ✓ Double shot peening process - 30 June 2022
- o Fatigue test with 5 mil cycle run-out - 30 September 2022
- o Evaluation of the results - 15 October 2022

(*) The second shot peening involves a reduction of the Almen intensity in order to significantly reduce the surface roughness.

Conclusions

- **CONTROLLED SHOT PEENING**

Guarantees an increase in fatigue performance of about 25%; since this is the last operation, the previous parameters of the supply chain are not changed (mechanical processing / hardening treatments).

- **STREAM FINISHING POST SHOT PEENING**

It combines the low surface roughness with the residual compressive stresses of the shot peening, improving both the resistance to both pitting and bending.
Comparative fatigue tests on test samples representative of the application have shown increases in fatigue resistance that can reach up to 70%.

- **DOUBLE SHOT PEENING**

It represents an alternative to stream-finishing, where the first shot peening determines residual tensions in depth and the second, with a lower intensity, refines the surface with a reduction in roughness.

Characterization of blasted surfaces with mesh-based parameters

Stefanie Stöckel[1], Sophie Gröger[1], Frank Segel[1]

[1]Professorship of Production Measuring Technology, Chemnitz University of Technology, Germany

Abstract
This paper presents an approach for the determination of surface parameters based on a mesh calculated from identified features. The resulting mesh-based parameters are used to assess the homogeneity of the feature distribution on the surface. The preparation of the data is a prerequisite for creating the mesh. The input data of a surface has to be processed in such a way that the features can be identified, which are the valleys for the examined blasted surfaces. For this, areal measurements and the application of filters are necessary to separate the features. In the next step a mesh is created. The simplest form is a triangular structure, which has already been used to describe the homogeneity of surfaces with particles. Based on this, various characteristics are evaluated and analyzed statistically. The method has been applied to four surfaces blasted with different blast media as an example.

Keywords areal surface characterization, mesh-based parameters, surface topography.

Introduction

To study the effect of manufacturing processes or the function, e.g. wear behaviour or optical appearance, the geometrical microstructure can be characterized. The new ISO 21920-2 in 2D and the ISO 25178-2 in 3D offer parameters to assess the properties of surfaces. [1, 2] Many of these parameters are based on the evaluation of amplitudes. In the context of blasted surfaces additional information can be gained from the distribution of areal surface characteristics and the homogeneity of surface features.

From a mathematical point of view there is "no accepted metric for the level of inhomogeneity in a given node pattern". [3] There exists no generalization of results and so they are only valid for the special investigated scenario. [3] In engineering, there are different applications where homogeneity is important, for example for composite materials.

Various papers have been published where the Delaunay triangulation is used to describe the distribution of particles in composites. Zhang et al. give an overview of different methods to evaluate the inclusion homogeneity of composites, including Delaunay triangulation as one method based on the inter-distance of points. [4] In general, the method consists of two main process steps: detecting the features (points) and calculating the mesh with the resulting parameters. Bray et al. applied the Delaunay network to describe the nanoparticle dispersion of composites. [5] They processed the image and identified the hull of particles. Then calculated the center of mass points and meshed those. Sul et al. include the Motif analysis in the image processing. [6]

Two more examples are the application of this approach for the classification of polyps from endoscopy pictures in medicine and the analysis of particles in cement in construction engineering. [7, 8] Both use image processing to extract the features. In this paper, we describe a process how to adopt the existing method for blasted surfaces and derive an approach to calculate parameters for the evaluation of homogeneity.

Experimental Methods

Materials

As examples, four different surfaces created with different blast media are used. The blast media is varied in two sizes and two different geometrical designs. The letter S stands for spherical media named CHRONITAL, which is highly break-resistant and durable. The other media is GRITTAL GM with angular broken shape. The size is indicated by the number (60 = 0,7 - 1,25 mm and 10 = 0,05 - 0,2 mm).

A special developed blasting machine, called Twister, has been used to machine the specimens. There is a spin wheel located in the middle of the machine that distributes the blast media in a defined way for a uniform surface finish. The workpiece carrier rotates around itself and around the spin wheel. For the examples a speed of 9000 min^{-1} and a process time of 10 min has been applied.

Measurement

A confocal microscope with a 20 times objective has been used to extract a part of the blasted surface. The generated data is the basis for the identification of the features, which are the holes resulting from the impact of the blast media. The aim is to describe these features as points and to create the mesh in the following step. Figure 1 shows the blast media on the left and the measurement result for the four different surfaces.

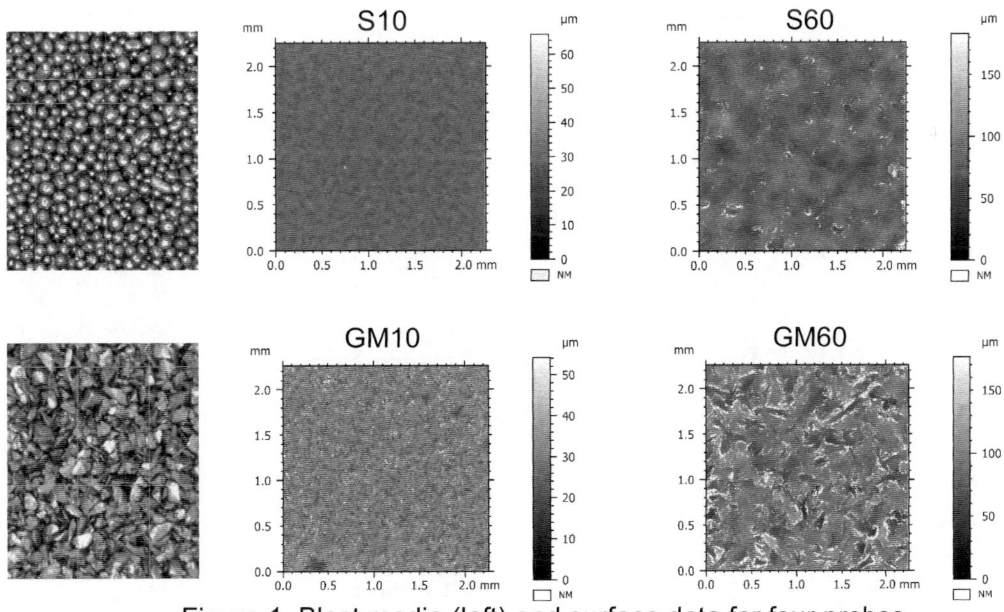

Figure 1. Blast media (left) and surface data for four probes

Data processing and filtering

The software MountainsMap® has been used to calculate all pre-processing steps and generate surface pictures. All four surfaces are prepared and then filtered with the same strategy to get comparable results. The data preparation follows three steps (Figure 2):

1. Extract topography layer, since the data of the measurement is multichannel data consisting of the topography and the intensity
2. Crop the data to 1x1 mm
3. Fill non-measured points

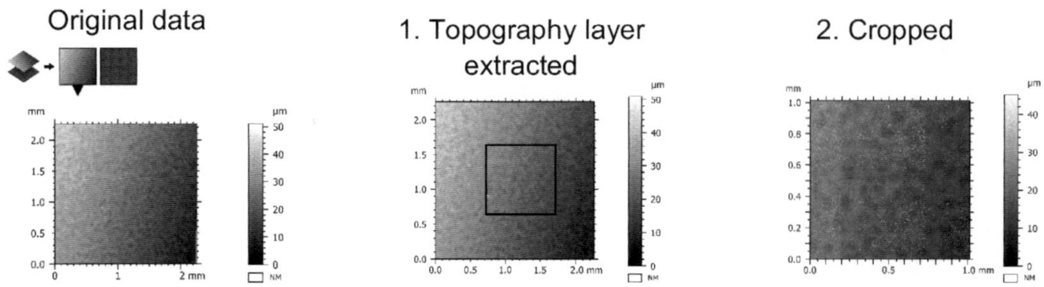

Figure 2. Data preparation of the surface (example S10)

After that, filters have been applied to detect the valleys as features. The filtering process is a combination of Gaussian filter to create the S-F-surface [2] and the Motif Analysis with Wolf-Pruning. [9] In the first step the alignment with LS-plane removes the tilt of the probes and so the form (F) by cutting long wavelengths. The S-Filter removes short wavelengths (roughness) with a Gauss filter of 8 µm. The result is the extracted waviness of the surface.

Motifs identify structures on surfaces as watershed segmentation method. [9] The method fills insignificant valleys slowly with water. On the saddle point, the water will flow into the adjacent valley. If that valley is significant they are combined, otherwise the filling is continued. A change tree shows the results and is the base for wolf pruning (Figure 3). Therefore, the difference in height between valley or peak and the adjacent saddle point is calculated. The pruning method finds those with the smallest height difference and combines with the adjacent saddle point. The process continues until a specific threshold is reached. Here the threshold is a percentage of the value Sz and as second criteria a percentage of the surface area.
Table 1 gives an overview of the applied filters for each probe.

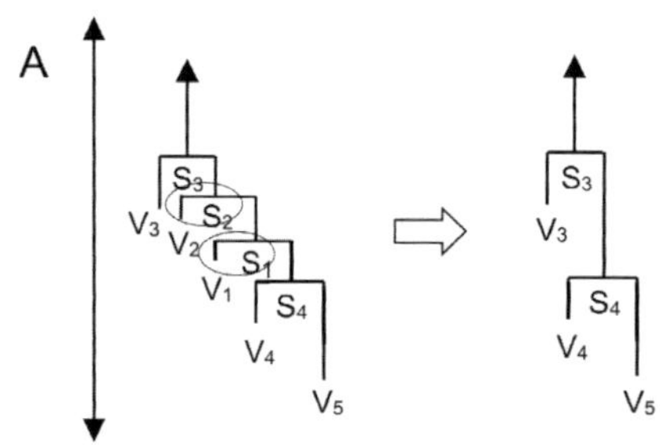

Figure 3. Change tree with A (height in µm), V (Valleys), S (saddle point) (after [2])

Table 1. Summary of filters

probe	Gaussian filter (in µm)	Motif analysis		
		Size of smoothing filter	Pruning (%Sz)	Pruning (%Surface area)
S10	8	5x5	5	0,1
S30	8	5x5	5	0,17
GM10	8	5x5	5	0,1
GM60	8	5x5	5	0,15

The point cloud represents the extreme values (x and y-coordinate) for each motif. Points which lie within a distance from 0 to 0,01 mm from the four edges are deleted. They are on or near the edge and it is impossible to know if these points are really the extreme points of the valleys or the valley is deeper outside the section.

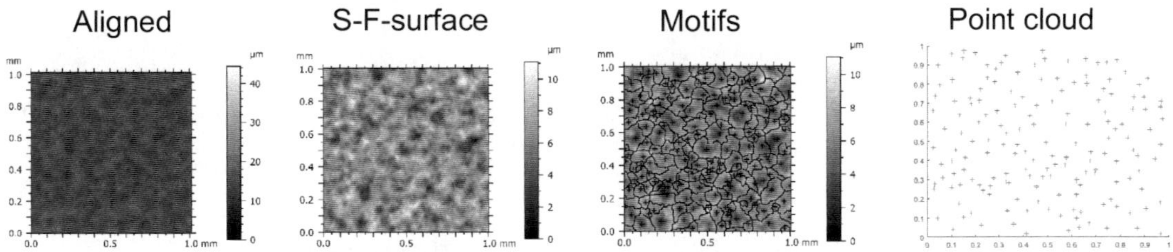

Figure 4. Filtering for example S10

Meshing

The next step is to create a mesh of the determined points of the surface (2D). The simplest structure are triangles, since they connect the points in the smallest possible distance. The Delaunay-Triangulation creates a mesh in such a way that no point is inside the circumcircle of any triangle, therefore it maximizes the minimum of all angles of the triangles. As stated in [4] these properties are suitable to describe homogeneity in general. Therefore, this algorithm is suitable and here applied for blasted surfaces. The algorithm and the following calculation of parameters are executed with MatLab.

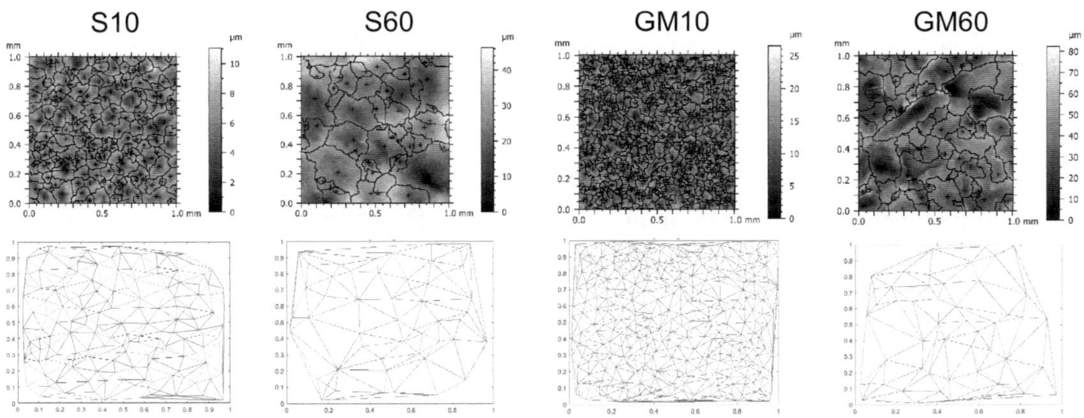

Figure 5. Surfaces with Motifs and the created Delaunay triangulation

Parameter

The Delaunay method creates triangles that can be geometrically described by side length and area. These two parameters are statistically analysed by calculating the mean value and standard deviation. Regular patterns show low values for standard deviation and irregular ones high values. In reference to other papers, the Area Disorder AD is calculated by

$$AD = 1 - \left(1 + \frac{\sigma_A}{\bar{A}}\right)^{-1} \tag{1}$$

and the variation index DT by

$$DT = \frac{\sigma_A}{\bar{A}} \tag{2}$$

where σ_A is the standard deviation of the triangle area and $\bar{A}$ the mean area. [4, 10] AD has a range from 0 to 1. The value 0 describes perfect order. A high value AD or DT means clustering and poor homogeneity. The histograms of side length and area show the distribution of these parameters.

Experimental Results

Table 2 presents the results for Sa and the number of Motifs as an example of standard parameters from the ISO standards. Comparing the results does not provide an information about the distribution of the features. The parameters in Table 3 on the other hand are a new possibility to describe the homogeneity.

Table 2. Results for selected standard parameters

Probe	Sa (µm)	Number of Motifs
S10	1,09	185
S60	5,66	54
GM10	2,2	388
GM60	9,73	68

Table 3. New parameters for four different blasted surfaces

Probe	Mean side length (mm)	Standard Deviation of side length (mm)	Mean Area (mm^2)	Standard Deviation of Area (mm^2)	DT	AD
S10	0,093	0,056	0,00275	0,00157	0,57	0,36
S60	0,177	0,104	0,00859	0,00580	0,68	0,40
GM10	0,065	0,050	0,00130	0,00078	0,60	0,37
GM60	0,164	0,094	0,00796	0,00611	0,77	0,43

The values show a bigger difference between the 10 and 60 than between the blast media S and GM. This could also be due to an influence of the size of the considered section, since there are fewer points for the bigger size blast media. The side length is influenced by the edge of the mesh where big ones can occur in long pointed triangles. The values DT and AD without units give similar results as expected.

Figure 6 displays the histograms for the area.

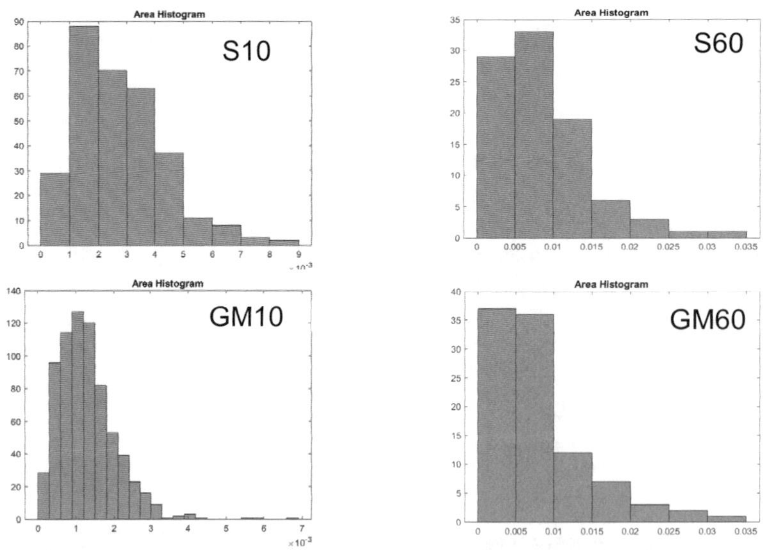

Figure 6. Histograms of the triangle area distribution

All four surfaces show a regular mesh pattern and together with the parameters can be categorized as homogeneous.

Discussion and Conclusions

The general approach with image processing, generating a mesh and calculating parameters is successfully applied. To validate the findings the method has to be applied on a wider range of surfaces. First for a larger number of surfaces manufactured the same way. Secondly for different process parameters for example with different process time. The shown examples are all homogeneous and it is of great interest to compare the results also to inhomogeneous surfaces. The aim is to define the limit of at least one parameter for a homogeneous surface (Figure 7).

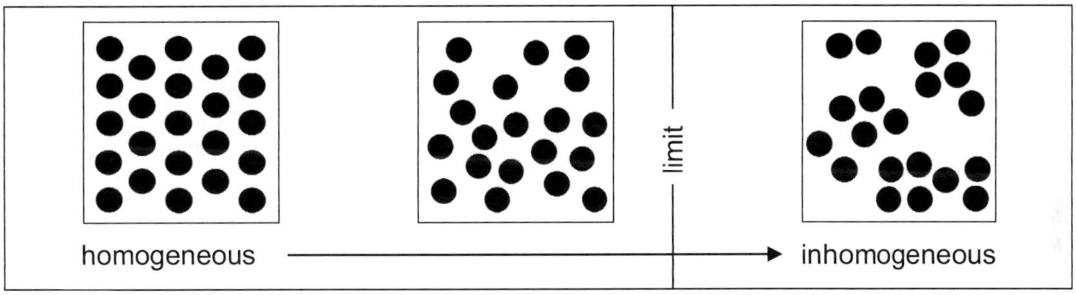

Figure 7. (In)homogenous surface

In general, the approach only focusses on the homogeneity of features and lacks the complete description of the surface as 2,5/3D topography. Therefore, the dimension of the features, for example the z-values or volume, will be described as a next step.

References

[1] *ISO 21920-2 Geometrical product specifications (GPS) — Surface texture: Profile — Part 2: Terms, definitions and surface texture parameters* (2021)

[2] *DIN EN ISO 25178-2 Geometrische Produktspezifikation (GPS) – Oberflächenbeschaffenheit: Flächenhaft - Teil 2: Begriffe und Oberflächen-Kenngrößen* (2012)

[3] U. Schilcher, G. Brandner and C. Bettstetter: *Quantifying inhomogeneity of spatial point patterns*, Computer Networks, Vol. 115 (2017), pp 65–81.

[4] L. Zhang, Z. Chen, J. Mao, S. Wang and Y. Zheng, *Quantitative evaluation of inclusion homogeneity in composites and the applications*, Journal of Materials Research and Technology, Vol. 3 (2020) issue 9, pp 6790–6807.

[5] D. J. Bray, S. G. Gilmour, F. J. Guild, T. H. Hsieh, K. Masania, and A. C. Taylor, *Quantifying nanoparticle dispersion: application of the Delaunay network for objective analysis of sample micrographs*, Journal of Materials Science, Vol. 19 (2011), pp 6437–6452.

[6] I. H. Sul, J. R. Youn, and Y. S. Song, *Quantitative dispersion evaluation of carbon nanotubes using a new analysis protocol*, Carbon, Vol. 4 (2011), pp 1473–1478.

[7] Y. Sargam and K. Wang, *Quantifying dispersion of nanosilica in hardened cement matrix using a novel SEM-EDS and image analysis-based methodology*, Cement and Concrete Research, Vol. 147 (2021).

[8] M. Häfner, M. Liedlgruber, A. Uhl, A. Vécsei and F. Wrba, *Delaunay triangulation-based pit density estimation for the classification of polyps in high-magnification chromo-colonoscopy*, Computer methods and programs in biomedicine, Vol. 107 (2012).

[9] *DIN EN ISO 16610-85 Geometrische Produktspezifikation (GPS) – Filterung – Teil 85: Morphologische Flächenfilterung: Segmentierung,* (2013)
[10] D. J. Bray, S. G. Gilmour, F. J. Guild and A. C. Taylor, *Quantifying nanoparticle dispersion by using the area disorder of Delaunay triangulation*, Journal of the Royal Statistical Society: Series C (Applied Statistics), Vol. 61 (2012), pp 253–275

AUTHOR INDEX